Autism

CURRENT CLINICAL NEUROLOGY

Daniel Tarsy, MD, SERIES EDITOR

Autism

Current Theories and Evidence

Andrew W. Zimmerman
Editor

Kennedy Krieger Institute and Johns Hopkins University,
Baltimore, Maryland

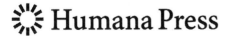 Humana Press

Editor
Andrew W. Zimmerman
Kennedy Krieger Institute and
Johns Hopkins University
707 N. Broadway
Baltimore MD 21205
USA

Series Editor
Daniel Tarsy
Beth Israel-Deaconess Medical Center and
Harvard Medical School
Boston, MA

ISBN: 978-1-60327-488-3   e-ISBN: 978-1-60327-489-0
DOI: 10.1007/978-1-60327-489-0

Library of Congress Control Number: 2008935824

Printed on acid-free paper

9 8 7 6 5 4 3 2 1

springer.com

DOI 10.1007/978-1-60327-489-0

ERRATA

Autism: Current Theories and Evidence

Andrew W. Zimmerman

Page iv: Cover illustration credits were inadvertently omitted:

Illustration: fetal brain (26 weeks gestational age), courtesy of Barbara J. Crain, M.D., Ph.D., Johns Hopkins University School of Medicine, Baltimore, MD and Hannah C. Kinney, M.D., Children's Hospital Boston and Harvard Medical School, Boston, MA. Cover design: Sharon N. Blackburn, B.A., Johns Hopkins University School of Medicine, Baltimore, MD.

Page v: Dedication text was inadvertently omitted:

Dedicated to those who live with and care for children and adults with autism, and those who strive through science to understand and treat them.

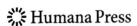 Humana Press

Preface

Andrew W. Zimmerman

Moving Beyond Hypotheses: A Call for Theories and Evidence

Creative thinking and collaborative scientific research have advanced our understanding of autism, and we are now beginning to synthesize the data into evidence and theories. Our lack of knowledge about the causes and pathogenesis of autism has been associated with widely divergent approaches to diagnosis and management, most of which have not been subjected to rigorous scientific scrutiny. The consequence for affected families is that while they recognize the importance of early intervention for their children, they are often left with indecision and confusion as to how to proceed. This book presents current theories about autism and the evidence that supports them. The goal is to show how the scientific method is revealing the biological bases of this spectrum of disorders, thereby leading the way to their treatment and prevention using evidence-based medicine.

Quotations from Leo Kanner

> In 1943: We must, then, assume that these children have come into the world with innate inability to form the usual, *biologically provided* affective contact with people, just as other children come into the world with innate physical or intellectual handicaps [1].

> In 1971: At long last, there is reason to believe that some answers to these questions (about autism) seem to be around the corner. *Biochemical explorations*, pursued vigorously in the very recent past, may open a new vista about the fundamental nature of the autistic syndrome [2].

Since Leo Kanner's 1943 description of autism as a biological disorder and follow-up of his original 11 patients in 1971, we have amassed large amounts of descriptive data. However, as in many areas of neuroscience, we are "data rich and theory poor" [3]. Furthermore, despite Kanner's optimism in 1971, substantive neurobiological answers were not "around the corner."

A plethora of hypotheses has been proposed to explain the various manifestations of autism, and many untested medical treatments being utilized are inspired by the needs of affected persons, yet are based on limited empirical data. Psychodynamic approaches to autism prevailed in the past, but in the last decade, there has been explosive growth in autism science as new researchers have been willing to discard or revise earlier concepts. Armed with an understanding of brain development from a neuroscience perspective and with the tools of molecular genetics, imaging, and environmental science, they are providing novel insights, energy, and fresh approaches, thereby ushering in a new era that would have amazed Kanner and his contemporaries.

Autism research is now a fast-moving field in which terminology changes frequently and challenges the classical definitions of the phenotypes used in experimental designs and patient care. Not surprisingly, scientific interest in autism has increased in parallel with similar increases in its prevalence, along with public awareness and concern. Thus, the annual numbers of original publications as well as review articles in the field have increased fourfold during the past decade (Fig. 1).

Now is the time to examine how we approach scientific discovery in autism. The clinical heterogeneity of this disorder, together with the inherent dynamic, changes during children's growth and development, confound static, linear models and simplistic, unilateral approaches. Global thinking and collaborative approaches are needed. *Dedicated efforts using logic and the scientific method are leading many researchers and funding agencies to move beyond the descriptive phase of observations, to form hypotheses, collect data systematically, define the evidence, and establish testable theories.*

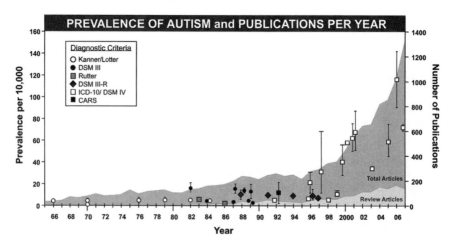

Fig. 1 Prevalence of autism per 10,000 (left) and autism publications (right) by year, 1966–2007. Prevalence figures are from separate publications, according to different diagnostic criteria used for autism assessment (inset), with confidence intervals (Courtesy of Lisa A. Croen, PhD). Total original research and review articles with key word "autism" are from PubMed [4]. Diagnostic criteria: Kanner [1]/Lotter [5]; DSM, Diagnostic and Statistical Manual III, III-R, IV [6]; ICD, International Classification of Diseases-10 [7]; CARS, Childhood Autism Rating Scale [8]

Definition

Hypothesis: a proposition that attempts to explain facts in a unified way and forms the basis of experiments designed to establish its plausibility.

Theory: a set of statements, including hypotheses, that explains observations in terms of those hypotheses [9].

In the scientific sense, a theory is a more advanced form of knowledge that emerges from hypotheses, empirical data, and evidence. The process of forming theories leads to discoveries, fitting Kuhn's description of how researchers observe "anomalies" that lead to a "paradigm shift" and eventually, "scientific revolution" [10]. The drive to understand autism and related neurodevelopmental disorders has motivated observations of clinical anomalies in autism for decades, but in recent years, important new advances in neuroscience, molecular biology, and related fields have offered the means to move beyond observations, develop hypotheses, and acquire experimental evidence. The application of scientific methodologies allows for translation of findings from basic science to the development of evidence-based clinical medicine in autism. This process was slow to start due to autism's inherent clinical heterogeneity and a lack of adequate funding for research. Now, however, consensus has been developing on the description of phenotypes, funding for research has increased, and investigators are actively communicating and validating various theories.

Theories result from observations of phenomena that are the basis for hypotheses, which lead to experimentation and empirical data, and with analysis, evidence. In turn, theories generate new and more refined observations and hypotheses, repeating the cycle and thereby confirming or modifying the original theory. This process represents a natural cycle in the development of new knowledge, which is especially relevant to autism research today (Fig.2).

Fig. 2 Cycle of progress in autism science. Observations lead to testable hypotheses; through collection of data and analysis of evidence, investigators develop theories. In turn, the cycle is repeated, informing new hypotheses and expanding generalizable knowledge

In developing theories, it is important, first and foremost, to distinguish science from "pseudo-science" or scientific from nonscientific theories [11]. Real progress will require that we engage the scientific method to develop theories. Valid theories emerge from hypotheses that can be refuted by empirical evidence. Einstein's work was an inspiring example of genuine science, in which theories have been validated and improved over time, because they are open to "falsification". In 1952, he reflected on the development of theories, noting that "The deeper we penetrate and the more extensive our theories become, the less empirical knowledge is needed to determine those theories"[12].

Increasingly focused investigation in autism research is now leading to theories that are maturing due to new evidence being generated by investigators to support them. This process is still in its infancy in autism but is maturing well, nourished by families affected by autism and private and public funding for science, with researchers working in a "balance between competition and cooperation, and between criticism and trust" [13].

The authors of *Autism: Current Theories and Evidence* have been instrumental in developing important biological theories about autism. All are contributing to autism research by constantly refining their theories through rigorous interdisciplinary collaboration and writing. They have generously accepted the invitation to describe the theories they have been developing and the evidence that supports them.

Autism: Current Theories and Evidence has 20 chapters divided into six sections: Molecular and Clinical Genetics; Neurotransmitters and Cell Signaling; Endocrinology, Growth, and Metabolism; Immunology, Maternal-Fetal Effects, and Neuroinflammation; Neuroanatomy, Imaging, and Neural networks; and Environmental Mechanisms and Models. The subjects cover a wide range of current scientific work in the field of autism, with strong and growing evidence to support them and demonstrate both the breadth and the depth of current autism research.

These theories support Kanner's optimism that we will find the biological bases for autism, resulting in improved treatment and prevention. Although it is clear that there is still much to be done, the work presented here demonstrates substantive progress in autism research. It is likely that no unifying theory will explain the myriad processes and forms of autism; rather, interrelated theories will weave together interacting causal pathways to explain its diverse manifestations.

This volume is by no means complete, as there are many other researchers who are developing valid theories in the field of autism. This is a challenging yet rewarding field that is engaging talented investigators. The reader is encouraged to consider how theories and the scientific method, in the hands of these and other dedicated researchers, are leading to greater knowledge and continued progress in autism research.

Acknowledgments I thank Alexander H. Hoon Jr., MD, Susan L. Connors, MD, and Pam K. Gillin, MSN, for their thoughtful reviews and helpful comments. Lisa A. Croen, PhD, provided the innovative graph on the prevalence of autism in Fig. 1. Sharon N. Blackburn, BA contributed creative graphic designs.

References

1. Kanner, L. (1943) Autistic disturbances of affective contact. *Nerv Child* **2**, 217–50.
2. Kanner, L. (1971) Follow-up study of eleven autistic children originally reported in 1943. *J Autism Child Schizophr* **1**, 119–145.
3. Rose, S. (2005) *The Future of the Brain: The Promise and Perils of Tomorrow's Neuroscience*. New York: Oxford University Press, p. 5.
4. PubMed. Online database of the National Library of Medicine (*http://www.ncbi.nlm.nih. gov/sites/entrez*).
5. Lotter, V. (1966). Epidemiology of autistic conditions in young children. *Social Psychiatry* **1**, 124–37.
6. *Diagnostic and Statistical Manual III (1980); IIIR (1987); and IV (1994)*. Washington, D.C.: American Psychiatric Association
7. *International Classification of Diseases – 10th Revision* (ICD-10). (2004). Geneva, Switzerland: World Health Organization.
8. Schopler, E., Reichler, R.J., DeVellis, R.F., Daly, K. (1980) Toward objective classification of childhool autism: Childhood Autism Rating Scale (CARS). *J Autism Dev Disord.* 10:91–103.
9. The American Heritage Science Dictionary. (2002) Boston: Houghton Mifflin Co.
10. Kuhn, T.S. (1996) The *Structure of Scientific Revolutions*, 3rd ed. Chicago: University of Chicago Press, pp. 52–65.
11. Popper, K. *The Logic of Scientific Discovery*. London: Routledge Classics, 1996, p. 57.
12. Einstein, A., Dec. 9, 1952, Albert Einstein Archives 36-549. In: Isaacson, W. *Einstein: His Life and Universe*. New York: Simon and Schuster, 2007, p. 118.
13. Godfrey-Smith, P. (2003) *Theory and Reality*. Chicago: University of Chicago Press, p. 228.

Contents

Contributors

Matthew P. Anderson, MD, PhD
Departments of Neurology & Pathology, Harvard Medical School, Harvard Institutes of Medicine, Boston, MA, USA

Alka Aneja, MD, MA
Department of Psychiatry, Kennedy Krieger Institute, Baltimore, MD, USA

Paul Ashwood, PhD
Department of Medical Microbiology and Immunology, The M.I.N.D. Institute, NIEHS Center for Children's Environmetal Heallth, University of California at Davis, Davis, CA, USA

Bonnie Auyeung, MA
Autism Research Centre, Department of Psychiatry, University of Cambridge, Cambridge, UK

Simon Baron-Cohen, PhD
Autism Research Centre, Dept of Psychiatry, University of Cambridge, Cambridge, UK

Filippo Biamonte, PhD
Laboratory of Developmental Neuroscience, Università Campus Bio-Medico, Rome, Italy

Mary E. Blue, PhD
Kennedy Krieger Institute, Departments of Neurology and Neuroscience, The Johns Hopkins University School of Medicine, Baltimore, MD, USA

Dejan B. Budimirovic, MD
Center for Genetic Disorders of Cognition & Behavior, Kennedy Krieger Institute, Baltimore, MD, USA

Steven Buyske, PhD
Departments of Statistics and Genetics, Rutgers University, New Brunswick, NJ, USA

George T. Capone, MD
Center for Genetic Disorders of Cognition and Behavior, Kennedy Krieger
Institute, Baltimore, MD, USA

Manuel F. Casanova, MD
Dept of Psychiatry & Behavioral Sciences, University of Louisville, Louisville,
KY, USA

Megan Clarke, BA
Center for Genetic Disorders of Cognition & Behavior, Kennedy Krieger
Institute, Baltimore, MD, USA

Susan L. Connors, MD
Department of Neurology and Developmental Medicine, Kennedy Krieger
Institute, Baltimore, MD, USA

Stephen R. Dager, MD
Department of Radiology, Interim Director, University of Washington Autism
Center, University of Washington, Seattle, WA, USA

Benjamin E. Deverman, PhD
Biology Division, California Institute of Technology, Pasadena, CA, USA

Emanuel DiCicco-Bloom, MD
Department of Neuroscience and Cell Biology, Graduate School of Biomedical
Sciences, Department of Pediatrics, UMDNJ-Robert Wood Johnson Medical
School, Piscataway, NJ, USA

Martin M. Evers, MD
Department of Psychiatry, Psychiatrist at Northern Westchester Hospital in
Mt. Kisco, New York, NY, USA

Seth D. Friedman, PhD
Dept of Radiology, University of Washington, Seattle, WA, USA

Martha R. Herbert, MD, PhD
Department of Neurology, Harvard Medical School, Martinos Center for
Biomedical Imaging, Charlestown, MA, USA

Luke Heuer, BS
Division of Rheumatology, Allergy and Clinical Immunology, The M.I.N.D.
Institute, NIEHS Center for Children's Environmental Health, University of
California at Davis, Davis, CA, USA

Christine F. Hohmann, PhD
Department of Biology, Morgan State University, Baltimore, MD, USA

Eric Hollander, MD
Seaver and New York Autism Center of Excellence, Department of Psychiatry,
Mount Sinai School of Medicine, New York, NY, USA

S. Jill James, PhD
University of Arkansas for Medical Sciences, Department of Pediatrics,
Arkansas Children's Hosp Research Institute, Little Rock, AR, USA

Michael V. Johnston, MD
Department of Neurology and Developmental Medicine, Kennedy Krieger
Institute, Baltimore, MD, USA

William G. Johnson, MD
Dept of Neurology, UMDNJ-Robert Wood Johnson Medical School,
Piscataway, NJ, USA

Walter E. Kaufmann, MD
Center for Genetic Disorders of Cognition and Behavior, Kennedy Krieger
Institute, Baltimore, MD, USA

Flavio Keller, MD
Laboratory of Developmental Neuroscience, Università Campus Bio-Medico,
Rome, Italy

Pamela J. Lein, PhD
Center for Research on Occupational and Environmental Toxicology, Oregon
Health and Science University, Portland, OR, USA

Kathryn McFadden, MD
Department of Pathology, University of Pittsburgh School of Medicine,
Pittsburgh, PA, USA

Nancy J. Minshew, MD,
Departments of Psychiatry and Neurology, University of Pittsburgh School of
Medicine, Pittsburgh, PA, USA

Carolyn Boylan Moloney, MD, PhD
Department of Pediatrics, The Johns Hopkins University School of Medicine,
Baltimore, MD, USA

Christina M. Morris, MS
Division of Pediatric Neurology, The Johns Hopkins University School of
Medicine, Baltimore, MD, USA

Roger Panteri, PhD
Laboratory of Developmental Neuroscience, Università Campus Bio-Medico,
Rome, Italy

Carlos A. Pardo-Villamizar, MD
Dept of Neurology, Division of Neuroimmunology and Infectious Disease,
The Johns Hopkins University School of Medicine, Baltimore, MD, USA

Paul H. Patterson, PhD
Biology Division, California Institute of Technology, Pasadena, CA, USA

Issac N. Pessah, PhD
Department of Molecular Bioscience, School of Veterinary Medicine,
Center for Children's Environmental Health Sciences and Disease Prevention,
University of California, Davis, CA, USA

Helen Petropoulos, BE
Dept of Radiology, University of Washington, Seattle, WA, USA

Mikhail Pletnikov, MD, PhD
Department of Psychiatry and Behavioral Sciences, The Johns Hopkins
University School of Medicine, Baltimore, MD, USA

Raili Riikonen, MD, PhD
Children's Hospital, University of Kuopio, Kuopio, Finland

Ian T. Rossman, PhD
Department of Neuroscience and Cell Biology, Graduate School
of Biomedical Sciences, UMDNJ-Robert Wood Johnson Medical School,
Piscataway, NJ, USA

Dennis W. W. Shaw, MD
Department of Radiology, University of Washington, Seattle, WA, USA

Harvey S. Singer, MD
Division of Pediatric Neurology, The Johns Hopkins University School of
Medicine, Baltimore, MD, USA

Stephen E. P. Smith, BS
Biology Division, California Institute of Technology, Pasadena, CA, USA

Madhura Sreenath, BA
Department of Neurology, UMDNJ-Robert Wood Johnson Medical School,
Piscataway, NJ, USA

Edward S. Stenroos, BS
Department of Neurology, UMDNJ-Robert Wood Johnson Medical School,
Piscataway, NJ, USA

Elaine Tierney, MD
Department of Psychiatry, Kennedy Krieger Institute, Baltimore, MD, USA

Judy Van de Water, PhD
Division of Rheumatology, Allergy and Clinical Immunology, The M.I.N.D.
Institute, NIEHS Center for Children's Environmetal Heallth, University of
California at Davis, Davis, CA, USA

Diane L. Williams, PhD
Department of Speech Language Pathology, Duquesne University,
Pittsburgh, PA, USA

Wensi Xu, BS
Biology Division, California Institute of Technology, Pasadena, CA, USA

Andrew W. Zimmerman, MD
Department of Pediatrics and Developmental Medicine, Kennedy Krieger
Institute, Baltimore, MD, USA

List of Color Plates

Part I
Molecular and Clinical Genetics

Chapter 1
ENGRAILED2 and Cerebellar Development in the Pathogenesis of Autism Spectrum Disorders

Ian T. Rossman and Emanuel DiCicco-Bloom

Abstract Autism and autism spectrum disorders (ASD) are complex neurodevelopmental diseases with strong genetic etiologies. While many brain regions are implicated in ASD pathogenesis, many studies demonstrate that the cerebellum is consistently abnormal in ASD patients, both neuroanatomically and functionally. Recently, the human gene *ENGRAILED2* (*EN2*), an important regulator of cerebellar development, was identified as an ASD-susceptibility gene, a finding replicated in three data sets. We review much of the literature implicating an abnormal cerebellum in ASD, as well as the molecular and cellular development of the cerebellum with respect to possible pathogenetic mechanisms that contribute to the ASD phenotype. In addition, we explore the role of *EN2* in normal cerebellar development as well as animal models in which abnormal *En2* expression produces ASD-like behavior and neuropathology. We also share preliminary data from our laboratory that suggest that *En2* promotes postnatal cerebellar development via cell cycle regulation and interactions with extracellular growth signals. After reviewing these data from different disciplines, it is our hope that the reader will better understand how abnormal cerebellar development contributes to ASD pathophysiology and pathogenesis. Further, we underscore the importance of multidisciplinary approaches to identifying ASD-associated genes and their functions during development of brain structures known to be abnormal in ASD patients.

Keywords Autism · cerebellum · genes · *engrailed2* · patterning · neurodevelopment

Introduction

Autism spectrum disorders (ASD) are highly heritable, neurodevelopmental disorders that affect 1:150 children in the United States [1]. Public and scientific interest in ASD has exploded over the last two decades, spurring intense

Emanuel DiCicco-Bloom
675 Hoes Lane, RWJSPH Room 362, Piscataway, NJ 08854, USA
e-mail: diciccem@umdnj.edu

A.W. Zimmerman (ed.), *Autism*, DOI: 10.1007/978-1-60327-489-0_1,
© Humana Press, Totowa, NJ 2008

3

investigations into specific genetic and environmental factors that contribute to ASD susceptibility. The result of these efforts is a broadened clinical definition that better describes the heterogeneity of ASD phenotypes and a profounder understanding of the molecular and neuropathologic complexities that underlie the disease. For the first time in history, we can assay the human genome at the level of the individual nucleotide to look for molecular patterns in ASD patients and families. At the same time, neuroimaging provides windows into the living brain of autistic children and adults, allowing comparisons that were once only possible postmortem. To put together all these data from genetics, biochemistry, neuroimaging, and epidemiology is daunting and, given the numerous brain regions and molecular pathways implicated in ASD pathogenesis, is beyond the scope of this text. Thus, our goal in this chapter is to focus on one brain structure, the cerebellum, and the evidence supporting its role as the most consistently abnormal region of the ASD brain. To do so, we will review the evidence from neuropathology and neuroimaging studies demonstrating reproducible deficits in the cerebella of ASD patients. Additionally, we will examine cerebellar development and pay attention to specific genes that coordinate cerebellar growth and cellular organization, while considering the possible consequences of disrupting these ontogenetic processes. Finally, we will demonstrate that one cerebellar gene, *ENGRAILED2*, is not only vital for normal cerebellar development but is indeed an ASD-susceptibility gene that functions to regulate postnatal cerebellar growth during critical periods at which ASD symptoms may first become noticeable. Thus, it is our goal that the reader will come away with a better appreciation for the role of the cerebellum in normal brain function, as well as ASD pathogenesis, and an understanding of how molecular genetics and neurodevelopmental biology can be utilized together to investigate a complex disease such as ASD.

History of Autism

Autism was first described by Leo Kanner in 1943 [2] as a developmental disease, present from birth, in which reciprocal social behavior, language, and communication are impaired and patients display restricted interests and repetitive behaviors. Today, these deficits make up the core DSM-IV criteria of ASD, which includes diagnoses of classical autism, Asperger syndrome, and pervasive developmental disorder-not otherwise specified (PDD-NOS). Though no closer to a cure for ASD than Dr. Kanner was in the 1940s, our clinical and basic science understanding of ASD symptoms and disease pathogenesis has improved dramatically, fueled by ever-advancing genetic and biomedical technologies. Further, increases in prevalence and public awareness of ASD have stimulated financial and intellectual movements to make ASD research a priority at all academic and government levels. As a result of these efforts, we now appreciate ASD as a complex genetic disorder and have begun to identify ASD-associated genes that regulate the development of brain structures known to be abnormal in humans with ASD. Uncovering these relationships between genes and normal neurodevelopmental processes provides insight into the mechanisms underlying abnormal brain development, and this information will bring us closer to fully understanding, and hopefully preventing, ASD.

ASD Neuropathology and Pathogenesis

Though extensively characterized clinically, ASD pathogenesis remains a mystery. Thus, pathologic abnormalities in brain structure and function may provide clues about the etiologies of this disease process. Postmortem and imaging studies have demonstrated a wide range of underlying neuropathology, involving a multiplicity of cortical and subcortical regions (reviewed by Palmen et al. 2004 [3]). For example, abnormal brain growth, and overgrowth, has been demonstrated by several neuroimaging studies comparing autistic children to normal age-matched controls [4, 5, 6, 7]. Further, increased brain size and weight have been described in postmortem analyses of autistic brains [3, 8, 9, 10, 11]. These studies suggest overall brain development is abnormal in ASD; however, it is currently unknown whether these pathologic abnormalities are causes or symptoms of ASD. More focused analyses of specific brain regions have found developmental abnormalities grossly, as well as microscopically. Several studies demonstrated cytoarchitectural abnormalities within cortical minicolumns, suggesting that ASD symptomology may result from abnormal connectivity within and between functional domains [12, 13, 14]. Additionally, several studies found reduced corpus callosal size in children with ASD, as well as abnormal communication between cortical structures during cognitive tasks, further suggesting that underconnectivity contributes to behavioral and cognitive abnormalities of autism ([15, 16, 17, 18, 19], see Minshew Chapter 18). Cortical dysgenesis, including cortical thickening, ectopic gray matter, disorganized lamina, and poor differentiation of the gray-white matter boundary, was described in six case studies by Bailey et al. (1998) [11]. In another study of subcortical brain structures, abnormally dense cell packing was described in hippocampi, amygdalae, and entorhinal cortices from 9 of 14 autistic brains [3, 9, 20]. A more recent study found amygdalar overgrowth during early postnatal development in ASD patients [21]. While compelling, based on the role of limbic structures, these studies await replication by other researchers and may not be generalizable to the ASD population at large. Collectively, though, these data describe the ASD brain as abnormal in structure, and accordingly in function, thus highlighting the importance of normal brain development. It is likely that multiple brain abnormalities exist within and between ASD patients and that the heterogeneity of clinical presentation is mediated by these brain differences. Therefore, the reader is encouraged to consider how hindbrain and cerebellar development and dysfunction, discussed below, could interact with these brain abnormalities in ASD.

Cerebellar Development, Function, and ASD

The cerebellum remains the most consistently abnormal brain region studied in humans with ASD. Cerebellar size, morphology, and function have all been found to be abnormal, and irregular patterns of cerebellar growth were found to begin from early infancy and continue into adolescence. In fact, 21 of 29 postmortem studies have found reduced numbers of Purkinje neurons, the primary cerebellar efferent neurons, without evidence of gliosis, suggesting these deficits may have occurred early in development [3, 9, 22, 23]. Purkinje neurons develop prenatally and are necessary for coordinating late gestation and postnatal cerebellar development via mechanical and neurochemical support of another cell population, the cerebellar granule neurons. Therefore, genetic or biochemical insults affecting early Purkinje neuron development could have sustained effects that disrupt cerebellar growth and function, contributing to ASD pathogenesis. Before discussing the body of evidence implicating cerebellar dysfunction in ASD, it is important to review how the cerebellum develops, keeping in mind that ASD pathogenesis may stem from perturbations at the genetic, biochemical, or cell biological level. The develop-ing mouse brain serves as one of the best models to understand these processes.

Cerebellar Development

Following neural tube closure in early development, the anterior region is subdivided into three vesicles: the prosencephalon, which gives rise to the telencephalon and diencephalon of the forebrain; the mesencephalon, which develops into the midbrain; and the rhombencephalon, which gives rise to the metencephalon (rostral) and myencephalon (caudal) of the hindbrain [24]. The anterior portion of the rhombencephalic metencephalon abuts the mesen-cephalon, at a genetically distinct boundary known as the mid–hindbrain junction. The identification and assignment of cells to this region (a process known as patterning) is achieved through induction of transcription factors in spatially and temporally restricted domains, which will be discussed later (Fig. 1.1 and Color Plate 1, following p. XX).

At this early stage (Embryonic day (E)8–9 in the mouse), the cerebellar anlage is delineated just caudal to the mid–hindbrain junction, along the dorsolateral aspect of the first rhombomere. As development progresses, morphogenetic bending of the neural tube creates the pontine flexure ventrally, causing a wide gap to develop dorsally and producing the distinctive diamond shape of the fourth ventricle [25]. Neuroepithelial proliferation in the cerebellar primordium occurs along the ventricular zone (VZ), similar to the forebrain, allowing the cerebellar primordium to thicken and take shape along the fourth ventricle. Between E10 and 14 in the mouse, cerebellar neuron progenitors begin to exit the cell cycle, migrate out of the proliferative VZ, and differentiate;

Patterning genes compartmentalize the neural tube into functionally distinct domains.

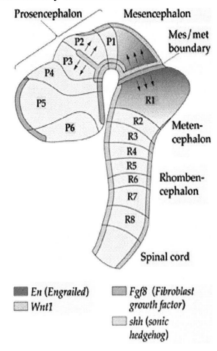

Fig. 1.1 Shortly after neural tube closure, patterning genes become expressed along the anterior-posterior, dorsal-ventral, and medial-lateral axes. These genes help compartmentalize the neural tube, giving rise to the forebrain (prosencephalon), midbrain (mesencephalon), and hindbrain (rhombencephalon). Depicted here is the neuromeric structure of the brain at approximately E8.5 in the mouse, when diffusible morphogens Wnt1, Fgf8, and *En* patterning genes are delineating the mid–hindbrain junction, and Shh, expressed by the floorplate, is involved in dorsal-ventral patterning. Note: *italicized* print denotes genes, whereas non-italicized print denotes gene products (i.e., proteins). From: Gilbert, S. F. 1997, http://8e.devbio.com/about.php, April 18, 2003

the first cells to do so between E10 and 11 will become neurons of the deep cerebellar nuclei [25, 26]. Between E12 and 14, cells that exit the cell cycle migrate out of the VZ and form a layer under the presumptive deep nuclei cells, giving rise to Purkinje neuron precursors, one of the two major types of cerebellar cortical neurons. Over the next several days of development, these Purkinje cell precursors continue to migrate out through deep nuclei precursors, forming a primitive Purkinje cell layer (PCL) that is several cells thick. A secondary site of proliferation develops as early as E10–12 at the posterior (or dorsal) edges of the cerebellar primordial neuroepithelium, the rhombic lip. These rhombic lip cells migrate out rostrally to form a superficial layer over the differentiated deep nuclei and Purkinje cell precursors; this layer is known as the external germinal layer (EGL) and will remain a secondary site of proliferation into postnatal development [25, 26, 27]. To achieve these early cytoarchitectural arrangements, cerebellar progenitor cells rely on radial glial- and extracellular matrix-supported migration away from the VZ [28]. In fact, mutations in mice that disrupt Purkinje precursor migration, such as *reeler*, disrupt the growth and trilaminar organization of the adult cerebellar cortex. Thus, early in cerebellar development, three major neuronal precursor populations develop: deep cerebellar nuclei, Purkinje, and rhombic lip precursors; the production and organization of these cells form a blueprint for later cerebellar patterning and histogenesis.

Postnatal Cerebellar Development

In mammals, following prenatal hindbrain development, the cerebellum undergoes a major postnatal expansion during which neurogenesis continues through infancy into early childhood. In humans, this period encompasses the ages (1–3 years) when ASD symptoms first appear, long after forebrain development is completed. Cerebellar granule neurons develop from a genetically distinct population of cells from the dorsal rhombic lip as early as E10.5 [29]. These cells express the bHLH transcription factor *Math1*, deletion of which eliminates the entire population of granule neuron precursors (GNPs) [27]. Throughout embryogenesis, these *Math1*-expressing cells migrate rostrally along the subpial surface of the developing cerebellum to form the EGL, a strongly proliferative niche overlying the Purkinje cells, as described above. Postnatal GNP development is spatially restricted, such that proliferating and differentiating cells are compartmentalized, even within domains. In the EGL, for example, proliferation occurs superficially in the outer EGL while cell cycle exit and inward migration along Bergmann glia begin in the inner EGL. Differentiation begins as GNPs migrate from the inner EGL into the molecular layer, where granule neuron parallel fibers are extended, and form contacts with Purkinje cell dendrites. While parallel fibers are left in the molecular layer, differentiating granule neuron cell bodies continue to migrate inwardly, extending their axons behind them to form the inner granule layer (IGL), below the PCL. Having arrived in the IGL, fully differentiated granule neurons terminate their migration and become integrated into cerebellar circuitry as glutamatergic interneurons (Fig. 1.2).

Proliferating GNPs in the EGL rely on the underlying neurons of the PCL to secrete extracellular growth factors, such as Sonic hedgehog (Shh) [30, 31, 32], pituitary adenylate cyclase activating peptide (PACAP) [33, 34, 35, 36], and insulin-like growth factor-1 (IGF-1) [37, 38, 39]; these growth factors, each acting through their own receptor(s), provide trophic support and survival of the developing EGL cells, as well as mitogenic (cell cycle promoting) and antimitogenic (cell cycle inhibiting) regulation. While homotypic cell–cell contacts

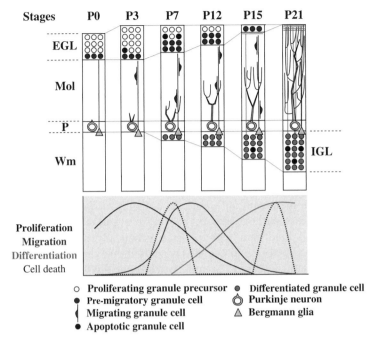

Fig. 1.2 Granule neuron precursors proliferate superficially in the external germinal layer (EGL), then migrate inwardly along Bergmann glia. During migration, the granule neuron precursors (GNPs) differentiate, producing an axon with parallel fibers that communicate with Purkinje cell dendrites, thereby forming the molecular layer. GNP migration terminates below the Purkinje cells, forming the IGL, where they are functionally integrated into the cerebellar circuitry. EGL, external granule cell layer; Mol, molecular layer; P, Purkinje cell layer; IGL, internal granule cell layer; Wm, white matter. Illustrated by Anthony Falluel-Morel, PhD (falluean@umdnj.edu) (*see* Color Plate 1)

between GNPs has been found to promote proliferation in vitro [40], in vivo proliferation is severely compromised by Purkinje cell loss [41]. Indeed, Shh, IGF-1, and PACAP mRNA and proteins have all been localized to Purkinje cells during development [30, 31, 33, 36, 38]. Further underscoring the developmental relationship between Purkinje cells and GNPs, adult granule neuron numbers were found to correlate to the number of Purkinje cells [42, 43, 44], and reciprocally, EGL integrity is required for normal Purkinje cell migration [45]. Genetic deletions targeted to Purkinje neurons that disrupt their development are deleterious to the future production of granule neurons, underlying the importance of the Purkinje-granule neuron relationship during development [32, 41].

Cerebellar Patterning Genes

Constructing regionally distinct boundaries in the neural tube so that future structures develop properly requires spatially and temporally regulated patterning gene expression. The neuroanatomical and functional abnormalities described in ASD may result from aberrant genetic events during crucial developmental timepoints. Further, many developmental patterning genes are evolutionarily conserved orthologs of *Drosophila* genes that encode positional information in developing fly larvae; *En1* and *En2*, for example, are two mammalian orthologs of the *Drosophila* gene *en*. These genes belong to a family of homeodomain containing transcription factors that bind to DNA to regulate the expression of other genes. The mammalian *En* genes operate similarly in mammals to specify and pattern the mid–hindbrain junction and the developing cerebellum. While mutations in essential patterning genes are often incompatible with life, variations in non-coding sequences may consequently alter the regulation of that gene's expression. These subtle changes in genetic code (as well as functional mutations) are responsible for the evolution of species but may also contribute to neuro-developmental disease pathogenesis by altering brain development. As alluded to earlier, the mid–hindbrain junction is a genetically distinct boundary in which transcription factors and secreted signals induce and cross-repress one another to delineate the cerebellar anlage from the midbrain, rostrally, and the hindbrain caudally [25] (see Fig. 1.1). The strategy of cross-repression and induction between transcription factors is not limited to the mid–hindbrain but is a common phenomenon at anatomical boundaries during development. In preventing ectopic expression of a gene within a neighboring population of neural precursors, specialized populations of neurons can develop side by side, a strategy well characterized in motor neuron development of the ventral spinal cord [46, 47, 48, 49, 50]. These epistatic relationships between patterning genes are essential, and abnormalities in their spatiotemporal expression patterns may be pathogenetic events in developmental disorders such as ASD.

As early as E8 in the mouse, the first murine ortholog of *Drosophila en*, *En1* is expressed across the mid–hindbrain junction, and its expression extends caudally demarcating the entire cerebellar anlage [51, 52]. *En2* is expressed 6–12 h later just caudal to the mid–hindbrain junction and overlaps with *En1* to demarcate the cerebellar anlage [51, 53]. The *En* genes are highly homologous [52] and functionally redundant [54, 55]; however, their expression patterns are spatially distinct, therefore they demonstrate divergent roles during development, as demonstrated by phenotypic differences between *En* mutants. Thus, normal brain development demands gene expression to occur in the right places, but also at the right times, to avoid aberrant gene interactions; given the dynamic interplay between patterning genes, it is not hard to imagine that dysfunction in one signal could have a "ripple effect" by altering expression in otherwise normal systems.

Highlighting the importance of interacting molecular systems in brain development, *En* signaling at the mid–hindbrain border was found to depend on the secreted factor *Wnt1*. Though *Wnt1* demarcates the mid–hindbrain border rostrally, and is not expressed in the cerebellar anlage, it was found to induce *En* expression. In fact, deletion of *Wnt1* results in loss of the entire cerebellum, presumably through lost induction of *En* [56]. Further, En proteins were found to work in concert with Fgf8 to repress *Pax6.1*, an anterior forebrain patterning gene, to maintain midbrain integrity [57]. Consistent with this, ectopic *Fgf8b* expression in either the forebrain or other hindbrain regions was found to induce *En1*, *En2*, *Pax5*, and *Gbx2* expression, conferring an anterior hindbrain phenotype to these ectopic sites [58]. Thus, the power of a gene to alter a tissue's fate is not limited to a particular group of cells but may be manifested in any neural population in which its protein can be biochemically active (e.g., in the case of FGF8b to bind to its receptor). Additionally, this group also demonstrated that *En2* and *Gbx2* are the first genes induced by Fgf8b in E9.5 mice, and that while *Fgf8*, *Wnt1*, and *Pax5* can initially be expressed independent of *En*, intact *En* expression is required for the maintenance of these genes' expression [59]. Thus, the cerebellar anlage arises from complex gene–gene, and gene–growth factor interactions in which induction and cross-repression are utilized to specify boundaries and confer distinct mesencephalic and metencephalic identities.

Patterning Gene Expression

As cerebellar development progresses and populations of cerebellar neurons emerge, patterning gene expression remains integral, and as the reader will see, unique combinations of genes are utilized to direct neural precursors to become a specific type of cerebellar neuron. Recently, Morales and Hatten [60] identified different populations of cells in the developing cerebellar VZ at E12.5. These populations were identified as either proliferating progenitor cells or postmitotic deep nuclear precursor cells via their combinatorial patterns of transcription factor expression. Both cell groups expressed three amino acid loop (TALE) family transcription factors *MEIS1* and *IRX3;* however, only postmitotic deep cerebellar nuclei precursors that had migrated superficially were found to express *MEIS2* and the LIM homeodomain containing transcription factors *LHX2/9* [60]. These transcription factors remain expressed in deep cerebellar nuclear precursors through mid-gestation as they migrate rostrally and inwardly below developing Purkinje cell precursors. Thus, gene expression allows identification of emerging cell populations but may also allow their tracking from birth through functional integration into neuronal circuitry.

A similar approach to that of Morales and Hatten (2006) [60] was used to demarcate deep cerebellar nuclear precursors originating from the rhombic lip. Fink et al. (2006) [61] found sequential expression of transcription factors *Pax6*, *Tbr1*, and *Tbr2* identified rhombic lip cells that migrated from the subpial stream through a nuclear transitory zone to join their VZ-derived counterparts. Interestingly, these rhombic lip cells gave rise exclusively to glutamatergic projection neurons, whereas VZ-derived deep cerebellar nuclear precursors expressed *Pax2* and were fated to become GABAergic interneurons [61]. Further, subpopulations of deep cerebellar nuclear cells found to be molecularly distinct in development remained so into adulthood and would later be found to occupy predictable medial-lateral positions within the deep cerebellar nuclei [61]. These data suggest that early gene expression functionally segregates some cerebellar neurons as progenitors; therefore, developmental abnormalities that disrupt gene expression could have predictable consequences on cerebellar function.

Combinatorial Gene Expression Confers Cell-Specific Identity

Given that combinatorial gene expression can specify multiple neuronal subtypes from a population of neural precursors, it follows that combinatorial stimuli (intrinsic and extracellular) can have a similar effect on a seemingly homogeneous population of differentiating cells: Purkinje neuron precursors. These cells give rise to a heterogeneous population of mature neurons that express positionally defined combinations of genes and proteins during development and adulthood. Given that Purkinje neuron deficits are the most consistent neuropathologic finding in ASD, these subtle differences in Purkinje cell identities (and possibly also functions?) may be very important in ASD pathogenesis and heterogeneity. To understand these differences in Purkinje neuron identity, we can examine developmental timepoints at which Purkinje neuron precursors genetically diverge from other neural precursors and from themselves. Beginning around E10.5, *En2* is expressed in all proliferating cells of the midbrain and cerebellar primordia [53]. By E12.5, though, some Purkinje neuron precursors exit the cell cycle, migrate out of the VZ, and begin to express the transcription factors *LHX1/5*. Thus, at this early stage, postmitotic Purkinje neuron precursors are homogeneously *LHX1/5* positive and can be differentiated from both proliferating (*En2*-expressing) VZ cells as well as the postmitotic deep nuclei precursors [60]. Interestingly, from E12.5 to 17.5, Purkinje cell development is regulated by *En2*, which works antagonistically with other homeodomain proteins, *Hoxa5* and *Hoxb7*, by repressing a Purkinje cell-specific gene, *Pcp2* [62]. However, by E16.5, the majority of Purkinje neurons express both *LHX1/5* and the differentiation marker calbindin, irrespective of individual cell *En2*, *Hox*, or *Pcp2* expression, suggesting that different genetic mechanisms are employed simultaneously to achieve a Purkinje cell fate. Thus, heterogeneity within the Purkinje neuron population is further achievable by augmenting the level of activity of each of these components and by other genetic and extracellular signals expressed later during development.

In addition to promoting cerebellar neuronal fates and subpopulation specialization, patterning genes, including *En2*, influence the rate of cerebellar growth as well as the organization of the organ along several spatial axes. To achieve these patterning events, *En2* expression during late embryogenesis (mouse E17.5-P0) becomes restricted to longitudinally oriented stripes, in which all cell types within the anterior-posterior/dorsal-ventral stripe express the gene. This striped pattern of *En2* expression sets up a cytoarchitectural map across the developing cerebellum, providing positional information to

Patterning gene expression in the prenatal WT and *En2* KO cerebellum.

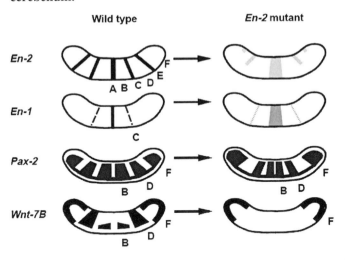

Fig. 1.3 Several patterning genes are expressed complementary to *En2* expression. While expression of cerebellar patterning genes is preserved in the *En2* knock-out (KO) mouse mutant, discussed below, their spatial expression patterns are disrupted. This suggests that positional information necessary for developing functional compartments and neuronal subpopulations may be lost, contributing to abnormal cerebellar morphology and function. WT, wild type. Adapted from Millen et al. (1995) [111]

innervating fibers and migrating neurons to assist finding their targets and regulates where other important patterning genes are expressed [63, 64, 65]. These other genes include *En1*, *Wnt7b*, and *Pax2*, which are also expressed in spatially restricted patterns of stripes, complementary to those of *En2* expression (Fig. 1.3). These patterning gene expression profiles are reminiscent of their phylogenetically older *Drosophila* orthologs, suggesting that mammalian compartmentalization may be the evolutionarily conserved process of larval segmentation in the fly [66].

Biochemical and Molecular Signals in Cerebellar Development

As cerebellar development progresses, the organ becomes less homogeneous, forming a recognizable ultrastructure composed of the hemispheres, the vermis, and the deep cerebellar nuclei. Within these structures, biochemical and molecular signals are expressed in parasagittal banding patterns that compartmentalize the developing cerebellum into reproducible topographies and functional domains [63, 67, 68, 69, 70, 71, 72]. This process begins mid-gestation and may be important for conferring identities to subpopulations of otherwise homogeneous cell types, such as Purkinje neurons. Though the functional consequence of having bio-chemically distinct Purkinje neuron subpopulations is not fully understood, patterns of Purkinje cell loss in ASD may reflect differential susceptibility between these neurons to genetic or biochemical insults. Characterizing the biochemical differences between surviving Purkinje neurons in postmortem samples from ASD patients may provide insight into the mechanisms of cell loss in these patients. For example, Leclerc et al. [71] demonstrated that Zebrin II (aldolase-C) and P-path (9-*O*-acetylated glycolipids) immunopositivity marked parasagittal bands of complementary compartments across the cerebellum with medial-lateral symmetry. Further, these markers identified three distinct populations of Purkinje neurons: Zebrin+, P-path−; Zebrin−, P-path+; and Zebrin+, P-path+. Addi-tionally, olivocerebellar and mossy fiber afferents were found to largely respect these compartments [71]. Further, expression of the previously mentioned Purkinje cell gene *L7/Pcp2* similarly follows parasagittal banding during late embryogenesis, and olivocerebellar afferents again project to specific clusters of immunopositive and negative Purkinje cells [73, 74]. Thus the cerebellum is functionally and biochemically, as well as anatomically, compartmentalized during embryo-genesis by genetic signals [72], suggesting that genetic perturbations could disrupt biochemical expression patterns and functional compartments.

Cerebellar Abnormalities in ASD

Given that almost all social or cognitive behaviors that are abnormal in ASD require some degree of attention regulation, it is possible that several functionally abnormal brain regions communicate with a common dysfunctional brain structure: the cerebellum. Though once believed to be solely involved in motor coordination, balance, and motor memory, the human cerebellum participates in higher order functions, including speech and attention regulation. In fact, shifting attention between stimuli requires an intact cerebellum, because even slight cerebellar abnormalities have been found to disrupt rapid shifting of attention. Interestingly, cerebellar activation patterns in ASD patients have been found to be abnormal, with reduced cerebellar activation in tasks requiring attentional shifting and increased activation in motor tasks [75]. In fact, ASD patients were found to perform similarly to patients who sustained cerebellar injury, with both groups showing reduced accuracy in shifted attention as a function of reaction time, compared to age-matched controls [76]. Cerebellar dysfunction at the behavioral and imaging level may have correlates at the molecular level. Purkinje neurons are activated by glutamatergic (i.e., excitatory) inputs from granule neu-rons and olivary climbing fibers but are themselves GABAergic (i.e., inhibitory). The rate-limiting enzyme glutamic acid decarboxylase (GAD67) that converts glutamate to GABA is highly expressed in adult Purkinje neurons. Recently, Yip et al. (2007) [77] found GAD67 expression by Purkinje neurons in postmortem cerebella of humans with ASD was reduced by 40% compared to normal controls.

This same group previously reported an abnormal increase in GAD67 expression by cerebellar basket cells [78], which function to inhibit Purkinje neuron firing by modulating action potential propagation at the axon hillock. These findings together suggest that surviving Purkinje neuron function is depressed in ASD by reduced neurotransmitter production and increased synaptic inhibition [77]. That deficits in attention regulation correlate with abnormal cerebellar activation, together with consistent cerebellar neuropathological findings and recent evidence of molecular dysregulation of Purkinje neurons suggests that ASD pathogenesis may be linked to abnormal cerebellar development in many patients.

Effects of Cerebellar Abnormalities in ASD

With even a basic understanding of how the cerebellum develops, one can appreciate how ASD-associated cerebellar abnormalities may arise. Given their prenatal development, the Purkinje neuron deficits described in ASD suggest that a prenatal insult may disrupt survival or maturation of these cells. If so, then cerebellar growth rates could be altered from birth, with growth retardation depending upon the severity of Purkinje cell loss or dysfunction, whereas overgrowth may be due to compensatory growth factor production by olivocerebellar afferent fibers; additionally, cerebellar size may appear grossly normal though underlying cellular losses have begun. However, Purkinje neuron maturation continues concurrently with GNP proliferation and the formation of functional circuitry; therefore, later Purkinje cell loss could also occur with minimal effects on cerebellar size, while greatly altering its function. Neuroimaging studies of cerebellar grey and white matter growth from early infancy through adolescence have indeed demonstrated dysregulated cerebellar growth in ASD patients, with overgrowth in early childhood correlating with increased head circumference [4, 5]. At 2–3 years old, cerebellar grey matter volumes were identical between ASD and normal children; however, by age 6–9 normal children's grey matter volumes increased whereas ASD children's did not; grey matter volumes remained reduced in ASD into adulthood. White matter volumes were 40% larger in ASD infants than normal children at age 2–3, normal children experienced a 50% increase in volume from ages 2–3 to 12–16 years, whereas ASD children's white matter volume growth was almost flat (7%). Naturally, these disparate grey and white matter growth patterns translate to vastly different developmental gray : white matter ratios, a possible indication of gross cerebellar dysfunction. While compelling, these data reflect changes in a subpopulation of children with ASD and may not be generalizable to the ASD community at large.

Numerous studies have measured overall cerebellar volume in children, adolescents and adults with ASD, with differing results. While initial studies suggested posterior cerebellar vermi were smaller in ASD than normal adults and exhibited abnormal foliation, these findings have not been replicated

[79, 80]; in fact, methodological irregularities, such as failure to control for IQ, may have accounted for these changes [81]. That said, when matched for IQ and age and corrected for total brain volume, the majority of neuroimaging and postmortem studies have reported an increase in cerebellar volume in ASD [4, 5, 81, 82, 83, 84, 85, 86]. Deep cerebellar nuclei neurons were also found to undergo changes in size and appearance, beginning large in childhood and becoming small and pale later in life. Though these changes remain to be replicated, similar changes were noted in the presynaptic neurons of the inferior olive, suggesting that olivocerebellar circuitry may be functionally abnormal in ASD. Similar to Purkinje-granule neuron numbers, there exists a linear relationship between brainstem olive and cerebellar Purkinje neuron numbers [68, 87]. Indeed, olivocerebellar projections begin early in brain development, and the olivary climbing fibers are known to provide trophic support for developing Purkinje neurons [88, 89, 90, 91]. Purkinje cell losses described in ASD ought to correlate with similar losses in correlating brainstem nuclei; however, only sporadic ASD case studies have found inferior olive cell losses [11, 92]. Thus, the period of ASD pathogenesis appears to begin prenatally, beginning early enough for neuronal populations to compensate for abnormalities in their target cells, though the exact timing and causes remain unknown.

Extracerebellar Functional Deficits in ASD

The maxim "structure dictates function" is often repeated in biology classes; therefore, structural brain abnormalities observed in ASD brains may underscore the behavioral deficits observed in humans with ASD. Functional neuroimaging in live patients suggests that functionality or connectivity within and between specific brain loci is abnormal [93]. For example, social cognition and emotion recognition deficits correlated with reduced amygdalar activation in patients with ASD [94, 95, 96, 97]. In several other studies, ASD patients showed hypoactivation and abnormal lateralization of the posterior superior temporal sulcus in tasks that infer mental intent by tracking the eye movements of others, as well as reduced activation in response to human voices [98, 99, 100, 101]. Further, abnormal processing of faces by patients with ASD is well characterized and correlates with reduced activation of the fusiform gyrus and inappropriate activation of adjacent cortical structures not associated with processing face stimuli [102, 103, 104, 105]. However, abnormal face perception and fusiform activation in ASD may be a byproduct of inappropriate focus, as these patients attended away from eyes and toward other facial features; in fact two studies found a correlation between time spent focusing on eye regions and magnitude of fusiform gyrus activation in ASD [104, 106]. Together, these studies suggest abnormal socialization and behavior in ASD may be due to processing of stimuli by inappropriate brain regions. Further, abnormal function in one structure may disrupt global processes in which that abnormal region communicates with other brain loci to code and respond to environmental cues. Thus, abnormal amygdalar coding of emotional information in facial expressions may disrupt higher order face recognition, leading to disrupted social behaviors associated with ASD.

En2 Transgenic Mice Phenocopy ASD Cerebellar Neuropathology

Animal models are useful tools for studying complex human diseases such as ASD. However, finding an appropriate animal model for ASD is difficult because it is a uniquely human disease characterized by behavior deficits in modalities absent from rodents (i.e. language, emotional reciprocity, etc.). However, several mouse mutants have been generated over the last several decades demonstrating a range of neuropathological deficits that appear similar to (or phenocopy) human disease states. Specifically, transgenic animal models in which *En2* expression has been altered, either by knock out (KO) or by overexpression, exhibit cerebellar neuropathology similar to that of ASD [63, 64, 107, 108, 109, 110, 111]. *En2* KO cerebella are approximately 30% smaller than the wild type (WT), resulting from a 30% reduction in numbers of all cell types, including Purkinje and granule neurons [63, 64, 110, 111]. Differences between WT and *En2*-deficient cerebellar development are evident as early as E15–16, when mutant cerebella fail to fuse medially and overall cerebellar size is diminished [64]. By E17.5, though, these differences have disappeared, suggesting that the absence of *En2* expression imparts a delay in cerebellar maturation during embryogenesis. In addition to gross morphological changes during late embryogenesis, *En2* mutant cerebella maintain expression of patterning genes although abnormalities are found [111]. Defects in hemispheric and vermal foliation are apparent in *En2* KO mice early in the postnatal period; while overall morphology is similar, there is an apparent delay in foliation such that mutant cerebella resemble 1–2 days younger WT cerebella. In addition to retarded growth and foliation, the formation of fissures that separate lobules is also delayed, and often incomplete, leaving normally separated lobules to appear fused in the KO hemispheres and vermis. These morphological deficits are more severe in the vermis where *En2* expression is greatest and mostly limited to the posterior lobules; in fact, early postnatal development of the anterior cerebellum appears morphologically normal though significantly delayed [64]. As a consequence of abnormal fissure formation, posterior lobes of the vermis are abnormally positioned, such that lobe VIII becomes a branch of IX and the sizes of lobes VII and VIII are reduced; interestingly, specific hypomorphic vermal phenotypes have also been described in groups of ASD patients [5, 64, 79, 80, 111].

Molecular Abnormalities in En2 Mutant Mice

The gross morphological abnormalities observed in postnatal cerebella of the *En2* KO mouse are punctuated by changes in the normal biochemical bands discussed above, and at the molecular level, these posterior lobules appear transformed so that they express characteristics of more anterior cerebellar molecular phenotypes [63, 111]. These patterns of molecular expression create compartments that are critical for establishing proper pre- and postsynaptic connections. Thus, an improper molecular environment may lead to abnormal cerebellar circuitry in these mice, similar to the functionally abnormal cerebella of human ASD patients. As demonstrated embryonically, postnatal expression of patterning genes are maintained in the absence of *En2* although their spatial boundaries are abnormal; for example, *Wnt7b* expression extends further posteriorly in *En2* KO cerebella, suggesting that these lobules have taken on an anterior genetic phenotype. Further, Zebrin II- and P-path banding patterns are overall maintained in the KO cerebella though there are several regions displaying abnormal or absent banding patterns [63]. Though gross abnormalities are largely absent from the anterior lobes, KO cerebella display abnormal Zebrin II banding anteriorly in lobule III and dorsomedially in lobule V, as well as posteriorly in lobules VIII and IX; in the lateral hemisphere, there is precocious banding along the fused crus II-paramedian lobule at P2 that gives way to an abnormally small band in the adult. In addition to abnormal biochemical markers of compartmentalization, the *En2* KO mouse cerebellum exhibits absent bands of expression of the Purkinje cell-specific *L7/Pcp*, most notably in the hypomorphic lobes of the posterior cerebellum. These disordered and lost compartmental markers suggest that subpopulations of Purkinje cells may be more susceptible to abnormal *En2* expression; these subpopulation differences may provide insight into patterns of Purkinje cell loss in humans with ASD [3]. By extension, this would suggest that functionally distinct cerebellar domains may be more susceptible to genetic perturbation and thus contribute to an ASD phenotype. In support of this idea, spinocerebellar afferents were found to form abnormal terminal fields in *En2* KO mouse posterior vermal lobules VIII and IX [65]. The KO afferent pattern was simpler than observed for the WT and formed a similar pattern of absent banding found by Zebrin II and *L7* expression though lobule identity was maintained. Several studies of humans with ASD have found abnormal functional circuitry by neuroimaging, as well as deficits in cerebellar-mediated attentional regulation [76, 79, 96, 98, 106, 112, 113, 114], thus the abnormal connectivity of the posterior cerebellar vermis observed in *En2* KO may have clinical correlates in humans with ASD.

Overexpression of En2 also Results in ASD-Like Cerebellar Pathology

In addition to the complete deletion of *En2* function in the KO, abnormal expression of the gene (specifically, ectopic overexpression in postmitotic Purkinje neurons) has been found to alter cerebellar development, suggesting that proper levels as well as spatiotemporal regulation are required to achieve cerebellar patterning. Similar to the KO, cerebellar development is retarded, overall size is reduced in the rostrocaudal and dorsoventral axes but relatively normal mediolaterally, and Purkinje cell numbers are reduced by 30% when *En2* is ectopically overexpressed in differentiating Purkinje cells [107]. Purkinje cell losses are consistent across all regions of the cerebellum; however, there is a selective loss of Purkinje cells underlying the reduced, or absent, fissures that normally separate sublobules. Further, *En2* overexpression in Purkinje cells delays their expression of parvalbumin and dendritic arborization, causes Purkinje cell soma to be smaller, and increases Purkinje cell ectopia outside the normal monolayer [115]. Accompanying Purkinje cell loss in fissures is a selective reduction in size of the EGL, once more demonstrating the functional relationship between Purkinje cells and EGL development. Given these disruptions in Purkinje and granule neuron development, as well as morphological abnormalities, it follows that biochemical compartmentalization and spinocerebellar afferent inputs also are disrupted by *En2* overexpression. In contrast to the *En2* KO mouse, early banding patterns corresponding to *L7* or cadherin-8 expression are normal; however, Zebrin II and P-path sagittal bands are disrupted in all lobules and severely fractured in posterior lobules VIII and IX. Whereas banding borders are sharply delineated in the WT, borders are diffuse in the transgenic, with mottling in the normally unmarked interband regions. Further, sagittal banding of a granule neuron marker, NADPH-diaphorase, is disrupted anteriorly and more severely disrupted in posterior lobules VI and VII [116]. As sagittal banding is thought to specify distinct Purkinje cell subpopulations for the purpose of functional compartmentalization [71], abnormal *En2* expression in transgenic mice disrupts this specification process, thereby disrupting topographical mapping of the cerebellum during development. In light of these patterning disruptions, spinocerebellar afferents are found to innervate transgenic cerebella in similar patterns to WT cerebella. However, like the biochemical banding patterns, spinocerebellar afferents are more diffusely spread with less distinct banding borders. Thus, two different genetic perturbations of *En2*, complete absence and ectopic overexpression, disrupt normal cerebellar development, resulting in similar morphologic, biochemical, cytological, and patterning abnormalities; further, these transgenic mouse cerebella phenocopy some of the most consistently cited neuropathologies reported in humans with ASD [3, 108, 109].

Functional Deficits in **En2** *Mutant Mice*

Given the relatively severe cerebellar deficits exhibited by *En2* KO mice, it is remarkable that they are not ataxic although they performed worse than WT on rotorod tests, suggesting that motor learning is impaired in the absence of *En2* [117]; these findings are in contrast to other cerebellar mouse mutants, including reeler, staggerer, and weaver, that display severe ataxia [118]. Humans with ASD, however, are also not ataxic even though they display cerebellar deficits and perform similarly to patients with cerebellar lesions on tests of attention regulation [76]. A recent study characterized social and neurocognitive behaviors in *En2* KO mice and found a reduction in social interactions with normal mice across developmental ages, as well as impaired learning and memory, in the absence of gross locomotor abnormalities [119]. In addition, this study found a cerebellum-specific increase in serotonin and its metabolites, a neurotransmitter system often found to be dysregulated in ASD [119]. A previous genetic analysis in humans found association between specific haplotypes of the serotonin transporter gene (*5HTT*) and autism in a homogeneous Irish population [120]. These haplotypes were later found to alter mRNA transcription of *5HTT*, which could alter brain serotonin levels [121]; thus in addition to neuropathology, *En2* KO mice behaviorally and neurochemically phenocopy ASD. These similarities between humans with ASD and *En2* transgenic mouse phenotypes were the impetus for exploring human *EN2* as a possible autism-susceptibility gene.

ASD Genetic and Environmental Etiology

The etiology of the ASD, in the absence of known cytogenetic abnormalities such as Fragile X, tuberous sclerosis, or duplication of 15q, remains unknown. Although environmental factors, including neurotoxicants such as pesticides and heavy metals, have been raised as possible disease-causing agents, presently they remain unsubstantiated [122]. However, several teratogens, including the anti-emetic thalidomide, anti-convulsant/mood stabilizer valproic acid (VPA), and the abortificant misoprostol, are associated with ASD; therefore, understanding their period of teratogenicity may provide insight into the timing, if not the mechanisms, of ASD pathogenesis. On the contrary, there is strong evidence for the heritability of ASD, including twin studies, duplex and multiplex families, and recent genetic association studies of specific genes. Further, there is an approximate 4:1 male : female ratio in ASD, and autistic females are more likely than autistic males to have very low IQs [1, 123]; there is currently no explanation for this sex ratio inequality.

Thalidomide and VPA

Prenatal exposure to thalidomide and VPA produce very similar dysmorphic features including neural tube defects, congenital heart defects, craniofacial abnormalities, abnormally shaped or posteriorly rotated ears, genital abnormalities, and limb defects [124], consistent with an early period of teratogenicity during the first 3–4 weeks postconception. Some of these dysmorphic features have been reported in idiopathic ASD, further suggesting that early embryonic insults increase ASD susceptibility. While thalidomide is not teratogenic in animal models, in utero-VPA administration in rodents has duplicated many of the skeletal and organ abnormalities (brain, heart), and most importantly has resulted in ASD-like behavioral abnormalities. Varying the embryonic day (E10–12.5) of VPA administration to pregnant rats produced different brainstem abnormalities in their offspring, suggesting that neurogenesis was disrupted at the time of exposure; importantly, Purkinje neuron deficits and reduced deep cerebellar nuclei volumes were found following VPA administration on and after E12.5 [125]. Behaviorally, both humans with ASD and rats exposed to VPA during early embryogenesis show enhanced eyeblink conditioning [126, 127]. Compared to late gestational or postnatal destruction of Purkinje neurons in which eyeblink conditioning is disrupted, early exposure has been shown to enhance this learning, suggesting that compensatory reorganization of brainstem–cerebellar circuitry may occur following early insults. Additionally, brain and systemic hyperserotonemia has also been described in both VPA animal models and humans with ASD; in fact, brain serotonin synthesis, which decreases after age 5, remains higher in autistic children as they age than their non-autistic siblings [128]. Further, both autistic children and adults demonstrate functionally asymmetric serotonin synthesis in pathways important for language and integration of sensory stimuli that correlate with specific language deficits [129, 130, 131]. The mechanism of VPA teratogenicity is not fully characterized though it has been shown to activate the retinoic acid response element (RARE), an important regulator of HOX gene expression [132]; in fact, VPA was found to up-regulate *Hoxa1* expression before and after its critical period of expression during hindbrain development, possibly via inhibition of histone deacetylase [133]. While mutations in HOXA1 have been found to cause syndromes of craniofacial abnormalities associated with mental retardation and ASD [134], it has failed to be associated with ASD in several populations [135, 136, 137, 138, 139]. Finally, rates of ASD among *in utero* thalidomide- and VPA-exposed children is approximately 4 and 11% [124, 140], respectively, suggesting that environmental exposures may increase ASD risk in genetically susceptible individuals but are unlikely to be causative of the developmental disease itself.

Twin Studies in ASD

The first ASD twin studies were published in 1977, and though they included a relatively small number of twins (21 pairs), they demonstrated a large imbalance in concordance between monozygotic (MZ) and dizygotic (DZ) twins [141, 142]. Further, these studies raised the possibility that broader phenotypes beyond classically defined autism may exist and that there would be strong genetic liability between MZ twins for these phenotypes as well. These results and assertions were validated by further twin and family studies published during the 1980s and 1990s, demonstrating that MZ concordance rates were 60–90%, whereas DZ concordance rates were 0–10%; a quantitative assessment of concordance inequality between MZ and DZ twins estimated greater than 90% heritability for autism [143, 144, 145, 146]. Unexpectedly, these studies also found that clinical heterogeneity extended to MZ twins, suggesting that shared gene expression does not necessarily dictate similar behavioral phenotypes [147]. Additionally, family studies conducted during this period extended analysis to non-twin siblings and other first- and second-degree relatives of ASD patients to better ascertain the genetic liability of the disease [143, 148, 149]. These studies found 2–6% autism rates among siblings of autistic patients compared to no autism found in siblings of Down syndrome patients; when a broader phenotype was considered (what is now considered ASD), the rate increased to 10–20% ASD incidence in autism siblings compared to 2–3% in Down syndrome siblings. The pattern of increased risk for first-degree versus second-degree relatives, in conjunction with the MZ–DZ concordance inequality led researchers to believe that autism was a complex heritable disease that resulted from epistatic interactions between approximately 2–10 genes [143]. It is currently believed that between 3 and 15 genes may confer susceptibility to ASD, with complex epistatic interactions contributing to the heterogeneity of clinical presentations [109, 150]. Further, epigenetic regulation of gene expression and interactions is now recognized to be essential to normal brain development, and several neurodevelopmental disorders are linked to abnormal epigenetics.

Rett Syndrome and Epigenetic Regulation

One such disease is Rett Syndrome, a progressive developmental disorder similar to ASD but almost exclusively affecting females. Unlike ASD in which several abnormal genes may be inherited, Rett syndrome results from mutations of methyl CpG-binding protein 2 (*MECP2*), an epigenetic effector molecule that represses gene expression by binding to methylated CpG islands (151; see Chapter 4 by Kaufmann et al.). Mutations that alter function of Mecp2 in mice, as well as humans with Rett, ASD, and Angelman syndromes, were recently demonstrated to disrupt expression of both imprinted and non-imprinted genes in the Angelman/Prader-Willi region of 15q11–q13 [152]. Thus, disruptions of complex epigenetic interactions may contribute to developmental disease pathogenesis, making individual genetic contributions difficult to ascertain. Further, novel mutations and polymorphic copy number variants within *MECP2* exons, introns, and the 3′-untranslated region (3′-UTR) have been identified in both males and females in several autistic populations [153, 154, 155]. This suggests that slight alterations in *MECP2* that are non-fatal can still disrupt brain function, possibly contributing to ASD pathogenesis. Interestingly, a recent study found that restoration of *Mecp2* expression fully reversed the disease phenotype in null mice exhibiting neurological impairments similar to humans with Rett syndrome [156]. These results suggest that PDD phenotypes may be reversible in humans and that future therapies aimed at correcting abnormal genetic and epigenetic interactions may prevent or attenuate the severity of developmental diseases such as ASD.

Genes of Interest in ASD

Once a genetic etiology of ASD was established, the search for ASD-associated genes began, first by identifying regions in the human genome containing marker variants [157] and then by identifying candidate genes within these regions [147, 158]. Ideal candidate genes were not only found in or near ASD-susceptibility loci but were expressed during brain development and/or brain function. Given the clinical and genetic heterogeneity of ASD, candidate gene association has been difficult to establish, and replication across populations nearly impossible. For example, a functional variant in the promoter of the tyrosine kinase receptor *MET* (7q31) was recently found to be associated with ASD [159] though this finding is yet to be replicated. *MET* signaling is found in various peripheral organs, as well as the developing brain, thus a functional variant that possibly alters receptor activity may contribute to abnormal growth and ASD pathophysiology. Several other genes involved in brain growth and function have been found to be associated with ASD in selected populations. Mutations in postsynaptic cell adhesion molecules neuroligins (NGLN) 3 and 4 on the X-chromosome [160, 161, 162] and a member of their receptor family, neurexin-beta [163] have been identified in ASD families, suggesting that disrupted synaptogenesis may contribute to ASD dysfunction. Further, the hormone oxytocin is implicated in regulating social behaviors including fear and pair bonding and is commonly administered during labor to stimulate parturition; recently, several allelic polymorphisms in the oxytocin receptor (*OXTR*) were found to be associated with ASD in Chinese and Caucasian populations though different alleles at the same locus were differentially associated between the two studies [164, 165]. Other genetic associations have been identified in case studies or subpopulations, some demonstrating a particular phenotype, such as macrocephaly and the tumor suppressor gene *PTEN* on chromosome 10q23 [166, 167]; the cerebellar Ca^{2+}-dependent secretion protein CAPS2 on human chromosome 7q that promotes neurotrophin release [168–170]; and the extracellular glycoprotein Reelin (*RELN*) on distal chromosome 7q, which is required for normal neural migration and cytoarchitecture during development [171, 172, 173]. Each of these genes alone may confer risk or may interact with other genetic or environmental factors that increase an individual's susceptibility to ASD. Our understanding of these genes' roles in ASD will be improved by replication across large populations; however, the heterogeneity of ASD makes replication very difficult.

Endophenotypes and ASD Genes

As our understanding of the underlying causes of ASD improves, we may be able to subdivide the disease into endophenotypes, which focus on select features of the disorder, such as language impairment or repetitive movements. Such features by themselves are not unique to the disorder but may exhibit greater genetic heritability than the entire spectrum of signs and symptoms comprising ASD. For example, *PTEN* abnormalities may segregate with a macrocephalic subpopulation of ASD but may not be associated with ASD in the general population. Recently, a quantitative trait loci (QTL) analysis of non-verbal communication in ASD families suggested that there are chromosomal regions that harbor ASD-susceptibility genes segregating selectively with severe language impairment but not other endophenotypes [174]. Further, despite lack of association between ASD and *HOXA1* in numerous studies, one polymorphism was found to be associated with increased head circumference in ASD and was reported to explain 5% of the variance of head size in the data set studied [175]. Thus, links between candidate genes and symptomology may help simplify the genetic and clinical heterogeneity of ASD into parsable domains and may also be better prognostic indicators of behavioral and pharmacological interventions.

ENGRAILED-2 is an Autism-Associated Gene

The pervasiveness of ASD across races, ethnicities, continents, and socioeconomic status may seem at odds with its heterogeneity; however, our current diagnostic criteria are inclusive and symptom-based, thus multiple modes of inheritance may all give rise to a similar clinical label [82, 109, 150, 176, 177, 178, 179, 180]. Thus, it is increasingly difficult to identify ASD-associated genes and replicate these associations across large, heterogeneous populations. That said, we have recently identified human *ENGRAILED-2* (*EN2*) as an ASD-susceptibility locus [108, 109]. As stated earlier, genome-wide analyses have identified chromosomal regions linked to ASD, thus candidate genes in these regions can be explored further for association with the disease. One region of interest has been the long arm of human chromosome 7 [157, 181]. The human *EN2* gene is found distally on the long arm of human chromosome 7, just outside an ASD-linkage region [182, 183, 184]. As described above, *EN2* is expressed during brain development and has been found to be essential for normal cerebellar function; further, *En2* mutant mice exhibit cerebellar neuropathology, as well as neurochemical and behavioral abnormalities that phenocopy humans with ASD. Thus, *EN2* was selected as a possible candidate gene to be screened for association with ASD.

The human *EN2* gene is relatively simple, spanning approximately 8.0 Kb and containing only two exons separated by a 3.3 Kb intron; four single nucleotide polymorphisms (SNPs) that span most of the *EN2* gene were chosen for analysis. SNPs are common (occur in at least 1% of the population) heritable variations found across the entire genome and can be used to track parental transmission of disease alleles to their children. Gharani et al. (2004) [109] genotyped four SNPs that span the *EN2* gene in 138 parent–child triads and a total of 167 extended pedigrees (including other affected and/or unaffected siblings) from the autism genetic research exchange (AGRE). They assessed allelic transmission from heterozygous parents to children with ASD and their unaffected siblings, using the transmission-disequilibrium test (TDT) [185]. Two intronic SNP alleles were each found to be overtransmitted to children diagnosed narrowly with autism as well as broadly with any ASD diagnosis and undertransmitted to their unaffected siblings. This suggests that inheritance of each allele is associated with autism and more broadly with ASD. Further, these two alleles were found to be overtransmitted together as a haplotype to both narrow and broadly diagnosed children; the exonic SNPs were not found to be associated with ASD, alone or even as haplotypes with the intronic SNPs, suggesting that the disease locus is either located within the *EN2* intron or is located elsewhere but is in tight linkage disequilibrium (LD) with the intronic SNPs [109].

These findings were replicated in two more non-overlapping, unrelated data sets of 222 additional AGRE pedigrees and 129 families from the National Institute of Mental Health (NIMH) [108]. All three data sets were combined to give a total of 518 families, and again the individual intronic SNPs, as well as the SNP haplotypes, were significantly associated with both the narrow diagnosis of classical autism and the broader spectrum of ASD; importantly, these associations were found to increase in statistical significance in the large, combined data set. This study also expanded analysis to 14 more SNPs spanning the entire *EN2* gene; however, no other SNPs individually, or in a haplotype with the two associated intronic SNPs, were found to be associated with ASD; in fact no other markers within two megabases 5' or 3' of *EN2* were found to be associated with ASD, suggesting that the intronic SNPs 972 and 973 are the ASD-susceptibility locus. Furthermore, *EN2* was found to confer disease risk in approximately 40% of the ASD cases studied [108], supporting *EN2* as an ASD-susceptibility locus. To our knowledge, this is the first time an ASD association has withstood replication across heterogeneous populations. Further, preliminary evidence suggests that the ASD-associated intron alters *En2* promoter-driven luciferase expression compared to the non-ASD-associated intron, suggesting that abnormal *cis*-regulation of *EN2* expression during development contributes to ASD pathogenesis (Dr. J. H. Millonig, personal communication).

En2 Function During Cerebellar Development

Despite its well characterized expression, the function of *En2* during cerebellar development remains largely unknown. As mentioned previously, *En2* expression prenatally plays an important part in specifying Purkinje neuron subpopulations and creating a topographical map of the embryonic cerebellum. Within the last decade, studies of Engrailed proteins have revealed expanded developmental roles for these presumptive transcription factors. Although traditionally viewed as DNA-binding proteins, the *Engrailed* homeoproteins have been empirically shown to participate in non-transcriptional activities, including translational regulation [186, 187, 188]. Amino acid sequence analysis has demonstrated that an 11-residue sequence in the Engrailed homeodomain is necessary and sufficient to function as a putative nuclear export signal (NES) [189, 190]. Indeed, Engrailed1 and 2 proteins were found to localize outside of the nucleus, in subplasmalemmal and intracellular vesicles of primary midbrain neurons [191]. These data suggested that homeoproteins are made available for secretion and correlated with previous findings that other homeodomain proteins, including Emx1 and HoxA7, localize to axons [191, 192]. In fact, Engrailed proteins were found to contain an unconventional secretion sequence that allowed their export from the cell, as well as internalization by other cells, through non-endocytotic mechanisms [190]. This secretion was blocked by the serine-threonine kinase CK2-phosphorylation of specific serine residues within the Engrailed secretion sequence [193]. Engrailed protein is also phosphorylated on a serine in its homeodomain by protein kinase A (PKA) that reduces its DNA-binding affinity [194]. Therefore, beyond *cis*- and *trans*-regulation of *Engrailed* by other developmental transcription factors, it is important to consider posttranslational modifications as important regulators of Engrailed functions and localization during development. Very recent work by Brunet et al. (2005) [186] found that Engrailed2 could differentially regulate retinal axon guidance in the *Xenopus* optic tectum, simultaneously attracting nasal growth cones yet repelling temporal growth cones. These functions were independent of transcription but rather involved internalized En2 binding to eukaryotic initiation factor 4E (eIF4E) and promoting the phosphorylation of eIF4E and its binding protein (eIF4EB), leading to local axonal protein translation [186]. Engrailed1 protein, however, has been found to be increased in an activity-dependent manner in the dendrites of adult midbrain neurons; further, hemizygous *En1* leads to midbrain neuron death in adult rodents [188, 195]. Thus, Engrailed proteins play many roles in the development and homeostasis of brain structures beyond their traditional roles as homeodomain transcription factors.

Preliminary Data

Our laboratory's ongoing research is aimed at defining *En2*'s cell biological function in cerebellar neural precursors, with particular emphasis on GNPs and postnatal cerebellar development. We demonstrated previously that ectopic overexpression of *En2* in embryonic cortical neuron precursors alters neurogenesis [108]. This is important given that ASD-associated SNPs within *EN2* may alter the timing or localization of its own expression, and ectopic expression has been shown to produce ASD-like neuropathology [107, 115]. Thus, in its native population (cerebellar GNPs), we hypothesized that *En2* could modulate cell cycle, thereby participating in granule neurogenesis. Postnatal *En2* expression begins in postmitotic, premigratory GNPs of the inner EGL [64]; therefore, we further hypothesized that *En2* promotes GNP cell cycle exit and differentiation. By comparing *En2* KO and WT mouse GNPs in vitro as well in vivo, we have preliminary evidence that in the absence of *En2*, GNPs remain more prolifera-tive and fail to differentiate. This difference in proliferation was greatest in KO GNPs from the central vermis, the region of greatest *En2* expression; in fact, there were no statistical differences in DNA synthesis (a measure of prolifera-tion) between KO and WT GNPs isolated from the lateral hemispheres, a region of very low or absent *En2* expression [64]. These regional differences in *En2*-regulated DNA synthesis further underscore the heterogeneity of cerebel-lar neuronal populations. Additionally, these data suggest that abnormal *En2* regulation during development may differentially disrupt medial cerebellar pathways required for attention regulation and speech production, requiring compensation from lateral cerebellar circuitry involved in higher order cogni-tive functions [24], thereby contributing to ASD phenotypes.

En2 and IGF-1

Although there appears to be increased GNP proliferation in the KO, we found that even in the absence of *En2,* overall cell cycle regulation by extracellular growth factors remains intact; however, *En2* is found to modulate signaling through one developmentally relevant growth factor and its receptor, IGF1 [196]. IGF-1 is produced in high concentrations by Purkinje neurons, which are supplied by anterograde transport from the inferior olive afferent fibers, and is released into the EGL during postnatal GNP proliferation [37, 41, 88, 91] (Riikonen, Chapter 10). IGF-1, acting through its tyrosine-kinase receptor, IGF-1R, stimulates proliferation, survival, and/or differentiation, depending upon which of several possible downstream messengers are activated. In the absence of *En2*, IGF-1 stimulates a greater increase in DNA synthesis compared to WT cells; this effect was demonstrable in cultured GNPs, as well as in vivo through peripheral IGF-1 injection. Further, in the absence of *En2*, GNPs demonstrated reduced neurite outgrowth (a marker of neuronal differentiation) in response to several differentiation signals (including IGF-1). Taken together, these data support a role for *En2* in regulating granule neurogenesis by promot-ing cell cycle exit and differentiation, possibly via modulation of IGF-1 signaling. Although preliminary, these data provide a model in which ASD-associated changes in *EN2* expression would have predictable consequences on postnatal cerebellar growth. Abnormalities in cerebellar size, gray : white matter ratios, and growth rates are well characterized in ASD; therefore, understanding postnatal

En2 expression modulates the effects of IGF1 on cerebellar granule neurogenesis.

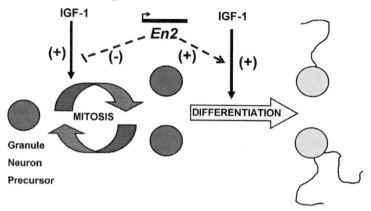

Fig. 1.4 This cartoon depicts our current hypothesis regarding the modulation of insulin-like growth factor-1 (IGF-1) signaling by *En2* expression. En2 interacts with undefined downstream signals to attenuate IGF-1-stimulated granule neuron precursor (GNP) proliferation while promoting GNP differentiation

En2 functions may shed light into the molecular pathways underlying these developmental abnormalities. On-going and future experiments will aim to modulate *En2* expression levels in cultured GNPs via cDNA overexpression and siRNA knockdown. Based upon our existing model (Fig. 1.4), we expect reduced *En2* expression to result in increased proliferation with a concomitant reduction in differentiation while overexpression will facilitate cell cycle exit and granule neuron differentiation. Through these experiments, we aim to functionally characterize *En2* to better understand the molecular and cellular complexities of normal cerebellar development. Further, it is our hope that these experiments will aid our understanding of ASD pathogenesis with respect to *En2*, which, at the time of writing, is the only ASD-associated gene to withstand replication across several data sets.

Conclusion

ASD are complex neurodevelopmental diseases with strong genetic etiologies and are highly variable among individuals. This clinical heterogeneity is symbolic of the complex genetic etiology and neuropathology observed across the ASD population. Despite this heterogeneity, many studies demonstrate that the cerebellum is consistently abnormal in ASD patients, both neuroanatomically and functionally. For example, humans with ASD demonstrate abnormal brain growth, with some patients exhibiting cerebellar gray : white matter ratios that deviate from age-matched controls throughout childhood and into adolescence.

Further, cerebellar-dependent cognitive tasks, including attention regulation and speech, are disrupted in humans with ASD while motor coordination is preserved, suggesting that the cerebellum is grossly intact in these patients. Recently, the human gene *ENGRAILED2*, an important regulator of cerebellar development, was identified as an ASD-susceptibility gene, a finding replicated in three data sets. Together, these data suggest that abnormal brain, and in particular, abnormal cerebellar development, underlie ASD pathogenesis. The role of *EN2* in normal cerebellar development is still under investigation; however, disruption of this gene in animal models produces ASD-like behavior and neuropathology. Thus, we aim to better understand normal functions of *EN2* as a possible window into the pathogenesis of ASD. Our preliminary data suggest that *En2* promotes postnatal GNP cell cycle exit and differentiation. Further, *En2* interacts with a developmentally important extracellular growth factor, IGF-1, to regulate GNP proliferation and differentiation in the postnatal period. Future studies will further explore the biochemical function of *En2*, as well as compare possible differences between *cis*-regulation of *EN2* expression by normal and ASD-associated introns.

References

1. CDC, Prevalence of Autism Spectrum Disorders–Autism and Developmental Disabilities Monitoring Network, Six Sites, United States, 2000. *MMWR Surveill. Summ.*, 2007. **56**(1): p. 1–11.
2. Kanner, L. and Eisenberg, L., Early Infantile Autism, 1943–1955. *Psychiatric Research Reports*, 1957. **7**: p. 55–65.
3. Palmen, S. J., van Engeland, H., Hof, P. R., and Schmitz, C., Neuropathological Findings in Autism. *Brain*, 2004. **127**(Pt 12): p. 2572–83.
4. Courchesne, E., Carper, R., and Akshoomoff, N., Evidence of Brain Overgrowth in the First Year of Life in Autism. *JAMA*, 2003. **290**(3): p. 337–44.
5. Courchesne, E., Karns, C. M., Davis, H. R., Ziccardi, R., Carper, R. A., Tigue, Z. D., Chisum, H. J., Moses, P., Pierce, K., Lord, C., Lincoln, A. J., Pizzo, S., Schreibman, L., Haas, R. H., Akshoomoff, N. A., and Courchesne, R. Y., Unusual Brain Growth Patterns in Early Life in Patients with Autistic Disorder: An MRI Study. *Neurology*, 2001. **57**(2): p. 245–54.
6. Hazlett, H. C., Poe, M., Gerig, G., Smith, R. G., Provenzale, J., Ross, A., Gilmore, J., and Piven, J., Magnetic Resonance Imaging and Head Circumference Study of Brain Size in Autism: Birth Through Age 2 Years. *Arch Gen Psychiatry*, 2005. **62**(12): p. 1366–76.
7. Sparks, B. F., Friedman, S. D., Shaw, D. W., Aylward, E. H., Echelard, D., Artru, A. A., Maravilla, K. R., Giedd, J. N., Munson, J., Dawson, G., and Dager, S. R., Brain Structural Abnormalities in Young Children with Autism Spectrum Disorder. *Neurology*, 2002. **59**(2): p. 184–92.
8. Bauman, M. L. and Kemper, T. L., Neuroanatomic Observations of the Brain in Autism: A Review and Future Directions. *Int J Dev Neurosci*, 2005. **23**(2–3): p. 183–7.
9. Kemper, T. L. and Bauman, M. L., The Contribution of Neuropathologic Studies to the Understanding of Autism. *Neurol Clin*, 1993. **11**(1): p. 175–87.
10. Kemper, T. L. and Bauman, M. L., Neuropathology of Infantile Autism. *Mol Psychiatry*, 2002. **7**(Suppl 2): p. S12–3.

11. Bailey, A., Luthert, P., Dean, A., Harding, B., Janota, I., Montgomery, M., Rutter, M., and Lantos, P., A Clinicopathological Study of Autism. *Brain*, 1998. **121**(Pt 5): p. 889–905.

12. Casanova, M. F., Buxhoeveden, D. P., and Brown, C., Clinical and Macroscopic Correlates of Minicolumnar Pathology in Autism. *J Child Neurol*, 2002. **17**(9): p. 692–5.

13. Casanova, M. F., Buxhoeveden, D. P., Switala, A. E., and Roy, E., Asperger's Syndrome and Cortical Neuropathology. *J Child Neurol*, 2002. **17**(2): p. 142–5.

14. Casanova, M. F., Buxhoeveden, D. P., Switala, A. E., and Roy, E., Minicolumnar Pathology in Autism. *Neurology*, 2002. **58**(3): p. 428–32.

15. Boger-Megiddo, I., Shaw, D. W., Friedman, S. D., Sparks, B. F., Artru, A. A., Giedd, J. N., Dawson, G., and Dager, S. R., Corpus Callosum Morphometrics in Young Children with Autism Spectrum Disorder. *J Autism Dev Disord*, 2006. **36**(6): p. 733–9.

16. Egaas, B., Courchesne, E., and Saitoh, O., Reduced Size of Corpus Callosum in Autism. *Arch Neurol*, 1995. **52**(8): p. 794–801.

17. Hardan, A. Y., Minshew, N. J., and Keshavan, M. S., Corpus Callosum Size in Autism. *Neurology*, 2000. **55**(7): p. 1033–6.

18. Just, M. A., Cherkassky, V. L., Keller, T. A., Kana, R. K., and Minshew, N. J., Functional and Anatomical Cortical Underconnectivity in Autism: Evidence from an Fmri Study of an Executive Function Task and Corpus Callosum Morphometry. *Cereb Cortex*, 2007. **17**(4): p. 951–61.

19. Minshew, N. J. and Williams, D. L., The New Neurobiology of Autism: Cortex, Connectivity, and Neuronal Organization. *Arch Neurol*, 2007. **64**(7): p. 945–50.

20. Bauman, M. and Kemper, T. L., Histoanatomic Observations of the Brain in Early Infantile Autism. *Neurology*, 1985. **35**(6): p. 866–74.

21. Schumann, C. M. and Amaral, D. G., Stereological Analysis of Amygdala Neuron Number in Autism. *J Neurosci*, 2006. **26**(29): p. 7674–9.

22. Williams, R. S., Hauser, S. L., Purpura, D. P., DeLong, G. R., and Swisher, C. N., Autism and Mental Retardation: Neuropathologic Studies Performed in Four Retarded Persons with Autistic Behavior. *Arch Neurol*, 1980. **37**(12): p. 749–53.

23. Ritvo, E. R., Freeman, B. J., Scheibel, A. B., Duong, T., Robinson, H., Guthrie, D., and Ritvo, A., Lower Purkinje Cell Counts in the Cerebella of Four Autistic Subjects: Initial Findings of the UCLA-NSAC Autopsy Research Report. *Am J Psychiatry*, 1986. **143**(7): p. 862–6.

24. Kandel, E. R., Schwartz, J. H., and Jessell, T. M., *Principles of Neural Science*. 4th ed. 2000, New York: McGraw-Hill, Health Professions Division. xli, 1414 p.

25. Hatten, M. E. and Heintz, N., Mechanisms of Neural Patterning and Specification in the Developing Cerebellum. *Annu Rev Neurosci*, 1995. **18**: p. 385–408.

26. Altman, J. and Bayer, S. A., *Development of the Cerebellar System: In Relation to Its Evolution, Structure, and Functions*. 1997, Boca Raton: CRC Press. 783 p., [16] p. of plates.

27. Ben-Arie, N., Bellen, H. J., Armstrong, D. L., McCall, A. E., Gordadze, P. R., Guo, Q., Matzuk, M. M., and Zoghbi, H. Y., Math1 Is Essential for Genesis of Cerebellar Granule Neurons. *Nature*, 1997. **390**(6656): p. 169–72.

28. Hatten, M. E., Central Nervous System Neuronal Migration. *Annu Rev Neurosci*, 1999. **22**: p. 511–39.

29. Alder, J., Cho, N. K., and Hatten, M. E., Embryonic Precursor Cells from the Rhombic Lip Are Specified to a Cerebellar Granule Neuron Identity. *Neuron*, 1996. **17**(3): p. 389–99.

30. Wallace, V. A., Purkinje-Cell-Derived Sonic Hedgehog Regulates Granule Neuron Precursor Cell Proliferation in the Developing Mouse Cerebellum. *Curr Biol*, 1999. **9**(8): p. 445–8.

31. Dahmane, N. and Ruiz i Altaba, A., Sonic Hedgehog Regulates the Growth and Patterning of the Cerebellum. *Development*, 1999. **126**(14): p. 3089–100.

32. Lewis, P. M., Gritli-Linde, A., Smeyne, R., Kottmann, A., and McMahon, A. P., Sonic Hedgehog Signaling Is Required for Expansion of Granule Neuron Precursors and Patterning of the Mouse Cerebellum. *Dev Biol*, 2004. **270**(2): p. 393–410.

33. Gonzalez, B. J., Basille, M., Vaudry, D., Fournier, A., and Vaudry, H., Pituitary Adenylate Cyclase-Activating Polypeptide Promotes Cell Survival and Neurite Outgrowth in Rat Cerebellar Neuroblasts. *Neuroscience*, 1997. **78**(2): p. 419–30.

34. Vaudry, D., Gonzalez, B. J., Basille, M., Fournier, A., and Vaudry, H., Neurotrophic Activity of Pituitary Adenylate Cyclase-Activating Polypeptide on Rat Cerebellar Cortex During Development. *Proc Natl Acad Sci USA*, 1999. **96**(16): p. 9415–20.

35. Nielsen, H. S., Hannibal, J., and Fahrenkrug, J., Expression of Pituitary Adenylate Cyclase Activating Polypeptide (PACAP) in the Postnatal and Adult Rat Cerebellar Cortex. *Neuroreport*, 1998. **9**(11): p. 2639–42.

36. Nicot, A., Lelievre, V., Tam, J., Waschek, J. A., and DiCicco-Bloom, E., Pituitary Adenylate Cyclase-Activating Polypeptide and Sonic Hedgehog Interact to Control Cerebellar Granule Precursor Cell Proliferation. *J Neurosci*, 2002. **22**(21): p. 9244–54.

37. Lin, X. and Bulleit, R. F., Insulin-Like Growth Factor I (IGF-I) Is a Critical Trophic Factor for Developing Cerebellar Granule Cells. *Brain Res*, 1997. **99**(2): p. 234–42.

38. Bartlett, W. P., Li, X. S., Williams, M., and Benkovic, S., Localization of Insulin-Like Growth Factor-1 mRNA in Murine Central Nervous System During Postnatal Development. *Dev Biol*, 1991. **147**(1): p. 239–50.

39. Ye, P., Xing, Y., Dai, Z., and D'Ercole, A. J., In Vivo Actions of Insulin-Like Growth Factor-I (IGF-I) on Cerebellum Development in Transgenic Mice: Evidence That IGF-I Increases Proliferation of Granule Cell Progenitors. *Brain Res*, 1996. **95**(1): p. 44–54.

40. Gao, W. O., Heintz, N., and Hatten, M. E., Cerebellar Granule Cell Neurogenesis Is Regulated by Cell-Cell Interactions in Vitro. *Neuron*, 1991. **6**(5): p. 705–15.

41. Smeyne, R. J., Chu, T., Lewin, A., Bian, F., Sanlioglu, S., Kunsch, C., Lira, S. A., and Oberdick, J., Local Control of Granule Cell Generation by Cerebellar Purkinje Cells. *Mol Cell Neurosci*, 1995. **6**(3): p. 230–51.

42. Herrup, K. and Sunter, K., Numerical Matching During Cerebellar Development: Quantitative Analysis of Granule Cell Death in Staggerer Mouse Chimeras. *J Neurosci*, 1987. **7**(3): p. 829–36.

43. Wetts, R. and Herrup, K., Direct Correlation between Purkinje and Granule Cell Number in the Cerebella of Lurcher Chimeras and Wild-Type Mice. *Brain Res*, 1983. **312**(1): p. 41–7.

44. Williams, R. W. and Herrup, K., The Control of Neuron Number. *Annu Rev Neurosci*, 1988. **11**: p. 423–53.

45. Jensen, P., Zoghbi, H. Y., and Goldowitz, D., Dissection of the Cellular and Molecular Events That Position Cerebellar Purkinje Cells: A Study of the Math1 Null-Mutant Mouse. *J Neurosci*, 2002. **22**(18): p. 8110–6.

46. Bertrand, N., Castro, D. S., and Guillemot, F., Proneural Genes and the Specification of Neural Cell Types. *Nat Rev Neurosci*, 2002. **3**(7): p. 517–30.

47. Briscoe, J., Pierani, A., Jessell, T. M., and Ericson, J., A Homeodomain Protein Code Specifies Progenitor Cell Identity and Neuronal Fate in the Ventral Neural Tube. *Cell*, 2000. **101**(4): p. 435–45.

48. Briscoe, J., Sussel, L., Serup, P., Hartigan-O'Connor, D., Jessell, T. M., Rubenstein, J. L., and Ericson, J., Homeobox Gene Nkx2.2 and Specification of Neuronal Identity by Graded Sonic Hedgehog Signalling. *Nature*, 1999. **398**(6728): p. 622–7.

49. Edlund, T. and Jessell, T. M., Progression from Extrinsic to Intrinsic Signaling in Cell Fate Specification: A View from the Nervous System. *Cell*, 1999. **96**(2): p. 211–24.

50. Gotz, M. and Huttner, W. B., The Cell Biology of Neurogenesis. *Nat Rev*, 2005. **6**(10): p. 777–88.

51. Davis, C. A. and Joyner, A. L., Expression Patterns of the Homeo Box-Containing Genes En-1 and En-2 and the Proto-Oncogene Int-1 Diverge During Mouse Development. *Genes Dev*, 1988. **2**(12B): p. 1736–44.

52. Joyner, A. L. and Martin, G. R., En-1 and En-2, Two Mouse Genes with Sequence Homology to the Drosophila Engrailed Gene: Expression During Embryogenesis. *Genes Dev*, 1987. **1**(1): p. 29–38.

53. Davis, C. A., Noble-Topham, S. E., Rossant, J., and Joyner, A. L., Expression of the Homeo Box-Containing Gene En-2 Delineates a Specific Region of the Developing Mouse Brain. *Genes Dev*, 1988. **2**(3): p. 361–71.

54. Hanks, M., Wurst, W., Anson-Cartwright, L., Auerbach, A. B., and Joyner, A. L., Rescue of the En-1 Mutant Phenotype by Replacement of En-1 with En-2. *Science*, 1995. **269**(5224): p. 679–82.

55. Hanks, M. C., Loomis, C. A., Harris, E., Tong, C. X., Anson-Cartwright, L., Auerbach, A., and Joyner, A., Drosophila Engrailed Can Substitute for Mouse Engrailed1 Function in Mid-Hindbrain, but Not Limb Development. *Development*, 1998. **125**(22): p. 4521–30.

56. Wassef, M., Bally-Cuif, L., and Alvarado-Mallart, R. M., Regional Specification During Cerebellar Development. *Perspect Dev Neurobiol*, 1993. **1**(3): p. 127–32.

57. Scholpp, S., Lohs, C., and Brand, M., Engrailed and Fgf8 Act Synergistically to Maintain the Boundary between Diencephalon and Mesencephalon. *Development*, 2003. **130**(20): p. 4881–93.

58. Liu, A., Losos, K., and Joyner, A. L., Fgf8 Can Activate Gbx2 and Transform Regions of the Rostral Mouse Brain into a Hindbrain Fate. *Development*, 1999. **126**(21): p. 4827–38.

59. Liu, A. and Joyner, A. L., En and Gbx2 Play Essential Roles Downstream of Fgf8 in Patterning the Mouse Mid/Hindbrain Region. *Development*, 2001. **128**(2): p. 181–91.

60. Morales, D. and Hatten, M. E., Molecular Markers of Neuronal Progenitors in the Embryonic Cerebellar Anlage. *J Neurosci*, 2006. **26**(47): p. 12226–36.

61. Fink, A. J., Englund, C., Daza, R. A., Pham, D., Lau, C., Nivison, M., Kowalczyk, T., and Hevner, R. F., Development of the Deep Cerebellar Nuclei: Transcription Factors and Cell Migration from the Rhombic Lip. *J Neurosci*, 2006. **26**(11): p. 3066–76.

62. Sanlioglu, S., Zhang, X., Baader, S. L., and Oberdick, J., Regulation of a Purkinje Cell-Specific Promoter by Homeodomain Proteins: Repression by Engrailed-2 Vs. Synergistic Activation by Hoxa5 and Hoxb7. *J Neurobiol*, 1998. **36**(4): p. 559–71.

63. Kuemerle, B., Zanjani, H., Joyner, A., and Herrup, K., Pattern Deformities and Cell Loss in Engrailed-2 Mutant Mice Suggest Two Separate Patterning Events During Cerebellar Development. *J Neurosci*, 1997. **17**(20): p. 7881–9.

64. Millen, K. J., Wurst, W., Herrup, K., and Joyner, A. L., Abnormal Embryonic Cerebellar Development and Patterning of Postnatal Foliation in Two Mouse Engrailed-2 Mutants. *Development*, 1994. **120**(3): p. 695–706.

65. Vogel, M. W., Ji, Z., Millen, K., and Joyner, A. L., The Engrailed-2 Homeobox Gene and Patterning of Spinocerebellar Mossy Fiber Afferents. *Brain Res*, 1996. **96**(1–2): p. 210–8.

66. Oberdick, J., Baader, S. L., and Schilling, K., From Zebra Stripes to Postal Zones: Deciphering Patterns of Gene Expression in the Cerebellum. *Trends Neurosci*, 1998. **21**(9): p. 383–90.

67. Ozol, K., Hayden, J. M., Oberdick, J., and Hawkes, R., Transverse Zones in the Vermis of the Mouse Cerebellum. *J Comp Neurol*, 1999. **412**(1): p. 95–111.

68. Wassef, M., Cholley, B., Heizmann, C. W., and Sotelo, C., Development of the Olivocerebellar Projection in the Rat: II. Matching of the Developmental Compartmentations of the Cerebellum and Inferior Olive through the Projection Map. *J Comp Neurol*, 1992. **323**(4): p. 537–50.

69. Sotelo, C. and Wassef, M., Cerebellar Development: Afferent Organization and Purkinje Cell Heterogeneity. *Philos Trans R Soc Lond*, 1991. **331**(1261): p. 307–13.

70. Wassef, M., Sotelo, C., Thomasset, M., Granholm, A. C., Leclerc, N., Rafrafi, J., and Hawkes, R., Expression of Compartmentation Antigen Zebrin I in Cerebellar Transplants. *J Comp Neurol*, 1990. **294**(2): p. 223–34.

71. Leclerc, N., Schwarting, G. A., Herrup, K., Hawkes, R., and Yamamoto, M., Compartmentation in Mammalian Cerebellum: Zebrin II and P-Path Antibodies Define Three Classes of Sagittally Organized Bands of Purkinje Cells. *Proc Natl Acad Sci USA*, 1992. **89**(11): p. 5006–10.

72. Hawkes, R. and Leclerc, N., Purkinje Cell Axon Collateral Distributions Reflect the Chemical Compartmentation of the Rat Cerebellar Cortex. *Brain Res*, 1989. **476**(2): p. 279–90.

73. Paradies, M. A. and Eisenman, L. M., Evidence of Early Topographic Organization in the Embryonic Olivocerebellar Projection: A Model System for the Study of Pattern Formation Processes in the Central Nervous System. *Dev Dyn*, 1993. **197**(2): p. 125–45.

74. Paradies, M. A., Grishkat, H., Smeyne, R. J., Oberdick, J., Morgan, J. I., and Eisenman, L. M., Correspondence between L7-Lacz-Expressing Purkinje Cells and Labeled Olivocerebellar Fibers During Late Embryogenesis in the Mouse. *J Comp Neurol*, 1996. **374**(3): p. 451–66.

75. Allen, G. and Courchesne, E., Differential Effects of Developmental Cerebellar Abnormality on Cognitive and Motor Functions in the Cerebellum: An fMRI Study of Autism. *Am J Psychiatry*, 2003. **160**(2): p. 262–73.

76. Courchesne, E., Townsend, J., Akshoomoff, N. A., Saitoh, O., Yeung-Courchesne, R., Lincoln, A. J., James, H. E., Haas, R. H., Schreibman, L., and Lau, L., Impairment in Shifting Attention in Autistic and Cerebellar Patients. *Behav Neurosci*, 1994. **108**(5): p. 848–65.

77. Yip, J., Soghomonian, J. J., and Blatt, G. J., Decreased GAD67 mRNA Levels in Cerebellar Purkinje Cells in Autism: Pathophysiological Implications. *Acta Neuropathol*, 2007. **113**(5): p. 559–68.

78. Yip, J., Soghomonian, J. J., Bauman, M., Kemper, T., and Blatt, G. J., Functional Status of the Cerebellar Purkinje Cells in Autistic Brains. *International Meeting for Autism Research (IMFAR)*, 2006. PS 4.47, p. 142.

79. Courchesne, E., Yeung-Courchesne, R., Press, G. A., Hesselink, J. R., and Jernigan, T. L., Hypoplasia of Cerebellar Vermal Lobules VI and VII in Autism. *N Engl J Med*, 1988. **318**(21): p. 1349–54.

80. Murakami, J. W., Courchesne, E., Press, G. A., Yeung-Courchesne, R., and Hesselink, J. R., Reduced Cerebellar Hemisphere Size and Its Relationship to Vermal Hypoplasia in Autism. *Arch Neurol*, 1989. **46**(6): p. 689–94.

81. Palmen, S. J. and van Engeland, H., Review on Structural Neuroimaging Findings in Autism. *J Neural Transm*, 2004. **111**(7): p. 903–29.

82. DiCicco-Bloom, E., Lord, C., Zwaigenbaum, L., Courchesne, E., Dager, S. R., Schmitz, C., Schultz, R. T., Crawley, J., and Young, L. J., The Developmental Neurobiology of Autism Spectrum Disorder. *J Neurosci*, 2006. **26**(26): p. 6897–906.

83. Hardan, A. Y., Minshew, N. J., Harenski, K., and Keshavan, M. S., Posterior Fossa Magnetic Resonance Imaging in Autism. *J Am Acad Child Adolesc Psychiatry*, 2001. **40**(6): p. 666–72.

84. Hardan, A. Y., Minshew, N. J., Mallikarjuhn, M., and Keshavan, M. S., Brain Volume in Autism. *J Child Neurol*, 2001. **16**(6): p. 421–4.

85. Piven, J., Nehme, E., Simon, J., Barta, P., Pearlson, G., and Folstein, S. E., Magnetic Resonance Imaging in Autism: Measurement of the Cerebellum, Pons, and Fourth Ventricle. *Biol Psychiatry*, 1992. **31**(5): p. 491–504.

86. Piven, J., Saliba, K., Bailey, J., and Arndt, S., An MRI Study of Autism: The Cerebellum Revisited. *Neurology*, 1997. **49**(2): p. 546–51.

87. Herrup, K., Shojaeian-Zanjani, H., Panzini, L., Sunter, K., and Mariani, J., The Numerical Matching of Source and Target Populations in the CNS: The Inferior Olive to Purkinje Cell Projection. *Brain Res*, 1996. **96**(1–2): p. 28–35.

88. Torres-Aleman, I., Pons, S., and Arevalo, M. A., The Insulin-Like Growth Factor I System in the Rat Cerebellum: Developmental Regulation and Role in Neuronal Survival and Differentiation. *J Neurosci Res*, 1994. **39**(2): p. 117–26.

89. Torres-Aleman, I., Pons, S., and Garcia-Segura, L. M., Climbing Fiber Deafferentation Reduces Insulin-Like Growth Factor I (IGF-I) Content in Cerebellum. *Brain Res*, 1991. **564**(2): p. 348–51.

90. Torres-Aleman, I., Pons, S., and Santos-Benito, F. F., Survival of Purkinje Cells in Cerebellar Cultures Is Increased by Insulin-Like Growth Factor I. *Eur J Neurosci*, 1992. **4**(9): p. 864–869.

91. Torres-Aleman, I., Villalba, M., and Nieto-Bona, M. P., Insulin-Like Growth Factor-I Modulation of Cerebellar Cell Populations Is Developmentally Stage-Dependent and Mediated by Specific Intracellular Pathways. *Neuroscience*, 1998. **83**(2): p. 321–34.

92. Rodier, P. M., Ingram, J. L., Tisdale, B., Nelson, S., and Romano, J., Embryological Origin for Autism: Developmental Anomalies of the Cranial Nerve Motor Nuclei. *J Comp Neurol*, 1996. **370**(2): p. 247–61.

93. Belmonte, M. K., Allen, G., Beckel-Mitchener, A., Boulanger, L. M., Carper, R. A., and Webb, S. J., Autism and Abnormal Development of Brain Connectivity. *J Neurosci*, 2004. **24**(42): p. 9228–31.

94. Pelphrey, K., Adolphs, R., and Morris, J. P., Neuroanatomical Substrates of Social Cognition Dysfunction in Autism. *Ment Retard Dev Disabil Res Rev*, 2004. **10**(4): p. 259–71.

95. Baron-Cohen, S., Ring, H. A., Wheelwright, S., Bullmore, E. T., Brammer, M. J., Simmons, A., and Williams, S. C., Social Intelligence in the Normal and Autistic Brain: An fMRI Study. *Eur J Neurosci*, 1999. **11**(6): p. 1891–8.

96. Critchley, H. D., Daly, E. M., Bullmore, E. T., Williams, S. C., Van Amelsvoort, T., Robertson, D. M., Rowe, A., Phillips, M., McAlonan, G., Howlin, P., and Murphy, D. G., The Functional Neuroanatomy of Social Behaviour: Changes in Cerebral Blood Flow When People with Autistic Disorder Process Facial Expressions. *Brain*, 2000. **123**(Pt 11): p. 2203–12.

97. Wang, A. T., Dapretto, M., Hariri, A. R., Sigman, M., and Bookheimer, S. Y., Neural Correlates of Facial Affect Processing in Children and Adolescents with Autism Spectrum Disorder. *J Am Acad Child Adolesc Psychiatry*, 2004. **43**(4): p. 481–90.

98. Castelli, F., Frith, C., Happe, F., and Frith, U., Autism, Asperger Syndrome and Brain Mechanisms for the Attribution of Mental States to Animated Shapes. *Brain*, 2002. **125**(Pt 8): p. 1839–49.

99. Boddaert, N., Belin, P., Chabane, N., Poline, J. B., Barthelemy, C., Mouren-Simeoni, M. C., Brunelle, F., Samson, Y., and Zilbovicius, M., Perception of Complex Sounds: Abnormal Pattern of Cortical Activation in Autism. *Am J Psychiatry*, 2003. **160**(11): p. 2057–60.

100. Boddaert, N., Chabane, N., Gervais, H., Good, C. D., Bourgeois, M., Plumet, M. H., Barthelemy, C., Mouren, M. C., Artiges, E., Samson, Y., Brunelle, F., Frackowiak, R. S., and Zilbovicius, M., Superior Temporal Sulcus Anatomical Abnormalities in Childhood Autism: A Voxel-Based Morphometry MRI Study. *Neuroimage*, 2004. **23**(1): p. 364–9.

101. Gervais, H., Belin, P., Boddaert, N., Leboyer, M., Coez, A., Sfaello, I., Barthelemy, C., Brunelle, F., Samson, Y., and Zilbovicius, M., Abnormal Cortical Voice Processing in Autism. *Nat Neurosci*, 2004. **7**(8): p. 801–2.

102. Hubl, D., Bolte, S., Feineis-Matthews, S., Lanfermann, H., Federspiel, A., Strik, W., Poustka, F., and Dierks, T., Functional Imbalance of Visual Pathways Indicates Alternative Face Processing Strategies in Autism. *Neurology*, 2003. **61**(9): p. 1232–7.

103. Schultz, R. T., Gauthier, I., Klin, A., Fulbright, R. K., Anderson, A. W., Volkmar, F., Skudlarski, P., Lacadie, C., Cohen, D. J., and Gore, J. C., Abnormal Ventral Temporal Cortical Activity During Face Discrimination among Individuals with Autism and Asperger Syndrome. *Arch Gen Psychiatry*, 2000. **57**(4): p. 331–40.

104. Hadjikhani, N., Joseph, R. M., Snyder, J., Chabris, C. F., Clark, J., Steele, S., McGrath, L., Vangel, M., Aharon, I., Feczko, E., Harris, G. J., and Tager-Flusberg, H., Activation of the Fusiform Gyrus When Individuals with Autism Spectrum Disorder View Faces. *Neuroimage*, 2004. **22**(3): p. 1141–50.

105. Pierce, K., Muller, R. A., Ambrose, J., Allen, G., and Courchesne, E., Face Processing Occurs Outside the Fusiform 'Face Area' in Autism: Evidence from Functional MRI. *Brain*, 2001. **124**(Pt 10): p. 2059–73.

106. Dalton, K. M., Nacewicz, B. M., Johnstone, T., Schaefer, H. S., Gernsbacher, M. A., Goldsmith, H. H., Alexander, A. L., and Davidson, R. J., Gaze Fixation and the Neural Circuitry of Face Processing in Autism. *Nat Neurosci*, 2005. **8**(4): p. 519–26.

107. Baader, S. L., Sanlioglu, S., Berrebi, A. S., Parker-Thornburg, J., and Oberdick, J., Ectopic Overexpression of Engrailed-2 in Cerebellar Purkinje Cells Causes Restricted Cell Loss and Retarded External Germinal Layer Development at Lobule Junctions. *J Neurosci*, 1998. **18**(5): p. 1763–73.

108. Benayed, R., Gharani, N., Rossman, I., Mancuso, V., Lazar, G., Kamdar, S., Bruse, S. E., Tischfield, S., Smith, B. J., Zimmerman, R. A., Dicicco-Bloom, E., Brzustowicz, L. M., and Millonig, J. H., Support for the Homeobox Transcription Factor Gene Engrailed 2 as an Autism Spectrum Disorder Susceptibility Locus. *Am J Hum Genet*, 2005. **77**(5): p. 851–68.

109. Gharani, N., Benayed, R., Mancuso, V., Brzustowicz, L. M., and Millonig, J. H., Association of the Homeobox Transcription Factor, Engrailed 2, with Autism Spectrum Disorder. *Mol Psychiatry*, 2004. **9**(5): p. 474–84.

110. Joyner, A. L., Herrup, K., Auerbach, B. A., Davis, C. A., and Rossant, J., Subtle Cerebellar Phenotype in Mice Homozygous for a Targeted Deletion of the En-2 Homeobox. *Science*, 1991. **251**(4998): p. 1239–43.

111. Millen, K. J., Hui, C. C., and Joyner, A. L., A Role for En-2 and Other Murine Homologues of Drosophila Segment Polarity Genes in Regulating Positional Information in the Developing Cerebellum. *Development*, 1995. **121**(12): p. 3935–45.

112. McAlonan, G. M., Cheung, V., Cheung, C., Suckling, J., Lam, G. Y., Tai, K. S., Yip, L., Murphy, D. G., and Chua, S. E., Mapping the Brain in Autism. A Voxel-Based MRI Study of Volumetric Differences and Intercorrelations in Autism. *Brain*, 2005. **128**(Pt 2): p. 268–76.

113. Courchesne, E., Saitoh, O., Yeung-Courchesne, R., Press, G. A., Lincoln, A. J., Haas, R. H., and Schreibman, L., Abnormality of Cerebellar Vermian Lobules VI and VII in Patients with Infantile Autism: Identification of Hypoplastic and Hyperplastic Subgroups with Mr Imaging. *AJR Am J Roentgenol*, 1994. **162**(1): p. 123–30.

114. Kuemerle, B., Gulden, F., Cherosky, N., Williams, E., and Herrup, K., The Mouse Engrailed Genes: A Window into Autism. *Behav Brain Res*, 2007 Jan 10. **176**(1): p. 121–32.

115. Jankowski, J., Holst, M. I., Liebig, C., Oberdick, J., and Baader, S. L., Engrailed-2 Negatively Regulates the Onset of Perinatal Purkinje Cell Differentiation. *J Comp Neurol*, 2004. **472**(1): p. 87–99.

116. Baader, S. L., Vogel, M. W., Sanlioglu, S., Zhang, X., and Oberdick, J., Selective Disruption Of "Late Onset" Sagittal Banding Patterns by Ectopic Expression of Engrailed-2 in Cerebellar Purkinje Cells. *J Neurosci*, 1999. **19**(13): p. 5370–9.

117. Gerlai, R., Millen, K. J., Herrup, K., Fabien, K., Joyner, A. L., and Roder, J., Impaired Motor Learning Performance in Cerebellar En-2 Mutant Mice. *Behav Neurosci*, 1996. **110**(1): p. 126–33.

118. Goldowitz, D. and Koch, J., Performance of Normal and Neurological Mutant Mice on Radial Arm Maze and Active Avoidance Tasks. *Behav Neural Biol*, 1986. **46**(2): p. 216–26.

119. Cheh, M. A., Millonig, J. H., Roselli, L. M., Ming, X., Jacobsen, E., Kamdar, S., and Wagner, G. C., En2 Knockout Mice Display Neurobehavioral and Neurochemical Alterations Relevant to Autism Spectrum Disorder. *Brain Res*, 2006. **1116**(1): p. 166–76.

120. Conroy, J., Meally, E., Kearney, G., Fitzgerald, M., Gill, M., and Gallagher, L., Serotonin Transporter Gene and Autism: A Haplotype Analysis in an Irish Autistic Population. *Mol Psychiatry*, 2004. **9**(6): p. 587–93.

121. Bradley, S. L., Dodelzon, K., Sandhu, H. K., and Philibert, R. A., Relationship of Serotonin Transporter Gene Polymorphisms and Haplotypes to mRNA Transcription. *Am J Med Genet B Neuropsychiatr Genet*, 2005. **136**(1): p. 58–61.

122. Muhle, R., Trentacoste, S. V., and Rapin, I., The Genetics of Autism. *Pediatrics*, 2004. **113**(5): p. e472–86.

123. Volkmar, F. R., Szatmari, P., and Sparrow, S. S., Sex Differences in Pervasive Developmental Disorders. *J Autism Dev Disord*, 1993. **23**(4): p. 579–91.

124. Arndt, T. L., Stodgell, C. J., and Rodier, P. M., The Teratology of Autism. *Int J Dev Neurosci*, 2005. **23**(2–3): p. 189–99.

125. Ingram, J. L., Peckham, S. M., Tisdale, B., and Rodier, P. M., Prenatal Exposure of Rats to Valproic Acid Reproduces the Cerebellar Anomalies Associated with Autism. *Neurotoxicol Teratol*, 2000. **22**(3): p. 319–24.

126. Sears, L. L., Finn, P. R., and Steinmetz, J. E., Abnormal Classical Eye-Blink Conditioning in Autism. *J Autism Dev Disord*, 1994. **24**(6): p. 737–51.

127. Stanton, M. E., Peloso, E., Brown, K. L., and Rodier, P., Discrimination Learning and Reversal of the Conditioned Eyeblink Reflex in a Rodent Model of Autism. *Behav Brain Res*, 2007. **176**(1): p. 133–40.

128. Chugani, D. C., Muzik, O., Behen, M., Rothermel, R., Janisse, J. J., Lee, J., and Chugani, H. T., Developmental Changes in Brain Serotonin Synthesis Capacity in Autistic and Nonautistic Children. *Ann Neurol*, 1999. **45**(3): p. 287–95.

129. Chandana, S. R., Behen, M. E., Juhasz, C., Muzik, O., Rothermel, R. D., Mangner, T. J., Chakraborty, P. K., Chugani, H. T., and Chugani, D. C., Significance of Abnormalities in Developmental Trajectory and Asymmetry of Cortical Serotonin Synthesis in Autism. *Int J Dev Neurosci*, 2005. **23**(2–3): p. 171–82.

130. Chugani, D. C., Muzik, O., Rothermel, R., Behen, M., Chakraborty, P., Mangner, T., da Silva, E. A., and Chugani, H. T., Altered Serotonin Synthesis in the Dentatothalamocortical Pathway in Autistic Boys. *Ann Neurol*, 1997. **42**(4): p. 666–9.

131. Muller, R. A., Chugani, D. C., Behen, M. E., Rothermel, R. D., Muzik, O., Chakraborty, P. K., and Chugani, H. T., Impairment of Dentato-Thalamo-Cortical Pathway in Autistic Men: Language Activation Data from Positron Emission Tomography. *Neurosci Lett*, 1998. **245**(1): p. 1–4.

132. Stodgell, C. J., Ingram, J. L., O'Bara, M., Tisdale, B. K., Nau, H., and Rodier, P. M., Induction of the Homeotic Gene Hoxa1 through Valproic Acid's Teratogenic Mechanism of Action. *Neurotoxicol Teratol*, 2006. **28**(5): p. 617–24.

133. Phiel, C. J., Zhang, F., Huang, E. Y., Guenther, M. G., Lazar, M. A., and Klein, P. S., Histone Deacetylase Is a Direct Target of Valproic Acid, a Potent Anticonvulsant, Mood Stabilizer, and Teratogen. *J Biol Chem*, 2001. **276**(39): p. 36734–41.

134. Tischfield, M. A., Bosley, T. M., Salih, M. A., Alorainy, I. A., Sener, E. C., Nester, M. J., Oystreck, D. T., Chan, W. M., Andrews, C., Erickson, R. P., and Engle, E. C., Homozygous Hoxa1 Mutations Disrupt Human Brainstem, Inner Ear, Cardiovascular and Cognitive Development. *Nat Genet*, 2005. **37**(10): p. 1035–7.

135. Sen, B., Sinha, S., Ahmed, S., Ghosh, S., Gangopadhyay, P. K., and Usha, R., Lack of Association of Hoxa1 and Hoxb1 Variants with Autism in the Indian Population. *Psychiatr Genet*, 2007. **17**(1): p. 1.

136. Gallagher, L., Hawi, Z., Kearney, G., Fitzgerald, M., and Gill, M., No Association between Allelic Variants of Hoxa1/Hoxb1 and Autism. *Am J Med Genet B Neuropsychiatr Genet*, 2004. **124**(1): p. 64–7.

137. Collins, J. S., Schroer, R. J., Bird, J., and Michaelis, R. C., The Hoxa1 A218g Polymorphism and Autism: Lack of Association in White and Black Patients from the South Carolina Autism Project. *J Autism Dev Disord*, 2003. **33**(3): p. 343–8.

138. Romano, V., Cali, F., Mirisola, M., Gambino, G., D'Anna, R., Di Rosa, P., Seidita, G., Chiavetta, V., Aiello, F., Canziani, F., De Leo, G., Ayala, G. F., and Elia, M., Lack of Association of Hoxa1 and Hoxb1 Mutations and Autism in Sicilian (Italian) Patients. *Mol Psychiatry*, 2003. **8**(8): p. 716–7.

139. Talebizadeh, Z., Bittel, D. C., Miles, J. H., Takahashi, N., Wang, C. H., Kibiryeva, N., and Butler, M. G., No Association between Hoxa1 and Hoxb1 Genes and Autism Spectrum Disorders (Asd). *J Med Genet*, 2002. **39**(11): p. e70.

140. Miller, M. T., Stromland, K., Ventura, L., Johansson, M., Bandim, J. M., and Gillberg, C., Autism Associated with Conditions Characterized by Developmental Errors in Early Embryogenesis: A Mini Review. *Int J Dev Neurosci*, 2005. **23**(2–3): p. 201–19.

141. Folstein, S. and Rutter, M., Genetic Influences and Infantile Autism. *Nature*, 1977. **265**(5596): p. 726–8.

142. Folstein, S. and Rutter, M., Infantile Autism: A Genetic Study of 21 Twin Pairs. *J Child Psychol Psychiatry*, 1977. **18**(4): p. 297–321.
143. Pickles, A., Bolton, P., Macdonald, H., Bailey, A., Le Couteur, A., Sim, C. H., and Rutter, M., Latent-Class Analysis of Recurrence Risks for Complex Phenotypes with Selection and Measurement Error: A Twin and Family History Study of Autism. *Am J Hum Genet*, 1995. **57**(3): p. 717–26.
144. Ritvo, E. R., Freeman, B. J., Mason-Brothers, A., Mo, A., and Ritvo, A. M., Concordance for the Syndrome of Autism in 40 Pairs of Afflicted Twins. *Am J Psychiatry*, 1985. **142**(1): p. 74–7.
145. Bailey, A., Le Couteur, A., Gottesman, I., Bolton, P., Simonoff, E., Yuzda, E., and Rutter, M., Autism as a Strongly Genetic Disorder: Evidence from a British Twin Study. *Psychol Med*, 1995. **25**(1): p. 63–77.
146. Le Couteur, A., Bailey, A., Goode, S., Pickles, A., Robertson, S., Gottesman, I., and Rutter, M., A Broader Phenotype of Autism: The Clinical Spectrum in Twins. *J Child Psychol Psychiatry*, 1996. **37**(7): p. 785–801.
147. Rutter, M., Genetic Studies of Autism: From the 1970s into the Millennium. *J Abnorm Child Psychol*, 2000. **28**(1): p. 3–14.
148. Bolton, P., Macdonald, H., Pickles, A., Rios, P., Goode, S., Crowson, M., Bailey, A., and Rutter, M., A Case-Control Family History Study of Autism. *J Child Psychol Psychiatry*, 1994. **35**(5): p. 877–900.
149. Ritvo, E. R., Mason-Brothers, A., Jenson, W. P., Freeman, B. J., Mo, A., Pingree, C., Petersen, P. B., and McMahon, W. M., A Report of One Family with Four Autistic Siblings and Four Families with Three Autistic Siblings. *J Am Acad Child Adolesc Psychiatry*, 1987. **26**(3): p. 339–41.
150. Bartlett, C. W., Gharani, N., Millonig, J. H., and Brzustowicz, L. M., Three Autism Candidate Genes: A Synthesis of Human Genetic Analysis with Other Disciplines. *Int J Dev Neurosci*, 2005. **23**(2–3): p. 221–34.
151. Amir, R. E., Van den Veyver, I. B., Wan, M., Tran, C. Q., Francke, U., and Zoghbi, H. Y., Rett Syndrome Is Caused by Mutations in X-Linked Mecp2, Encoding Methyl-CpG-Binding Protein 2. *Nat Genet*, 1999. **23**(2): p. 185–8.
152. Samaco, R. C., Hogart, A., and LaSalle, J. M., Epigenetic Overlap in Autism-Spectrum Neurodevelopmental Disorders: Mecp2 Deficiency Causes Reduced Expression of Ube3a and Gabrb3. *Hum Mol Genet*, 2005. **14**(4): p. 483–92.
153. Coutinho, A. M., Oliveira, G., Katz, C., Feng, J., Yan, J., Yang, C., Marques, C., Ataide, A., Miguel, T. S., Borges, L., Almeida, J., Correia, C., Currais, A., Bento, C., Mota-Vieira, L., Temudo, T., Santos, M., Maciel, P., Sommer, S. S., and Vicente, A. M., MECP2 Coding Sequence and 3'UTR Variation in 172 Unrelated Autistic Patients. *Am J Med Genet B Neuropsychiatr Genet*, 2007 June 5. **144B**(4): p. 473–83.
154. Li, H., Yamagata, T., Mori, M., Yasuhara, A., and Momoi, M. Y., Mutation Analysis of Methyl-CpG Binding Protein Family Genes in Autistic Patients. *Brain Dev*, 2005. **27**(5): p. 321–5.
155. Xi, C. Y., Ma, H. W., Lu, Y., Zhao, Y. J., Hua, T. Y., Zhao, Y., and Ji, Y. H., Mecp2 Gene Mutation Analysis in Autistic Boys with Developmental Regression. *Psychiatr Genet*, 2007. **17**(2): p. 113–6.
156. Guy, J., Gan, J., Selfridge, J., Cobb, S., and Bird, A., Reversal of Neurological Defects in a Mouse Model of Rett Syndrome. *Science*, 2007. **315**(5815): p. 1143–7.
157. Yang, M. S. and Gill, M., A Review of Gene Linkage, Association and Expression Studies in Autism and an Assessment of Convergent Evidence. *Int J Dev Neurosci*, 2007. **25**(2): p. 69–85.
158. Tabor, H. K., Risch, N. J., and Myers, R. M., Candidate-Gene Approaches for Studying Complex Genetic Traits: Practical Considerations. *Nat Rev*, 2002. **3**(5): p. 391–7.
159. Campbell, D. B., Sutcliffe, J. S., Ebert, P. J., Militerni, R., Bravaccio, C., Trillo, S., Elia, M., Schneider, C., Melmed, R., Sacco, R., Persico, A. M., and Levitt, P., A Genetic Variant That Disrupts Met Transcription Is Associated with Autism. *Proc Natl Acad Sci USA*, 2006. **103**(45): p. 16834–9.

160. Blasi, F., Bacchelli, E., Pesaresi, G., Carone, S., Bailey, A. J., and Maestrini, E., Absence of Coding Mutations in the X-Linked Genes Neuroligin 3 and Neuroligin 4 in Individuals with Autism from the Imgsac Collection. *Am J Med Genet B Neuropsychiatr Genet*, 2006. **141**(3): p. 220–1.

161. Jamain, S., Quach, H., Betancur, C., Rastam, M., Colineaux, C., Gillberg, I. C., Soderstrom, H., Giros, B., Leboyer, M., Gillberg, C., and Bourgeron, T., Mutations of the X-Linked Genes Encoding Neuroligins Nlgn3 and Nlgn4 Are Associated with Autism. *Nat Genet*, 2003. **34**(1): p. 27–9.

162. Laumonnier, F., Bonnet-Brilhault, F., Gomot, M., Blanc, R., David, A., Moizard, M. P., Raynaud, M., Ronce, N., Lemonnier, E., Calvas, P., Laudier, B., Chelly, J., Fryns, J. P., Ropers, H. H., Hamel, B. C., Andres, C., Barthelemy, C., Moraine, C., and Briault, S., X-Linked Mental Retardation and Autism Are Associated with a Mutation in the Nlgn4 Gene, a Member of the Neuroligin Family. *Am J Hum Genet*, 2004. **74**(3): p. 552–7.

163. Feng, J., Schroer, R., Yan, J., Song, W., Yang, C., Bockholt, A., Cook, E. H., Jr., Skinner, C., Schwartz, C. E., and Sommer, S. S., High Frequency of Neurexin 1beta Signal Peptide Structural Variants in Patients with Autism. *Neurosci Lett*, 2006. **409**(1): p. 10–3.

164. Jacob, S., Brune, C. W., Carter, C. S., Leventhal, B. L., Lord, C., and Cook, E. H., Jr., Association of the Oxytocin Receptor Gene (OXTR) in Caucasian Children and Adolescents with Autism. *Neurosci Lett*, 2007. **417**(1): p. 6–9.

165. Wu, S., Jia, M., Ruan, Y., Liu, J., Guo, Y., Shuang, M., Gong, X., Zhang, Y., Yang, X., and Zhang, D., Positive Association of the Oxytocin Receptor Gene (OXTR) with Autism in the Chinese Han Population. *Biol Psychiatry*, 2005. **58**(1): p. 74–7.

166. Butler, M. G., Dasouki, M. J., Zhou, X. P., Talebizadeh, Z., Brown, M., Takahashi, T. N., Miles, J. H., Wang, C. H., Stratton, R., Pilarski, R., and Eng, C., Subset of Individuals with Autism Spectrum Disorders and Extreme Macrocephaly Associated with Germline Pten Tumour Suppressor Gene Mutations. *J Med Genet*, 2005. **42**(4): p. 318–21.

167. Buxbaum, J. D., Cai, G., Chaste, P., Nygren, G., Goldsmith, J., Reichert, J., Anckarsater, H., Rastam, M., Smith, C. J., Silverman, J. M., Hollander, E., Leboyer, M., Gillberg, C., Verloes, A., and Betancur, C., Mutation Screening of the Pten Gene in Patients with Autism Spectrum Disorders and Macrocephaly. *Am J Med Genet B Neuropsy Chiatr Genet*, 2007 June 5. **144B**(4): p. 484–91.

168. Sadakata, T., Kakegawa, W., Mizoguchi, A., Washida, M., Katoh-Semba, R., Shutoh, F., Okamoto, T., Nakashima, H., Kimura, K., Tanaka, M., Sekine, Y., Itohara, S., Yuzaki, M., Nagao, S., and Furuichi, T., Impaired Cerebellar Development and Function in Mice Lacking Caps2, a Protein Involved in Neurotrophin Release. *J Neurosci*, 2007. **27**(10): p. 2472–82.

169. Sadakata, T., Washida, M., and Furuichi, T., Alternative Splicing Variations in Mouse Caps2: Differential Expression and Functional Properties of Splicing Variants. *BMC Neurosci*, 2007. **8**: p. 25.

170. Sadakata, T., Washida, M., Iwayama, Y., Shoji, S., Sato, Y., Ohkura, T., Katoh-Semba, R., Nakajima, M., Sekine, Y., Tanaka, M., Nakamura, K., Iwata, Y., Tsuchiya, K. J., Mori, N., Detera-Wadleigh, S. D., Ichikawa, H., Itohara, S., Yoshikawa, T., and Furuichi, T., Autistic-Like Phenotypes in Cadps2-Knockout Mice and Aberrant Cadps2 Splicing in Autistic Patients. *J Clin Invest*, 2007. **117**(4): p. 931–43.

171. Fatemi, S. H., Snow, A. V., Stary, J. M., Araghi-Niknam, M., Reutiman, T. J., Lee, S., Brooks, A. I., and Pearce, D. A., Reelin Signaling Is Impaired in Autism. *Biol Psychiatry*, 2005. **57**(7): p. 777–87.

172. Skaar, D. A., Shao, Y., Haines, J. L., Stenger, J. E., Jaworski, J., Martin, E. R., DeLong, G. R., Moore, J. H., McCauley, J. L., Sutcliffe, J. S., Ashley-Koch, A. E., Cuccaro, M. L., Folstein, S. E., Gilbert, J. R., and Pericak-Vance, M. A., Analysis of the Reln Gene as a Genetic Risk Factor for Autism. *Mol Psychiatry*, 2005. **10**(6): p. 563–71.

173. Persico, A. M., D'Agruma, L., Maiorano, N., Totaro, A., Militerni, R., Bravaccio, C., Wassink, T. H., Schneider, C., Melmed, R., Trillo, S., Montecchi, F., Palermo, M., Pascucci, T., Puglisi-Allegra, S., Reichelt, K. L., Conciatori, M., Marino, R., Quattrocchi, C. C., Baldi, A.,

Zelante, L., Gasparini, P., and Keller, F., Reelin Gene Alleles and Haplotypes as a Factor Predisposing to Autistic Disorder. *Mol Psychiatry*, 2001. **6**(2): p. 150–9.

174. Chen, G. K., Kono, N., Geschwind, D. H., and Cantor, R. M., Quantitative Trait Locus Analysis of Nonverbal Communication in Autism Spectrum Disorder. *Mol Psychiatry*, 2006. **11**(2): p. 214–20.

175. Conciatori, M., Stodgell, C. J., Hyman, S. L., O'Bara, M., Militerni, R., Bravaccio, C., Trillo, S., Montecchi, F., Schneider, C., Melmed, R., Elia, M., Crawford, L., Spence, S. J., Muscarella, L., Guarnieri, V., D'Agruma, L., Quattrone, A., Zelante, L., Rabinowitz, D., Pascucci, T., Puglisi-Allegra, S., Reichelt, K. L., Rodier, P. M., and Persico, A. M., Association between the Hoxa1 A218g Polymorphism and Increased Head Circumference in Patients with Autism. *Biol Psychiatry*, 2004. **55**(4): p. 413–9.

176. Chen, W., Landau, S., Sham, P., and Fombonne, E., No Evidence for Links between Autism, MMR and Measles Virus. *Psychol Med*, 2004. **34**(3): p. 543–53.

177. Fombonne, E., The Prevalence of Autism. *JAMA*, 2003. **289**(1): p. 87–9.

178. Fombonne, E., Epidemiological Surveys of Autism and Other Pervasive Developmental Disorders: An Update. *J Autism Dev Disord*, 2003. **33**(4): p. 365–82.

179. Fombonne, E., Modern Views of Autism. *Can J Psychiatr*, 2003. **48**(8): p. 503–5.

180. Fombonne, E., Epidemiology of Autistic Disorder and Other Pervasive Developmental Disorders. *J Clin Psychiatry*, 2005. **66**(Suppl 10): p. 3–8.

181. Folstein, S. E. and Rosen-Sheidley, B., Genetics of Autism: Complex Aetiology for a Heterogeneous Disorder. *Nat Rev*, 2001. **2**(12): p. 943–55.

182. Alarcon, M., Cantor, R. M., Liu, J., Gilliam, T. C., and Geschwind, D. H., Evidence for a Language Quantitative Trait Locus on Chromosome 7q in Multiplex Autism Families. *Am J Hum Genet*, 2002. **70**(1): p. 60–71.

183. Auranen, M., Vanhala, R., Varilo, T., Ayers, K., Kempas, E., Ylisaukko-Oja, T., Sinsheimer, J. S., Peltonen, L., and Jarvela, I., A Genomewide Screen for Autism-Spectrum Disorders: Evidence for a Major Susceptibility Locus on Chromosome 3q25-27. *Am J Hum Genet*, 2002. **71**(4): p. 777–90.

184. Liu, J., Nyholt, D. R., Magnussen, P., Parano, E., Pavone, P., Geschwind, D., Lord, C., Iversen, P., Hoh, J., Ott, J., and Gilliam, T. C., A Genomewide Screen for Autism Susceptibility Loci. *Am J Hum Genet*, 2001. **69**(2): p. 327–40.

185. Spielman, R. S., McGinnis, R. E., and Ewens, W. J., Transmission Test for Linkage Disequilibrium: The Insulin Gene Region and Insulin-Dependent Diabetes Mellitus (IDDM). *Am J Hum Genet*, 1993. **52**(3): p. 506–16.

186. Brunet, I., Weinl, C., Piper, M., Trembleau, A., Volovitch, M., Harris, W., Prochiantz, A., and Holt, C., The Transcription Factor Engrailed-2 Guides Retinal Axons. *Nature*, 2005. **438**(7064): p. 94–8.

187. Nedelec, S., Foucher, I., Brunet, I., Bouillot, C., Prochiantz, A., and Trembleau, A., Emx2 Homeodomain Transcription Factor Interacts with Eukaryotic Translation Initiation Factor 4e (Eif4e) in the Axons of Olfactory Sensory Neurons. *Proc Natl Acad Sci USA*, 2004. **101**(29): p. 10815–20.

188. Di Nardo, A. A., Nedelec, S., Trembleau, A., Volovitch, M., Prochiantz, A., and Montesinos, M. L., Dendritic Localization and Activity-Dependent Translation of Engrailed1 Transcription Factor. *Mol Cell Neurosci*, 2007. **35**(2): p. 230–6.

189. Maizel, A., Bensaude, O., Prochiantz, A., and Joliot, A., A Short Region of Its Homeodomain Is Necessary for Engrailed Nuclear Export and Secretion. *Development*, 1999. **126**(14): p. 3183–90.

190. Joliot, A., Maizel, A., Rosenberg, D., Trembleau, A., Dupas, S., Volovitch, M., and Prochiantz, A., Identification of a Signal Sequence Necessary for the Unconventional Secretion of Engrailed Homeoprotein. *Curr Biol*, 1998. **8**(15): p. 856–63.

191. Joliot, A., Trembleau, A., Raposo, G., Calvet, S., Volovitch, M., and Prochiantz, A., Association of Engrailed Homeoproteins with Vesicles Presenting Caveolae-Like Properties. *Development*, 1997. **124**(10): p. 1865–75.

192. Briata, P., Di Blas, E., Gulisano, M., Mallamaci, A., Iannone, R., Boncinelli, E., and Corte, G., Emx1 Homeoprotein Is Expressed in Cell Nuclei of the Developing Cerebral Cortex and in the Axons of the Olfactory Sensory Neurons. *Mech Dev*, 1996. **57**(2): p. 169–80.

193. Maizel, A., Tassetto, M., Filhol, O., Cochet, C., Prochiantz, A., and Joliot, A., Engrailed Homeoprotein Secretion Is a Regulated Process. *Development*, 2002. **129**(15): p. 3545–53.

194. Hjerrild, M., Stensballe, A., Jensen, O. N., Gammeltoft, S., and Rasmussen, T. E., Protein Kinase A Phosphorylates Serine 267 in the Homeodomain of Engrailed-2 Leading to Decreased DNA Binding. *FEBS Lett*, 2004. **568**(1–3): p. 55–9.

195. Sonnier, L., Le Pen, G., Hartmann, A., Bizot, J. C., Trovero, F., Krebs, M. O., and Prochiantz, A., Progressive Loss of Dopaminergic Neurons in the Ventral Midbrain of Adult Mice Heterozygote for Engrailed1. *J Neurosci*, 2007. **27**(5): p. 1063–71.

196. Rossman, I. T., Kamdar, S., Millonig, J., DiCicco-Bloom, E., Extracellular Growth Factors Interact with Engrailed2 (En2), an Autism-Associated Gene, to Control Cerebellar Development, in *Society for Neuroscience*. 2005, Washington, DC. p. Program No. 596.2.

Chapter 2
Teratogenic Alleles in Autism and Other Neurodevelopmental Disorders

William G. Johnson, Madhura Sreenath, Steven Buyske, and Edward S. Stenroos

Abstract Genes for neurodevelopmental disorders have proven difficult to find. In the case of autism, even though a number of genes associated with the disorder have been identified, the cause of the disorder is far from clear. We discuss here a new category of genetic contribution to human disorders, i.e., gene variants (alleles) that act in mothers during pregnancy to contribute to the neurodevelopmental disorder in her offspring. We have termed these maternally acting alleles "teratogenic alleles" because they act in the fetus in a way somewhat analogous to chemical teratogens or drug teratogens ingested by a mother during pregnancy that act to damage her fetus. Although two examples of this mechanism have been known for some time, new examples have been found and since 2003 the number of examples has more than doubled. Even though this number is growing rapidly, the present number, *33*, is not large. It is not clear whether this is so because teratogenic alleles are not yet well known and therefore not looked for, whether the possibility of a teratogenic allele is not considered when studies are designed, or whether there are in fact very few teratogenic alleles in comparison with alleles that act in the fetus. All of the teratogenic alleles reported so far have been in neurodevelopmental and developmental disorders. Again it is not clear whether this is simply because these are the disorders that happen to have been studied so far by suitable study designs, or whether teratogenic alleles are actually more commonly encountered in these disorders, perhaps as a regular feature or even as a characteristic feature of these disorders.

In autism, some of the brain abnormalities are known to occur very early in pregnancy. A number of teratogenic alleles have already been reported in autism and more will be reported. Demonstration of the action of teratogenic alleles in autism opens up new opportunities for therapy and prevention.

Keywords Autism · teratogenesis · genes · alleles · maternal · pregnancy

W.G. Johnson
Department of Neurology, UMDNJ-Robert Wood Johnson Medical School,
Piscataway, NJ, USA
e-mail: wjohnson@umdnj.edu

A.W. Zimmerman (ed.), *Autism*, DOI: 10.1007/978-1-60327-489-0_2,
© Humana Press, Totowa, NJ 2008

Introduction

Genes that determine or contribute to human disorders are usually thought of as acting in the affected individual, and most of them do. However, an important class of genes acts in the mother of the affected individual to contribute to a developmental or neurodevelopmental disorder in her affected offspring. This model, the gene–teratogen model, consists of the following:

(1) Maternal alleles, called "teratogenic alleles" because they act in a fashion somewhat analogous to ordinary chemical teratogens ingested by a mother during pregnancy that alter the fetal environment and affect fetal development. Teratogenic alleles can be alleles of specific autosomal or X-linked genes acting in dominant, recessive, codominant, multifactorial, or interactive fashion, including low penetrance alleles of functional polymorphisms.
(2) Fetal alleles, called "modifying or specificity alleles" because they modulate the effect of teratogenic alleles with respect to severity or tissue distribution of the developmental effect.
(3) Environmental factors, which may act at different times during gestation. Because some of these terms may be challenging for the nonspecialist reader, a general glossary is included at the end of the chapter.

Thirty-three reports of suspected or documented teratogenic alleles have been found to date (Table 2.1), a number that has more than doubled over the past four years [1]. Still, this number, though growing, is small. It is not clear whether this is a rare and unusual mechanism or whether the small number reflects the fact that action of a teratogenic allele is not often considered in genetic studies.

Most of the reported teratogenic alleles have been found in neurodevelopmental disorders (Table 2.1). Again, it is not clear how to interpret this. It is possible that this observation results solely from an artifact of ascertainment; that is, it may be somewhat more natural to consider maternal effects in neurodevelopmental disorders, especially those that are diagnosed in childhood. It is also possible that most of the reported teratogenic alleles occur in neurodevelopmental disorders because they play a role more frequently in these disorders. It is even possible that teratogenic alleles are a common and characteristic feature of neurodevelopmental disorders and that they would be recognized as if they were more frequently sought. The evidence is currently lacking for this possibility, but it cannot be excluded at present.

Another reason to consider the possible action of teratogenic alleles more frequently in designing genetic studies is that genes for some neurodevelopmental disorders have been difficult to find. An increasing number of susceptibility genes of small effect have been reported but it seems unclear yet as to how these act or interact to contribute to the neurodevelopmental disorders being studied. Teratogenic alleles could be an important part of the overall picture of the etiology of neurodevelopmental disorders, with multiple teratogenic alleles

Table 2.1 Reports of teratogenic alleles

Teratogenic allele	Disease	Analysis by	References
Maternal TDT			
1. *MTR*2756G*	NTD/spina bifida	Mat TDT, log-linear	[13]
2. *MTRR*66G*	NTD/spina bifida	Mat TDT, log-linear	[13]
3. *GSTP1*val105*	Autism	Mat TDT	[5]
Maternal TDT equivalent			
4. *PAH* mutations	Maternal phenylketonuria	Mat TDT equivalent	[48]
5. Rh *d*	Erythroblastosis fetalis	Mat TDT equivalent	[18]
6. Rh *d*	Schizophrenia	Mat TDT equivalent, log-linear	[19, 49]
Case–parent log-linear analysis without maternal TDT or equivalent			
7. *MTHFD1*653Q*	Spina bifida	Log-linear	[21]
8. *CYP1A1*6235C*	Low birth weight	Log-linear	[22]
9. *NAT1* compos. genotypes	NTD/spina bifida	Log-linear	[23]
10. *CCL-2*2518AA*	NTD/spina bifida	Log-linear	[24]
Regression analysis			
11. *APOE*E2*	Lower LDL-C, apoB, higher HDL-C, apoA1 in newborns	Forward stepwise regression analysis	[25]
12. *APOC3*S2*	Lower newborn LDL-C, apoB, HDL-C, apoA1	Forward stepwise regression analysis	[25]
13. *LPL*S447X*	Lower newborn LDL-C, apoB, TG	Forward stepwise regression analysis	[25]
14. *GSTP1*Val105,*Val114*	Asthma	Multiple linear regression	[26]
Case–control plus case TDT			
15. *MTHFR*677T, CT, TT*	Oral–facial clefting, cleft lip ± cleft palate	Case–control, case TDT negative	[27]
16. *MTHFR 677T*	Congenital heart disease	Case–control, case TDT negative	[28]
17. *MTHFR*677TT*	Down syndrome	Case–control, case TDT negative	[29]
18. *HLA-DR*4*	Autism	Case–control, case TDT negative	[30]

Table 2.1 (continued)

Teratogenic allele	Disease	Analysis by	References
Case–control			
19. *C4B*0*	Autism	Case–control	[31]
20. *HLA-DR4*	Autism	Case–control	[34]
21. *GSTP1-1b*	Recurrent early pregnancy loss	Case–control	[37]
22. *MTHFR*1298C*	NTD/spina bifida	Case–control	[38]
23. *MTRR*66G *GG*	Down syndrome	Case–control	[39]
24. *MTHFR*677T/ MTRR*66GG* " "	Down syndrome	Case–control	[39]
25. *GSTT1*0*	Oral–facial clefting	Case–control	[40]
26. *MTHFR*1298C*	NTD/spina bifida	Case–control	[42]
27. *GSTM1*0*	Recurrent pregnancy loss	Case–control	[43]
28. *DHFR*19 bp deletion*	Spina bifida	Case–control	[44]
29. *CYP1A1*2A*	Recurrent pregnancy loss	Case–control	[45]
30. *DHFR*19 bp deletion*	Preterm delivery	Case–control	[46]
31. *MTHFR*1298C, CC*	Down syndrome	Case–control	[29]
32. *MTHFR*1298C, CC*	Down syndrome	Case–control	[50]
33. *RFC1*A80G, GG*	Down syndrome	Case–control	[50]

The table lists the reports of teratogenic alleles (1) according to the method of analysis used and (2) according to the date of the report beginning with the earliest. The specific teratogenic allele is listed on the left, next the specific disease or disorder studied, then the study design, and finally the literature reference. The specific teratogenic allele is given in the nomenclature: gene symbol*allele designation. The names of the genes corresponding to the gene symbol are given in the text and correspond to the designation in OMIM (Online Mendelian Inheritance in Man). Gene symbols are given in upper case letters and italicized, while the corresponding protein symbol is given in the same upper case letters but not italicized. Sometimes a genotype is given, as in numbers 31 and 32, where "*MTHFR*1298C*" refers to the C-allele at position 1298 of the gene *MTHFR*" and *CC*, the genotype, refers to a double dose of the C-allele, i.e., homozygosity. Human genes are given in upper case italics, corresponding proteins in the same upper case letters that are not italicized. *MTR*: methionine synthase gene; *MTRR*: methionine synthase reductase gene; *GSTP1*: glutathione *S*-transferase P1 gene; NTD: neural tube defect; *PAH*: phenylalanine hydroxylase gene; Rh: Rhesus factor, a protein that individuals either have or do not have on the surface of their red blood cells; *MTHFD1*: methylenetetrahydrofolate dehydrogenase gene; *CYP1A1*: Cytochrome P450 1A1 gene; *NAT1*: N-acetyltransferase gene; *CCL*: monocyte chemoattractant protein gene; APOE: Apolipoprotein E; APOC3: Apolipoprotein C3; LPL: Lipoprotein lipase; *MTHFR*: methylenetetrahydrofolate reductase gene; HLA: human leukocyte antigen system; OFC: Oral–facial clefting; CHD: congenital heart defect; C4B: complement component 4B, a blood group antigen; *GSTT1*: Glutathione *S*-transferase T1 gene; *GSTM1*: Glutathione *S*-transferase M1 gene; *DHFR*: Dihydrofolate reductase gene; *RFC1*: reduced folate carrier 1 gene.

interacting with multiple fetal genes, both interacting with environmental factors. But again, at present there is no evidence to support this possibility.

Teratogenic alleles have some unexpected features. First, since they act in the mother to affect the fetus, the mother and not the fetus is the genetic patient. From the point of view of the fetus, a teratogenic allele is an environmental effect, not a genetic effect. For teratogenic alleles acting in autosomal recessive fashion, the fetus will nonetheless be an obligate heterozygote for the harmful allele (and possibly a homozygote depending upon the paternal contribution) even though the allele may not act in the fetus at all. For a dominant teratogenic allele, the fetus may not even carry a single copy of the maternal allele that affects its environment adversely. Interestingly, teratogenic alleles are not readily studied in inbred mouse strains, since both mother and fetus are homozygous for the same allele at nearly all loci.

The model does not include the effects of maternal dietary constituents, drug/toxin/teratogen ingestion, smoking, endocrine or metabolic effects, infection or inflammation or other maternal effects on the fetus, if these maternal effects are solely environmental and not related to maternal genotype. Nor does the model include the direct interaction of a maternal environmental effect with a fetal genotype. The model does not include mitochondrial genes since they do not segregate. Alleles originating from fetal or maternal microchimerism or from genomic imprinting are not included since these will not be inherited through many generations.

Approaches to Documenting Action of a Teratogenic Allele

Locating or identifying a candidate gene for a modifying/specificity allele is straightforward because the ordinary methods of linkage mapping are suitable and well developed, e.g., the lod score method, sib pair methods, or standard case-parent methods. For this reason, genes for neurodevelopmental disorders identified by these methods are likely to be of this type.

However, documenting a teratogenic allele is not straightforward since the mother of the proband with the neurodevelopmental disorder is the one who is affected, i.e., the "genetic patient." Pedigrees collected for use in the lod score method often do not contain the right individuals for study if the pedigree is simply recoded so that the individuals with the neurodevelopmental disorder are now "unaffected" and their mothers are "affected." Sib pair methods are also problematic in the study of a teratogenic locus, even if parents are included, because a sib pair contains only one "affected" individual, the mother. Likewise, in standard transmission disequilibrium test (TDT) carried out with neurodevelopmental probands and their parents, the only affected individuals in the group are the mothers.

The presence of a teratogenic allele may first be suspected when examination of a dataset containing probands, mothers, and fathers for a neurodevelopmental disorder shows a significantly increased frequency for an allele of interest among mothers but not fathers compared to controls (Table 2.2). Increased allele frequency in probands is also to be expected by descent from mothers. Table 2.3 shows an example of apparent relative risks for cases and mothers for one or two copies of a risk allele. For a teratogenic allele, the expected disease

Table 2.2 Origin of changes in allele frequencies and allele transmissions for genes acting in mothers only, cases only or through maternal imprinting

	Gene acting in mothers only	Gene acting in cases only	Maternal imprinting
Elevated allele frequency in cases	By enrichment	Causative	Causative
Elevated allele frequency in mothers	Causative	By enrichment	By enrichment
Elevated allele frequency in fathers	No	By enrichment	No
Overtransmission of maternal allele	No	Yes	Yes
Overtransmission paternal allele	No	Yes	No

The table characterizes the effects of genes acting solely in mothers, solely in offspring, or via maternal imprinting in terms of the susceptibility allele frequencies in cases and parents. The table also shows whether the allele would be expected to be overtransmitted from parents to the case. The table distinguishes between allele frequencies elevated because the allele in question is causative in the subjects and allele frequencies that are necessarily elevated by enrichment—for example, higher frequencies by descent in offspring for a teratogenic allele acting maternally only. "Causative" means that the allelic frequency is higher than in the population because the allele itself is a susceptibility allele or is in tight linkage disequilibrium with such an allele. "By enrichment" means that the allelic frequency is increased simply because of Mendelian inheritance, with no causative effect. "By enrichment" increases will be attenuated relative to "causative" increases.

allele frequencies from highest to lowest are mothers > neurodevelopmental probands > fathers = controls. However, allelic association studies are problematic since the allele may be subject to population stratification. In this situation, the case families may not be matched to the controls and the study may be invalid. Since there is no data on population stratification for that allele, it may not be realized that the study is invalid.

If maternal allele frequencies are increased and TDT analysis shows increased frequency of transmissions from mothers but not fathers to probands, then genomic imprinting or mitochondrial transmission is likely; the allele should be investigated to determine if it is subject to imprinting. However, if no increased frequency of transmissions is found, then the action of a teratogenic allele is possible.

There are at present three approaches for documenting the action of a suspected teratogenic allele. The first approach is to use TDT analysis for transmissions from maternal grandparents to mothers as we and others have suggested [2,3]. Since the mother is the genetic patient for a teratogenic locus, the expectation is that such transmission/disequilibrium will occur. This approach has been successful in a study of spina bifida [4] and a study of autism [5] (see Table 2.1). This approach works best for disorders where the individuals with the neurodevelopmental disorder are young, since living maternal grandparents can be more readily found. This approach is direct and strongly supports

Table 2.3 Illustration of allelic enrichment in cases due to a maternal genetic effect and allelic enrichment among mothers due to a case genetic effect

Effects of gene acting in mothers but not in cases			
Allele frequency, r_1 and r_2 in mothers	0.13, 2.00, 2.00	0.25, 1.50, 2.00	0.50, 1.00, 2.00
Apparent r_1 in offspring	1.38	1.22	1.25
Apparent r_2 in offspring	1.76	1.44	1.50
Effects of gene acting in cases but not in mothers			
Allele frequency, r_1 and r_2 r_2 in cases	0.13, 2.00, 2.00	0.25, 1.50, 2.00	0.50, 1.00, 2.00
Apparent r_1 in mothers	1.38	1.22	1.25
Apparent r_2 in mothers	1.76	1.44	1.50
Maternal imprinting effect			
Allele frequency, r_{mi}	0.25, 2.00		
Apparent r_1 in offspring	1.50		
Apparent r_2 in offsping	2.00		
Apparent r_1 in mothers	1.50		
Apparent r_2 in mothers	2.00		

Calculations based on Buyske (2007) [51].
Note: r_1 = relative risk of one copy of risk allele, r_2 = relative risk of two copies of risk allele, r_{mi} = relative risk of maternally imprinted allele. Allele frequencies have been selected to give a genetic attributable fraction of 25%.
The table gives an illustration of how allele frequencies can be enriched in subjects other than the "genetic patient." General formulas are given in Buyske (2007) *s1*. Each column begins with an example of a risk allele frequency and relative risks (as compared to zero copies) of one or two copies of that allele in the genetic patient. The rows below then show the apparent relative risk in the offspring or mothers. For example, consider a locus with a recessive teratogenic allele acting only in the mothers such that a mother homozygous for the risk allele has offspring with double the risk of a disorder (the last column of the first panel of the table). A case–control study ignoring mothers would be expected to find that the disorder followed an additive model with relative risks of 1.25 and 1.5 for one and two copies of the allele, respectively.

the action of a teratogenic allele, but it does not address interaction between maternal and fetal genes.

The second approach is to use the case-parent log-linear method [6, 7, 8, 9, 10]. Since this method requires only the trio consisting of the individual with the neurodevelopmental disorder and the two parents, it may be the most suitable approach if living maternal grandparents are difficult to find. It also addresses the question of interaction between maternal and fetal genes. With this method, the data is stratified by parental mating type and a modeling term is included for maternal genotype. For example, a mother with AA genotype and a father with aa genotype represent the same mating type as a mother with aa genotype and a father with AA genotype, but if the former pair occurs more often in parents of affected offspring, then that is evidence that the AA genotype in mothers is a risk genotype for offspring. There are reasons to believe that the case-parent log-linear method may have less power than maternal TDT for the same number of families studied. The one report that used both methods on the same dataset is discussed below.

The third approach uses "pents," that is, families with neurodevelopmental proband, parents and maternal grandparents (five individuals per family) [11]. The pent approach has the advantage of estimating both maternal and off-spring genetic effects, and offers increased power, per proband, compared with other family-based approaches. As a practical matter, the difficulty in obtaining DNA from the proband, parents and maternal grandparents generally rules out the pent analysis for all but early onset disorders. Five genotyped subjects per proband will also strike those used to case–control designs as a heavy genotyping load.

Other approaches have also been presented [12].

Examples of Teratogenic Alleles

At least *33* known examples can be cited, most of them recently reported, for which this mechanism has been shown to play a role in causing neurodevelopmental disorders (see Table 2.1). The studies reported varying sample sizes, used varying analytical methods and documented the action of a teratogenic allele with proofs of varying certainty.

Documentation of a Teratogenic Allele by Maternal Transmission Disequilibrium Test

*MTR*2756G* and *MTRR*66G* as Teratogenic Alleles for Spina Bifida

The clearest method of documenting the presence of a teratogenic allele is through maternal TDT. Thus far, action of a teratogenic allele has been reported in two diseases, both neurodevelopmental disorders. In the case of spina bifida, Doolin et al. [13] (see Table 2.1, #1 & 2) showed that polymorphisms of the methionine synthase gene, *MTR* (*MTR*2756G*), and the methionine synthase reductase gene, *MTRR* (*MTRR*66G*), were associated with spina bifida in maternal trios, or mother trios, consisting of mothers of affected individuals and their parents. Maternal TDT was significant in both cases ($p = 0.004$ for *MTR*2756G* and $p = 0.05$ for *MTRR*66G*). The log-linear method was also used; although these results were not statistically significant, they were suggestive ($p = 0.10$ for *MTR*2756G* and $p = 0.08$ for *MTRR*66G*). Modeling with the log-linear approach suggested that the risk increased with the number of maternal high-risk alleles. This result was particularly interesting and important because a number of previous studies had attempted without success to associate *MTR*2756* with spina bifida using either case–control studies of individuals with spina bifida or a case TDT study design (affected individuals and parents) because the rationale for suspecting this as a candidate gene was extremely strong. Maternal deficiencies of both folate and of vitamin B12 have been associated with spina bifida in the fetus and MTR is the only enzyme in humans requiring both folate and vitamin B12 for its action. The presumption in the earlier negative studies was that these alleles acted in the individuals affected with spina bifida rather than in the mothers. When the possibility of a teratogenic allele was considered, this excellent candidate gene was indeed found to be associated with spina bifida.

GSTP1*313A and GSTP1*341C as Teratogenic Alleles for Autism

Recently, polymorphisms of the glutathione S-transferase P1 (GSTP1) gene (GSTP1*313A and GSTP1*341C, i.e., GSTP1*val^{105} and GSTP1*ala^{114}, respectively) were associated with autism as teratogenic alleles, using maternal TDT in trios consisting of mothers of individuals with autism and their parents [5] (see Table 2.1, #3). GSTP1*val^{105} contributed all or most of the haplotype effect observed. GSTP1 is a major enzyme that contributes to detoxifying xenobiotics and reducing oxidative stress. Another function of GSTP1 is regulating c-Jun N-terminal kinase (JNK) by binding to JNK, in which capacity it may also regulate oxidative stress. Interestingly, GSTP1*val^{105} lies within the H-site of the GSTP1 protein, the region where electrophilic toxins, xenobiotics or metabolites bind to GSTP1 for conjugation with GSH, detoxification and excretion. Also, both GSTP1 polymorphisms lie within the region contributing to binding of the GSTP1 protein to the JUN–JNK complex. This finding raises the question of whether oxidative stress originating in mothers could contribute to or potentiate the autism phenotype in affected fetuses.

Documentation of a Teratogenic Allele by Maternal Transmission Disequilibrium Test Equivalent

Teratogenic alleles have been shown to contribute to two neurodevelopmental disorders, maternal phenylketonuria (maternal PKU) and Rh incompatibility, using a method that is, in a sense, equivalent to maternal TDT.

Maternal Phenylketonuria

Maternal PKU results from the major intrauterine effect on fetuses of mothers with PKU. Phenylketonuria itself is a recessive postnatal disorder. Both untreated homozygous PKU mothers and fathers have elevated blood phenylalanine. However, heterozygous offspring of untreated PKU mothers may develop maternal PKU, a disorder different from PKU, that has an abnormal developmental and neurodevelopmental phenotype [14–16]. Nearly all cases of PKU and hence maternal PKU are known to result from mutations in the phenylalanine hydroxylase (PAH) gene. Thus the mutations in the maternal

(*PAH*) gene, through the mechanism of the resulting elevation of maternal blood phenylalanine or other metabolite(s) in the untreated PKU mother, act during pregnancy as a teratogen for the fetus (see Table 2.1, #4).

Infants with PKU [17] are normal at birth and develop a progressive metabolic disorder postnatally characterized by vomiting, eczema, mental retardation and infantile spasms with hypsarrythmia on the electroencephalogram. In contrast, infants with maternal PKU [17] have a congenital nonprogressive disorder of fetal onset characterized by microcephaly, abnormal facies, mental retardation, congenital heart disease, and prenatal and postnatal growth retardation. The teratogenic effect in maternal PKU is not dependent upon the fetal genotype, although the fetus is an obligate heterozygote since the mother is a homozygote for phenylketonuria and the father (usually) has the normal genotype.

Since PKU mothers of maternal PKU offspring are homozygotes for *PAH* mutations, each maternal grandparent is a heterozygote who has transmitted the mutant allele to the mother. If a maternal TDT was carried out in these maternal trios (maternal PKU mothers and their parents), it would show transmission disequilibrium. Thus, this documentation of action of a teratogenic allele in maternal PKU is, in a sense, the equivalent of a maternal TDT. However, since the maternal PKU mothers have PKU and are thus known to be homozygotes for PAH mutations, such a TDT is unnecessary.

Rh Incompatibility

That Rh incompatibility disorders in the fetus result from the action of a teratogenic allele is (see Table 2.1, #5) readily shown by the same reasoning given above for maternal PKU. For Rh incompatibility to occur, the mother must lack the Rh D antigen on the surface of her erythocytes and thus be Rh-negative (RhD-negative), i.e., a homozygote for Rh *d*-allele with *d/d* genotype. An additional requirement is that the fetus should carry an Rh *D*-allele that can be only of paternal origin (father is RhD-positive). Thus, during pregnancy, the Rh *d/d* mother is exposed to the Rh D antigen produced by the fetus. Since the mother lacks this antigen, she makes antibodies to Rh D antigen as the pregnancy progresses. During a subsequent pregnancy with an RhD-positive fetus, maternal antibodies are again produced but in greater amount and in an accelerated fashion. With further RhD-positive fetuses, the mother mounts an immunological attack upon the fetus, who may develop erythroblastosis fetalis and a developmental disorder. This consists of three clinical syndromes in the fetus and neonate: anemia of the newborn, neonatal jaundice that can lead to the severe fetal encephalopathy of kernicterus, and the generalized neonatal edema of hydrops fetalis with massive anasarca, pleural effusions, and ascites. The teratogenic allele here is Rh *d*, for which the mother is a homozygote. Each of her parents carries an Rh *d* allele and has transmitted an Rh *d* allele to her. Thus, transmission disequilibrium is present. The mechanisms of fetal damage are unclear, but cytokine abnormalities have been observed [18].

Rh incompatibility has been linked to one neurodevelopmental disorder, schizophrenia (see Table 2.1, #6). An increased incidence of schizophrenia has been found in offspring of pregnancies with Rh incompatibility [19], but this was not the case for autism in a similar study [20].

Documentation of a Teratogenic Allele by the Log-Linear Method

At least four teratogenic alleles have been documented by the log-linear method without using maternal TDT.

The *MTHFD1*653Q* Allele and Spina Bifida

The *MTHFD1*653Q* allele of the folate-related gene for the trifunctional enzyme, methylenetetrahydrofolate dehydrogenase, was shown in the Irish population to be a teratogenic allele for spina bifida and possibly a risk factor for decreased embryonic survival (see Table 2.1, #7). The Q-allele may act in cases as well as mothers [21]. The authors speculated that the allele could act through the function of MTHFD1 enzyme as a source of 10-formyltetrahydrofolate, which is essential for rapidly dividing cells. Another possible action was less efficient purine or pyrimidine synthesis.

The *CYP1A1*6235C* Allele and Low Birth Weight

The *CYP1A1*6235C* allele of the P450 cytochrome oxidase CYP1A1 was shown [22] in a Chinese population to be a teratogenic allele for low birth weight (see Table 2.1, #8). The *6235C* allele acted in cases as well as mothers. There was no interaction between the case effect and the maternal effect of the allele. Case TDT was negative. There was neither any effect of another CYP1A1 polymorphism nor a polymorphism of CYP2E1. CYP1A1 is an important phase II enzyme in the detoxification and excretion of xenobiotics. The authors speculated that a possible mechanism of action was interaction of the allele in mothers and maternal passive smoking. The *6235C* allele in mothers had been previously shown to modify the association between maternal smoking and infant birth weight. Although few of these women were smokers, most of their husbands were smokers, thereby subjecting these mothers to passive smoking.

NAT1 Alleles and Spina Bifida Risk

The log-linear method has been used to associate alleles of the gene for *N*-acetyltransferase, *NAT1* [23], with risk of spina bifida (see Table 2.1, #9). Using the log-linear method, the authors analyzed genotypes of five single nucleotide polymorphisms (SNPs), for which the less common allele was associated with reduced or no enzyme activity, after constructing a composite genotype from the five SNP genotypes. Both case and maternal genotypes influenced spina bifida risk such that *NAT1* alleles that reduced or abolished enzyme activity appeared to decrease spina bifida risk. These results suggested that *NAT1* alleles act as teratogenic alleles for spina bifida. In addition, these alleles acted in the spina bifida cases themselves. NAT1 is a detoxification enzyme that contributes to folate catabolism and acetylation of aromatic and heterocyclic amines. *N*-Acetyltransferases also contribute to the formation of mercapturic acids from glutathione conjugates, produced by glutathione *S*-transferases, for their detoxification and excretion. It would not be surprising that decreasing folate catabolism in spina bifida mothers and cases would decrease spina bifida risk. The possibility that the detoxification function of NAT1 could also be involved is intriguing because it suggests that exogenous maternal environmental factors such as maternal smoking could play a role.

The *CCL-2*(-2518)A* Promoter Allele and Spina Bifida Risk

A fourth teratogenic allele, the *-2518A* allele of the monocyte chemoattractant protein 1 gene (*CCL-2*), was documented [24] using the log-linear method in spina bifida (see Table 2.1, #10). The maternal AA genotype was hypothesized to increase spina bifida risk by allowing the mother to mount a less vigorous systemic and/or local response to infection.

Documentation of a Teratogenic Allele by Regression Methods

Maternal Alleles Influence Newborn Lipoprotein Concentrations

Forward stepwise regression analysis was used to document the influence of three maternal alleles of genes related to lipoproteins on newborn lipoprotein concentrations (see Table 2.1, #11, 12, 13). The presence of the *APOE*E2* allele in mothers was found to be associated with lower LDL-C and apoB concentrations and higher HDL-C and apoA1 concentrations in their newborn offspring. The maternal *APOC3*S2* allele was found to be associated with lower LDL-C, apoB, HDL-C, and apoA1 concentrations in their newborn offspring. The *LPL*S447X* allele in mothers was found to be associated with lower LDL-C, apoB, and TG levels in their offspring [25].

Maternal *GSTP1* Alleles and the Asthma Phenotype in Children

Since maternal factors were known to affect heritability and expression of asthma and atopy, and since the glutathione *S*-transferase P1 (GSTP1) gene *GSTP1*val^{105}* allele was known to influence asthma phenotypes, genotypes of two GSTP1 gene alleles (*GSTP1*val^{105}* and *GSTP1*ala^{114}*) in children with asthma and their parents were correlated with data on lung function in the affected children using multiple regression analysis (see Table 2.1, #14). Paternal genotypes did not influence lung function in the child. However, maternal $GSTP1$ val^{105}/val^{105} and maternal $GSTP1$ ala^{114}/val^{114} genotypes were significantly associated with certain parameters of lung function in the children: forced expiratory volume in 1second (FEV$_1$) and forced vital capacity (FVC). Maternal $GSTP1$ val^{105}/val^{105} genotype was significantly and strongly predictive of the FEV$_1$/FVC ratio. Maternal $GSTP1$ ala^{114}/val^{114} genotype was significantly and strongly associated with higher FEV$_1$ and FEV$_1$/FVC ratios. These findings remained significant after correction for atopic status in child and mother, smoking during pregnancy, passive smoke exposure, and case and paternal $GSTP1$ genotype [26].

The authors noted that GSTP1 detoxifies products of oxidative stress and agents that contribute to oxidative stress. They interpreted the findings as suggesting that the maternal $GSTP1$ genotypes influenced the mother's ability to manage environmental oxidative stress and modulate transmission of oxidative stress to the fetus during pregnancy and that this effect contributes to the observed postnatal pulmonary function changes in the offspring.

Documentation of a Teratogenic Allele by Case–Control Methods Plus Case Transmission Disequilibrium Test

Case–control studies suggest the presence of a teratogenic allele if maternal allele frequencies are higher than paternal allele frequencies and control allele frequencies of the allele in question. However, this would also be seen in the case of maternal imprinting. Showing lack of transmission disequilibrium for transmissions from mothers to cases or transmissions from parents to cases makes imprinting unlikely and strengthens the case for action of a teratogenic allele as an explanation for elevated maternal allele frequencies.

Maternal *MTHFR 677T* Allele as a Risk Factor for Cleft Lip With or Without Cleft Palate (CL/P)

Since periconceptual folate supplementation in the mother may contribute to the prevention of CL/P in her offspring, the *MTHFR C677T* polymorphism was studied in 64 CL/P cases and their parents. Members of the 64 trios were genotyped for the polymorphism, and the data were analyzed for transmission from parents to children (see Table 2.1, #15). No transmission disequilibrium was documented. Next, genotypes of this polymorphism for the 64 mothers and 106 controls were compared in a case–control study design using 2×2 contingency tables, and odds ratios were determined. Mothers had a significantly higher frequency of the T-allele compared to controls. Odds ratios for CL/P in offspring of mothers with the CT genotype, the TT genotype, and overall were significantly increased: 2.75, 2.51, and 2.68, respectively [27].

Thus, the maternal *MTHFR* *T*-allele appears to be a teratogenic allele for CL/P. The finding of a negative case TDT decreases the possibility that imprinting might be an explanation for the findings since transmission disequilibrium for transmissions from mothers to CL/P cases would be observed if maternal imprinting were present [27].

Maternal *MTHFR 677T* Allele as a Risk Factor for Congenital Heart Disease

Since periconceptual folate supplementation in the mother may prevent congenital heart defects (CHD) in her offspring, the important folate-related enzyme methylenetetrahydrofolate reductase (MTHFR) was studied in 133 case trios, i.e., families consisting of a CHD proband and the parents (see Table 2.1, #16). Trios were genotyped for the *MTHFR C677T* polymorphism and studied by case TDT. There was no transmission disequilibrium for transmissions from parents to proband. Next 158 mothers, their CHD-affected children, and 261 control women were genotyped for this polymorphism. Using logistic regression, maternal genotypes CT and TT combined were associated with a threefold increased risk of CHD and a sixfold increased risk for the conotruncal subtype of CHD in the offspring. The same maternal genotypes also showed a gene–environment interaction with periconceptual folate supplementation, suggesting that this might contribute to the prevention of CHD [28].

Thus, the maternal *MTHFR* *T*-allele appears to be a teratogenic allele for CHD. The finding of a negative case TDT decreases the possibility that imprinting might be an explanation for the findings since transmission disequilibrium for transmissions from mothers to CHD cases would be observed if maternal imprinting were present [28].

Maternal *MTHFR 677T* Allele as a Risk Factor for Down Syndrome

To study the question of whether abnormality of folate metabolism predisposes to chromosomal nondisjunction, two folate-related gene polymorphisms, *MTHFR C677T* and *MTHFR A1298C*, were studied in 149 mothers of Down syndrome (DS) cases and 165 control mothers in a case–control study design (see Table 2.1, #17, 31). The *677T* allele was significantly more frequently found in case mothers compared with control mothers ($p = 0.0001$) as was the *677 TT* genotype ($p = 0.003$). Interestingly, the excess of *677TT* homozygotes was confined to case mothers aged less than 31 years; in case mothers who were more than 31 years old, the distribution of this genotype was comparable with controls. In mothers homozygous for *677T* and mothers homozygous for *1298C*, the relative risk of a DS-affected pregnancy was significantly increased: 7.67 and 4.40, respectively.

In 84 DS families in which cases and parents were genotyped, the *677T* and -*C* alleles were transmitted equally from the mothers to DS offspring. Interestingly, the *677T* allele was preferentially transmitted from fathers to DS offspring. Transmission disequilibrium test was not carried out for the 1298C allele and this is therefore listed in a different section of Table 2.1 (#31).

Thus, the maternal *MTHFR 677T* allele appears to be a teratogenic allele for DS, since the negative case TDT decreases the possibility that imprinting might be an explanation for the findings. Transmission disequilibrium for transmissions from mothers to DS cases would be observed if maternal imprinting were present [29]. The C allele of *MTHFR A1298C* also appears to be a teratogenic allele for DS. If members of these DS case trios were genotyped for the *MTHFR A1298C* polymorphism and TDT was not found, the evidence for the C allele as a teratogenic allele would be stronger.

HLA-DR4 as a Risk Factor for Autism

Because there was substantial evidence for an association between specific antigens of the HLA system and autoimmune disorders, because autoimmune disorders had been shown to have a higher incidence among members of autism families, especially mothers, and because HLA-DR4 had been previously associated with autism, the *DR4* allele was studied in autism cases and their parents in a case–control study design (see Table 2.1, #18). The autism families were ascertained from two groups: 16 families (16 cases and mothers and 14 fathers) from Eastern Tennessee, which had diagnosis made by an experienced observer by DSM-IV criteria and by the Childhood Autism Rating Scale; and 33 families with multiple affected males consisting of anonymous DNA samples from the Autism Genetic Resource Exchange (AGRE), selected to be evenly distributed from all parts of the United States and with case diagnoses made by the Autism Diagnostic Interview-Revised. *DR4* genotyping was by DNA analysis [30].

In the East Tennessee families, cases and mothers had a significantly elevated frequency of *DR4* compared with controls using low-resolution genotyping: 62.50% (odds ratio 5.54, 95% CI 1.74–18.67) and 68.75% (odds ratio 4.20, 95% CI 1.37–13.27), respectively. With high-resolution testing, *DRβ1*040101* frequency in mothers was again higher for the East Tennessee families but not the AGRE families than that in controls. There was no transmission disequilibrium for transmissions from mothers to cases, no sharing of alleles, and no significant change in the distribution of *DR4* alleles among those with *DR4* [30].

The authors hypothesized that a prenatal maternal–fetal immune interaction in the East Tennessee families could affect fetal brain development. This interaction could result from immune stimulation by infections, allergy, or some yet unknown environmental factor that could perhaps be specific for the East Tennessee region [30].

Suggestion of a Teratogenic Allele by Case–Control Methods Without Case Transmission Disequilibrium Test

Complement *C4B*0* and *HLA-DR4* as Possible Teratogenic Alleles for Autism

Since a number of immune abnormalities had been reported in autism raising the question of an immune contribution to autism, it was hypothesized that one or more genes within the major histocompatibility complex (MHC) could contribute to autism. The null allele (*C4B*0*) of the *C4B* gene (see Table 2.1, #19) showed significantly increased frequency in individuals with autism and their mothers but not their fathers [31]. The frequency of certain MHC-extended haplotypes, especially *B44-SC30-DR4* [a haplotype containing allele *44* at the *HLA-B* locus (*B*44*), allele *S* at *BF* (*BF*S*), allele *C* at *C2* (*C2*C*), allele *3* at *C4A* (*C4A*3*), the null allele at *C4B* (*C4B*0*), and allele *4* at *HLA-DRβ1* (*DR*4*)], was significantly increased both in individuals with autism and their mothers but not in their fathers [32,33]. However, with the exception of *C4B*0*, none of the individual components of the extended haplotype was associated with autism [33]. In a later study [34] of the hypervariable region-3 (HVR-3) within *HLA-DRβ1*, significantly increased frequency of certain sequences was reported, including *DRβ1*0401*, within *DR*4*, in autism probands and their mothers but not their fathers (see Table 2.1, #20). Although they had found specific evidence for association of *C4B*0* and *HLA-DR*4* with autism cases and mothers, the authors noted the high degree of linkage disequilibrium within the MHC region and emphasized that while the gene associated with autism could be *C4B*0* or *HLA-DR*4*, another gene within the MHC or multiple genes within the MHC could be the source of their findings. As an explanation of their striking maternal findings, the question was raised of whether a gene acting in the mother during pregnancy might contribute to autism in her fetus [34].

A repeat study confirmed association of *C4B*0* in individuals with autism but parents were not studied [35]. In a repeat study of autism families that included *HLA-DR4*, an increased frequency of case mothers was noted compared with published control frequencies [36]. The transmission disequilibrium test was negative for transmissions of *DR4* from mothers to cases making maternal imprinting unlikely as an explanation for the increased frequency in maternal *DR4*. However, correction for multiple comparisons was not made, and the large number of comparisons that were made in the study rendered the increased frequency of *DR4* among autism mothers not statistically significant. Thus a teratogenic allele could not be documented from these promising data. A repeat study with a stronger study design is needed.

The *GSTP1*1b* Allele as a Possible Teratogenic Allele for Recurrent Early Pregnancy Loss

Since lifestyle factors, e.g., smoking and alcohol or coffee use, have been associated with recurrent early pregnancy loss (REPL), enzymes detoxifying potentially toxic substances in alcoholic beverages, coffee, or cigarettes were considered as risk factors for REPL. GSTP1 has widespread tissue distribution unlike most GSTs, which are chiefly localized in liver. The *GSTP1*105 val* allele of the *GSTP1 ile105val* polymorphism is associated with lower enzyme activity. In a case–control study, the *GSTP1*105 val* allele was studied in 187 women with REPL and 109 women with uncomplicated obstetrical history (see Table 2.1, #21). Odds ratios for *GSTP1*105 val* homozygosity in cases versus controls were elevated in gravidas who used coffee, tobacco, and alcohol and were statistically significant for the first two cases but not for alcohol use. Fetal genotypes were not determined [37].

There are some problems in suggesting that *GSTP1*105 val* is a teratogenic allele based on these data. First, the authors speculated that impaired placental detoxification might play a role. Since the placenta consists largely of fetal tissue whose growth is determined by fetal genes, this mechanism would imply that detoxification controlled by fetal genes would be impaired. However, their data suggest a maternal effect, not a fetal effect, and the question of significant abnormality of fetal genes is not addressed. Second, no TDT was carried out for transmissions from mothers to fetuses to address the possibility of genomic imprinting [37].

These data certainly raise the interesting question of whether *GSTP1*105 val* is a teratogenic allele for REPL. In that case, the allele would be acting in maternal tissue, not fetal tissue, a possibility that is quite plausible. The data were statistically significant but significance was lost after correction for multiple comparisons. The study needs to be replicated using a stronger study design.

The *MTHFR*1298C* Allele as a Possible Teratogenic Allele for Spina Bifida

Since the *MTHFR*677T* allele had been associated with spina bifida, the authors tested alleles of two additional polymorphisms, the *MTHFR A1298C* polymorphism and the *A80G* polymorphism of the reduced folate carrier gene, *RFC-1* in an Italian population [38]. They found increased frequency of the *MTHFR*1298C* and *RFC-1*80G* alleles in mothers and cases. Although the authors did not do statistical analysis, 2×2 tests of their data showed that the increase in mothers homozygous for the *MTHFR*1298C* allele was highly significant, even allowing for multiple comparisons, for the *CC* genotype as well as for the *C* allele frequency (see Table 2.1, #22). The *RFC-A80G* polymorphism data did not show a significant increase in mothers with the *GG* genotype compared to controls and showed a marginally significant increase for the *G* allele frequency in mothers compared to controls that was lost after correction for multiple comparisons.

Thus, the *MTHFR*1298C* allele can be suggested as a possible teratogenic allele for spina bifida (see Table 2.1, #22). No TDT was carried out and maternal imprinting was not excluded as an alternative explanation for the increased frequency in spina bifida mothers of the *MTHFR*1298C* allele. Confirmation with a stronger study design is required to document action of *MTHFR*1298C* as a teratogenic allele.

Methylenetetrahydrofolate Reductase *(MTHFR)*677T* and Methionine Synthase Reductase *(MTRR)*66G* as Possible Teratogenic Alleles for Down Syndrome

Since the folate-related gene polymorphisms *MTRR* A66G and *MTHFR* C677T had been linked to DS, the prevalence of these genotypes was studied in a case–control study design comparing DS mothers with control mothers [39] (see Table 2.1, #23, 24). The *66G* allele in DS mothers in single dose was associated with significantly increased risk of DS in her affected offspring compared with control mothers, OR 7.97 (95% CI 1.04–61.2, $p = 0.017$); in double dose, the risk from the *66G* allele was further increased, OR 15.0 (1.94–116, $p = 0.0005$).

The *MTHFR C677T* polymorphism in DS mothers showed no such increased risk of DS in her affected offspring compared with control mothers when analyzed by itself. However, the *MTHFR*677T* allele in DS mothers, as the *CT* or *TT* genotype, was associated with increased risk of DS in her offspring in those mothers who also possessed the *MTRR*66G* allele in double dose compared to control mothers, OR 2.98 (1.19–7.46), $p = 0.02$. The mechanism of action of the *MTRR*66G* and *MTHFR*677T* alleles was unclear. Although the *MTHFR*677T* allele was associated with higher homocysteine, *MTRR* alleles were not.

Thus, it is possible that the *MTRR*66G* and also the *MTHFR*677T* alleles may act as teratogenic alleles for DS based upon this case–control study. However, there are alternative explanations for these data. Fetal genotypes were not examined. No TDT was carried out for transmissions from mothers to affected children, and imprinting was not excluded. Further studies are required to document the action of these alleles as teratogenic alleles.

Glutathione S-Transferase Deletion Allele (*GSTT1*0*) as a Possible Teratogenic Allele for Oral–Facial Clefting

Although smoking during pregnancy appears to be related to oral–facial clefting (OFC), the results have been unclear. Since susceptibility to cigarette smoke could result from polymorphic variation in enzymes that carry out biotransformation of xenobiotics, a case–control study was carried out in 113 infants affected with nonsyndromic oral clefting, 104 control infants, and their mothers using polymorphisms of genes for GSTT1 and CYP1A1, important biotransformation enzymes [40]. Neither gene showed association with oral clefting. However, when mothers who smoked and carried *GSTT1*0* were compared with nonsmoking mothers with the wild-type genotype, an increase in risk of having an affected child was found in smoking mothers who were homozygotes for *GSTT1*0*, odds ratio 3.2 (95% CI 0.9–11.6) (see Table 2.1, #25). If both the smoker mother and her affected offspring were homozygotes for *GSTT1*0*, then the risk of oral clefting in the infant would be still greater, odds ratio 4.9 (0.7–36.9) [40].

Thus, *GSTT1*0* may be a teratogenic allele for oral clefting in mothers who smoke, an example of gene–environment interaction. *GSTT1*0* homozygotes have no GSTT1 activity, since they have a double dose of the polymorphic whole-gene deletion allele. *GSTT1*0* may also be a modifying/specificity allele (i.e., an allele that acts in the fetus to modify or potentiate the phenotype caused by a teratogenic allele) since the odds ratio increased if both the mother who smokes and her fetus were GSTT1*0 homozygotes. However, since the two confidence intervals overlapped, this is suggested but not documented [40].

There are limitations to this study and alternate explanations for the data. One limitation is that the confidence intervals included 1.0. Thus the results were not statistically significant despite their interest and coherence. Also, no TDT was carried out, and imprinting was not excluded. The transmission disequilibrium test can be carried out with such a deletion polymorphism using a newer method [41] or using genotyping methods that detect heterozygotes. Thus, *GSTT1*0* is a possible teratogenic allele for oral clefting, but further studies are needed.

*MTHFR*1298C* as a Possible Teratogenic Allele for Spina Bifida

Since two common folate-related polymorphisms of methylenetetrahydrofolate (MTHFR), *MTHFR C677T* and *A1298C*, had been associated with spina bifida in some populations, these polymorphisms were studied in 74 cases and 102 parents of cases and compared with 110 control individuals from the State of Yucatan, Mexico [42]. Neither polymorphism was associated with cases or case fathers compared with controls. However, compared with controls, case mothers had significantly increased frequencies of the *1298C* allele, odds ratio 2.91 (95% CI 1.02–8.68, $p = 0.025$), increased frequencies of the heterozygous *1298* genotype, *AC*, odds ratio 2.78 (0.90–8.88), and marginally significantly increased frequency of the doubly heterozygous *677/1298* haplotype, *CT/AC*, odds ratio 4.0 (0.68–26.19, $p = 0.054$) [42].

Thus the *1298C* allele alone (see Table 2.1, #26) or in combination with the *677T* allele is a possible teratogenic allele for spina bifida in the Yucatan population. Since TDT was not carried out for transmissions from mothers to their affected offspring, imprinting was not excluded. Further studies are required to confirm this interesting finding, published thus far only in abstract form.

The Glutathione *S*-Transferase Gene Deletion Allele, *GSTM1*0*, as a Possible Risk Factor for Recurrent Pregnancy Loss

Since prior studies had suggested that both genetic and environmental factors might contribute to the etiology of recurrent pregnancy loss (RPL), gene deletion polymorphisms of two key enzymes that detoxify environmental chemicals, glutathione S-transferase (GST) GSTM1 and GSTT1, were used in a case–control of 115 case mothers with RPL and 160 control individuals [43]. A significantly greater proportion of case mothers, 65.2%, were homozygotes for GSTM1*0 compared with 45.6% of controls, odds ratio 2.23 (95% CI 1.36–3.66) (see Table 2.1, #27); however the proportions of case mothers and controls homozygous for GSTT1*0 did not differ [43].

Thus, GSTM1*0 is a possible teratogenic allele for RPL. Neither fetal genotypes nor TDT was carried out for transmissions from mothers to fetuses. Thus, imprinting remains a possible explanation for these findings [43]. Newer methods allow TDT to be carried out with gene deletion polymorphisms as discussed earlier.

The Dihydrofolate Reductase 19-Basepair Deletion Allele, A Possible Teratogenic Allele for Spina Bifida

Dihydrofolate reductase (DHFR) is a key enzyme in folate metabolism. It is necessary to reduce folic acid in supplements to tetrahydrofolate, the form useful for the cell. More importantly, it is also necessary to reduce the DHF produced in the cell by the action of thymidylate synthase, an enzyme necessary for DNA synthesis. DNA synthesis is a necessity for cell division, especially in development in utero. The DHFR 19-bp deletion allele is highly polymorphic and may limit rapid thymidylate synthesis and hence cell division. The DHFR deletion polymorphism was studied in spina bifida cases and parents in a case–control study design [44]. Deletion homozygotes were significantly more frequent among spina bifida mothers but not fathers or cases (see Table 2.1, #28). This raised the question of whether the allele could act in case mothers to limit the metabolism of folic acid supplements and to limit rapid DNA synthesis during development in utero.

Thus, the DHFR 19-bp deletion allele is a possible teratogenic allele for spina bifida. No TDT was carried out and imprinting remains an alternate explanation for the data [44]. Further studies are required to document the action of the DHFR deletion allele as a teratogenic allele.

The CYP1A1*2A Allele as a Possible Teratogenic Allele for Recurrent Pregnancy Loss

Since the risk of miscarriage is known to be increased by environmental and lifestyle factors, e.g., stress, smoking, and alcohol consumption, that increase oxidative stress, and since cytochrome P450 (CYP) enzymes reduce oxidative stress by detoxifying xenobiotics producing oxidative stress and also by detoxifying the products of oxidative stress, a case–control study was carried out in 160 women with RPL compared with 63 healthy women with successful reproductive history using five polymorphisms of detoxification enzymes [45]. A significant association between the CYP1A1*A1 allele and RPL was found in mothers with RPL compared with controls, odds ratio 1.93 (95% CI 1.10–3.38, $p = 0.023$) [45] (see Table 2.1, #29).

Thus *CYP1A1*A1* may be a teratogenic allele for RPL. Limitations of the study are loss of statistical significance if multiple comparisons are applied as well as the lack of genotyping of fetuses. The transmission disequilibrium test was not carried out for transmissions between mothers and fetuses, and imprinting was not excluded [45]. Further studies are required to document the action of *CYP1A1*A1* as a teratogenic allele.

The *DHFR* 19-bp Deletion Allele as a Possible Teratogenic Allele for Preterm Delivery

The *DHFR* 19-bp deletion allele was studied in women delivering preterm infants compared to woman delivering healthy infants in a case–control study design, since the allele may contribute to decreased availability of folic acid supplements and decrease DNA synthesis during stress periods when rapid DNA synthesis is required such as during development *in utero* [46].

The data were analyzed by logistic regression and were used to examine the effect of the deletion allele by deletion homozygosity and heterozygosity on the preterm delivery, very preterm delivery and low birth weight, controlling for potential confounding variables (age, parity, ethnicity, smoking, and body mass index). Odds ratios were significantly elevated for preterm delivery, 3.03 (95% CI 1.02–8.95), and elevated for very preterm delivery, 6.27 (0.82–48.19), and low birth weight, 2.47 (0.92–6.64) [46] (see Table 2.1, #30).

Thus, the *DHFR* 19-bp deletion allele may be a teratogenic allele for preterm delivery, very preterm delivery, and low birth weight. However, the elevated odds ratio was significant only for the first of these. More importantly, off-spring were not genotyped and TDT for transmissions from mother to affected offspring was not carried out [46]. Thus, imprinting remains an alternative explanation for the data.

The Methylenetetrahydrofolate Reductase (MTHFR)*1298C Allele as a Possible Teratogenic Allele for Down Syndrome

Since the mechanisms of meiotic nondysjunction and the contribution of mater-nal age to trisomy 21 DS were poorly understood, since genomic DNA hypo-methylation may be associated with chromosomal instability and abnormal segregation, and since impairment of folate metabolism has been causally related to DNA hypomethylation and abnormal purine synthesis, seven folate gene polymorphisms (including the methylenetetrahydrofolate reductase, *MTHFR*, and the reduced folate carrier1, *RFC1*, genes) were studied in a case–control study design in 94 DS mothers and 264 control women [50] (see Table 2.1, #32, 33). Increased risk of DS in the affected offspring was found for the *MTHFR*1298C* allele, odds ratio 1.46 (95% CI 1.02–2.10), and the *RFC1*80G* allele, odds ratio 1.48 (1.05–2.10). Increased risk of DS in the affected offspring was found for genotype *CC* of *MTHFR A1298C*, odds ratio 2.29 (1.06–4.96), and genotype *GG* of *RFC1 A80G*, odds ratio 2.05 (1.03–4.07). Interestingly, significant interactions were found between maternal age at conception and either the *MTHFR*1298C* or the *RFC1*80G* alleles. Interactions between a number of genotype pairs were also found [50].

Thus, the *MTHFR*1298C* allele and the *RFC1*80G* allele are possible teratogenic alleles for DS. The DS offspring genotypes were not determined and TDT was not carried out for transmissions from DS mothers to DS cases [50]. Thus, imprinting remains an alternate explanation of the data.

Potential Examples of Teratogenic Alleles

A few other potential examples of teratogenic alleles have been encountered that were not well enough described to be considered here. One of the more interesting of these is diabetic embryopathy. This is a developmental disorder seen in infants of diabetic mothers. Since there are a growing number of genetic risk factors for diabetes mellitus, it would be reasonable to determine whether some of these, if present in the mother, are more likely than others to be associated with diabetic embryopathy in the fetus. Some mouse genetic studies have been carried out but we have been unable to find a report of such a study in humans. Alternatively, diabetic embryopathy could result solely from the level, timing, and duration of maternal hyperglycemia during pregnancy.

A Word About Mechanisms by Which Teratogenic Alleles Act

There are obviously an unlimited number of possible mechanisms by which teratogenic alleles might act. However, some mechanisms might be more common than others. With over 30 teratogenic alleles reported thus far, it may be possible to begin to identify some underlying themes, with the understanding that this collection of teratogenic alleles identified thus far is not necessarily representative.

A previous examination of this question identified folate metabolism, the immune system, DNA synthesis, and oxidative stress as central to the action of the teratogenic alleles that had been reported at that time [1]. Inspection of Table 2.1 confirms that these areas continue to be well represented.

What additional mechanisms besides these could be important to consider in searching for teratogenic alleles in the future? Maternally acting genes that affect the fetus could act in a number of different cells including the ovum, the ovarian granulosa cells, cells affecting ovulation, the fallopian tubes, the uterine implantation mechanism, and portions of the uterus interacting with the placenta, among others. One interesting area of maternal interaction with the fetus is the spiral arteries of the uterus that are invaded by endovascular trophoblast and interstitial trophoblast cells from the fetal placenta. This cellular invasion seems to be important for a successful pregnancy. In order to carry out a successful invasion of the maternal spiral arteries, the fetal endovascular trophoblasts must bind to the maternal endothelial cells of the spiral arteries. A number of ligands are required for the binding of these two cell types. This particularly close and important interaction of maternal and fetal cells requires the action of a number of genes that could be considered candidates for teratogenic alleles.

It remains unknown whether teratogenic alleles will remain a small but interesting class of agents affecting fetal development or whether their number will expand greatly. It also remains unknown whether the association of teratogenic alleles so far with neurodevelopmental disorders and outcomes of pregnancy results simply from the fact that this is an obvious group of disorders in which to look for them or whether these teratogenic alleles may be an important and characteristic mechanism of central importance for these disorders. It is hoped that future research will contribute to answering these questions.

Teratogenic Alleles and Autism

Teratogenic alleles are highly relevant to autism. Four of the 33 reports of teratogenic alleles are in autism and more will be reported soon. In autism, some of the brain abnormalities are known to occur very early in pregnancy. Thus, postnatal therapy of autism, while valuable and important, may not completely prevent or reverse the processes begun prenatally by the genetic and environmental factors that contribute to autism. Placental villous abnormalities have also been noted in autism, another indicator of prenatal pathology in autism [47].

Since teratogenic alleles contribute to the autism phenotype during pregnancy, they open up an opportunity to prevent and ameliorate processes contributing to autism during pregnancy. Identification of new teratogenic alleles for autism is likely to shed light on the processes and mechanisms by which they contribute to autism. Study of these processes and mechanisms is important because it could lead to approaches to detect difficulties that could lead to autism in utero. Identification of these processes and mechanisms could ultimately lead to approaches to prevent or reverse damage to the fetus that occurs in utero and even postnatally.

The first step is to understand the concept of teratogenic alleles and how studies can be designed to detect them. Furthering this goal is a major purpose of this chapter.

General Glossary

Human genes are given in upper case italics; corresponding proteins are given in the same upper case letters that are not italicized.

Microchimerism: is the presence of a small number of cells, genetically distinct from those of the host individual, e.g., maternal cells found in her fetus or fetal cells found in its mother.

Maternal imprinting: imprinting is a genetic phenomenon by which specific genes are "imprinted," i.e., modified, during gametogenesis without any base pair change (e.g., by methylation) so that they are expressed differently from the same gene that is not imprinted. Such a gene, for example, may not be expressed. If a maternal gene is imprinted, e.g., in the egg, then the maternal gene of the fetus may not be expressed while the paternal gene will still be expressed normally.

Transmission disequilibrium test (TDT): is a family-based association method to test for the presence of genetic linkage between a genetic marker and a trait. Transmission disequilibrium test detects genetic linkage only in the presence of genetic association. The TDT measures the overtransmission of an allele from one or both parents to their offspring.

Polymorphism: is a "common" DNA variation that occurs in at least 1% of chromosomes, i.e., 2% of individuals in the population.

Mutation: is a "rare" DNA variation occurring in less than 1% of chromosomes.

Oxidative stress: refers to the state of oxidative damage in a cell, tissue, or organ that can occur as a result of increased reactive oxygen species (ROS) production or a decrease in antioxidant defenses.

Hypsarrhythmia: abnormal and irregular electroencephalogram (EEG) commonly found in patients with infantile spasms.

Erythroblastosis fetalis: also known as hemolytic disease of the newborn (HDN), is an immune condition that develops in a fetus when the IgG antibodies that have been produced by the mother have passed through the placenta including antibodies that attack the red blood cells in the fetal circulation.

Kernicterus: is bilirubin staining and damage to brain centers of infants caused by elevated levels of bilirubin. Rh incompatibility between mother and fetus may cause hemolysis of fetal red blood cells, thereby releasing unconjugated bilirubin into the fetal blood. Since the fetal blood–brain barrier is not fully formed, some of this released bilirubin enters the brain and interferes with normal neuronal development.

Hydrops fetalis: is a condition in the fetus characterized by an accumulation of fluid, or edema, in at least two fetal compartments.

Anasarca: (or "extreme generalized edema") is widespread swelling of the skin due to effusion of fluid into the extracellular space.

Ascites: is the accumulation of fluid in the abdominal cavity.

Low-density lipoprotein (LDL): is a type of lipoprotein. Low-density lipoproteins transport cholesterol to the arteries and can be retained there, starting the formation of plaques. Increased levels of LDL are associated with artherosclerosis and thus heart attack, stroke, and peripheral vascular disease. Low-density lipoprotein is often referred to as "bad cholesterol."

Apolipoprotein B (APOB): is the primary apolipoprotein (protein portion of a lipoprotein) of LDLs. Apolipoprotein B binds to LDL receptors in various cells throughout the body delivering cholesterol to these cells.

High-density lipoprotein (HDL): is a type of lipoprotein. High-density lipoproteins transport cholesterol from the body's tissues to the liver. High levels of HDLs seem to protect against cardiovascular disease, while low levels of

HDLs increase the risk for heart disease. High-density lipoprotein is often referred to as "good cholesterol."

Apolipoprotein A-I (ApoA-I): is the primary apolipoprotein (protein portion of a lipoprotein) of HDLs. The protein promotes cholesterol efflux from tissues to the liver for excretion. As a major component of HDL, ApoA-I helps to clear cholesterol from arteries.

TG: Thyroglobulin, the carrier protein for thyroid hormone.

Atopy: is an allergic hypersensitivity affecting parts of the body not in direct contact with the allergen. It may involve eczema (atopic dermatitis), allergic conjunctivitis, allergic rhinitis, and/or asthma.

Human leukocyte antigen system: The human leukocyte antigen (HLA) system is the name of the human major histocompatibility complex (MHC). The major HLA antigens are essential elements in immune function. The HLA region includes a large number of genes, e.g., the *HLA-DR* genes, including the *HLA-DR*4* allele, and the *C4B* gene and its null allele, *C4*0*.

DSM-IV: the Diagnostic and Statistical Manual of Mental Disorders (DSM) IV is a handbook that lists different categories of mental disorders and the criteria for diagnosing and identifying them. It is published by the American Psychiatric Association.

Null-allele: is a mutant copy of a gene that completely lacks that gene's normal function.

Haplotype: is a combination of alleles at multiple closely linked loci that are usually transmitted together.

DNA methylation: is a type of chemical modification of DNA that can be inherited without changing the DNA sequence. It involves the addition of a methyl group to DNA and can involve hypo- or hypermethylation.

References

1. Johnson WG. Teratogenic alleles and neurodevelopmental disorders. *BioEssays* 2003; 25: 464–477.
2. Johnson WG. The DNA polymorphism–diet–cofactor–development hypothesis and the gene-teratogen model for schizophrenia and other developmental disorders. Am J *Med Genet B (Neuropsychiatr Genet)* 1999; 88: 311–323.
3. Mitchell LE. Differentiating between fetal and maternal genotypic effects, using the transmission test for linkage disequilibrium. *Am J Hum Genet* 1997; 60: 1006–1007.
4. Doolin MT, Barbaux S, McDonnell M, Hoess K, Whitehead AS, Mitchell LE. Maternal genetic effects, exerted by genes involved in homocysteine remethylation, influence the risk of spina bifida. *Am J Hum Genet* 2002; 71(5): 1222–1226.
5. Williams TA, Mars AE, Buyske SG, Stenroos ES, Wang R, Factura-Santiago MF et al. Risk of autistic disorder in affected offspring of mothers with a glutathione S-transferase P1 haplotype. *Arch Pediatr Adolesc Med* 2007; 161(4): 356–361.
6. Weinberg CR, Wilcox AJ, Lie RT. A log-linear approach to case-parent-triad data: Assessing effects of disease genes that act either directly or through maternal effects and that may be subject to parental imprinting. *Am J Hum Genet* 1998; 62(4): 969–978.

7. Wilcox AJ, Weinberg CR, Lie RT. Distinguishing the effects of maternal and offspring genes through studies of "case-parent triads". *Am J Epidemiol* 1998; 148(9): 893–901.
8. Weinberg CR, Wilcox AJ. Re: "Distinguishing the effects of maternal and offspring genes through studies of 'case-parent triads' " and "a new method for estimating the risk ratio in studies using case-parental control design". *Am J Epidemiol* 1999; 150(4): 428–429.
9. Weinberg CR. Methods for detection of parent-of-origin effects in genetic studies of case-parents triads. *Am J Hum Genet* 1999; 65(1): 229–235.
10. Starr JR, Hsu L, Schwartz SM. Assessing maternal genetic associations: A comparison of the log-linear approach to case-parent triad data and a case–control approach. *Epidemiol* 2005; 16(3): 294–303.
11. Mitchell LE, Weinberg CR. Evaluation of offspring and maternal genetic effects on disease risk using a family-based approach: The "pent" design. *Am J Epidemiol* 2005; 162(7): 676–685.
12. Mitchell LE, Starr JR, Weinberg CR, Sinsheimer JS, Mitchell LE, Murray JC. Maternal Genetic Effects. Concurrent Invited Sessions I, #14, American Society of Human Genetics, Annual Meeting, Wed Oct 26 8–10 pm, Salt Lake City, UT, 2005.
13. Doolin MT, Barbaux S, McDonnell M, Hoess K, Whitehead AS, Mitchell LE. Maternal genetic effects, exerted by genes involved in homocysteine remethylation, influence the risk of spina bifida. *Am J Hum Genet* 2002; 71(5): 1222–1226.
14. Koch R, Levy HL, Matalon R, Rouse B, Hanley WB, Trefz F et al. The international collaborative study of maternal phenylketonuria: Status report 1994. *Acta Paediatr Suppl* 1994; 407: 111–119.
15. Allen RJ, Brunberg J, Schwartz E, Schaefer AM, Jackson G. MRI characterization of cerebral dysgenesis in maternal PKU. *Acta Paediatr Suppl* 1994; 407: 83–85.
16. Abadie V, Depondt E, Farriaux JP, Lepercq J, Lyonnet S, Maurin N et al. [Pregnancy and the child of a mother with phenylketonuria]. *Archives Pediatr* 1996; 3: 489–486.
17. Menkes JH *Textbook of Child Neurology*. 4 ed. Philadelphia: Lea & Febiger 1990.
18. Westgren M, Ek S, Remberger M, Ringden O, Stangenberg M. Cytokines in fetal blood and amniotic fluid in Rh-immunized pregnancies. *Obstet Gynecol* 1995; 86(2): 209–213.
19. Hollister JM, Laing P, Mednick SA. Rhesus incompatibility as a risk factor for schizophrenia in male adults. *Arch Gen Psychiatry* 1996; 53: 19–24.
20. Zandi PP, Kalaydjian A, Avramopoulos D, Shao H, Fallin MD, Newschaffer CJ. Rh and ABO maternal-fetal incompatibility and risk of autism. *Am J Med Genet B Neuropsychiatr Genet* 2006; 141(6): 643–647.
21. Brody LC, Conley M, Cox C, Kirke PN, McKeever MP, Mills JL et al. A polymorphism, R653Q, in the trifunctional enzyme methylenetetrahydrofolate dehydrogenase/methenyltetrahydrofolate cyclohydrolase/formyltetrahydrofolate synthetase is a maternal genetic risk factor for neural tube defects: Report of the Birth Defects Research Group. *Am J Hum Genet* 2002; 71(5): 1207–1215.
22. Chen D, Hu Y, Yang F, Li Z, Wu B, Fang Z et al. Cytochrome P450 gene polymorphisms and risk of low birth weight. *Genet Epidemiol* 2005; 28(4): 368–375.
23. Jensen LE, Hoess K, Mitchell LE, Whitehead AS. Loss of function polymorphisms in NAT1 protect against spina bifida. *Hum Genet* 2006; 120(1): 52–57.
24. Jensen LE, Etheredge AJ, Brown KS, Mitchell LE, Whitehead AS. Maternal genotype for the monocyte chemoattractant protein 1 A(-2518)G promoter polymorphism is associated with the risk of spina bifida in offspring. *Am J Med Genet A* 2006; 140(10): 1114–1118.
25. Descamps OS, Bruniaux M, Guilmot PF, Tonglet R, Heller FR. Lipoprotein concentrations in newborns are associated with allelic variations in their mothers. *Atherosclerosis* 2004; 172(2): 287–298.
26. Carroll WD, Lenney W, Child F, Strange RC, Jones PW, Fryer AA. Maternal glutathione S-transferase GSTP1 genotype is a specific predictor of phenotype in children with asthma. *Pediatr Allergy Immunol* 2005; 16(1): 32–39.

27. Martinelli M, Scapoli L, Pezzetti F, Carinci F, Carinci P, Stabellini G et al. C677T variant form at the MTHFR gene and CL/P: A risk factor for mothers? *Am J Med Genet* 2001; 98(4): 357–360.

28. van Beynum IM, Kapusta L, den Heijer M, Vermeulen SH, Kouwenberg M, Daniels O et al. Maternal MTHFR 677C>T is a risk factor for congenital heart defects: Effect modification by periconceptional folate supplementation. *Eur Heart J* 2006; 27(8): 981–987.

29. Rai AK, Singh S, Mehta S, Kumar A, Pandey LK, Raman R. MTHFR C677T and A1298C polymorphisms are risk factors for down's syndrome in Indian mothers. *J Hum Genet* 2006; 51(4): 278–283.

30. Lee LC, Zachary AA, Leffell MS, Newschaffer CJ, Matteson KJ, Tyler JD et al. HLA-DR4 in families with autism. *Pediatr Neurol* 2006; 35(5): 303–307.

31. Warren RP, Singh VK, Cole P, Odell JD, Pingree CB, Warren WL et al. Increased frequency of the null allele at the complement C4b locus in autism. *Clin Exp Immunol* 1991; 83: 438–440.

32. Warren RP, Singh VK, Cole P, Odell JD, Pingree CB, Warren WL et al. Possible association of the extended MHC haplotype B44-SC30-DR4 with autism. *Immunogenetics* 1992; 36: 203–207.

33. Daniels WW, Warren RP, Odell JD, Maciulis A, Burger RA, Warren WL et al. Increased frequency of the extended or ancestral haplotype B44- SC30-DR4 in autism. *Neuropsychobiology* 1995; 32: 120–123.

34. Warren RP, Odell JD, Warren WL, Burger RA, Maciulis A, Daniels WW et al. Strong association of the third hypervariable region of HLA-DR beta 1 with autism. *J Neuroimmunol* 1996; 67: 97–102.

35. Odell D, Maciulis A, Cutler A, Warren L, McMahon WM, Coon H et al. Confirmation of the association of the C4B null allelle in autism. *Hum Immunol* 2005; 66(2): 140–145.

36. Torres AR, Maciulis A, Stubbs EG, Cutler A, Odell D. The transmission disequilibrium test suggests that HLA-DR4 and DR13 are linked to autism spectrum disorder. *Hum Immunol* 2002; 63(4): 311–316.

37. Zusterzeel PL, Nelen WL, Roelofs HM, Peters WH, Blom HJ, Steegers EA. Polymorphisms in biotransformation enzymes and the risk for recurrent early pregnancy loss. *Mol Hum Reprod* 2000; 6(5): 474–478.

38. De Marco P, Calevo MG, Moroni A, Arata L, Merello E, Cama A et al. Polymorphisms in genes involved in folate metabolism as risk factors for NTDs. *Eur J Pediatr Surg* 2001; 11(Suppl 1): S14–S17.

39. O'Leary VB, Parle-McDermott A, Molloy AM, Kirke PN, Johnson Z, Conley M et al. MTRR and MTHFR polymorphism: Link to down syndrome? *Am J Med Genet* 2002; 107(2): 151–155.

40. van Rooij IA, Wegerif MJ, Roelofs HM, Peters WH, Kuijpers-Jagtman AM, Zielhuis GA et al. Smoking, genetic polymorphisms in biotransformation enzymes, and nonsyndromic oral clefting: A gene–environment interaction. *Epidemiology* 2001; 12(5): 502–507.

41. Buyske S, Williams TA, Mars AE, Stenroos ES, Wong R, Ming X et al. Analysis of case–parent trios at a locus with a deletion allele: Association of GSTM1 with autism. *BMC Genet* 2006; 7: 8.

42. Gonzalez-Herrera LJ, Flores-Machado MP, Castillo-Zapata IC, Garcia-Escalante MG, Pinto-Escalante D, Gonzalez-Del Angel A. Interaction of C677T and A1298C polymorphisms in the MTHFR gene in association with neural tube defects in the State of Yucatan, Mexico. *Am J Hum Genet* 2002; 71(4): 367.

43. Sata F, Yamada H, Kondo T, Gong Y, Tozaki S, Kobashi G et al. Glutathione S-transferase M1 and T1 polymorphisms and the risk of recurrent pregnancy loss. *Mol Hum Reprod* 2003; 9(3): 165–169.

44. Johnson WG, Stenroos ES, Spychala J, Buyske S, Chatkupt S, Ming X. A new 19 bp deletion polymorphism in intron-1 of dihydrofolate reductase (DHFR)-A risk factor for spina bifida acting in mothers during pregnancy? *Am J Med Genet* 2004; 124A(4): 339–345.

45. Suryanarayana V, Deenadayal M, Singh L. Association of CYP1A1 gene polymorphism with recurrent pregnancy loss in the south Indian population. *Hum Reprod* 2004; 19(11): 2648–2652.

46. Johnson WG, Scholl TO, Spychala JR, Buyske S, Stenroos ES, Chen X. Common dihydrofolate reductase 19 bp deletion allele: A novel risk factor for preterm delivery. *Am J Clin Nutr* 2005; 81: 664–668.

47. Anderson GM, Jacobs-Stannard A, Chawarska K, Volkmar FR, Kliman HJ. Placental trophoblast inclusions in autism spectrum disorder. *Biol Psychiatry* 2007; 61(4): 487–491.

48. Rouse B, Azen C. Effect of high maternal blood phenylalanine on offspring congenital anomalies and developmental outcome at ages 4 and 6 years: The importance of strict dietary control preconception and throughout pregnancy. *J Pediatr* 2004; 144(2): 235–239.

49. Palmer CG, Turunen JA, Sinsheimer JS, Minassian S, Paunio T, Lonnqvist J et al. RHD maternal-fetal genotype incompatibility increases schizophrenia susceptibility. *Am J Hum Genet* 2002; 71(6): 1312–1319.

50. Scala I, Granese B, Sellitto M, Salome S, Sammartino A, Pepe A et al. Analysis of seven maternal polymorphisms of genes involved in homocysteine/folate metabolism and risk of Down syndrome offspring. *Genet Med* 2006; 8(7): 409–416.

51. Buyske S. Maternal genotype effects can alias case genotype effects in case–control studies. Submitted. 2007.

Chapter 3
Cholesterol Deficit in Autism: Insights from Smith–Lemli–Opitz Syndrome

Alka Aneja and Elaine Tierney

Abstract Cholesterol is necessary for neuroactive steroid production, growth of myelin membranes, and normal embryonic and fetal development. It also modulates the oxytocin receptor, as well as ligand activity and G-protein coupling of the serotonin-1A receptor. A deficit of cholesterol may perturb these biological processes and thereby contribute to autism spectrum disorders (ASD), as observed in Smith–Lemli–Opitz syndrome (SLOS) and some subjects with ASD in the autism genetic resource exchange (AGRE). A clinical diagnosis of SLOS can be confirmed by laboratory testing and is partially treatable by cholesterol supplementation. Thus, the threshold should be low for obtaining a blood sterol analysis for the biochemical diagnosis of SLOS.

Keywords Autism · SLOS · cholesterol · neurosteroids · treatment

Introduction

Autism spectrum disorders (ASD) are characterized by disturbances of brain function and defined by core areas of specific abnormalities in reciprocal social interaction, communication, and restrictive or repetitive interests and behaviors [1]. With an incidence of 1 in 150 [2], ASD manifest in early childhood and usually persist throughout life. In the majority of cases, specific underlying causes cannot be identified. However, a number of factors are being investigated including infectious, metabolic, genetic, and environmental, with specific causes being documented in generally less than 10–12% of cases [1]. In addition to the association of ASD with specific heritable disorders (e.g., fragile X syndrome, phenylketonuria, and tuberous sclerosis), evidence for a genetic contribution includes occurrence of cognitive, language, and behavioral disturbances in close relatives, increased recurrence risk in sibs, and increased concordance in monozygotic

E. Tierney
Department of Psychiatry, Kennedy Krieger Institute, 3901 Greenspring Ave.,
Baltimore, MD 21211, USA
e-mail: tierney@kennedykrieger.org

compared to dizygotic twins [1]. One of the genetic disorders with autism and sterol abnormalities is Smith–Lemli–Opitz syndrome (SLOS) [3].

Cholesterol Pathways and SLOS

Cholesterol is an important building block for the body's cell membranes, myelination of the central nervous system, the formation of all steroid hormones (e.g., cortisol, testosterone, estrogen), and formation of bile acids necessary for digestion of fats. Cholesterol deficiency during prenatal life can cause birth defects and postnatally can cause poor growth and developmental delays. SLOS is an autosomal recessive disorder due to an inborn error of cholesterol metabolism that is caused by mutations of the 7-dehydrocholesterol (7-DHC) reductase gene (DHCR7) [4], located on chromosome 11q12–13 [5, 6]. SLOS is not uncommon, with an estimated incidence among individuals of European ancestry of 1 in 20,000 to 1 in 60,000 births and a carrier frequency of at least 1% [7]. In persons with SLOS, this enzyme functions abnormally and, as a result, there is not enough cholesterol produced in the body and 7-DHC accumulates (Fig. 3.1).

A clinical diagnosis of SLOS can be confirmed by biochemical testing. An elevated plasma 7-DHC level relative to the total cholesterol level establishes the diagnosis. SLOS is not only identifiable but is also treatable by cholesterol supplementation. Thus, it is important to know the variations in presentation of SLOS so that individuals with ASD who have SLOS can be identified and started on treatment.

SLOS is characterized by a broad spectrum of phenotypic abnormalities including developmental delay, the characteristic facial anomalies of wide-set eyes (hypertelorism), posteriorly rotated ears, a prominent nasal bridge, a high

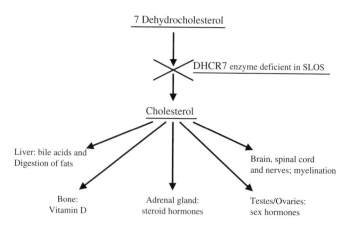

Fig. 3.1 The pivotal role of cholesterol

arched palate, upturned nares (Figs 3.2 and 3.3), and an abnormal amount of syndactyly (webbing) between the second and third toes (2–3 toe syndactyly) (Fig. 3.3). The lower the 7-dehydrocholesterol level in plasma, the more severe

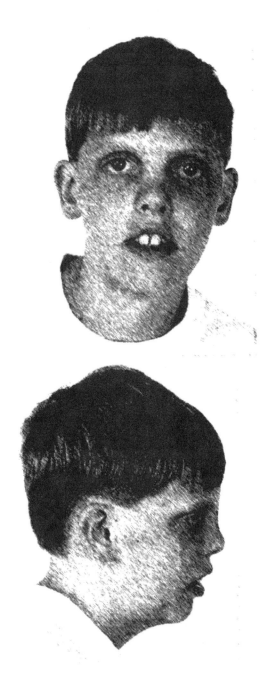

Fig. 3.2 (a and b) Facial features of SLOS: Ptosis, upturned nares, and micrognathia

A. Aneja, E. Tierney

Fig. 3.3 Abnormal degree of webbing between the second and third toes (2–3 toe syndactyly)

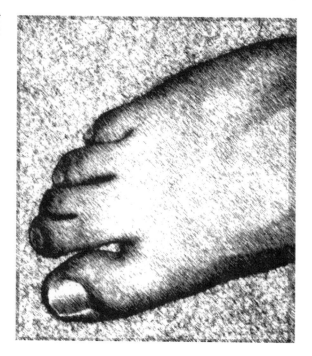

are the physical findings in SLOS; malformations of the heart, brain, lungs, and genitals are frequently observed in more severe cases [7]. Severely affected fetuses often are not viable or the infants die during the perinatal period [7]. In the mild SLOS cases, the physical phenotype may not be readily discernable and toe syndactyly may not be present. Often an individual with a mild SLOS phenotype is diagnosed after a relative, who has a greater degree of physical manifestations, receives the diagnosis of SLOS.

The SLOS behavioral phenotype includes many features that are also seen in ASD, such as social and language communication impairments as well as repetitive and ritualistic behaviors. Among 17 subjects with SLOS who were administered the algorithm questions of the Autism Diagnostic Interview-Revised (ADI-R), 53% met the criteria for autism [8], as diagnosed by both the DSM-IV and the ADI-R algorithm [9]. A recent study reported that approximately three-fourths of the children with SLOS in their study had some variant of ASD, suggesting the most consistent relationship with ASD of any single gene disorder [10]. Other behavioral characteristics include repeated self-injury (89%), self-biting (54%), head banging (48%), and opisthokinesis (54%; a highly characteristic upper body movement in which the child performs an arched backwards diving motion) [8]. Sensory hyper-reactivity, temperament dysregulation, and sleep disturbance are also commonly seen in SLOS [8]. Cognitive abilities in individuals with SLOS range from borderline intellectual functioning to profound mental retardation [3]. These physical and behavioral features of SLOS are summarized in Table 3.1.

Table 3.1 Phenotypic features of SLOS

Physical features	Behavioral features
Microcephaly	Autism spectrum disorder
Ptosis (drooping of eyelids)	Social impairment
Low-set and small ears	Communication disorder
Soft cleft palate or bifid uvula	Repetitive and ritualistic behaviors
Hand or foot malformations (2–3 toe syndactyly, clinodactyly)	Mental retardation
	Self-injury
Malformations of heart, brain, and lungs	Opisthokinesis (backward arched motion)
Malformations in gastrointestinal and genital tracts	Severe sleep disturbance
Failure to thrive/feeding difficulties	Attention deficit hyperactivity disorder
Growth retardation	Sensory hyperreactivity
Hypotonia	Anxiety

Although SLOS is associated with ASD [8, 10], the incidence of SLOS and other sterol disorders among individuals with ASD is unknown. One study investigated the incidence of biochemically diagnosed SLOS and other sterol abnormalities in blood samples from a cohort of subjects with ASD who were in multiplex families (families in which more than one child in the immediate family had ASD) [11]. Using gas chromatography/mass spectrometry, cholesterol and its precursor sterols were quantified in 100 samples from subjects with ASD obtained from the Autism Genetic Resource Exchange (AGRE) specimen repository [12]. Although no sample had sterol levels consistent with SLOS, 19 samples had total cholesterol levels lower than 100 mg/dl, which is below the fifth centile for children over age 2 years [11]. These findings suggest that, in addition to SLOS, there may be other disorders of sterol metabolism or homeostasis associated with ASD. Also, the unexpected finding in the above study, that 19% of children from a sample of mostly multiplex ASD sibships have substantial hypocholesterolemia, warrants further research. This is important since the study of hypocholesterolemia and its predicted effects on neurosteroid metabolism and its related cholesterol-dependent biomechanisms may offer important insights into the causes and treatment of ASD.

Other studies indicate that individuals with decreased dietary intake or increased intestinal losses of cholesterol will have *increased* levels of cholesterol precursors, especially lathosterol [13] which we did not find in comparing the subjects with cholesterol levels below 100 mg/dl versus the full group cohort. Rather, we found that individuals with a cholesterol level below 100 mg/dl had statistically *lower* levels of lathosterol compared to the entire group cohort [11], indicating that the cause of the hypocholesterolemia was decreased cholesterol synthesis rather than increased cholesterol losses from gastrointestinal disturbances or abnormal diets. The data argue that the subjects with ASD who have low cholesterol levels have intrinsically reduced cholesterol synthesis, similar to patients with SLOS.

Furthermore, although the majority of individuals with SLOS have lower than normal cholesterol levels, there are a number of individuals diagnosed clinically and biochemically with SLOS who have normal total cholesterol levels and only mildly elevated levels of 7-DHC. Thus, a normal value for plasma cholesterol does not exclude the diagnosis of SLOS. The specific test for SLOS—analysis of the ratio of 7-DHC to cholesterol—is required. Neuroleptic medications may impair the function of 7-DHC reductase; therefore, if a sample demonstrates a borderline abnormality, additional testing of sterol metabolism in cultured lymphoblasts should be performed to see whether the elevated 7-DHC finding is repeatable.

The recognition of the biochemical cause of SLOS not only significantly improved diagnostic accuracy by measuring the ratio of plasma levels of 7-DHC to total cholesterol but also offered a potential treatment strategy. The current standard for treatment is to begin dietary cholesterol supplementation as soon as the condition is diagnosed. There are reports that cholesterol supplementation may improve growth, speech articulation, and neurodevelopmental status [14] although Sikora et al. [15] found that developmental quotients did not improve over time with cholesterol supplementation. Yet, autism features may respond to treatment. Among 17 subjects with SLOS who were administered the algorithm questions of the ADI-R, of the 9 patients who began cholesterol supplementation before the age of 5 years, 22% satisfied the criteria at the age of 4–5 years (while on supplementation), whereas of the 8 subjects not supplemented with cholesterol before age of 5 years, 88% met the criteria for autism at the age of 4–5 years (prior to supplementation) [8]. The standard treatment is dietary cholesterol supplementation (150 mg/kg/day) using crystalline cholesterol suspended in OraPlus or pasteurized egg yolk.

Professional organizations are making recommendations for screening for biochemical and genetic disorders that are associated with ASD, as described in the American Academy of Neurology and the Child Neurology Society's Practice Parameter: Screening and Diagnosis of Autism [16]. The recommended diagnostic tests to evaluate a child for recognized causes of ASD (in the presence of mental retardation) include a lead level, high-resolution chromosome study and DNA analysis for fragile X. Based on additional symptoms and clinical findings, tests to be considered include EEG, quantitative urinary organic and plasma amino acids, and carbohydrate-deficient glycoprotein analysis [17].

Cholesterol-Related Mechanisms in ASD

The sterol abnormality associated with SLOS may affect the brain in multiple ways in the same person and may explain why there is a constellation of features in ASD. Current research suggests that alterations in cholesterol production secondary to SLOS affect the development of the brain and central nervous system through their effects on dendrite differentiation, the function of brain receptors, myelination, and the synthesis of steroid hormones. Disruption of these components may result in the presentation of various ASD features.

The following are examples of hypotheses that can be made related to sterol function in ASD and SLOS:

1. Cholesterol is essential for the growth of myelin membranes. In mice created to lack the ability to synthesize cholesterol in myelin-forming oligodendrocytes, it was shown that cholesterol is an indispensable component of myelin membranes and that cholesterol availability to oligodendrocytes is a rate-limiting factor for brain maturation [18]. The myelin oligodendrocyte glycoprotein (MOG) is a myelin-specific protein of the central nervous system that is known to be affected by MOG-specific demyelinating antibodies. This autoimmune process has been associated with obsessive-compulsive disorder [19]. In SLOS, obsessive-compulsive symptoms can be severe and autism-associated motor movements (stereotypies) are common [8]. When insufficient cholesterol is present, 7-DHC may be present in myelin where cholesterol would have existed and might cause abnormal myelin function.

2. Cholesterol is essential for normal embryonic and fetal development. CNS abnormalities, similar to those observed in individuals with ASD, have been noted in individuals with SLOS and are thought to result from hedgehog protein malfunction secondary to inadequate availability of cholesterol [7]. The SLOS mouse model has also been shown to exhibit commissural deficiencies, hippocampal abnormalities, and hypermorphic development of serotonin (5-HT) neurons that may help explain the ASD behavioral presentation seen in individuals with SLOS [20].

3. Cholesterol is a precursor for neurosteroid production, and abnormal neurosteroid function may be related to ASD features. Neurosteroids modulate neurotransmitter receptor activity and exhibit neurodevelopmental and neuroprotective effects. The defect in cholesterol synthesis in SLOS may lead to abnormal neurosteroid production [21]. A deficit in neurosteroids may be associated with both anxiety and mood disorders [22]. There may be decreased neurosteroid levels, dehydroepiandrosterone (DHEA), and DHEA-sulfate (DHEA-S) in adult patients with autistic disorder [23]. Parents of SLOS patients frequently report changes in their children's abnormal behaviors within days of supplementation, before there is a change in the plasma cholesterol or 7-DHC level. As cholesterol cannot cross the blood–brain barrier, the behavioral changes observed may be due to altered levels of cholesterol-derived steroid precursors or other compounds rather than the change in the level of circulating cholesterol.

4. Cholesterol has a modulatory role in the function of the oxytocin receptor [23]. Oxytocin itself has been found to be involved in social function [24]. Therefore, a low cholesterol level may impair the function of the oxytocin receptor in individuals who have a mutation in the gene for the oxytocin receptor and this in turn could lead to abnormal social functioning in individuals with SLOS and individuals with ASD from other etiologies.

5. Cholesterol functions as a modulator of the ligand binding activity and G-protein coupling of the serotonin1A (5-HT1A) receptor [25]. Cholesterol is also a component of the lipid rafts of the serotonin transporter [26]. Dysfunction in the serotonin system may lead to the social and behavioral difficulties seen in individuals with ASD, just as some symptoms of ASD are alleviated by treatment with medications that are known to interact with the serotonin receptor. Abnormal serotonergic neuron development has been demonstrated in a SLOS mouse model [20]. SLOS may present with impairing irritability, sleep disturbance and obsessive-compulsive disorder, and these symptoms may improve with cholesterol supplementation or treatment with serotonin reuptake inhibitors.

Treatment Approaches

In individuals with SLOS, cholesterol therapy can significantly decrease the number of autistic behaviors, as well as the number of infections, and may result in increased growth, weight gain, and improved sleep [7, 27]. It can help reduce irritability and hyperactivity and lead to happier affect and attention span [28, 29, 30]. Other behaviors that improve with cholesterol supplementation include self-injury [28], aggressive behaviors [29, 31], temper outbursts, trichotillomania, and tactile defensiveness [29]. Individuals with SLOS treated with cholesterol have also been reported to be more sociable, including initiating hugs and being more active than passive [28, 31].

Although the effect of cholesterol supplementation on cognitive development is disappointing [7], it is possible that early administration of cholesterol therapy in infancy or in childhood may improve developmental outcomes in SLOS patients. Furthermore, no side effects have been reported with cholesterol therapy to date. Cholesterol is supplied in natural form (eggs, cream, liver) or as purified cholesterol; the starting dose for purified cholesterol is 40–50 mg/kg/day. Tube feeding is often required in infants and younger children because of feeding difficulties. Medical and surgical management of gastroesophageal reflux may be required. When necessary, fresh frozen plasma is required as a source of cholesterol for rapid management of infections or for surgical procedures. On occasions, when acutely ill, patients with SLOS might develop overt adrenal insufficiency requiring treatment. There is evidence that statins (a class of medications used to lower cholesterol in those with abnormally high levels) may improve DHCR7 activity and thus lead to an increase in cholesterol levels in individuals with mild DHCR7 deficiency [32], in human fibroblasts from mildly affected individuals [33] and in a SLOS mouse model [34].

Certain medications used to treat behavioral and psychiatric disorders lower the production of cholesterol by interfering with the enzyme DHCR7. However, the benefits of these medications may outweigh the potential risks of lowering cholesterol production. Mildly to moderately elevated levels of

7-DHC have been observed in three psychiatric patients without SLOS who were treated with haloperidol [35]; 7-DHC levels were directly proportional to the dose of haloperidol and decreased to normal upon discontinuation of haloperidol therapy.

Conclusion

The importance of performing biochemical analyses for SLOS in individuals with ASD who have specific physical or behavioral features is critical to its early diagnosis [36]. The study of phenotype–genotype relationships in single-gene disorders such as fragile X syndrome, Rett syndrome, and SLOS may provide insights as to how the disruptions of biological mechanisms of these disorders correlate to the features of ASD observed in these disorders.

Research on SLOS also leads to the question whether abnormalities in cholesterol metabolism not due to SLOS may exist in patients with "typical" ASD. A cholesterol deficit might lead to physical structural abnormalities; abnormal structure of sterol-rich membranes, such as myelin; dysfunction of serotonin and other brain receptors; and impairment in the synthesis and metabolism of sterols, including neurosteroids. In some forms of ASD, the symptoms of ASD may be due to interaction of components that are sterol dependent. In addition, further study of SLOS and the resulting abnormal cholesterol conditions might also help us understand mechanisms of importance to research and treatment of ASD in patients without defects in cholesterol metabolism.

Recommendations

Because a specific treatment is available for the biochemical defect of SLOS, the threshold should be low for obtaining a sterol analysis for SLOS (total cholesterol and 7-DHC) for an individual with ASD. The association of ASD with other physical or behavioral manifestations that warrant testing include 2–3 toe syndactyly, ptosis, soft cleft palate/bifid uvula, failure to thrive or feeding difficulties (beyond food selectivity), growth retardation, postnatal onset of microcephaly, hand or foot malformation, abnormal genitalia, hypotonia, mental retardation, severe sleep disturbance, self-injury, or opisthokinesis (fast backward arching motion).

Acknowledgments We gratefully acknowledge the resources provided by the Autism Genetic Resource Exchange (AGRE) consortium and the participating AGRE families. AGRE is a program of Autism Speaks and is supported, in part, by grant 1u2 4MH081810 from the National Institute of Mental Health to Clara M. Lajonchere (PI). This work was also supported by the Smith–Lemli–Opitz Advocacy and Exchange, Autism Speaks, the National Institute of Mental Health Studies to Advance Autism Research and Treatment (STAART)

Center 154MH066417 (PI: Rebecca Landa), the National Institute for Child Health and Human Development Mental Retardation and Developmental Disability Research Center P30HD24061 (PI: Michael Cataldo), the National Institute of Mental Health Research Units on Pediatric Psychopharmacology N01MH80011 (PI: Michael Aman), and the KKI Center for Genetic Disorders of Cognition and Behavior.

References

1. Muhle R, Trentacoste SV, Rapin I. The genetics of autism. Pediatrics 2004; 113: e472–e486.
2. Centers for Disease Control (CDC). (February 9, 2007). Morbidity and Mortality Weekly Report: Surveillance Summaries. http://www.cdc.gov/MMWR/pdf/ss/ss5601.pdf.
3. Smith DW, Lemli L, Opitz JM. A newly recognized syndrome of multiple congenital anomalies. J Pediatr 1964; 64:210–217.
4. Tint GS, Irons M, Elias ER, Batta AK, Frieden R, Chen TS, et al. Defective cholesterol biosynthesis associated with the Smith-Lemli-Opitz syndrome. N Eng J Med1994; 330:107–113.
5. Wassif CA, Maslen C, Kachilele-Linjewile S, et al. Mutations in the human sterol delta7-reductase gene at 11q12-13 cause Smith-Lemli-Opitz syndrome. Am J Hum Genet 1998; 63(1):55–62.
6. Fitzky BU, Witsch-Baumgartner M, Erdel M, et al. Mutations in the Delta7-sterol reductase gene in patients with the Smith-Lemli-Opitz syndrome. Proc Natl Acad Sci USA 1998; 95(14):8181–8186.
7. Kelley RI, Hennekam RCH. Smith-Lemli-Opitz Syndrome, In: The metabolic and molecular basis of inherited disease, 8th edition. Edited by Scriver CR, Beaudet AL, Sly WS, Valle D. New York, McGraw Hill, 2000, pp. 6183–6201.
8. Tierney E, Nwokoro NA, Porter FD, Freund LS, Ghuman JK, Kelley RI. Behavior phenotype in the RSH/Smith-Lemli-Opitz syndrome. Am J Med Genet 2001; 98:191–200.
9. Lord C, Rutter M, Le Couteur A. Autism diagnostic interview-revised: A revised version of a diagnostic interview for caregivers of individuals with possible pervasive developmental disorders. J Autism Dev Disord 1994; 24:659–685.
10. Sikora DM, Pettit-Kekel K, Penfield J, Merkens LS, Steiner RD. The near universal presence of autism spectrum disorders in children with Smith-Lemli-Opitz syndrome. Am J Med Genet Part A 2006; 140:1511–1518.
11. Tierney E, Bukelis I, Thompson RE, et al. Abnormalities of Cholesterol Metabolism in Autism Spectrum Disorders. Am J Med Genet B Neuropsychiatr Genet 2006; 141B:666–668.
12. Geschwind DH, Sowinski J, Lord C, et al. AGRE Steering Committee. The autism genetic resource exchange: a resource for the study of autism and related neuropsychiatric conditions. Am J Hum Genet 2001; 69(2):463–466.
13. Lund E, Sisfontes L, Reihner E, Bjorkhem I. Determination of serum levels of unesterified lathosterol by isotope dilution-mass spectrometry. Scand J Clin Lab Invest 1989; 49(2):165–171.
14. Irons M, Elias ER, Abuelo D, Bull MJ, Greene CL, Johnson VP, Keppen L, Schanen C, Tint GS, Salen G. Treatment of Smith-Lemli-Opitz syndrome: Results of a multicenter trial. Am J Med Genet 1997; 68:311–314.
15. Sikora DM, Ruggiero M, Petit-Kekel K, Merkens LS, Connor WE, Steiner RD. Cholesterol supplementation does not improve developmental progress in Smith-Lemli-Opitz syndrome. J Pediatr 2004; 144(6):783–791.
16. Filipek PA, Accardo PJ, Ashwal S, et al. Practice parameter: screening and diagnosis of autism: Report of the Quality Standards Subcommittee of the American Academy of Neurology and the Child Neurology Society. Neurology 2000; 55:468–479.

17. Zimmerman AW. Autism Spectrum Disorders, In: Treatment of pediatric neurologic disorders. Edited by Singer HS, Kossof EH, Hartman AL, Crawford TO. Boca Raton, Florida, Taylor and Francis, 2005, pp. 489–494.
18. Saher G, Brugger B, Lapper-Siefke C, et al. High cholesterol level is essential for myelin membrane growth. Nat Neurosci 2005; 8:468–475.
19. Zai G, Bezchlibnyk YB, Richter MA, et al. Myelin oligodendrocyte glycoprotein (MOG) gene is associated with obsessive-compulsive disorder. Am J Med Genet B Neuropsychiatr Genet 2004; 129:64–68.
20. Waage-Baudet H, Lauder JM, Dehart DB, et al. Abnormal serotonergic development in a mouse model for the Smith-Lemli-Opitz syndrome: implications for autism. Int J Dev Neurosci 2003; 21:451–459.
21. Marcos J, Guo LW, Wilson WK, Porter FD, Shackleton C. The implications of 7-dehydrosterol-7-reductase deficiency (Smith-Lemli-Opitz syndrome) to neurosteroid production. Steroids 2004; 69:51–60.
22. Strous RD, Golubchik P, Maayan R, et al. Lowered DHEA-S plasma levels in adult individuals with autistic disorder. Eur Neuropsychopharmacol 2005; 15:305–309.
23. Gimpl G, Wiegand V, Burger K, Fahrenholz F. Cholesterol and steroid hormones: modulators of oxytocin receptor function. Prog Brain Res 2002; 139:43–55.
24. Hollander E, Bartz J, Chaplin W, et al. Oxytocin increases retention of social cognition in autism. Biol Psychiatry 2007; 61(4):498–503.
25. Chattopadhyay A, Jafurulla M, Kalipatnapu S, Pucadyil TJ, Harikumar KG. Role of cholesterol in ligand binding and G-protein coupling of serotonin1A receptors solubilized from bovine hippocampus. Biochem. Biophys Res Commun 2005; 327:1036–1041.
26. Magnani F, Tate CG, Wynne S, Williams C, Haase J. Partitioning of the serotonin transporter into lipid.microdomains modulates transport of serotonin. J Biol Chem 2004; 279(37):38770–38778.
27. Tierney E, Nwokoro NA, Kelley RI. Behavioral phenotype of RSH/Smith-Lemli-Opitz syndrome. Ment Retard Dev Disabil Res Rev 2000; 6(2):131–134.
28. Irons M., Elias ER., Abuelo D, Tint GS, Salen G. Clinical features of the Smith-Lemli-Opitz syndrome and treatment of the cholesterol metabolic defect. Internat Pediatr 1995; 10:28–32.
29. Nwokoro NA, Mulvihill JJ. Cholesterol and bile acid replacement therapy in children and adults with Smith-Lemli-Opitz (SLO/RSH) syndrome. Am J Med Genet B Neuropsychiatr Genet 1997; 68:315–321.
30. Opitz JM. RSH (so called Smith-Lemli-Opitz) syndrome. Curr Opin Pediatr 1999; 11:353–362.
31. Ryan AK, Bartlett K, Clayton P, et al. Smith-Lemli-Opitz syndrome: A variable clinical and biochemical phenotype. J Med Genet 1998; 35:558–565.
32. Jira PE, Wevers RA, de Jong J, et al. Simvastatin. A new therapeutic approach for Smith-Lemli-Opitz syndrome. J Lipid Res 2000; 41(8):1339–1346.
33. Wassif CA, Krakowiak PA, Wright BS, et al. Residual cholesterol synthesis and simvastatin induction of cholesterol synthesis in Smith-Lemli-Opitz syndrome fibroblasts. Mol Genet Metab 2005; 85(2):96–107. Epub 2005 Feb 5.
34. Correa-Cerro LS, Wassif CA, Kratz L, et al. Development and characterization of a hypomorphic Smith-Lemli-Opitz syndrome mouse model and efficacy of simvastatin therapy. Hum Mol Genet 2006; 15(6):839–851. Epub 2006 Jan 30.
35. Nowaczyk MJM, Tierney E. Smith-Lemli-Opitz syndrome: Demystifying genetic syndromes. Kingston, New York, National Association for the Dually Diagnosed Publishing, 2004: 207–223.
36. Nowaczyk MJM, Waye JS. The Smith-Lemli-Opitz syndrome: A novel metabolic way of. understanding developmental biology, embryogenesis and dysmorphology. Clin. Genet. 2001; 59: 75–386.

Chapter 4
Autism in Genetic Intellectual Disability

Insights into Idiopathic Autism

**Walter E. Kaufmann, George T. Capone, Megan Clarke,
and Dejan B. Budimirovic**

Abstract Despite early controversy, it is currently accepted that a substantial proportion of children with intellectual disability of genetic origin meet criteria for autism spectrum disorders (ASD). This has led to an increased interest in studying conditions such as Fragile X syndrome (FXS) as genetic models of idiopathic ASD. Here, largely based on our own studies, we expand this notion to propose that the study of ASD in genetic intellectual disability can provide important clues about many aspects of idiopathic ASD including its core behavioral features. Thus, FXS could reveal a molecular–neurobiological–behavioral continuum for deficits in complex social interactions in ASD. Down syndrome (DS) could disclose similar bases for repetitive and stereotypic behaviors in ASD, while DS and Rett syndrome are likely to share commonly affected molecular–neurobiological–behavioral pathways with individuals with idiopathic ASD who experience developmental regression. Consequently, the in-depth characterization of ASD in genetic intellectual disability could be doubly rewarding by improving the clinical management of severely affected individuals with these disorders and by shedding light into key aspects of idiopathic ASD.

Keywords Autism · genetics · intellectual disability · fragile X · down · Rett

Introduction

Autistic features have long been recognized in multiple genetic disorders associated with intellectual disability. However, autism has only recently been characterized in a comprehensive manner in these disorders. The comorbidity of intellectual disability and autism raises significant methodological and clinical issues. The first is whether severe cognitive impairment, present in a substantial proportion of patients with disorders such as Down syndrome (DS), precludes a confident diagnosis of autism. This is an issue of relevance not only to genetic

W.E. Kaufmann
Kennedy Krieger Institute, 3901 Greenspring Ave., Baltimore, MD 21211, USA
e-mail: Kaufmann@kennedykrieger.org

A.W. Zimmerman (ed.), *Autism*, DOI: 10.1007/978-1-60327-489-0_4,
© Humana Press, Totowa, NJ 2008

intellectual disability, but also to other severely impaired individuals who fulfill Diagnostic and Statistical Manual of Mental Disorders, 4th edition, Text Revision (DSM-IV-TR) [1] criteria for autism without a recognizable etiology. A second, overlapping, issue is the contribution of delayed or impaired communication, a frequent feature of genetic disorders associated with intellectual disability, to the diagnosis of autism or autism spectrum disorders (ASD). Despite these concerns, studies in the last decade have reported that a significant proportion of patients with Fragile X syndrome (FXS), DS, and Velocardiofacial syndrome (VCFS) meet DSM-IV criteria for ASD and demonstrate behavioral profiles consistent with the DSM-IV label [2, 3]. Other disorders that display features overlapping with those of idiopathic ASD or social impairments of relevance to autism include Rett syndrome (RTT), Angelman syndrome, Prader–Willi syndrome, Smith–Magenis syndrome, Williams syndrome, Turner syndrome, tuberous sclerosis, San Filippo syndrome, phenylketonuria, adenylosuccinate lyase deficiency, Cohen syndrome, and Smith–Lemli–Opitz syndrome [2, 3].

General Issues

The study of genetic disorders associated with ASD has several implications. In addition to addressing the particular diagnostic and therapeutic issues affecting a subset of individuals with severe phenotypes and complex medical and educational needs within FXS [4], DS [5], VCFS [6], and other conditions, this line of research could identify important genetic and neurobiological mechanisms in idiopathic ASD. A second, less well-recognized rationale is that the aforementioned genetic disorders may lead to a more complete characterization of the heterogeneous behavioral features of idiopathic autism. The latter issue is in line with the recent interest in subdividing idiopathic ASD into discrete clinical groups or endophenotypes [7, 8]. This approach is already demonstrating its value in the characterization of genetic abnormalities underlying idiopathic autism. It is our opinion that detailed phenotyping in idiopathic ASD and genetic disorders associated with ASD will also be essential for the development of specific clinical guidelines and novel treatment trials for all autistic disorders. The following sections describe the current and potential contributions of genetic disorders associated with intellectual disability to the behavioral and clinical features, the neurobiology, and finally the genetics and molecular biology of idiopathic ASD. This chapter will focus on our and others' work on FXS, DS, and RTT.

Background on Genetic Disorders Associated with Idiopathic Autism

Fragile X Syndrome

Fragile X syndrome is the most prevalent form of inherited intellectual disability, and the second most common genetic etiology of intellectual disability,

Table 4.1 Characteristic features of Fragile X syndrome

Physical features	Neurobehavioral features
Large ears	Mild to moderate mental retardation
Thick nasal bridge	Language delay, predominantly expressive
Prominent jaw	Rapid/burst-like speech
High-arched/narrow palate	Attentional-organizational dysfunction
Pale blue irises	Visuospatial impairment
Strabismus	Hyperactivity
Pectus excavatum	Hyperarousal
Kyphoscoliosis	Anxiety, particularly social
Lax joints	Autistic-like features
Single palmar crease	Aggressive behavior
Flat feet	Stereotypic/preservative behavior
Cutis laxa	Hypotonia
Mitral valve prolapse	Nystagmus
	Seizures

From: Adapted from Kaufmann WE, Reiss AL. Molecular and cellular genetics of Fragile X syndrome. *Am J Med Genet* 1999; 88: 11–24 [10].

affecting 1:4000 males and 1:6000 females [9]. The disorder is linked to the expansion of a CGG polymorphism in the (5′UTR) regulatory region of the *FMR1* gene. When the normal number (~30) of CGG repeats increases to >200 (full mutation), *FMR1* promoter hypermethylation, *FMR1* transcriptional silencing [i.e., no production of Fragile X mental retardation protein (Fmrp), an RNA-binding protein and translational regulator], and the FXS phenotype occur. Intermediate level expansions (60–200 CGG repeats), which are termed permutations, are not associated with FXS [10] but with a carrier status or other clinical phenotypes (e.g., mild cognitive/behavioral problems, Fragile X Tremor Ataxia Syndrome or FXTAS) [11]. In addition to mild to moderate mental retardation (MR), FXS is characterized by dysmorphic features, connective tissue abnormalities (e.g., lax joints), and other non-CNS phenotypical anomalies (e.g., macroorchidism after puberty) (Table 4.1). However, variable cognitive and language impairments and associated neurobehavioral problems, including attentional difficulties, hyperactivity, anxiety, and autistic disorder, constitute the major medical and educational concerns for patients with FXS [10, 4]. Similar to other genetic disorders, because of ascertainment bias and variable diagnostic approaches, the exact proportion of individuals with certain phenotypical features in FXS is unclear [4]. Nonetheless, as expected in an X-linked condition, the characteristic features of FXS are more prominent in affected males, and ASD is almost exclusively a clinical issue in the latter group.

Down Syndrome

Down syndrome is the most common genetic cause of intellectual disability, occurring in an estimated 1:1740 live births [12, 13]. The DS phenotype

Table 4.2 Characteristic features of Down syndrome

Physical features (variable)	Neurobehavioral features (variable)
Flat facial profile	**Early development**
Poor primitive and deep tendon reflexes	Intellectual speech impairment
Hypotonia	Hypotonia-motor delay
Hyperlaxity of joints	Infantile seizures
Excessive skin on neck	**Childhood**
Slanted palpebral fissures	Inattention-hyperactivity
Pelvic dysplasia	Aggressive–disruptive behaviors
Anomalous auricles	Stereotypies–autistic features
Dysplastic midphalanx 5th finger	Unusual sensory preferences-responding
Single palmar crease	**Adolescence and adulthood**
Adolescence and adulthood	Depression
Down syndrome newborns demonstrating more than four features (100%)	Anxiety
Down syndrome newborns demonstrating more than six features (90%)	Obsessive–compulsive features
	Late adulthood
	Alzheimer-type pathology and dementia
	Seizures
	Parkinsonian features

From: Adapted from Roizen NJ, Patterson D. Down's syndrome. *Lancet* 2003; 361: 1281–1289 [13] & Capone GT, Roizen NJ, Rogers PT. Down Syndrome. In: Accardo PJ and Johnston MV (Eds). *Developmental Disabilities in Infancy and Childhood*. Baltimore: Paul H. Brookes Publishing Co, 2007: 285–308 [15].
From: Hagberg B. Clinical manifestations and stages of Rett syndrome. *Ment Retard Dev Disabil Res Rev* 2002; 8:61–65 [17]. Reprinted by permission.

results from trisomy 21. In addition to characteristic dysmorphic features, anatomical abnormalities include cardiac and gastrointestinal malformations. Neurological abnormalities include cognitive impairment, neuromusclar hypotonia, and occasionally seizures [13]. Although most children with DS are described as sociable and affectionate [14], a relatively significant proportion (10–15%) manifest atypical neurobehavioral symptoms. These include hyperactivity and impulsivity, oppositional and disruptive behavior, stereotypic movement and autistic features. In contrast to FXS, emphasis on behavioral syndromes is a relatively new development in DS. Nevertheless, the complex management issues involving care of individuals with DS and abnormal behavior and the diagnostic challenges represented by the similarities between SMD and ASD highlight the importance of this research area in DS. Table 4.2 summarizes the most salient features of the DS phenotype.

Rett Syndrome

Rett syndrome is an X-linked condition that affects predominantly females, with an incidence of approximately 1:9000 girls by age 12 years. Rett syndrome is a severe disorder, lethal in most male cases [16] and the second leading cause of global developmental delay and severe intellectual disability in females, after DS [17]. The majority of RTT cases are associated with mutations in the coding region of *MECP2*, a gene located on Xq28, which encodes the transcriptional repressor methyl-CpG-binding protein 2 (Mecp2) [18, 19]. Rett syndrome is a complex condition because of its dynamic evolution (Table 4.3), particularly from the neurological viewpoint, and range of clinical presentations (i.e., classic vs. atypical RTT) [17]. Depending on the clinical stage (for stages, see Table 4.3), females with RTT could appear normal, although cognitive impairment typically becomes evident by 18 months of age. In addition to language delay or loss and frequent deceleration of head growth, individuals with RTT present with loss of motor skills and hand use as well as characteristic stereotypic hand-wringing movements. Additional manifestations include respiratory irregularities,

Table 4.3 Characteristic features and stages of Rett syndrome

Original staging system	Later additions
Stage I: Early-onset stagnation	
Onset age: 6 months to 1.5 year	Onset from 5 months of age
Developmental progress delayed	Early postural delay
Developmental pattern still not significantly abnormal	Dissociated Development "Bottom-shufflers"
Duration: weeks to months	
Stage II: developmental regression	
Onset age: 1–3 or 4 year	
Loss of acquired skills/communication	Loss of acquired skills: fine finger, babble/ words, active playing
Mental deficiency appears	Occasionally "in another world"
	Eye contact preserved
	Breathing problems still modest
	Seizures in only 15%
Duration: weeks to months, possibly 1 year	
Stage III: psuedostationary period	
Onset age: after passing Stage II	"Wake-up" period
Some communicative restitution	Prominent hand apraxia/dyspraxia
Apparently preserved ambulent ability	
Unapparent, slow neuromotor regression	
Duration: years to decades	
Stage IV: late motor deterioration	
Onset age: when Stage III ambulation ceases	Subgrouping introduced:
	Stage IV A: previous walkers, now nonambulent
Complete wheelchair dependency	Stage IV B: never ambulent
Severe disability: wasting and distal distortion	
Duration: decades	

From: Hagberg B. Clinical manifestations and stages of Rett syndrome. *Ment Retard Dev Disabil Res Rev* 2002, 8: 61–65 [17]. Reprinted by permission.

seizures, and notably impaired social interaction. The latter led to the unique inclusion of RTT in the DSM-IV-TR [1] as the only etiologically defined condition among ASD. Additional links between RTT and idiopathic ASD, to be discussed below, include the identification of individuals with RTT-like *MECP2* mutations and other neurological phenotypes such as nonspecific intellectual disability, Angelman syndrome, and ASD [17, 19]. Consistent with the widespread distribution of *Mecp2*, non-CNS features of RTT include disturbed gastrointestinal motility and abnormal autonomic vascular regulation, all apparently related to peripheral neuronal dysfunction. Table 4.3 depicts the most salient features of RTT.

Clinical and Behavioral Features of Genetic Disorders Associated With Autism: Implications for Idiopathic Autism or Autism Spectrum Disorders

With the exception of FXS, the characterization of autistic features in genetic disorders associated with intellectual disability is at an early stage. Despite this, available data suggest that each major genetic disorder presenting with ASD has a distinctive profile. Interestingly, these profiles highlight specific aspects of the autistic disorder in such a way that their study, as a whole, provides complementary insights into most key features of idiopathic ASD.

Autism in Fragile X syndrome: Deficit in Complex Social Interaction and Adaptive Socialization, with Severe Social Withdrawal but no Regression

Of all genetic disorders associated with ASD, the best characterized is FXS [2]. Despite initial controversy [20, 21], it is now accepted that a relatively large proportion (~20–45%) of boys with FXS meet DSM-IV criteria for a nonregressive type of ASD [4, 22, 23]. Concern about the validity of ASD diagnosis in FXS originates from the fact that most males with FXS display some autistic features (e.g., gaze avoidance, hand flapping) [4]. Nonetheless, we [24, 25] and others [26, 27] have demonstrated that it is possible to identify a group of boys with FXS who exhibit a core social interaction impairment in accordance with the DSM-IV definition of ASD. Indeed, these individuals show a neurobehavioral profile similar to that of their counterparts with idiopathic ASD [28, 29, 30]: severe social indifference [31]; a spectrum of social interaction deficits [23, 25] that is relatively independent of cognitive function [25, 27]; greater receptive than expressive language delay [27, 31, 32]; persistence of gaze avoidance during continuous social challenge [32]; and a fairly stable diagnosis over time [33, 31, 27]. Furthermore, the profile of autistic features on the

Autism Diagnostic Interview-Revised (ADI-R) [34] of boys with FXS indicates that diagnosis and severity of ASD are driven by impairment in complex socioemotional aspects and not in simple social behaviors [25]. Emphasizing the core social disturbance in males with FXS and ASD, we have shown that in statistical models including several measures of communication skills and adaptive socialization, the latter is the only significant predictor of ASD [25] (Fig. 4.1). The relevance of this central impairment in socialization skills is highlighted by the fact that adaptive socialization is considered a key reference measure for resolving diagnostic discrepancies between the two gold standard instruments in idiopathic ASD [34], the ADI-R [33] and the Autism Diagnostic Observation Schedule-Generic [36].

Extending the abovementioned findings, an examination of the role of dimensions of social behavior in diagnosis and severity of ASD in boys with FXS demonstrated that delay in adaptive socialization skills is the primary correlate and severity of social withdrawal is a close secondary factor [31]. It is important to note that among boys with FXS, the most severe ASD phenotype is linked to both impaired adaptive socialization and prominent social withdrawal [31]. An in-depth study of these two behavioral dimensions has led to an initial understanding of the relationship between ASD and social anxiety, the other major social disorder in FXS [4]. We have observed that social withdrawal behaviors, which include both avoidance and indifference, are distributed in a continuum of severity in boys with FXS [31]. As described years ago, a large proportion of boys with FXS display excessive shyness without apparent functional or clinical consequences; however, the rest show either marked avoidance or a more severe combination of severe avoidance and indifference. In line with severe avoidance, the intermediate social withdrawal phenotype is linked to the diagnosis of social anxiety, while the most affected group presents an extremely high frequency of severe ASD diagnosis. These clinical observations suggest that social anxiety and ASD have a common behavioral root in FXS, namely social withdrawal, and that the interaction between social withdrawal and impaired adaptive socialization and its cognitive

Fig. 4.1 Diagram of the relationship between skills and ASD in FXS. Note that delay in socialization skills is a selective contributor to the diagnosis and severity (measure as ADI-R/ADOS-G scores) of ASD in FXS. Abbreviations: Rec, receptive; Exp, expressive; lang, language skills; ADI-R, Autism Diagnostic Interview-Revised; ADOS-G, Autism Diagnostic Observation Schedule-Generic

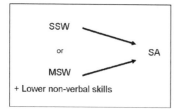 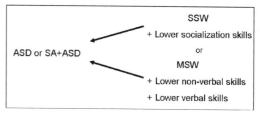

Fig. 4.2 Model of the relationships among social withdrawal, cognitive impairment, social anxiety, and ASD in FXS. Left panel: Note that either severe social withdrawal (SSW) per se or mild social withdrawal (MSW) in conjunction with lower nonverbal skills would lead to social anxiety (SA). Right panel: A more complex combination of deficits, specifically the addition of lower socialization or verbal skills, is required for ASD alone or comorbid with social anxiety. Abbreviations: SSW, severe social withdrawal; MSW, mild social withdrawal; SA, social anxiety

correlates (e.g., deficit in verbal reasoning) [31] will ultimately determine the ASD phenotype (Fig. 4.2). At the neurobiological level, we postulate that ASD in FXS has an obligatory cortical component, involving prefrontal and temporal regions, which when combined with limbic dysfunction leads to a severe ASD phenotype. Other cognitive deficits (e.g., severe nonverbal delay/parietal lobe dysfunction) would constitute variable components of ASD in FXS. In summary, the study of ASD in FXS could provide important clues about both core elements of impaired reciprocal social interaction and limbic (i.e., amygdalar) components of the social cognition system disrupted in idiopathic ASD [37, 38].

Autism in Down Syndrome: Complex and Simple Stereotypic Behaviors and High Prevalence of Regression

Depending upon the diagnostic criteria used and the method of ascertainment, the prevalence of ASD in individuals with DS is estimated to be between 5–10% [39, 40], which represents a 25-fold increase in risk for ASD compared to the general population. As with FXS in the 1990s, pediatricians and mental health providers have been reluctant to recognize or diagnose ASD in children with DS, resulting in uncertain educational placement, a missed opportunity for rational pharmacotherapy, and unnecessary hardship for parents [39, 41]. The controversy of DS + ASD has been influenced by stereotyped notions about DS, ASD, or severe cognitive impairment, as well as by the unique challenges of evaluating children with particularly low cognitive and adaptive skills and associated maladaptive behaviors. For these reasons, there has been considerably less research interest in DS + ASD when compared to other neurogenetic syndromes with severe intellectual disability. While young children with DS often have marked delay in speech production, this is well compensated for by the use of sign or gesture [42]). In contrast, individuals with DS at risk for developing ASD may display atypical behaviors during infancy or the toddler years [43]. Social indifference, lack of sustained joint attention, and disinterest in gesture or functional communication may also be noted. Other behaviors seen prior to 36 months may include stereotypies, irritability head banging or self-injury, fascination with lights or ceiling fans, episodic deviation of eye gaze, extreme food refusal, and unusual stereotyped play with toys or other objects. Associated auditory processing impairments may cause the child to act as if deaf. Children with DS and a history of infantile spasms or myoclonic seizures are at particularly high risk for developing ASD [44, 45].

Formal analyses of behavioral profiles of children with DS and ASD have confirmed clinical impressions that stereotypic behaviors are prominent [46, 47] (Table 4.4). These stereotypies include both simple motor and more complex behaviors [47]. Considering that stereotypic movement disorder (SMD) is an important comorbidity in DS, in our studies we have compared children with DS + ASD, not only in those with typical behavior but also in children with DS + SMD [46, 47]. Children with DS + ASD typically satisfied —three to four criteria under social impairment, compared to only—one to two criteria for the DS + SMD group. Using the highly informative Aberrant Behavior Checklist (ABC), we also demonstrated that ABC's lethargy/social withdrawal behavior, specifically items representing avoidance and indifference, better differentiated the DS + ASD and DS + SMD groups [47]) (Table 4.4). Analyses with the corresponding Relating scale of the Autism Behavior Checklist further support the distinction between DS + ASD and DS + SMD [47]. Notably, atypical social behaviors displayed by children with DS + SMD or disruptive behavior disorder (DB), though reminiscent of ASD, do not significantly impair social function.

To date, only one study has specifically employed the prelinguistic ADOS-G and the ADI-R, in addition to DSM-IV criteria "gold standard" instruments, in children with DS. In individuals with DS and severe-profound intellectual disability, ASD could be identified, but discordance between the two instruments

Table 4.4 Behavioral characteristics of DS + ASD, DS + SMD, and typical DS (ABC profiles)

Subscale	ASD	SMD	Typical	ANOVA	t-tests
Irritability	13.2±9.3	7.4±6.4	4.4±4.4	$F < 0.0001$	a
Lethargy	18.1±9.4	6.6±5.5	2.5±3.6	$F < 0.0001$	a,b,d
Stereotypy	12.5±4.1	7.2±3.1	0.5±1.5	$F < 0.0001$	a,b,c
Hyperactivity	20.8±10	15.4±7.1	8.5±8.7	$F < 0.0001$	a,e
Inappropriate speech	2.5±3.0	2.3±2.6	1.0±1.7	$F = 0.01$	ns

ASD versus Typical: (a) $P < 0.0001$; ASD versus SMD: (b) $P < 0.0001$; SMD versus Typical: (c) $P < 0.0001$, (d) $P < 0.0001$, (e) $P = 0.0002$. Post-hoc pairwise t-test statistically significant <0.003 after correcting for multiple comparisons using the Bonferroni procedure.
From: Capone GT, Grados M, Kaufmann WE, Bernad-Ripoll S, Jewell A. Down syndrome and comorbid autism-specturm disorder: characterization using the Aberrant Behavior Checklist. *Am J Med Genet* 134A: 373–380 [46]. Reprinted by permission.

raised both methodological and conceptual issues [48]. This investigation empha-sized that although lower cognitive performance is an important correlate of ASD (i.e., lower than typical DS or DS associated with SMD [46]), it is not an obligatory component, as we have recently verified in our own data. Nonetheless, we acknowledge that the relationship between profound cognitive impairment ($IQ < 25$), maladaptive behavior, ASD risk and severity remains controversial and additional studies on the subject are needed [49, 50, 51]. It is plausible that these complete neuro behavioral clusters reflect "overlapping-yet-distinct func-tional outcomes" resulting from underlying neurobiological impairment deter-mined by the approximately 350 genes mapping to chromosome 21. Thus, the "severe developmental delay" explanation becomes proxy for a "genetically neurobiologically mediated impairment in brain organization and function," which results in severe intellectual disability with variable expression of ASD.

In terms of temporal evolution, a large proportion of children with DS symptoms of ASD have a slow and insidious onset, progressing over many months or years (G.T. Capone, personal observation). However, in approxi-mately one-third of our cohort with DS + ASD, there is a history given of deterioration in cognitive-speech-language-social skills without motor dete-rioration, clinical seizures, or prior atypical (for DS) development (unpublished data). Many display loss of skills between 3 and 6 years when stereotypy, irritability, sensory aversions and maladaptive behaviors may appear or inten-sify (personal observation), leading to the diagnosis of late-onset autism or childhood disintegrative disorder (CDD) [46]. Interestingly, in terms of the behavioral profiles described below, there are no differences between children with DS and typical autism and those with CDD [47]) (Table 4.5). Clearly, descriptive studies and methods of investigation employing well-defined diag-nostic criteria are needed to better understand the phenomenon of regression in DS + ASD.

We conclude that ASD in DS is a good model for understanding the behavioral and neurobiological relationship between stereotypic behavior and

Table 4.5 Aberrant behavior checklist in down syndrome and autism spectrum disorders by DSM-IV Type

Subscale	Autism = 38	PDD = 8	CDD = 12	ANOVA	t-Tests
Irritability	14.2 ± 10.1	8.5 ± 6.7	13.3 ± 7.7	F = 0.30	ns
Lethargy	18.6 ± 9.4	8.6 ± 5.8	22.8 ± 7.4	F = 0.002	a,b
Stereotypy	12.3 ± 4.2	11.2 ± 4.0	14.7 ± 3.4	F = 0.12	ns
Hyperactivity	20.7 ± 10.8	20.8 ± 10.0	21.3 ± 8.0	F = 0.98	ns
Inappropriate speech	2.0 ± 2.4	4.4 ± 5.0	2.4 ± 2.9	F = 0.12	ns

PDD versus CDD: (a) $P < 0.001$; Autism versus PDD: (b) $P = 0.005$.
Post-hoc pairwise *t-test* statistical significance < 0.003 after correcting for multiple compar-isons using the Bonferroni procedure.
From: Capone GT, Grados M, Kaufmann WE, Bernad-Ripoll S, Jewell A. Down syndrome and comorbid autism-spectrum disorder: characterization using the Aberrant Behavior Checklist. *Am J Med Genet* 134A: 373–380 [46]. Reprinted by permission.

social reciprocity in idiopathic ASD, as well as for studying early and late regression phenomena. Finally, DS + ASD will provide unique insights into core social interaction impairments in individuals with severe cognitive impairment.

Autism in Rett Syndrome: Preserved Communication and Motor Function With Regression

The study of autistic features in RTT has focused on two different aspects: the differential diagnosis from idiopathic ASD in young females and the RTT + ASD comorbidity in higher functioning RTT patients. The issue of differentiating RTT from idiopathic ASD has been central to this genetic disorder since its initial descriptions emphasized the combination of "autism, dementia, ataxia, and loss of purposeful hand use" as key features of RTT [52]. Subsequent refinements of the RTT phenotype highlighted the diversity and severity of the motor and communication deficits, which contrasted with the relative preservation of cognitive and motor function in idiopathic ASD [53, 54, 55]. Another important distinction is that autistic features in RTT follow the dynamic course of the disorder (see Table 4.3 for clinical stages in RTT), in contrast to the relative stability of idiopathic ASD. Recognition of autistic manifestations in RTT typically coincides with the regressive phase (Stage II), between 1 and 4 years, in the most common "classic" form of the disorder [55, 56] that also includes loss of language and fine motor skills and is followed by clinical improvement to virtual disappearance of social interaction deficits by late childhood (see Table 4.3) [17]. In spite of the clinical differences between RTT and idiopathic ASD, it is clear that autistic symptomatology is more prevalent in RTT than in comparable samples of females with idiopathic severe intellectual disability [57] and that many young girls with RTT meet DSM-IV criteria for ASD [58]. Our recent large-scale study of 313 RTT patients concluded that even after several years following the identification of *MECP2* as the "RTT gene," a significant proportion of patients is diagnosed as having (idiopathic) ASD in early life (i.e., ~18%). A profile emerged for the girls with RTT and misdiagnosis; they tended to have a milder phenotype, particularly in terms of motor function, with relatively late appearance of typical RTT features (Table 4.6; [59]). Interestingly, R306C, a mutation typically associated with a milder RTT phenotype, and T158M, a mutation linked to a wide range of phenotypical outcomes [60], were overrepresented in girls with RTT and ASD misdiagnosis [61].

A more controversial issue is the possibility of RTT + ASD comorbidity beyond the regressive phase of the disorder. The strongest evidence comes from the study of a milder (atypical) RTT phenotype, the so-called preserved speech variant (PSV), which, similar to other milder forms of RTT, is commonly associated with the R133C mutation [61, 62]. Although girls with PSV have better communication skills, they also tend to have more severe autistic

Table 4.6 Early Clinical features in RTT patients with and without initial diagnosis of autism.

Early clinical symptoms	No autism diagnosis mean score	Autism diagnosis mean score	P value for autism diagnosis	Odds ratio (OR) for autism diagnosis	95% confidence interval (CI) LCI	UCI
Age at diagnosis of Rett syndrome in years	4.72	6.08	0.09	1.05	0.99	1.10
Age at loss of hand function in months	23.25	29.74	0.02	1.02	1.00	1.04
Age at loss of communication in months	20.25	25.69	0.04	1.02	1.00	1.04
Age at onset of hand stereotypies in months	26.02	31.83	0.05	1.02	1.00	1.03

From: Young DJ, Bebbington A, Anderson A, Ravine D, Ellaway C, Kulkarni A, de Klerk N, Kaufmann WE, Leonard H. The diagnosis of autism in a female: could it be Rett syndrome? *Eur J Pediatr* 2007: in press [59]. Reprinted by permission.

features; nonetheless, the autistic symptoms appear to regress by early adolescence [62]. These observations, in conjunction with several surveys demonstrating a low frequency of *MECP2* mutations among individuals with ASD [63, 64, 65, 66], lead to the conclusion that disruptions of the *MECP2* gene are a rare cause of a stable autistic phenotype. Overall, the study of the relationship between RTT and ASD is informative in that it emphasizes that the expression of autistic symptoms requires minimally preserved motor and communication systems (i.e., uncommon misdiagnosis in girls with severe RTT phenotype), a critical point for the diagnosis of ASD in individuals with severe intellectual disability. The close association of the emergence of autistic features and the loss of cognitive and fine motor skills in RTT indicates that this genetic disorder may be a good model of regression in ASD. It is interesting to notice that regardless of the specific phenotype (e.g., RTT, Angelman syndrome-like), individuals with *MECP2* mutations almost invariably present with loss of developmental skills [19]. Finally, although *MECP2* mutations per se may not be a common etiology of ASD, the association of RTT and ASD diagnosis signifies the potential role of imbalances in Mecp2 (*MECP2* product) in the pathogenesis of idiopathic autism.

Insights into the Neurobiology of Idiopathic Autism

Despite the detailed analyses of autistic features in FXS, DS, and other disorders with relatively well-understood neuroanatomy, the neurobiological correlates of ASD in these conditions have only been occasionally explored. This is due to a number of difficulties inherent to this type of research. First, a major source of information on the neurobiology of genetic intellectual disability is the study of mouse models that reproduce only to a limited extent the behavioral features, including autistic manifestations, of the disorder. Other experimental strategies, such as in vitro models, could disclose relevant data only if the cellular model is already linked to a genetic correlate of ASD (e.g., R133C *MECP2* mutation in some girls with RTT and autism). For all these reasons, most of the neurobiology of genetic intellectual disability and ASD has to rely on neuroimaging and neurophysiological investigations of affected subjects.

Neurobiology of Fragile X Syndrome and Autism Spectrum Disorder: Anomalies of the Cerebellar Vermis and Limbic Dysfunction

To our knowledge, only one study has directly examined affected neural systems in individuals with FXS who meet DSM-IV criteria for ASD. As shown in Fig. 4.3, we found that boys with FXS and autism proper (as opposed to milder forms of ASD) have, on MRI scans, relatively larger posterior–superior cerebellar vermi than their counterparts without autism, although they are smaller than in typically developing controls [67]. Interestingly, the abnormal region (i.e., lobules VI–VII) is the same one previously described as relatively smaller in individuals with idiopathic ASD [68], a finding confirmed in our study [67]. Despite these observations about the cerebellum, the profile of ASD in FXS, as in idiopathic ASD, suggests a major disturbance in limbic and adjacent temporal regions [38]. Although mild MRI volumetric increases in the hippocampus have been reported in subjects with FXS [69], so far no study has evaluated the relationship between these morphometric changes and ASD status [70]. Moreover, cortisol reactivity (i.e., variability in cortisol levels) to a social challenge, a measure of limbic–hypothalamic function, has been found to

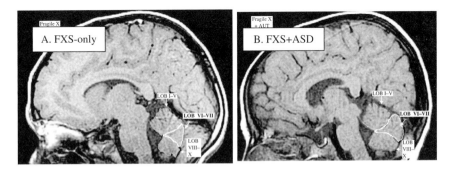

Fig. 4.3 Cerebellar vermis abnormality in FXS and ASD. Representative midsagittal MRIs of subjects with FXS with (FXS + ASD) and without (FXS-only) autism. The posterior–superior cerebellar vermis (lobules VI–VII) is delineated in white. Note in B the larger and protruding outline of this vermian region

be decreased in children with FXS and severe autistic behavior [71]. The opposite seems to be true for FXS subjects with prominent social avoidance [71, 72]. Although intuitively, decreased hormonal and behavioral responses to social stimuli are compatible with ASD, the precise mechanism of these cortisol changes in FXS is unknown. We have reported a higher frequency of acetylation of the glucocorticoid-negative regulator annexin-1 in males with FXS [73], particularly in those with severe social withdrawal [74]. However, annexin-1 is involved in the acute phase of cortisol modulation [75], and it is therefore unclear whether it has any role in the reported slow return to baseline observed in boys after a cognitive/social challenge [72]. Another candidate for abnormal cortisol regulation in FXS is the glucocorticoid receptor alpha; the synthesis of this low-affinity cortisol receptor is directly regulated by the deficient Fmrp [76], and its levels are decreased in dendrites of hippocampal neurons in a mouse model of FXS [77]. These animals also show increased cortisol levels after a stressful situation [78]; however, no behaviors of relevance to ASD appear to correlate with these cortisol anomalies.

Two general neuronal abnormalities have been described in FXS and/or corresponding mouse models: aberrant configuration of dendritic spines (i.e., long, tortuous, immature appearance; [79]) and enhanced activity of class I metabotropic glutamate receptors leading to increased long-term depression [80]. The latter is linked to the postulated negative regulatory role of Fmrp in protein synthesis triggered by metabotropic glutamate receptor activation [79, 80]. Although these anomalies involve brain regions implicated in idiopathic ASD, their ubiquitous nature and lack of formal comparisons involving individuals with FXS and ASD preclude the establishment of meaningful relationships. Nonetheless, the fact that one study showed reduced class I metabotropic glutamate receptor-dependent long-term potentiation in the lateral amygdala [81] (an area linked to both anxiety and ASD [38, 82]) in mice deficient in Fmrp, and that class I metabotropic glutamate receptor antagonists will be available for clinical use relatively soon [83] has increased the interest in metabotropic glutamate receptors as potential targets in ASD. In conclusion, the best-characterized neurobiological correlate of ASD in FXS is a relative enlargement of the posterior–superior vermis. However, data on FXS subjects and experimental models suggest that several limbic regions may also be functionally abnormal. Considering the close relationship between social anxiety and ASD in FXS, careful work will be required for differentiating limbic anomalies linked to either disorder.

Neurobiology of Down Syndrome and Autism Spectrum Disorders: Cerebellar Enlargement and Stereotypies

The neuroanatomical features of DS include microcephaly and decreased brain size, selective volumetric reduction of the frontal lobe, hippocampus and cerebellum, immature gyral patterns, abnormal neocortical lamination, and delayed cortical fiber myelination [13, 67, 70, 79], many of them confirmed

Table 4.7 Bilateral increase of WM volumes in the brainstem and cerebellum in DS + ASD

	DS only ($N = 15$)	DS + ASD ($N = 15$)
Brain	1022.7 ± 106.9	1041.6 ± 157.4
GM	641.8 ± 62.2	661.4 ± 105.8
WM	380.9 ± 55.0	380.2 ± 56.5
Brainstem	31.7 ± 4.0	32.1 ± 4.9
GM	19.6 ± 3.3	18.4 ± 4.3
WM	12.1 ± 2.0	13.7 ± 1.6*
Cerebellum	84.6 ± 9.5	88.5 ± 12.0
GM	67.1 ± 8.1	66.6 ± 11.9
WM	17.5 ± 2.4	21.8 ± 3.0*

*$p < 0.05$

by MRI morphometric studies [67, 70, 84, 85, 86, 87, 88]. However, little is known about the neuroanatomy of DS and comorbid ASD. In a recent study [89], we examined regional brain volume changes in children with DS and typical behavior (DS-only), DS + ASD, and controls. We found that when compared to the DS-only group, children with DS + ASD show significant bilateral increases in WM volumes in the brainstem and cerebellum (Table 4.7). Although still smaller than in age-matched controls, the relatively larger cerebellar volumes in DS + ASD correlated positively and selectively with severity of stereotypies (specifically with item number 11 on the ABC, "*stereotyped, repetitive movements*"). In addition, initial assessments of brain growth show that individuals with DS + ASD display an accelerated pattern between the ages of 2 and 5 years, when compared with controls. The preferential enlargement of cerebellar WM and the pattern of brain growth in subjects with DS + ASD mimic those observed in children with idiopathic ASD [90]. These results indicate that ASD in DS follows a similar pathogenetic course to idiopathic ASD, specifically pointing to cerebellar WM hyperplasia as a trademark of ASD against different genetic backgrounds [89].

Mice with segmental trisomy (Ts65Dn) have a dosage imbalance for genes corresponding to those on human chromosome 21q21–22.3 [91]. Although there are no reports of behavioral abnormalities resembling ASD in these mice, they exhibit cognitive deficits found in idiopathic ASD and postulated in DS + ASD (i.e., working memory and long-term memory; [92]). Additional characterizations of the Ts65Dn mice, including the synaptic bases of their neurobehavioral abnormalities [93,94], may contribute to better understanding and clinical approaches to ASD in DS [95].

MeCP2Deficiency and the Neurobiology of Autism Spectrum Disorders

There is no direct information on the neurobiology of RTT and autistic features. Nevertheless, postmortem and neuroimaging (magnetic resonance spectroscopy,

MRS) data relevant to the regressive stage of RTT (see Table 4.3), when autistic features are noticed, indicate that the basic process is glutamate-dependent toxicity. Both increases in glutamate concentration [96] and in the density of NMDA receptors [97, 98] in the cerebral cortex provide a solid basis for an excitotoxic phenomenon [99]. We recently showed increased glutamate concentrations in the frontal white matter [100], an area where reductions in axonal markers and elevations in astrocytic markers have also been described [101, 102]. These data suggest that the period of regression in RTT, with its associated emergence of autistic features, could also be linked to white matter disturbances, which have been recently implicated in idiopathic ASD [103]. Although to date no study has characterized patients with RTT and autistic features by neuroimaging or other methods, our recent work indicate that girls with RTT and *MECP2* mutations associated with either ASD misdiagnosis or comorbidity tend to have relatively greater preservation of anterior frontal cortex volumes [104].

Behavioral characterizations of mice deficient in Mecp2 indicate that features of relevance to ASD, especially anxiety, correlate with severity of Mecp2 (*MECP2* product) deficit and with volumetric reductions of the amygdala and hippocampus [105]. The same group reported that choline supplementation to nursing dams attenuated motor function but not fear conditioning (i.e., anxiety) in these Mecp2-deficient mice [106], suggesting that abnormal social behavior associated with Mecp2 deficit may have unique underlying mechanisms. Another study in mice carrying an RTT-like *MECP2* mutation demonstrated increased anxiety-like behavior and elevated serum glucocorticoid levels, which were associated with increased limbic expression of corticotropin-releasing hormone [107]. The same animals displayed deficits in contextual fear and social memories and different paradigms of social interaction [108, 109]. Altogether, the study of the neurobiology of *MECP2* mutations and RTT suggests that Mecp2 deficit leads to autistic features in the context of an excitotoxic process that could involve several forebrain structures. As in FXS, *MECP2* mutations could be associated with both anxiety- and ASD-like behaviors and disturbances in the hypothalamic–pituitary–adrenocortical (HPA) axis.

Genetic and Molecular Pathways Common to Autistic Disorders

This is probably the area in which most significant advances have been made toward understanding the bases of ASD. As expected of a heterogeneous condition, multiple molecular pathways have been implicated in the genesis of autistic manifestations. Despite these accomplishments, major methodological obstacles still remain. Although analyses of samples and cell lines of affected patients can lead to straightforward results, the links between in vitro measures in peripheral cells and ASD neurobiology are tenuous and, at present, require a combination of experimental approaches and/or analyses of postmortem brain samples.

Molecular Basis of Fragile X Syndrome and Autism Spectrum Disorders: Cytoplasmic FMR1-Interacting Protein 1 and Other Downstream Fmrp Targets

Since 1991, it has been known that mutations in *FMR1* are the cause of the vast majority of FXS cases and that the phenotypic manifestations of FXS are the result of a marked reduction in the levels of *FMR1*'s product, the Fragile X mental retardation protein (Fmrp) [10]. In contrast with consistent reports on correlations between magnitude of Fmrp decrease and severity of physical and cognitive phenotype [4, 110], lymphocytic Fmrp levels do not seem to predict behavioral abnormalities in FXS [111]. Only recently, large-scale studies have demonstrated a modest relationship between Fmrp deficit and severity of autistic behavior [112, 113]. These findings are not surprising considering that Fmrp is an RNA-binding protein that regulates the synthesis, particularly at synaptic sites, of a relatively large number of proteins (5–8% total mRNA; [10, 113, 114, 115]). Therefore, specific neurobehavioral features in FXS are more likely to depend on a relatively greater involvement of certain Fmrp targets and neuronal circuits that are not reflected in general measures of Fmrp. A recent publication by Nishimura and colleagues [116] examined gene expression profiles in lymphoblasts from boys with FXS and ASD, comparing them with typically developing controls and boys with duplication of chromosome 15 and ASD (dup15q; a recognized genetic abnormality associated with ASD; [2]). Of 120 differentially expressed genes, including 15 previously identified in neuronal [77]) and "phenotypically generic" lymphoblast [76] FXS/Fmrp-deficient samples, 68 were also dysregulated in the dup15q group (Table 4.8). Among them there was G protein-coupled receptor 155 (*GPR155*), a gene regulated by the cytoplasmic *FMR1*-interacting protein 1 (*CYFIP1*), an antagonist and binding partner of Fmrp that is a member of the Rac GTPase system involved in neurite development [79, 117]. Since *CYFIP1* and another one of its targets, [the janus kinase and microtubule-interacting protein 1 (*JAKMIP1* or *MARLIN-1*)], were also dysregulated in patients with dup15q; Jakmip1 was reduced in brains of *FMR1* knockout mice; and *JAKMIP1* and *GPR155* were differentially expressed in male sib pairs discordant for idiopathic ASD; it can be concluded that the *CYFP1* signaling pathway is implicated in different genetic forms of ASD. Although the abovementioned study [116] did not formally compare subjects with FXS with and without ASD, the comprehensive and comparative nature of the assays suggests that the study of peripheral cells from individuals with FXS and ASD may be highly informative for understanding mechanisms underlying idiopathic ASD. Additional analyses

Table 4.8 Genes dysregulated in lymphoblasts from patients with FXS and ASD

Gene name	Gene abbreviation	Levels
Nuclear receptor subfamily 3 group C member 1	NR3C1 (*)	Upregulated
Vimentin	VIM (*)	Downregulated
Iduronate 2-sulfatase	IDS (**)	Upregulated
Hairy and enhancer of split 1	HES1 (**&)	Upregulated
Immunoglobulin superfamily, member 3	IGSF3 (**)	Upregulated
CDK2-associated protein 2	CDK2AP2 (**)	Downregulated
Ubiquitin specific peptidase 8	USP8 (**)	Downregulated
MAX-like protein X	MLX (**)	Downregulated
Ribosomal protein S5	RPS5 (**)	Downregulated
C-terminal binding protein 1	CTBP1 (**)	Downregulated
Spleen tyrosine kinase	SYK (**)	Downregulated
F-box protein 6	FBXO6 (**)	Downregulated
Mitogen-activated protein kinase kinase kinase 11	MAP3K11 (**)	Downregulated
Sorting nexin 15	SNX15 (**)	Downregulated
CD44 antigen	CD44 (**)	Downregulated
G protein-coupled receptor 155	GPR155 (@)	Downregulated

(*) Reported by Miyashiro et al. [77]).
(**) Reported by Brown et al. [76].
(&) Associated with attention-deficit hyperactivity disorder (Brookes et al. [118]).
(@) Also found in patients with chromosome 15 duplication and ASD (Nishimura et al.; [116])

will have to determine whether the specific neural pathways affected by this gene expression dysregulation are those implicated in ASD of unknown cause.

Molecular Basis of Down Syndrome and Autism Spectrum Disorders: Genes Involved in Early Brain Development

The characteristic cognitive–behavioral profile of DS + ASD suggests that this phenotype is distinct. Although the DS critical region of chromosome 21 includes 360 unique genes [119], little is known about the dosage of these genes and phenotypical variability in DS (congenital heart defects; [120]). However, a recent study of 34 idiopathic autism-affected relative pairs with the regressive phenotype demonstrated genetic linkage to chromosome 21q between markers DS21S1432 and DS21S1899 [121]. Of the seven known genes that map to this region, two (BTG3 and CXADR) are expressed in fetal brain and a third one maps just outside this region and is also involved in early brain development (NCAM2). Of particular interest is BTG3 (also termed APRO4/ANA) since balanced expression of this gene appears to be critical for neuronal differentiation in the forebrain [122, 123], a fundamental process linked to both intellectual disability and ASD [124]. In conclusion, although no specific molecular data are available on ASD in DS, studies of idiopathic ASD indicate that genes in chromosome 21 may play a critical role in the pathogenesis of ASD.

Mecp2 Levels are Dysregulated in Idiopathic Autism Spectrum Disorders

In preceding sections, we have discussed the relationship between RTT, *MECP2* mutations, and the diagnosis of ASD. Although mutations in the coding region of *MECP2* are infrequently associated with ASD [63, 64, 65], reduced levels of Mecp2 secondary to either mutations in regulatory regions of the gene [125, 66] or as yet unknown physiological signals [126, 127] have been reported in both male and female subjects with ASD. Of particular interest is the demonstration of abnormal Mecp2 levels, particularly reductions, associated with aberrant promoter methylation in frontal cortex of subjects with ASD [127]. Although changes in brain Mecp2 levels are not specific to ASD, since they are also observed in other developmental disorders (e.g., Angelman syndrome), they may significantly contribute to neurological disturbances in ASD. For instance, reports suggest that Mecp2 abnormalities lead to allelic dysregulation and decreased expression of GABAA receptor gene subunits and reduced expression of UBE3A [128, 129]. As stated with regard to molecular abnormalities in FXS and ASD, the link between aberrant Mecp2 expression and affected neural pathways in ASD is yet unknown.

Data on the neuronal phenotype [130] and levels of Mecp2 [131] in the brain from subjects with RTT and other developmental disorders allowed us to propose a model for phenotypes secondary to Mecp2 deficit (Fig. 4.4). If mild reductions in Mecp2 levels (due to disrupted regulatory signals) occur during the early postnatal period when synaptic activity plays a critical role in modulating neuronal and synaptic development, the outcome is most likely ASD or Angelman syndrome. If the Mecp2 deficit is severe or persistent into late postnatal life, as expected from *MECP2* mutations, the most probable outcome would be a RTT phenotype.

Concluding Remarks

We have presented published and preliminary data, as well as some hypothetical models, supporting the notion that the study of genetic disorders associated with intellectual disability and ASD, namely FXS, DS, and RTT, has important implications not only for the individuals affected by these severe comorbidities, but also for the entity termed "idiopathic ASD." The behavioral features of ASD in genetic disorders are varied, and therefore informative of several key aspects of ASD, such as core social interaction impairment, stereotypic behaviors, and developmental regression. They also represent a contribution to the evaluation of the role of severe cognitive and motor impairment in the expression of autistic features. The emerging knowledge on neuroimaging of ASD in FXS, DS, and RTT emphasizes the involvement of brain areas already implicated in idiopathic ASD, in particular the cerebellum and limbic regions. These MRI morphometric approaches may eventually identify additional

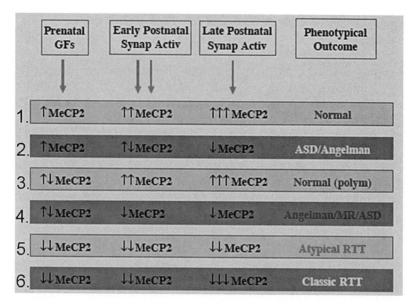

Fig. 4.4 Model of phenotypical outcomes secondary to Mecp2 deficiency.

(1) During normal development, onset of Mecp2 expression (↑) coincides with early neuronal differentiation directed by specific signals [i.e., growth factors (GFs)]. Levels of Mecp2 expression/function increase (↑↑), for most cortical and limbic regions, in early postnatal life and are strongly modulated by synaptic activity during the critical period of synaptic maturation. In the same regions, Mecp2 levels continue to increase (↑↑↑) into adulthood.

(2) If developmental synaptic activity or other factors (e.g., 15q11–13 abnormality) regulating Mecp2 expression in early postnatal life are disturbed, levels of Mecp2 could decrease and a phenotype of Angelman syndrome or ASD may develop.

(3) If Mecp2 polymorphisms or mild prenatal Mecp2 deficits (↑↓) occur, depending on the genetic compensatory capacity of the subject or X inactivation skewing, no phenotype (polymorphism [polym]) or (4) a non-RTT disorder with mild Mecp2 deficiency (↓) may arise. This situation will explain the majority of non-RTT phenotypes associated with MECP2 mutations, including Angelman syndrome, ASD, and nonsyndromic MR.

(5) If Mecp2 dysfunction takes place early and is severe (↓↓), as in most patients with pathogenic *MECP2* mutations, development of subcortical pathways will be affected. These secondary/compensatory neurotransmitter changes, in combination with insufficient Mecp2-dependent response to synaptic signals during the critical postnatal period, will perpetuate Mecp2 deficiency.

(6) If 'facilitating' factors (e.g., genetic polymorphisms, unfavorable X inactivation skewing) are also present, a more severe classic RTT phenotype with severe MeCP2 deficit (↓↓↓) will emerge. Otherwise, MeCP2 function will remain at a moderately low level (↓↓) and an atypical/variant RTT phenotype will develop.

Note that this model does not distinguish between RTT patients with or without *MECP2* mutations, since the postulates are based on Mecp2 function that could be impaired by other genes functionally associated with Mecp2. Abbreviations: GFs, growth factors; Synap Activ, synaptic activity. The intensity of the gray shading symbolizes the presence (darker) or absence (lighter) of negative or 'facilitating' factors that lead to a more severe phenotype neural circuits involved in both genetic and idiopathic ASD. It remains to be seen to what extent animal models of these genetic disorders will provide valuable data for idiopathic ASD. Nonetheless, recent progress on mouse models of autism [132] suggests that better tools will ultimately be available for the characterization of these experimental paradigms. Data on the molecular correlates of ASD in genetic disorders have recently been quite revealing. Undoubtedly, abnormalities in Mecp2 expression play a role in the development of ASD. The issue to be determined is at which level of the pathogenetic process they do so. Furthermore, gene expression analyses place Fmrp targets at the center of pathways common to several genetic forms of autism. Data are also suggestive of an important role for genes in the DS critical region of chromosome 21. The integration of all these pieces of data is a major challenge, to be better addressed when additional data become available.

The first goal in the field of ASD associated with genetic intellectual disability is, of course, to acquire more data on the aforementioned areas. However, it is also necessary to introduce new approaches. In terms of behavioral studies, experimental paradigms should complement findings derived from clinical measures. Our recent work on identifying dynamic behavioral features of ASD in FXS by the social approach scale is a good example [133]. Naturalistic observations may also be informative as revealed by our study of RTT's neurological and behavioral phenotype through video recordings by parents [134]. In terms of neuroimaging, there is the need for applying the entire spectrum of techniques. Given the close association of severe cognitive impairment and ASD in these disorders, functional MRI remains an elusive approach. Most likely, the study of gene expression profiles in lymphoid and postmortem samples from affected individuals will continue providing promising leads. The challenge here is the integration of molecular and neurobiological data; the comprehensive evaluation of *CYFIP1* and its targets in ASD associated with FXS and chromosome 15 duplication by Nishimura and colleagues [116] illustrates that such work is feasible.

The present review was based on our work; consequently, we did not intend for it to be comprehensive. The study of other genetic disorders associated with ASD (e.g., VCFS, tuberous sclerosis) could also be extremely valuable. The increasing number of genetic abnormalities reported in individuals with "idiopathic" ASD [135] will most likely change the view of the relationship between genetic disorders and autism and, perhaps also, the criteria for diagnosing ASD. Regardless, in our opinion, our sketchy knowledge on the molecular, neurobiological, and behavioral correlates of ASD in FXS, DS, and RTT already demonstrates that these disorders are valuable models for autism research.

Acknowledgments This work was partially supported by NIH grants HD33175, MH67092, HD24448, HD24061, RR00052, and by awards from FRAXA and IRSA to W.E.K. We are grateful to the families who have participated in our research projects at the Center for Genetic Disorders of Cognition & Behavior, Kennedy Krieger Institute, Baltimore, Maryland.

References

1. APA (American Psychiatric Association) *Diagnostic and Statistical Manual of Mental Disorders*, *4th Edition*, *4th Text Rev* Washington D.C: American Psychiatric Association Press 2000.
2. Cohen D, Pichard N, Tordjman S, Baumann C, Burglen L, Excoffier E, Lazar G, Mazet P, Pinquier C, Verloes A, Heron D. Specific genetic disorders and autism: clinical contribution towards their identification. *J Autism Dev Disord* 2005; 35: 103–116.
3. Feinstein C, Singh S. Social phenotypes in neurogenetic syndromes. *Child Adolesc Psychiatr Clin N Am* 2007; 16: 631–647.
4. Hagerman RJ. The Physical and Behavioral Phenotype. In: Hagerman RJ and Hagerman PJ (Eds) *Fragile X Syndrome: Diagnosis, Treatment, and Research*, 3rd ed., Baltimore, MD: Johns Hopkins University Press, 2002: 3–110.
5. Capone GT, Goyal P, Ares W, Lannigan E. Neurobehavioral disorders in children, adolescents, and young adults with down syndrome. *Am J Med Genet* 2006; 142C: 158–172.
6. Antshel KM, Aneja A, Strunge L, Peebles J, Fremont WP, Stallone K, Abdulsabur N, Higgins AM, Shprintzen RJ, Kates WR. Autistic spectrum disorders in velo-cardio facial syndrome (22q11.2 Deletion). *J Autism Dev Disord* 2006; in press 37(9): 1776–1786. Epub 2006 Dec 19.
7. Spence SJ, Cantor RM, Chung L, Kim S, Geschwind DH, Alarcon M. Stratification based on language-related endophenotypes in autism: attempt to replicate reported linkage. *Am J Med Genet* 2006; 141B: 591–598.
8. Viding E, Blakemore SJ. Endophenotype approach to developmental psychopathology: implications for autism research. *Behav Genet* 2007; 37: 51–60.
9. Sherman SL, Marsteller F, Abramowitz AJ, Scott E, Leslie M, Bregman J. Cognitive and behavioral performance among Fmr1 high-repeat allele carriers surveyed from special education classes. *Am J Med Genet* 2002; 114: 458–465.
10. Kaufmann WE, Reiss AL. Molecular and cellular genetics of Fragile X syndrome. *Am J Med Genet* 1999; 88: 11–24.
11. Hagerman PJ, Hagerman RJ. The Fragile-X premutation: a maturing perspective. *Am J Hum Genet* 2004; 74: 805–16.
12. CDC. Improved National Prevalence Estimates for 18 Selected Major Birth Defects – United States, 1999–2001. MMWR 2006; 54: 51–52.
13. Roizen NJ, Patterson D. Down's syndrome. Lancet 2003; 361: 1281–1289.
14. Evans DW, Gray FL. Compulsive-like behavior in individuals with down syndrome: its relation to mental age level, adaptive and maladaptive behavior. *Child Dev* 2000; 71: 288–300.
15. Capone GT, Roizen NJ, Rogers PT. Down Syndrome. In: Accardo PJ, and Johnston MV (Eds). *Developmental Disabilities in Infancy and Childhood*. Baltimore: Paul H. Brookes Publishing Co, 2007: 285–308.
16. Kankirawatana P, Leonard H, Ellaway C, Scurlock J, Mansour A, Makris CM, Dure LS 4th, Friez M, Lane J, Kiraly-Borri C, Fabian V, Davis M, Jackson J, Christodoulou J, Kaufmann WE, Ravine D, Percy AK. Early progressive encephalopathy in boys and Mecp2 mutations. *Neurology* 2006; 67: 164–166.
17. Hagberg B. Clinical manifestations and stages of Rett syndrome. *Ment Retard Dev Disabil Res Rev* 2002; 8: 61–65.
18. Akbarian S. The neurobiology of Rett syndrome. *Neuroscientist* 2003; 9: 57–63.
19. Kaufmann WE, Johnston MV, Blue ME. Mecp2 expression and function during brain development: implications for Rett syndrome's pathogenesis and clinical evolution. *Brain Dev* 2005; 27: S77–S87.
20. Wolff PH, Gardner J, Paccla J, Lappen J. The greeting behavior of Fragile X males. *Am J Ment Retard* 1989; 93: 406–411.

21. Cohen IL, Sudhalter V, Pfadt A, Jenkins EC, Brown WT, Vietze PM. Why are autism and the Fragile X syndrome associated? Conceptual and methodological issues. *Am J Hum Genet* 1991; 48: 195–202.

22. Baumgardner TL, Reiss AL, Freund LS, Abrams MT. Specification of the neuro-behavioral phenotype in males with Fragile X syndrome. *Pediatrics* 1995; 95: 744–752.

23. Bailey DB Jr, Hatton DD, Skinner M, Mesibov G. Autistic behavior, FMR1 protein, and developmental trajectories in young males with Fragile X syndrome. *J Autism Dev Disord* 2001; 31: 165–174.

24. Kau AS, Tierney E, Bukelis I, Stump MH, Kates WR, Trescher WH, Kaufmann WE. Social behavior profile in young males with Fragile X syndrome: characteristics and specificity. *Am J Med Genet* 2004; 126A: 9–17.

25. Kaufmann WE, Cortell R, Kau ASM, Bukelis I, Tierney E, Gray RM, Cox C, Capone GT, Standard P. Autism spectrum disorder in X syndrome: communication, social interaction, and specific behaviors. *Am J Med Genet* 2004; 129A: 225–234.

26. Rogers SJ, Wehner DE, Hagerman R. The behavioral phenotype in Fragile X: symptoms of autism in very young children with Fragile X syndrome, idiopathic autism and other developmental disorders. *J Dev Behav Pediatr* 2001; 22: 409–417.

27. Lewis P, Abbeduto L, Murphy M, Richmond E, Giles N, Bruno L, Schroeder S. Cognitive, language and social-cognitive skills of individuals with Fragile X syndrome with and without autism. J Intellect Disabil Res 2006; 50: 532–545.

28. Eaves LC, Ho HH, Eaves DM. Subtypes of autism by cluster analysis. *J Autism Dev Discord* 1994; 24: 3–22.

29. Spiker D, Lotspeich LJ, Dimiceli S, Myers RM, Risch N. Behavioral phenotypic variation in autism multiplex families: evidence for a continuous severity gradient. *Am J Med Genet* 2002; 114: 129–136.

30. Losh M, Piven J. Social-cognition and the broad autism phenotype: identifying genetically meaningful phenotypes. J Child Psychol Psychiatry 2007; 48: 105–112.

31. Budimirovic DB, Bukelis I, Cox C, Gray RM, Tierney E, Kaufmann WE. Autism spectrum disorder in Fragile X syndrome: differential contribution of adaptive socialization and social withdrawal. *Am J Med Genet* 2006; 140A: 1814–1826.

32. Philofsky A, Hepburn SL, Hayes A, Hagerman R, Rogers SJ. Linguistic and cognitive functioning and autism symptoms in young children with Fragile X syndrome. *Am J Ment Retard* 2004; 109: 208–218.

33. Sabaratnam M, Murthy NV, Wijeratne A, Buckingham A, Payne S. Autistic-like behaviour profile and psychiatric morbidity in Fragile X syndrome: a prospective ten-year follow-up study. *Eur Child Adolesc Psychiatry* 2003; 12: 172–177.

34. Lord C, Rutter M, Le Couteur A. Autism diagnostic interview–revised: a revised version of a diagnostic interview for caregivers of individuals with possible pervasive developmental disorders. *J Autism Dev Disord* 1994; 24: 659–685.

35. Tomanik SS, Pearson DA, Loveland KA, Lane DM, Bryant Shaw J. Improving the reliability of autism diagnoses: examining the utility of adaptive behavior. *J Autism Dev Disord* 2007; 37: 921–928.

36. Lord C, Rutter M, DiLavore PC. *Autism Diagnostic Observation Schedule-Generic.* Chicago, IL: University of Chicago 1998.

37. Eigsti IM, Shapiro T. A systems neuroscience approach to autism: biological cognitive, and clinical perspectives. *Ment Retard Dev Disabil Res Rev* 2003; 9: 205–215.

38. Pelphrey K, Adolphs R, Moris JP. Neuroanatomical substrates of social cognition dysfunction in autism. *Ment Retard Dev Disabil Res Rev* 2004; 10: 259–271.

39. Ghaziuddin M, Tsai LY, Ghaziuddin N. Autism in down's syndrome: presentation and diagnosis. *J Intellect Disabil Res* 1992; 36: 449–456.

40. Kent L, Evans J, Paul M, Sharp M. Comorbidity of autistic spectrum disorders in children with down syndrome. *Dev Med Child Neurol* 1999; 41: 153–158.

41. Howlin PL, Wing L, Gould J. The recognition of autism in children with down syndrome: implications for intervention and some speculations about pathology. *Dev Med Child Neurol* 1995; 37: 398–414.
42. Chapman RS, Hesketh LJ. Behavioral phenotype of individuals with down syndrome. *Ment Retard Dev Disabil Res Rev* 2000; 6: 84–95.
43. Capone GT. Down syndrome and autistic spectrum disorders. In: W. Cohen and M. Madnick (Eds). Down Syndrome: Visions of the 21st century. New York: John Wiley & Sons Inc, 2002: 327–338.
44. Eisermann MM, DeLaRaillere A, Dellatolas G, Tozzi E, Nabbout R, Dulac O, Chiron C. Infantile spasms in down syndrome: effects of delayed anticonvulsive treatment. *Epilepsy Res* 2003; 55: 21–27.
45. Goldberg-Stern H, Strawsburg RH, Patterson B, Hickey F, Bare M, Gadoth N, Degrauw TJ. Seizure frequency and characteristics in children with down syndrome. *Brain Dev* 2001; 23: 275–78.
46. Capone GT, Grados M, Kaufmann WE, Bernad-Ripoll S, Jewell A. Down syndrome and co-morbid autism spectrum disorder: characterization using the aberrant behavior checklist. *Am J Med Genet* 2005; 134A: 373–380.
47. Carter JC, Capone GT, Gray RM, Cox CC, Kaufmann WE. Autistic-spectrum disorders in down syndrome: further delineation and distinction from other behavioral abnormalities. *Am J Med Genet B* 2007; 144: 87–94.
48. Starr EM, Berument SK, Pickles A, Tomlins M, Bailey A, Papanikolaou K, Rutter M. Brief report: autism in individuals with down syndrome. *J Autism Dev Disord* 2005; 5: 66–73.
49. Vig S, Jedrysek E. Autistic features in young children with significant cognitive impairment: autism or mental retardation. *J Autism Dev Disord* 1999; 3: 235–248.
50. Wing L, Gould J. Severe impairments of social interaction and associated abnormalities in children: epidemiology and classification. *J Autism Dev Disord* 1979; 1: 11–29.
51. Waterhouse L, Morriss R, Allen D, Dunn M, Fein D, Feinstein C, Rappin I, Wing L. Diagnosis and classification in autism. *J Autism Dev Disord* 1996; 1: 59–86.
52. Hagberg B, Aicardi J, Dias K, Ramos O. A progressive syndrome of autism, dementia, ataxia, and loss of purposeful hand use in girls: Rett's syndrome: report of 35 cases. *Ann Neurol* 1983; 14: 471–479.
53. Olsson B, Rett A. Behavioral observations concerning differential diagnosis between the Rett syndrome and autism. *Brain Dev* 1985; 7: 281–289.
54. Olsson B, Rett A. Autism and Rett syndrome: behavioural investigations and differential diagnosis. *Dev Med Child Neurol* 1987; 29: 429–441.
55. Percy AK, Zoghbi HY, Lewis KR, Jankovic J. Rett syndrome: qualitative and quantitative differentiation from autism. *J Child Neurol* 1988; 3: S65–S67.
56. Trevathan E, Naidu S. The clinical recognition and differential diagnosis of Rett syndrome. *J Child Neurol* 1988; 3: S6–S16.
57. Mount RH, Charman T, Hastings RP, Reilly S, Cass H. Features of autism in Rett syndrome and severe mental retardation. *J Autism Dev Disord* 2003; 33: 435–442.
58. Tsai LY. Is Rett syndrome a subtype of pervasive developmental disorders. *J Autism Dev Disord* 1992; 22: 551–561.
59. Young DJ, Bebbington A, Anderson A, Ravine D, Ellaway C, Kulkarni A, de Klerk N, Kaufmann WE, Leonard H. The diagnosis of autism in a female: could it be Rett syndrome. *Eur J Pediatr* 2007; in press 167(6): 661–669. Epub 2007 Aug 8.
60. Colvin L, Leonard H, de Klerk N, Davis M, Weaving L, Williamson S, Christodoulou J. Refining the phenotype of common mutations in Rett syndrome. *J Med Genet* 2004; 41: 25–30.
61. Zappella M, Meloni I, Longo I, Hayek G, Renieri A. Preserved speech variants of the Rett syndrome: molecular and clinical analysis. *Am J Med Genet* 2001; 104A: 14–22.
62. Zappella M, Meloni I, Longo I, Canitano R, Hayek G, Rosaia L, Mari F, Renieri A. Study of MECP2 gene in Rett syndrome variants and autistic girls. *Am J Med Genet* 2003; 119B: 102–107.

63. Beyer KS, Blasi F, Bacchelli E, Klauck SM, Maestrini E, Poustka A. Mutation analysis of the coding sequence of the MECP2 gene in infantile autism. International molecular genetic study of autism consortium (IMGSAC). *Hum Genet* 2002; 111: 305–309.

64. Lobo-Menendez F, Sossey-Alaoui K, Bell JM, Copeland-Yates SA, Plank SM, Sanford SO, Skinner C, Simensen RJ, Schroer RJ, Michaelis RC. Absence of MECP2 mutations in patients from the South Carolina autism project. *Am J Med Genet* 2003; 117B: 97–101.

65. Carney RM, Wolpert CM, Ravan SA, Shahbazian M, Ashley-Koch A, Cuccaro ML, Vance JM, Pericak-Vance MA. Identification of MECP2 mutations in a series of females with autistic disorder. *Pediatr Neurol* 2003; 28: 205–211.

66. Coutinho AM, Oliveira G, Katz C, Feng J, Yan J, Yang C, Marques C, Ataide A, Miguel TS, Borges L, Almeida J, Correia C, Currais A, Bento C, Mota-Vieira L, Temudo T, Santos M, Maciel P, Sommer SS, Vicente AM. Mecp2 coding sequence and 3'UTR variation in 172 unrelated autistic patients. *Am J Med Genet* 2007; 144B: 475–483.

67. Kaufmann WE, Cooper KL, Mostofsky SH, Capone GT, Kates WR, Newschaffer CJ, Bukelis I, Stump MH, Jann AE, Lanham DC. Specificity of cerebellar vermian abnormalities in autism: a quantitative magnetic resonance imaging study. *J Child Neurol* 2003; 18: 463–470.

68. Brambilla P, Hardan A, di Nemi SU, Perez J, Soares JC, Barale F. Brain anatomy and development in autism: review of structural MRI studies. *Brain Res Bull* 2003; 61: 557–69.

69. Kates WR, Abrams MT, Kaufmann WE, Breiter SN, Reiss AL. Reliability and validity of MRI measurement of the amygdala and hippocampus in children with Fragile X syndrome. *Psychiatry Res* 1997; 75: 31–48.

70. Kates WR, Folley BS, Lanham DC, Capone GT, Kaufmann WE. Cerebral growth in Fragile X syndrome: review and comparison with down syndrome. *Microsc Res Tech* 2002; 57: 159–167.

71. Hessl D, Glaser B, Dyer-Friedman J, Blasey C, Hastie T, Gunnar M, Reiss AL. Cortisol and behavior in Fragile X syndrome. *Psychoneuroendocrinology* 2002; 27: 855–872.

72. Hessl D, Glaser B, Dyer-Friedman J, Reiss AL. Social behavior and cortisol reactivity in children with Fragile X syndrome. *J Child Psychol Psychiatry* 2006; 47: 602–610.

73. Sun H-T, Cohen S, Kaufmann WE. Annexin 1 is abnormally expressed in Fragile X syndrome: a two-dimensional electrophoresis study in lymphocytes. *Am J Med Genet* 2001; 103: 81–90.

74. Kaufmann WE, Danko CG, Kau ASM, Thevarajah S, Bukelis I, Tierney E, Neuberger I. Increased protein acetylation in lymphocytes predicts autistic behavior in Fragile X syndrome. *Ann Neurol* 2003; 54: S105–S106.

75. Buckingham JC, Solito E, John C, Tierney T, Taylor A, Flower R, Christian H, Morris J. Annexin 1: a paracrine/juxtacrine mediator of glucocorticoid action in the neuroendocrine system. *Cell Biochem Funct* 2003; 21: 217–221.

76. Brown V, Jin P, Ceman S, Darnell JC, O'Donnell WT, Tenenbaum SA, Jin X, Feng Y, Wilkinson KD, Keene JD, Darnell RB, Warren ST. Microarray identification of FMRP-associated brain mRNAs and altered mRNA translational profiles in Fragile X syndrome. *Cell* 2001; 107: 489–499.

77. Miyashiro KY, Beckel-Mitchener A, Purk TP, Becker KG, Barret T, Liu L, Carbonetto S, Weiler IJ, Greenough WT, Eberwine J. RNA cargoes associating with fmrp reveal deficits in cellular functioning in Fmr1 null mice. *Neuron* 2003; 37: 417–431.

78. Markham JA, Beckel-Mitchener AC, Estrada CM, Greenough WT. Corticosterone response to acute stress in a mouse model of Fragile X syndrome. *Psychoneuroendocrinology* 2006; 31: 781–785.

79. Kaufmann WE, Moser HW. Dendritic anomalies in disorders associated with mental retardation. *Cereb Cortex* 2000; 10: 981–991.

80. Bear MF. Therapeutic implications of the mGluR theory of Fragile X mental retardation. *Genes Brain Behav* 2005; 4: 393–398.

81. Zhao MG, Toyoda H, Ko SW, Ding HK, Wu LJ, Zhuo M. Deficits in trace fear memory and long-term potentiation in a mouse model for Fragile X syndrome. *J Neurosci* 2005; 25: 7385–7392.

82. Mathew SJ, Ho S. Etiology and neurobiology of social anxiety disorder. *J Clin Psychiatry* 2006; 67(Suppl 12): 9–13.

83. Spooren W, Gasparini F. mGlu5 receptor antagonists: a novel class of anxiolytics? *Drug News Perspect* 2004; 17: 251–257.

84. Weis S, Weber G, Neuhold A, Rett A. Down syndrome: MR quantification of brain structures and comparison with normal control subjects. *AJNR Am J Neuroradiol* 1991; 12: 1207–1211.

85. Raz N, Torres IJ, Briggs SD, Spencer WD, Thornton AE, Loken WJ, Gunning FM, McQuain JD, Driesen NR, Acker JD. Selective neuroanatomic abnormalities in down's syndrome and their cognitive correlates: evidence from MRI morphometry. *Neurology* 1995; 45: 356–366.

86. Aylward EH, Habbak R, Warren AC, Pulsifer MB, Barta PE, Jerram M, Pearlson GD. Cerebellar volume in adults with down syndrome. *Arch Neurol* 1997; 74: 73–82.

87. Aylward EH, Li Q, Honeycutt NA, Warren AC, Pulsifer MB, Barta PE, Chan MD, Smith PD, Herram M, Pearlson GD. MRI volumes of the hippocampus and amygdala in adults with down's syndrome with and without dementia. *Am J Psychiatry* 1999; 156: 564–568.

88. Pinter JD, Eliez S, Schmitt JE, Capone GT, Reiss AL. Neuroanatomy of down's syndrome: a high resolution MRI study. *Am J Psychiatry* 2001; 158: 1659–1665.

89. Carter JC, Capone GT, Kaufmann WE. Neuroanatomic correlates of autism and stereotypy in children with Down syndrome. Neuroreport 2008; 19: 653–656.

90. Carper RA, Moses P, Tigue ZD, Courchesne E. Cerebral lobes in autism: early hyperplasia and abnormal age effects. *Neuroimage* 2002; 16: 1038–1051.

91. Hyde LA, Crnic LS, Pollock A, Bickford PC. Motor learning in Ts65Dn mice, a model for down syndrome. *Dev Psychobiol* 2001; 38: 33–45.

92. Escorihuela RM, Vallina IF, Martinez-Cue C, Baamonde C, Dierssen M, Tobena A, Florez J, Fernandez-TeruelA. Impaired short- and long-term memory in ts65dn mice, a model for down syndrome. *Neurosci Lett* 1998; 247: 171–174.

93. Belichenko PV, Masliah E, Kleschevnikov AM, Villar AJ, Epstein CJ, Salehi A, Mobley WC. Synaptic structural abnormalities in the Ts65Dn mouse model of down syndrome. *J Comp Neurol* 2004; 480: 281–98.

94. Benavides-Piccione R, Ballesteros-Yanez I, de Lagran MM, Elston G, Estivill X, Fillat C, Defilipe J, Dierssen M. On dendrites in down syndrome and DS murine models: a spiny way to learn. *Prog Neurobiol* 2004; 74: 111–126.

95. Fernandez F, Morishita W, Zuniga E, Nguyen J, Blank M, Malenka RC, Garner CC. Pharmacotherapy for cognitive impairment in a mouse model of down syndrome. *Nat Neurosci* 2007; 10: 411–413.

96. Pan JW, Lane JB, Hetherington H, Percy AK. Rett syndrome: 1H spectroscopic imaging at 4.1 tesla. *J Child Neurol* 1999; 14: 524–528.

97. Blue ME, Naidu S, Johnston MV. Development of amino acid receptors in frontal cortex from girls with Rett syndrome. *Ann Neurol* 1999; 45: 541–545.

98. Colantuoni C, Jeon O-H, Hyder K, Chenchik A, Khimani AH, Narayanan V, Hoffman EP, Kaufmann WE, Naidu S, Pevsner J. Gene expression profiling in postmortem Rett syndrome brain: differential gene expression and patient classification. *Neurobiol Dis* 2001; 8: 847–865.

99. Kaufmann WE. Cortical development in Rett syndrome: molecular, neurochemical, and anatomical aspects. In: Kerr A and Engerström IW (Eds). *The Rett Disorder and the Developing Brain.* Oxford: Oxford University Press, 2001: 85–110.

100. Farage L, Nagae-Poetscher LM, Bibat G, Kaufmann W, Barker PB, Naidu S, Horská A. Single voxel proton MR spectroscopy in Rett syndrome: clinical and

genetic correlations. *ASNR 44th Annual Meeting*, 2006: May 15, San Diego, California.

101. Horská A, Naidu S, Herskovits EH, Wang PY, Kaufmann WE, Barker PB. Quantitative proton MR spectroscopic imaging in early Rett syndrome. *Neurology* 2000; 54: 715–722.

102. Khong P-L, Lam C-W, Ooi CGC, Ko C-H, Wong VCN. Magnetic resonance spectroscopy and analysis of MECP2 in Rett syndrome. *Pediatr Neurol* 2002; 26: 205–209.

103. Minshew NJ, Williams DL. The new neurobiology of autism: cortex, connectivity, and neuronal organization. *Arch Neurol* 2007; 64: 945–950.

104. Carter JC, Lanham DC, Pham D, Bibat G, Naidu S, Kaufmann WE. Selective cerebral volume reduction in Rett syndrome: a multiple approach MRI study. *AJNR Am J Neuroradiol* 2007; 29(3): 436–411. Epub 2007 Dec 7.

105. Stearns NA, Schaevitz LR, Bowling H, Nag N, Berger UV, Berger-Sweeney J. Behavioral and anatomic abnormalities in Mecp2 mutant mice: a model for Rett syndrome. *Neuroscience* 2007; 146: 907–921.

106. Nag N, Berger-Sweeney JE. Postnatal dietary choline supplementation alters behavior in a mouse model of Rett syndrome. *Neurobiol Dis* 2007; 26: 473–480.

107. McGill BE, Bundle SF, Yaylaoglu MB, Carson JP, Thaller C, Zoghbi HY. Enhanced anxiety and stress-induced corticosterone release are associated with increased crh expression in a mouse model of Rett syndrome. *Proc Natl Acad Sci USA* 2006; 103: 18267–18272.

108. Moretti P, Bouwknecht JA, Teague R, Paylor R, Zoghbi HY. Abnormalities of social interactions and home-cage behavior in a mouse model of Rett syndrome. *Hum Mol Genet* 2005; 14: 205–220.

109. Moretti P, Levenson JM, Battaglia F, Atkinson R, Teague R, Antalffy B, Armstrong D, Arancio O, Sweatt JD, Zoghbi HY. Learning and memory and synaptic plasticity are impaired in a mouse model of Rett syndrome. *J Neurosci* 2006; 26: 319–327.

110. Kaufmann WE, Abrams MT, Chen W, Reiss AL. Genotype, molecular phenotype, and cognitive phenotype: correlations in Fragile X syndrome. *Am J Med Genet* 1999; 83: 286–295.

111. Bailey DB Jr, Hatton DD, Skinner M, Mesibov G. Autistic behavior, FMR1 protein, and developmental trajectories in young males with Fragile X syndrome. *J Autism Dev Disord* 2001; 31: 165–174.

112. Loesch DZ, Huggins RM, Bui QM, Taylor AK, Pratt C, Epstein J, Hagerman RJ. Effect of Fragile X status categories and FMRP deficits on cognitive profiles estimated by robust pedigree analysis. *Am J Med Genet* 2003; 122A: 13–23.

113. Bagni C, Greenough WT. From mRNP trafficking to spine dysmorphogenesis: the roots of Fragile X syndrome. *Nat Rev Neurosci* 2005; 6: 376–387.

114. Duan R, Jin P. Identification of messenger RNAs and microRNAs associated with Fragile X mental retardation protein. *Methods Mol Biol* 2006; 342: 267–276.

115. Zalfa F, Achsel T, Bagni C. mRNPs, polysomes or granules: FMRP in neuronal protein synthesis. *Curr Opin Neurobiol* 2006; 16: 265–269.

116. Nishimura Y, Martin CL, Lopez AV, Spence SJ, Alvarez-Retuerto AI, Sigman M, Steindler C, Pellegrini S, Schanen NC, Warren ST, Geschwind DH. Genome-wide expression profiling of lymphoblastoid cell lines distinguishes different forms of autism and reveals shared pathways. *Hum Mol Genet* 2007; 16: 1682–1698.

117. Schenck A, Bardoni B, Langmann C, Harden N, Mandel JL, Giangrande A. CYFIP/Sra-1 controls neuronal connectivity in drosophila and links the Rac1 GTpase pathway to the Fragile X protein. *Neuron* 2003; 38: 887–898.

118. Brookes K, Xu X, Chen W, Zhou K, Neale B, Lowe N, Anney R. The analysis of 51 genes in DSM-IV combined type attention deficit hyperactivity disorder: association signals in DRD4, DAT1 and 16 other genes. *Mol Psychiatry* 2006; 10: 934–953.

119. Gardiner K, Fortuna A, Bechtel L, Davisson MT. Mouse models of down syndrome: how useful can they be? Comparison of the gene content of human chromosome 21 with orthologous mouse genomic regions. *Gene* 2003; 318: 137–147.

120. Tang Y, Schapiro MB, Franz DN, Patterson BJ, Hickey FJ, Schorry EK, Hopkin RJ, Wylie M, Narayan T, Glauser TA, Gilbert DL, Hershey AD, Sharp FR. Blood expression profiles for tuberous sclerosis complex 2, neurofibromatosis type 1, and down's syndrome. *Ann Neurol* 2004; 56: 808–814.

121. Molloy CA, Keddache M, Martin LJ. Evidence for linkage on 21q and 7q in a subset of autism characterized by developmental regression. *Mol Psychiatry* 2005; 10: 741–746.

122. Rahmani Z. APRO4 regulates scr tyrosine kinase activity in pc12 cells. *J Cell Sci* 2006; 119: 646–658.

123. Seibzehnrubl FA, Buslei R, Eyupoglu IY, Seufert S, Hahnen E, Blumcke I. Histone deacetlyase inhibitors increase neuronal differentiation in adult forebrain precursor cells. *Exp Brain Res* 2007; 176: 672–678.

124. Kaufmann WE, Carter JC, Bukelis I, Lieberman DN. Neurobiology of genetic mental retardation. In: Gilman S (Ed). *Neurobiology of Disease.* Burlington, MA: Elsevier Academic press, 2007: 563–580.

125. Shibayama A, Cook EH Jr, Feng J, Glanzmann C, Yan J, Craddock N, Jones IR, Goldman D, Heston LL, Sommer SS. MECP2 structural and 30-UTR variants in schizophrenia, autism and other psychiatric diseases: a possible association with autism. *Am J Med Genet* 2004; 128B: 50–53.

126. Samaco RC, Nagarajan RP, Braunschweig D, LaSalle JM. Multiple pathways regulate MECP2 expression in normal brain development and exhibit defects in autism-spectrum disorders. *Hum Mol Genet* 2004; 13: 629–639.

127. Nagarajan RP, Hogart AR, Gwye Y, Martin MR, Lasalle JM. Reduced MeCP2 expression is frequent in autism frontal cortex and correlates with aberrant MECP2 promoter methylation. *Epigenetics* 2006; 1: 172–182.

128. Samaco RC, Hogart A, LaSalle JM. Epigenetic overlap in autism-spectrum neurodevelopmental disorders: MECP2 deficiency causes reduced expression of UBE3A and GABRB3. *Hum Mol Genet* 2005; 14: 483–492.

129. Hogart A, Nagarajan RP, Patzel KA, Yasui DH, Lasalle JM. 15q11–13 GABAA receptor genes are normally biallelically expressed in brain yet are subject to epigenetic dysregulation in autism-spectrum disorders. *Hum Mol Genet* 2007; 16: 691–703.

130. Kaufmann WE, MacDonald SM, Altamura C. Dendritic cytoskeletal protein expression in mental retardation: an immunohistochemical study of the neocortex in Rett syndrome. *Cereb Cortex* 2000; 10: 992–1004.

131. LaSalle JM, Goldstine J, Balmer D, Greco CM. Quantitative localization of heterogeneous methyl-CpG-binding protein 2 (MeCP2) expression phenotypes in normal and Rett syndrome brain by laser scanning cytometry. *Hum Mol Genet* 2001; 10: 1729–1740.

132. Roberts JE, Weisenfeld LA, Hatton DD, Heath M, Kaufmann WE. Social approach and autistic behavior in children with Fragile X syndrome. *J Autism Dev Disord* 2007; 37: 1748–1760.

133. Ricceri L, Moles A, Crawley J. Behavioral phenotyping of mouse models of neurodevelopmental disorders: relevant social behavior patterns across the life span. *Behav Brain Res* 2007; 176: 40–52.

134. Fyfe S, Downs J, McIlroy O, Burford B, Lister J, Reilly S, Laurvick CL, Philippe C, Msall M, Kaufmann WE, Ellaway C, Leonard H. Development of a video-based evaluation tool in Rett syndrome. *J Autism Dev Disord* 2007; 37: 1636–1646.

135. Beaudet AL. Autism: highly heritable but not inherited. *Nat Med* 2007; 13: 534–536.

Part II
Neurotransmitters and Cell Signaling

Chapter 5
Serotonin Dysfunction in Autism

Mary E. Blue, Michael V. Johnston, Carolyn B. Moloney, and Christine F. Hohmann

Abstract This chapter reviews the evidence for the involvement of the neurotransmitter serotonin in the etiology of autism. Serotonin-containing neurons in the raphe nuclei of the brainstem are among the first neurons to be generated, and their axonal projections extend to widespread areas throughout the brain and spinal cord. Thus, the serotonergic system can influence early developmental events throughout the brain. Other evidence is the generally higher level of serotonin in the blood of patients with autism. In the brain of young children with autism, however, serotonin synthesis is decreased compared to that in normal siblings. Pharmacological studies show symptomatic improvement with agents that enhance serotonergic function and exacerbation of behaviors with drugs that decrease serotonin. Genetic analyses show mutations in the serotonin transporter in autism. Animal models that mimic the changes in serotonin show behavioral, structural and biochemical features that resemble autism. Conversely, models based on other features of autism often show changes in serotonergic markers. The role serotonin exerts in autism is likely through its interactions with other transmitter systems, neurotrophic and growth factors, and immunological factors. Taking all the studies together, the evidence suggests that serotonin is a critical player in the development of the autism phenotype.

Keywords Autism · serotonin · neurotrophins · cerebral cortex · neuroimmunity

Introduction

For many years, it has been postulated that the neurotransmitter serotonin (5-hydroxytryptophan or 5-HT) plays a significant role in the pathogenesis of autism. Both the peripheral and the central nervous systems (CNSs) contain

M.E. Blue
Neuroscience Laboratory Room 400R, Kennedy Krieger Institute, 707 North Broadway, Baltimore, MD 21205, USA
e-mail: blue@kennedykrieger.org

A.W. Zimmerman (ed.), *Autism*, DOI: 10.1007/978-1-60327-489-0_5,
© Humana Press, Totowa, NJ 2008

serotonergic neurons, and their axons are distributed to widespread areas of the brain and body, making it likely that 5-HT could impact development globally. During brain development, serotonergic neurons in the brainstem are among the first long-tract projection neurons to be "born" and are thus in a position to influence early brain development [1, 2, 3, 4, 5]. In fact, 5-HT influences both neurogenesis and pathway formation [6, 7, 8]. On the level of the synapse, 5-HT acts as a neuromodulator, setting the level of inhibition and excitation to other afferents [9, 10, 11, 12, 13]. At a "systems" level, 5-HT acting through its receptors and transporter modulates mood and affect [14, 15, 16, 17, 18]. For example, the selective serotonin reuptake inhibitors (SSRIs) have revolutionized the treatment of depression and other affective disorders in adults [19, 20, 21, 22, 23, 24, 25, 26]. Although the treatment of children with SSRIs has had more mixed results [26, 27, 28, 29, 30], they are effective in treating some of the behavioral features of autism [19, 31, 32, 33, 34, 35, 36, 37]. In this chapter, we will review what is known about the role of 5-HT in brain development as it relates to the development of the autism phenotype.

The first evidence that 5-HT may play a role in autism came from serum measures of 5-HT from patients with autism [38]. Blood serum and platelet 5-HT levels were elevated (reviewed in [39, 40, 41]) although these effects varied by age and ethnicity (for review see [42]). In addition, several studies suggested changes in blood levels of the 5-HT precursor amino acid tryptophan, which likely would result in altered 5-HT synthesis in the CNS [43, 44, 45]. Reports of 5-HT levels in cerebrospinal fluid also showed highly variable results [42]. Interestingly, a recent report measuring plasma levels of 5-HT in the family members of children with autism showed that levels of 5-HT in the plasma from mothers of children with autism were decreased significantly compared to normal control mothers [46]. The authors suggested that plasma 5-HT levels may more accurately reflect the changes in the CNS than the platelet values and this view has recently received support from experimental studies in mice [47]. Moreover, decreased maternal plasma serotonin during gestation appears to dramatically reduce the ability of the offspring to synthesize serotonin in the brain [47, 48].

Studies that assess 5-HT receptors in platelets or whole blood of individuals with autism show decreased 5-HT_2 receptor binding [49, 50]. A recent SPECT analysis also found significant reductions in 5-HT_{2A} binding in the cortex [51]. In addition, clinical trials with 5-HT_2 receptor-acting drugs show clinical benefit for impaired social behaviors and aggression in autism [52, 53].

Several genetic analyses suggest a role for 5-HT in autism. Promotor region polymorphisms for the serotonin transporter (5-HTT) and for the 5-HTT gene SLC6A4 are associated with a susceptibility to autism [54, 55, 56]. In monozygotic twins who are discordant for autism, Hu et al. recently demonstrated an association between increased 5-HTT expression in peripheral lymphocytes and the autism phenotype [57]. In addition, 5-HT homeostasis in the brain and the periphery is regulated by the hypothalamo–pituitary axis (HPA) [58, 59, 60] and thus is vulnerable to environmental factors like stress. HPA responsiveness is altered in autism, particularly after exposure to novel situations or psychosocial stress [61, 62].

Psychopharmacologic intervention studies further support a role for 5-HT in autism. Compounds that alter the serotonergic system produce moderate behavioral improvements in autism [31, 32, 63, 64] and is reviewed in [19]. The SSRIs lead to improvements in social behavior while decreasing aggressive and stereotyped behaviors in children with autism [33, 34, 35, 36, 37, 65]. Consistent with these therapeutic effects, decreasing CNS 5-HT via tryptophan depletion exacerbates symptoms in patients with ASD [42, 66]. A review of these studies by Kolevzon et al. suggests that the differences in the efficacy and tolerability of the SSRIs depend on the age at which they are given [19]. This is not surprising given the different role that 5-HT plays during brain development.

Perhaps the most compelling evidence for the developmental role of 5-HT in autism comes from landmark studies by Chugani and colleagues [67, 68], showing an altered developmental trajectory of serotonergic innervation to the cerebral cortex. Their studies, using positron emission tomography (PET) imaging of a tryptophan analog, demonstrated that young children with autism do not display the developmental peak in brain 5-HT synthesis capacity seen in typically developing children [67, 68]. Boys with autism have specific decreases in 5-HT synthesis in the dentatothalamocortical pathway [68]. Moreover, early developmental decreases in cortical serotonin appear to correlate with the severity of autistic symptoms as measured by speech development [69]. Most recently, Piven and colleagues have also observed a relationship between gray matter overgrowth in autistic brains and serotonin transporter variants [70]. As described below, a mouse model that mimics the selective 5-HT decreases in the developing cortex shares many of the phenotypic changes observed in autism.

The Role of Serotonin in Cortical Development and Plasticity

Serotonergic afferents from the brainstem raphe nuclei innervate cerebral cortex during a critical time in cortical morphogenesis. Similar to the peak in 5-HT synthesis at 2 years of age in humans, rodents show a transient peak in 5-HT levels in the first few days after birth [71, 72]. At this time, layer IV of sensory areas of cortex exhibits dense patches of staining for 5-HT and 5-HTTs, particularly in the "barrel field" in primary somatosensory cortex (SI) [73, 74, 75, 76]. This region of SI contains specialized columnar processing units with distinct clusters of layer IV neurons or "barrels" that receive sensory input from individual whiskers [77]. The serotonergic patches are in fact due to the glutamatergic thalamocortical afferents (TCAs), which transiently express 5-HTT along with the vesicular monoamine transporter (VMAT2), taking up 5-HT as a "borrowed transmitter" [78, 79, 80, 81, 82, 83, 84] and presumably releasing it onto developing target neurons in cortex. In addition, 5-HT$_{1B}$ receptors, also located on TCAs [80], profoundly inhibit the release of glutamate from these axons [85]. Thus, 5-HT has a developmentally transient, presynaptic inhibitory effect upon thalamocortical transmission in developing cerebral cortex.

In vitro studies also show that 5-HT in cultured neurons stimulates neurite outgrowth from thalamic neurons independent of their cortical targets [86, 87] and enhances survival and maturation of cortical glutamatergic neurons [88, 89]. In vivo, it appears that too little or too much 5-HT is detrimental to cortical development. Neonatal, systemic serotonergic depletions delay development of several cortical layers [90], the appearance of TCA patterning in the barrel field [76], and ultimately lead to decreases in the size of the barrel field [91, 92]. Conversely, in monoamine oxidase A (MAOA) and 5-HTT knock-out mice, which have dramatically increased levels of extracellular 5-HT, barrels are not visible due to overlap of TCAs from adjacent whiskers [82, 93, 94]. These serotonergic effects are at the root of altered dendritic and synaptic development [95]. Barrel formation is restored in MAOA and 5-HTT single and double knockouts by the blockade of 5-HT synthesis or the additional knockout of 5-HT$_{1B}$ receptors, which normally inhibit glutamate release [82, 96]. Thus, in vivo results confirm a U-shaped dose response to serotonin in development with too little and too much altering the normal course of morphogenesis.

Since TCAs use glutamate as their neurotransmitter [97], glutamate receptors (GluRs) likely play a role in the serotonergic modulation of cortical development. Experimental evidence suggests that GluRs play roles in the activity-dependent refinement of synaptic connectivity [98, 99, 100, 101]. GluRs are classified broadly into two groups: ionotropic sites, linked to ion channels, and metabotropic sites, linked to second messengers [102]. The ionotropic sites include those activated by the exogenous agonists, N-methyl-D-aspartate (NMDA), amino-3-hydroxy-5-methyl-4-isoxazole propionic acid (AMPA), and kainate (KA).

NMDA receptors influence both the retraction of incorrectly placed axon arbors and synapses and the elaboration of correctly positioned terminals [98, 99, 100, 101, 103, 104, 105, 106, 107]. Barrel formation is disrupted in transgenic mice that have knocked down NMDAR-1 receptor expression [106, 107, 108, 109]. Metabotropic glutamate receptors (mGluRs) also modulate glutamate release and neuron excitability in the whisker to barrel pathway [110, 111, 112] and have well documented roles in cortical development and activity-dependent plasticity [113, 114, 115, 116, 117, 118, 119, 120]. Taken together, the results of these developmental studies suggest that interactions between TCAs and serotonergic afferents in cortex, acting via 5-HT and glutamate receptors, shape cortical development.

Neurotrophins, BDNF, and Serotonin

Neurotrophins are key regulators of neuronal survival, differentiation, and synapse formation, and plasticity in the nervous system and some of their function appears to be intimately connected to serotonergic modulatory influences. Neurotrophins and their receptors are expressed in the neocortex and hippocampus [121, 122,

123, 124], and the patterns of neurotrophin expression are activity-dependent and regulated by sensory inputs and electrical activity [123, 125, 126, 127]. Interestingly, the neurotrophin brain derived neurotrophic factor (BDNF) and 5-HT show reciprocal regulation; both are responsive to environmental factors [126, 128]. For example, mice heterozygous for BDNF expression (BDNF+/–) display premature, age-associated loss in forebrain serotonergic innervation [129], and 5-HTT function is impaired in the brains of BDNF-deficient (BDNF+/–) mice [130]. Localized increases in BDNF expression promote serotonergic fiber sprouting after injury [131, 132]. In turn, 5-HT depletion via inhibition of synthesis is accompanied by decreases in BDNF levels in the mature hippocampus [133].

BDNF expression in the brain is sensitive to a variety of psychoactive drugs [134], and abnormalities in neurotrophin expression have been implicated in the etiology of several brain disorders that show altered cortical maturation and plasticity, such as schizophrenia and depression [135, 136, 137]. In autism, several studies show increased serum or blood levels of BDNF expression [138, 139, 140]. Whether this change reflects peripheral increases in 5-HT levels or central deficits in 5-HT is not known.

BDNF and its receptor, trkB, are densely expressed on cortical and hippocampal neurons and influence both axonal and dendritic growth in a highly neuron-specific and age-dependent manner [141, 142]. Cortical expression of the trkB receptor peaks in the first 2 weeks postnatally in the rat [143], but BDNF action on cortical plasticity continues into adulthood [141, 144]. With maturation, trkB becomes enriched at the site of glutamatergic synapses and therefore is uniquely able to modulate experience-dependent plasticity [145].

During brain development, factors such as perinatal stress or environmental enrichment promote long-term alterations in BDNF expression in brain and blood plasma [126, 146, 147, 148]. In rat, maternal infection moreover can cause long-term increases in BDNF within the cerebral cortex and other brain areas [149]. Interestingly, within the context of gender disparity in autism, BDNF regulation may be differentially regulated in males and females [150, 151]. BDNF expression may be mediated by serotonergic mechanisms in that 5-HT$_{2A}$ receptor antagonists have been shown to block stress-induced decreases in BDNF expression in the hippocampus and cortex [152]. Overall, these data suggest complex interactions among 5-HT homeostasis, neurotrophin expression, stress, and gender in autism.

Serotonin and Immune Activation

Immunological dysfunction is another emerging component of ASD pathology [153, 154, 155, 156]. Several reports suggest that up to 60% of patients with ASD have some type of systemic immune dysfunction, either as part of cellular or humoral immune responses [157, 158, 159]. Other pathological evidence of immunological reactions within the CNS is the presence of lymphocyte infiltration and microglial nodules [160, 161] and increases in pro-inflammatory cytokines in peripheral blood samples from patients with autism [157, 162, 163, 164, 165, 166]. Recently, Molloy et al. [167] reported that the production of Th2-associated cytokines in leukocytes is increased in patients with autism.

Postmortem neuropathological studies of brain tissues suggest an active and ongoing neuroinflammatory process in autism with astroglial and neuroglial activation in the cerebral cortex and white matter [168]. However, pathogenic mechanisms involving astroglia and microglia also are common to other brain disorders. In autism, it is possible that different factors (e.g., genetic susceptibility, maternal factors, prenatal environmental exposures) may trigger the development of these neuroglial reactions. Protein arrays show that cytokines/chemokines such as MCP-1, IL-6, and TGF-β1, which are mainly derived from activated neuroglia, are the most prevalent cytokines in brain tissues and CSF from patients with autism [168]. In addition, stress can activate pro-inflammatory cytokine release directly and, via altered serotonin homeostasis, further modulate pro-inflamatory cytokine activity [169]. Moreover, stressors and cytokines may share a common ability to impair neuronal plasticity via alterations of BDNF and other neurotrophins in the brain [170].

Serotonin influences immune cell function and pro-inflamatory cytokine release although the precise function and directionality of serotonergic effects are still inconclusive [169, 170, 171]. Recent studies suggest that peripheral 5-HT promotes the inflammatory response and cytokine release via 5-HT$_{1A}$ and 5-HT$_{1B}$ receptors [172, 173, 174]. Patients with major depressive disorders have increased cytokine expression [175], and prolonged treatment with antidepressants such as SSRIs can reduce IL6 and other pro-inflammatory cytokines [176].

Cytokines, in turn, affect both peripheral and central serotonergic systems. A number of inflammatory conditions are associated with central serotonergic insufficiencies [177, 178], and interferon alpha treatment leads to decreases in 5-HT [179–182]. Treatment of colonic mucosa cells with pro-inflammatory cytokines, including TNF-alpha, reduces 5-HTT function [183]. Pro-inflammatory cytokines also may affect CNS 5-HT availability by interacting with the catabolic enzyme indolamine-2,3-dyoxignesase (IDO) [177, 184]. This enzyme is inducible by pro-inflammatory cytokines as well as stress and diminishes tryptophan availability for 5-HT synthesis [177]. On the contrary, mitogen activation of pro-inflammatory cytokines leads to increased 5-HT uptake into raphe neurons [185]. Striatal 5-HT levels are increased after peripheral IL-6 injection [186], and IL-6 increases peripheral 5-HT levels in a rat model for pulmonary hypertension [187]. Thus, there is selectivity in the manner that pro-inflammatory cytokines modulate peripheral or central 5-HT homeostasis and vice versa.

Animal Models, Serotonin, and Autism

Animal models for autism that alter the serotonergic system provide additional evidence for a role of serotonin in autism. To mimic the findings of Chugani et al. [67, 68], C. Hohmann selectively lesioned serotonergic afferents to the

cortex and hippocampus in newborn mice. She did this by injecting the serotonin neurotoxin 5,7-dihydroxydopamine (5,7-DHT) bilaterally into the medial
forebrain bundle (MFB). The neonatal 5,7-DHT injections selectively lesioned
the serotonergic axons in the cortex while sparing those in other regions
(Fig. 5.1 and see Color Plate 2, following p. XX) [188, 189]. Although HPLC
measures of serotonin levels in brain indicated substantial recovery by

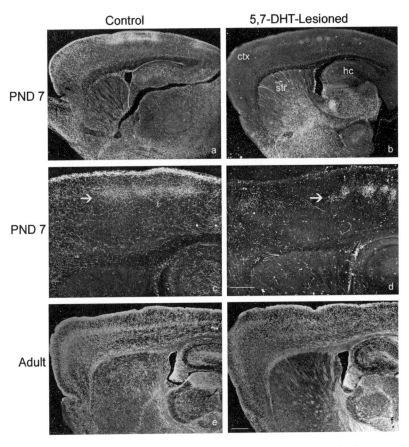

Fig. 5.1 Dark field photomicrographs of 5-hydroxytryptophan (5-HT) staining in parasagittal
sections from control [age normal (**a**, **e**), saline injected (**c**)] and 5,7-dihydroxydopamine
(5,7-DHT)-lesioned (**b**, **d**, **f**) mice at postnatal day (PND) 7 (**a**–**d**) and in the adult (**e**, **f**). The
density of serotonergic axons in cerebral cortex (ctx) and hippocampus (hc) is very depleted in
lesioned mice at PND 7. However, the dense serotonin positive patches are present in barrel
field cortex in control and lesioned mice [arrows; shown at higher magnification in (**c**) and (**d**)].
This is due to the transient expression of the serotonin transporter in thalamocortical axons
at this age. Over time, there is variable regrowth of the serotonergic axons into cortex
and hippocampus. The adult example in **f** shows the greatest degree of recovery observed.
Magnification bars = 500 μm; the bar in **d** is for **c** and **d**; the one in **f** is for **a**, **b**, **e**, **f**. Abbreviation:
str-striatum (*see* Color Plate 2)

adulthood [190], subsequent immunocytochemical studies revealed that while the 5-HT innervation of cortex recovered slowly, it was not fully complete by adulthood (Fig. 5.1). However, autoradiographic labeling of the serotonin transporter (5-HTT) showed an "overshoot" in 5-HTT binding of lesioned mice by the age of 1 month, indicating increased 5-HT uptake in the re-innervating axons [188]. Thus, the serotonergic innervation in this model may be more functionally recovered than is visible by immunocytochemistry.

Initial morphometric studies in young adult, lesioned mice revealed increases in overall cortical thickness attributable to changes in specific regions and layers of neocortex [190]. In older adult lesioned mice, layer IV in parietal cortex remains wider than in sham lesioned mice, and sex differences are apparent in layer II/III cortical width at the same age [189]. These data suggest that layer specific alterations in the cortical network persist into maturity, after the 5,7-DHT lesion. Interestingly, a recent study using magnetic resonance imaging (MRI) has shown that cortical thickness also is increased in autism with parietal and temporal lobes being preferentially affected [191].

Our recent unbiased stereologic analyses of cortical volume in the lesioned mice are consistent with the cortical thickness measures, in that we observed an increase in cortical volume in 1-week-old 5,7-DHT-lesioned mice compared to age-matched unlesioned normal mice (Fig. 5.2). The increase in total volume was principally due to the enlargement of the parietal cortex, where in human cases of autism significant increases in cortical thickness are present [191]. In adult lesioned mice, cortical volume is no longer different in the lesioned mice compared to normal mice. This age-specific increase in cortical volume mirrors the findings in children with autism, in which the substantial increases in cortical volume are primarily observed between 2 and 4 years of age [192–195], during the same period of development when significantly elevated head circumferences also are observed [196–198].

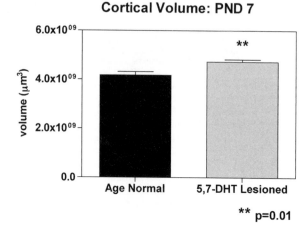

Fig. 5.2 Unbiased stereologic measures of cortical volume found that the volume of the cerebral cortex was significantly larger in 5,7-DHT-lesioned mice (n = 12) than in age normal control mice (n = 14) at postnatal day (PND) 7

Behavioral findings in the selective 5,7-DHT lesion model also show a number of similarities to autism. A battery of behavioral tests showed altered social and emotional behaviors in the 5,7-DHT-lesioned mice (see [188] for review). In a socially transmitted food preference task, the lesioned mice showed increases in stereotypic behaviors and less social interaction between demonstrator and observer mice [188]. One interesting finding of the behavioral studies is that in a variety of tests, the male lesioned mice exhibited altered behavior while female lesioned mice had responses more similar to age-matched normal mice. For example, in an open field object recognition task, only male lesioned mice exhibited altered exploration of displaced and novel objects [189], and in a cued and contextual fear conditioning test, male lesioned mice showed decreased auditory responsiveness to the conditioning stimulus despite an increased fear response [188]. These behavioral findings are interesting in light of the 3–4:1 male to female ratio in autism. Lesioned mice performed normally on a basic neurologic test battery with the exception of a small deficit in motor coordination in male 5,7-DHT-lesioned mice [189]. While it is unlikely that the motor deficits contributed significantly to the behavioral alterations in the lesioned mice, it is interesting to note mild motor deficits are observed in children with autism [199].

Another model for autism also has targeted the serotonergic system. Patricia Whitaker-Azmitia and colleagues set out to develop an animal model that mimics the hyperserotonemia in the serum of many patients with autism. In this model, pregnant rats are administered the serotonin agonist 5-methoxytryptamine (5-MT) beginning on gestation day 12 and continuing postnatally for various periods of time. This treatment results in increased serum levels of 5-HT [48], yet decreases in synaptosomal uptake of serotonin in the brainstem and forebrain of the offspring [200]. The authors hypothesize that the loss of serotonergic terminals, centrally, is due to negative feedback by 5-HT receptors to the increased levels of peripheral serotonin [48]. Interestingly, recent data from a genetic model show a reduction in central serotonergic neurotransmission resulting, as well, from decreased peripheral serotonin [47]. The 5-MT-treated offspring also show decreases in ultrasonic vocalizations, suggesting problems with maternal to pup bonding [48]. In addition, the treated offspring show some phenotypic features of autism such as seizures, hyper-responsiveness to sound and touch, and motoric changes that can be likened to stereotypies [201]. Magnetic resonance spectroscopy (MRS) showed no changes in cortical volume in treated animals [201], but that is not too surprising since the MRS studies were performed on adult animals. However, the MRS analysis demonstrated significantly lower NAA/Cr ratios but higher choline/Cr and myoinositol/Cr ratios in the treated animals [201]. This could be the result of metabolic disturbances, changes in maturation, or cell death [201] and is consistent with MRS studies showing metabolic disturbances in autism [202].

Prenatal exposure to the teratogen valproic acid (VPA) is another model for autism. The VPA model is based on the fact that prenatal exposure to VPA has been associated with autism (see [203] for review). VPA is administered to pregnant rats on gestational day (GD) 9, 11.5, or 12.5 depending on the study. VPA-exposed animals exhibit a number of behavioral alterations that resemble autism including decreased sensitivity to pain, diminished acoustic pre-pulse inhibition, locomotor and repetitive/stereotypic-like hyperactivity, and social behaviors [204, 205]. In fear conditioning experiments, VPA-treated rats are more anxious and have long lasting fear memories that are difficult to extinguish [205]. On a cellular level, VPA exposure leads to hyperreactivity and hyperplasticity of local circuits in the cortex and amygdala [205–207]. In terms of serotonin, VPA administered on GD 9 leads to significant increases in serotonin in the hippocampus [208], frontal cortex [203], and periphery [208]. Serotonergic cells in the dorsal raphe are displaced caudally in VPA-treated animals although the total number of serotonergic cells in the raphe is not different from normal controls [203, 209]. In vitro experiments indicate that VPA retards maturation of serotonergic neurons and that addition of the morphogen Sonic hedgehog, whose expression is important for the differentiation of raphe cells, is partially protective [209]. Thus, serotonin dysfunction also plays a key role in the VPA model of autism.

Additional models for autism are based on the hypothesis that defects in the development of the brainstem are responsible for some of the phenotypic features of autism [210–212]. Several reports indicate a strong association between *Hoxa1* and related homeobox genes and autism spectrum disorder [213–215]. These genes are involved in the segmental specification of the brain, particularly the brainstem, in early ontogeny [216]. Abnormally developed brainstem and cranial motor nuclei, during the first trimester, have been associated with autism [210], and mice mutants for Hoxa1 and Hoxb1 display craniofacial abnormalities in structures that are modulated by serotonin in early development [216, 217]. In addition, there is increasing evidence that another homeobox gene, *Engrailed 2* (*En2*), is also associated with autism [218–220]. The engrailed genes are also important in brainstem development, and *En2* knock-out mice show behavioral and cerebellar abnormalities that resemble those in autism [221, 222] (See Chapter 01 by Rossman et al.). Interestingly, the *En2* knock-out mice, when crossbred on a C57BL6J/129S2SV PAS background strain, have increased levels of 5-HT and 5-HIAA in the cerebellum [221]. Thus, En2 expression may regulate serotonergic function. It would be interesting to see whether serotonergic denervation would exacerbate the behavioral and neuropathological features in *En2* knock-out mice.

Summary

In summary, we have presented evidence for a key role of serotonin in the etiology of autism. We hypothesize that serotonin, through its reciprocal interactions with other neurotransmitter systems, as well as neurotrophic and neuroinflammatory factors, contributes to the behavioral impairments that define

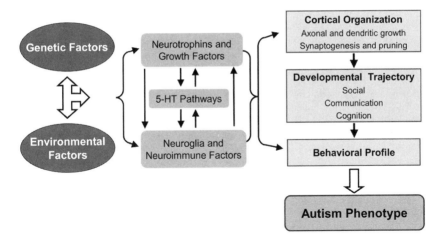

Fig. 5.3 Overview figure for summary

the autism phenotype (Fig. 5.3). We further hypothesize that cortical serotonergic imbalances, early in life, affect the establishment of cortical connections
and alter synaptic plasticity. Such serotonergic imbalances may be inborn,
genetic, or environmentally induced. It should not be overlooked that a number
of highly prevalent environmental toxins affect serotonergic neurotransmission
[223, 224, 225, 226, 227]. Other neurobiological factors such as neurotrophins
and neuroimmune factors (e.g., cytokines and chemokines) also influence the
establishment of neuronal connectivity and networks. Thus, we also hypothesize that abnormalities in neurobehavioral developmental trajectories (social,
communication, and cognitive) are influenced in part by neurotrophic and
neuroimmune pathways that lead to the behavioral profiles of the autism
phenotypes. The goal of future studies from our laboratories is to explore the
interactions among serotonin, neurotrophins, and neuroimmune factors in
autism and in animal models for autism.

References

1. Wallace J. A., Lauder J. M. (1983) Development of the serotonergic system in the rat
   embryo: an immunocytochemical study. *Brain Res Bull* **10**, 459–479.
2. Wallace J. A., Petrusz P., Lauder J. M. (1982) Serotonin immunocytochemistry in the adult
   and developing rat brain: methodological and pharmacological considerations. *Brain Res
   Bull* **9**, 117–129.
3. Lidov H. G., Molliver M. E. (1982) Immunohistochemical study of the development of
   serotonergic neurons in the rat CNS. *Brain Res Bull* **9**, 559–604.
4. Lidov H. G., Molliver M. E. (1982) An immunohistochemical study of serotonin neuron
   development in the rat: ascending pathways and terminal fields. *Brain Res Bull* **8**, 389–430.
5. Rubenstein J. L. (1998) Development of serotonergic neurons and their projections. *Biol
   Psychiatry* **44**, 145–150.

6. Lauder J. M., Wallace J. A., Krebs H. (1981) Roles for serotonin in neuroembryogenesis. *Adv Exp Med Biol* **133**, 477–506.

7. Hunt P. N., Gust J., Mccabe A. K., Bosma M. M. (2006) Primary role of the serotonergic midline system in synchronized spontaneous activity during development of the embryonic mouse hindbrain. *J Neurobiol* **66**, 1239–1252.

8. Laurent A., Goaillard J. M., Cases O., Lebrand C., Gaspar P., Ropert N. (2002) Activity-dependent presynaptic effect of serotonin 1B receptors on the somatosensory thalamo-cortical transmission in neonatal mice. *J Neurosci* **22**, 886–900.

9. Beique J. C., Chapin-Penick E. M., Mladenovic L., Andrade R. (2004) Serotonergic facilitation of synaptic activity in the developing rat prefrontal cortex. *J Physiol* **556**, 739–754.

10. Pardo C. A., Eberhart C. G. (2007) The neurobiology of autism. *Brain Pathol* **17**, 434–447.

11. Aghajanian G. K., Marek G. J. (1999) Serotonin, via 5-HT2A receptors, increases EPSCs in layer V pyramidal cells of prefrontal cortex by an asynchronous mode of glutamate release. *Brain Res* **825**, 161–171.

12. Aghajanian G. K., Marek G. J. (1997) Serotonin induces excitatory postsynaptic potentials in apical dendrites of neocortical pyramidal cells. *Neuropharmacology* **36**, 589–599.

13. Marek G. J., Wright R. A., Schoepp D. D. (2006) 5-Hydroxytryptamine2A (5-HT2A) receptor regulation in rat prefrontal cortex: interaction of a phenethylamine hallucinogen and the metabotropic glutamate2/3 receptor agonist LY354740. *Neurosci Lett* **403**, 256–260.

14. Kalia M. (2005) Neurobiological basis of depression: an update. *Metabolism* **54**, 24–27.

15. Gallinat J., Strohle A., Lang U. E., et al. (2005) Association of human hippocampal neurochemistry, serotonin transporter genetic variation, and anxiety. *Neuroimage* **26**, 123–131.

16. Aghajanian G. K., Marek G. J. (2000) Serotonin model of schizophrenia: emerging role of glutamate mechanisms. *Brain Res Brain Res Rev* **31**, 302–312.

17. Aghajanian G. K., Marek G. J. (1999) Serotonin and hallucinogens. *Neuropsychopharmacology* **21**, 16S–23S.

18. Marek G. J. (2007) Serotonin and dopamine interactions in rodents and primates: implications for psychosis and antipsychotic drug development. *Int Rev Neurobiol* **78**, 165–192.

19. Kolevzon A., Mathewson K. A., Hollander E. (2006) Selective serotonin reuptake inhibitors in autism: a review of efficacy and tolerability. *J Clin Psychiatry* **67**, 407–414.

20. Pae C. U., Patkar A. A. (2007) Paroxetine: current status in psychiatry. *Expert Rev Neurother* **7**, 107–120.

21. Carrasco J. L., Sandner C. (2005) Clinical effects of pharmacological variations in selective serotonin reuptake inhibitors: an overview. *Int J Clin Pract* **59**, 1428–1434.

22. Arroll B., Macgillivray S., Ogston S., et al. (2005) Efficacy and tolerability of tricyclic antidepressants and SSRIs compared with placebo for treatment of depression in primary care: a meta-analysis. *Ann Fam Med* **3**, 449–456.

23. Keller M. B. (2000) Citalopram therapy for depression: a review of 10 years of European experience and data from U.S. clinical trials. *J Clin Psychiatry* **61**, 896–908.

24. Bezchlibnyk-Butler K., Aleksic I., Kennedy S. H. (2000) Citalopram–a review of pharmacological and clinical effects. *J Psychiatry Neurosci* **25**, 241–254.

25. Geddes J. R., Freemantle N., Mason J., Eccles M. P., Boynton J. (2000) SSRIs versus other antidepressants for depressive disorder. *Cochrane Database Syst Rev* CD001851 (2).

26. Devane C. L., Sallee F. R. (1996) Serotonin selective reuptake inhibitors in child and adolescent psychopharmacology: a review of published experience. *J Clin Psychiatry* **57**, 55–66.

27. Hetrick S., Merry S., Mckenzie J., Sindahl P., Proctor M. (2007) Selective serotonin reuptake inhibitors (SSRIs) for depressive disorders in children and adolescents. *Cochrane Database Syst Rev* CD004851 (3).

28. Farley R. L. (2005) Pharmacological treatment of major depressive disorder in adolescents. *ScientificWorldJournal* **5**, 420–426.

29. Weller E. B., Tucker S., Weller R. A. (2005) The selective serotonin reuptake inhibitors controversy in the treatment of depression in children. *Curr Psychiatry Rep* **7**, 87–90.

30. Courtney D. B. (2004) Selective serotonin reuptake inhibitor and venlafaxine use in children and adolescents with major depressive disorder: a systematic review of published randomized controlled trials. *Can J Psychiatry* **49**, 557–563.

31. Moore M. L., Eichner S. F., Jones J. R. (2004) Treating functional impairment of autism with selective serotonin-reuptake inhibitors. *Ann Pharmacother* **38**, 1515–1519.

32. Posey D. J., Erickson C. A., Stigler K. A., Mcdougle C. J. (2006) The use of selective serotonin reuptake inhibitors in autism and related disorders. *J Child Adolesc Psychopharmacol* **16**, 181–186.

33. Baghdadli A., Gonnier V., Aussilloux C. (2002) [Review of psychopharmacological treatments in adolescents and adults with autistic disorders]. *Encephale* **28**, 248–254.

34. Buchsbaum M. S., Hollander E., Haznedar M. M., et al. (2001) Effect of fluoxetine on regional cerebral metabolism in autistic spectrum disorders: a pilot study. *Int J Neuropsychopharmacol* **4**, 119–125.

35. Delong G. R., Teague L. A., Mcswain Kamran M. (1998) Effects of fluoxetine treatment in young children with idiopathic autism. *Dev Med Child Neurol* **40**, 551–562.

36. Fatemi S. H., Realmuto G. M., Khan L., Thuras P. (1998) Fluoxetine in treatment of adolescent patients with autism: a longitudinal open trial. *J Autism Dev Disord* **28**, 303–307.

37. Ghaziuddin M., Tsai L., Ghaziuddin N. (1991) Fluoxetine in autism with depression. *J Am Acad Child Adolesc Psychiatry* **30**, 508–509.

38. Schain R., Freedman D. (1961) Studies on 5-hydroxyindole metablism in autistic and other mentally retarded children. *J Pediatr* **58**, 315–320.

39. Cook E. H. (1990) Autism: review of neurochemical investigation. *Synapse* **6**, 292–308.

40. Burgess N. K., Sweeten T. L., Mcmahon W. M., Fujinami R. S. (2006) Hyperserotoninemia and altered immunity in autism. *J Autism Dev Disord* **36**, 697–704.

41. Lam K. S., Aman M. G., Arnold L. E. (2006) Neurochemical correlates of autistic disorder: a review of the literature. *Res Dev Disabil* **27**, 254–289.

42. Cook E. H., Leventhal B. L. (1996) The serotonin system in autism. *Curr Opin Pediatr* **8**, 348–354.

43. D'eufemia P., Finocchiaro R., Celli M., Viozzi L., Monteleone D., Giardini O. (1995) Low serum tryptophan to large neutral amino acids ratio in idiopathic infantile autism. *Biomed Pharmacother* **49**, 288–292.

44. Buitelaar J. K., Willemsen-Swinkels S. H. (2000) Autism: current theories regarding its pathogenesis and implications for rational pharmacotherapy. *Paediatr Drugs* **2**, 67–81.

45. Klauck S. M., Poustka F., Benner A., Lesch K. P., Poustka A. (1997) Serotonin transporter (5-HTT) gene variants associated with autism? *Hum Mol Genet* **6**, 2233–2238.

46. Connors S. L., Matteson K. J., Sega G. A., Lozzio C. B., Carroll R. C., Zimmerman A. W. (2006) Plasma serotonin in autism. *Pediatr Neurol* **35**, 182–186.

47. Cote F., Fligny C., Bayard E., et al. (2007) Maternal serotonin is crucial for murine embryonic development. *Proc Natl Acad Sci U S A* **104**, 329–334.

48. Whitaker-Azmitia P. M. (2005) Behavioral and cellular consequences of increasing serotonergic activity during brain development: a role in autism? *Int J Dev Neurosci* **23**, 75–83.

49. Mcbride P. A., Anderson G. M., Hertzig M. E., et al. (1998) Effects of diagnosis, race, and puberty on platelet serotonin levels in autism and mental retardation. *J Am Acad Child Adolesc Psychiatry* **37**, 767–776.

50. Cook E. H., Jr., Arora R. C., Anderson G. M., et al. (1993) Platelet serotonin studies in hyperserotonemic relatives of children with autistic disorder. *Life Sci* **52**, 2005–2015.

51. Murphy D. G., Daly E., Schmitz N., et al. (2006) Cortical serotonin 5-HT2A receptor binding and social communication in adults with Asperger's syndrome: an in vivo SPECT study. *Am J Psychiatry* **163**, 934–936.

52. Buitelaar J. K., Willemsen-Swinkels S. H. (2000) Medication treatment in subjects with autistic spectrum disorders. *Eur Child Adolesc Psychiatry* **9**, 185–197.
53. Mcdougle C. J., Scahill L., Mccracken J. T., et al. (2000) Research Units on Pediatric Psychopharmacology (RUPP) Autism Network. Background and rationale for an initial controlled study of risperidone. *Child Adolesc Psychiatr Clin N Am* **9**, 201–224.
54. Cook E. H., Jr., Courchesne R., Lord C., et al. (1997) Evidence of linkage between the serotonin transporter and autistic disorder. *Mol Psychiatry* **2**, 247–250.
55. Devlin B., Cook E. H., Jr., Coon H., et al. (2005) Autism and the serotonin transporter: the long and short of it. *Mol Psychiatry* **10**, 1110–1116.
56. Sutcliffe J. S., Delahanty R. J., Prasad H. C., et al. (2005) Allelic heterogeneity at the serotonin transporter locus (SLC6A4) confers susceptibility to autism and rigid-compulsive behaviors. *Am J Hum Genet* **77**, 265–279.
57. Hu V. W., Frank B. C., Heine S., Lee N. H., Quackenbush J. (2006) Gene expression profiling of lymphoblastoid cell lines from monozygotic twins discordant in severity of autism reveals differential regulation of neurologically relevant genes. *BMC Genomics* **7**, 118.
58. Seckl J. R., Meaney M. J. (2004) Glucocorticoid programming. *Ann N Y Acad Sci* **1032**, 63–84.
59. Vazquez D. M., Lopez J. F., Van Hoers H., Watson S. J., Levine S. (2000) Maternal deprivation regulates serotonin 1A and 2A receptors in the infant rat. *Brain Res* **855**, 76–82.
60. Weaver I. C., Diorio J., Seckl J. R., Szyf M., Meaney M. J. (2004) Early environmental regulation of hippocampal glucocorticoid receptor gene expression: characterization of intracellular mediators and potential genomic target sites. *Ann N Y Acad Sci* **1024**, 182–212.
61. Corbett B. A., Mendoza S., Abdullah M., Wegelin J. A., Levine S. (2006) Cortisol circadian rhythms and response to stress in children with autism. *Psychoneuroendocrinology* **31**, 59–68.
62. Jansen L. M., Gispen-De Wied C. C., Wiegant V. M., Westenberg H. G., Lahuis B. E., Van Engeland H. (2006) Autonomic and neuroendocrine responses to a psychosocial stressor in adults with autistic spectrum disorder. *J Autism Dev Disord* **36**, 891–899.
63. Hazell P. (2007) Drug therapy for attention-deficit/hyperactivity disorder-like symptoms in autistic disorder. *J Paediatr Child Health* **43**, 19–24.
64. Hollander E., Phillips A., Chaplin W., et al. (2005) A placebo controlled crossover trial of liquid fluoxetine on repetitive behaviors in childhood and adolescent autism. *Neuropsychopharmacology* **30**, 582–589.
65. Cook E. H., Jr., Rowlett R., Jaselskis C., Leventhal B. L. (1992) Fluoxetine treatment of children and adults with autistic disorder and mental retardation. *J Am Acad Child Adolesc Psychiatry* **31**, 739–745.
66. Mcdougle C. J., Naylor S. T., Cohen D. J., Aghajanian G. K., Heninger G. R., Price L. H. (1996) Effects of tryptophan depletion in drug-free adults with autistic disorder. *Arch Gen Psychiatry* **53**, 993–1000.
67. Chugani D. C., Muzik O., Behen M., et al. (1999) Developmental changes in brain serotonin synthesis capacity in autistic and nonautistic children. *Ann Neurol* **45**, 287–295.
68. Chugani D. C., Muzik O., Rothermel R., et al. (1997) Altered serotonin synthesis in the dentatothalamocortical pathway in autistic boys. *Ann Neurol* **42**, 666–669.
69. Chandana S. R., Behen M. E., Juhasz C., et al. (2005) Significance of abnormalities in developmental trajectory and asymmetry of cortical serotonin synthesis in autism. *Int J Dev Neurosci* **23**, 171–182.
70. Wassink T. H., Hazlett H. C., Epping E. A., et al. (2007) Cerebral cortical gray matter overgrowth and functional variation of the serotonin transporter gene in autism. *Arch Gen Psychiatry* **64**, 709–717.
71. Hohmann C. F., Hamon R., Batshaw M. L., Coyle J. T. (1988) Transient postnatal elevation of serotonin levels in mouse neocortex. *Brain Res* **471**, 163–166.

72. Connell S., Karikari C., Hohmann C. F. (2004) Sex-specific development of cortical monoamine levels in mouse. *Brain Res Dev Brain Res* **151**, 187–191.

73. Fujimiya M., Hosoda S., Kitahama K., Kimura H., Maeda T. (1986) Early development of serotonin neuron in the rat brain as studied by immunohistochemistry combined with tryptophan administration. *Brain Dev* 8, 335–342.

74. D'amato R. J., Blue M. E., Largent B. L., et al. (1987) Ontogeny of the serotonergic projection to rat neocortex: transient expression of a dense innervation to primary sensory areas. *Proc Natl Acad Sci U S A* **84**, 4322–4326.

75. Rhoades R. W., Bennett-Clarke C. A., Chiaia N. L., et al. (1990) Development and lesion induced reorganization of the cortical representation of the rat's body surface as revealed by immunocytochemistry for serotonin. *J Comp Neurol* **293**, 190–207.

76. Blue M. E., Erzurumlu R. S., Jhaveri S. (1991) A comparison of pattern formation by thalamocortical and serotonergic afferents in the rat barrel field cortex. *Cereb Cortex* **1**, 380–389.

77. Woolsey T. A., Van Der Loos H. (1970) The structural organization of layer IV in the somatosensory region (SI) of mouse cerebral cortex. The description of a cortical field composed of discrete cytoarchitectonic units. *Brain Res* **17**, 205–242.

78. Bennett-Clarke C. A., Chiaia N. L., Rhoades R. W. (1996) Thalamocortical afferents in rat transiently express high-affinity serotonin uptake sites. *Brain Res* **733**, 301–306.

79. Bennett-Clarke C. A., Lane R. D., Rhoades R. W. (1995) Fenfluramine depletes serotonin from the developing cortex and alters thalamocortical organization. *Brain Res* **702**, 255–260.

80. Bennett-Clarke C. A., Leslie M. J., Chiaia N. L., Rhoades R. W. (1993) Serotonin 1B receptors in the developing somatosensory and visual cortices are located on thalamocortical axons. *Proc Natl Acad Sci U S A* **90**, 153–157.

81. Cases O., Lebrand C., Giros B., et al. (1998) Plasma membrane transporters of serotonin, dopamine, and norepinephrine mediate serotonin accumulation in atypical locations in the developing brain of monoamine oxidase A knock-outs. *J Neurosci* **18**, 6914–6927.

82. Lebrand C., Cases O., Adelbrecht C., et al. (1996) Transient uptake and storage of serotonin in developing thalamic neurons. *Neuron* **17**, 823–835.

83. Lebrand C., Cases O., Wehrle R., Blakely R. D., Edwards R. H., Gaspar P. (1998) Transient developmental expression of monoamine transporters in the rodent forebrain. *J Comp Neurol* **401**, 506–524.

84. Mansour-Robaey S., Mechawar N., Radja F., Beaulieu C., Descarries L. (1998) Quantified distribution of serotonin transporter and receptors during the postnatal development of the rat barrel field cortex. *Brain Res Dev Brain Res* **107**, 159–163.

85. Rhoades R. W., Bennett-Clarke C. A., Shi M. Y., Mooney R. D. (1994) Effects of 5-HT on thalamocortical synaptic transmission in the developing rat. *J Neurophysiol* **72**, 2438–2450.

86. Lieske V., Bennett-Clarke C. A., Rhoades R. W. (1999) Effects of serotonin on neurite outgrowth from thalamic neurons in vitro. *Neuroscience* **90**, 967–974.

87. Lotto B., Upton L., Price D. J., Gaspar P. (1999) Serotonin receptor activation enhances neurite outgrowth of thalamic neurones in rodents. *Neurosci Lett* **269**, 87–90.

88. Dooley A. E., Pappas I. S., Parnavelas J. G. (1997) Serotonin promotes the survival of cortical glutamatergic neurons in vitro. *Exp Neurol* **148**, 205–214.

89. Lavdas A. A., Blue M. E., Lincoln J., Parnavelas J. G. (1997) Serotonin promotes the differentiation of glutamate neurons in organotypic slice cultures of the developing cerebral cortex. *J Neurosci* **17**, 7872–7880.

90. Osterheld-Haas M. C., Hornung J. P. (1996) Laminar development of the mouse barrel cortex: effects of neurotoxins against monoamines. *Exp Brain Res* **110**, 183–195.

91. Bennett-Clarke C. A., Leslie M. J., Lane R. D., Rhoades R. W. (1994) Effect of serotonin depletion on vibrissa-related patterns of thalamic afferents in the rat's somatosensory cortex. *J Neurosci* **14**, 7594–7607.

92. Persico A. M., Altamura C., Calia E., et al. (2000) Serotonin depletion and barrel cortex development: impact of growth impairment vs. serotonin effects on thalamocortical endings. *Cereb Cortex* **10**, 181–191.
93. Vitalis T., Cases O., Callebert J., et al. (1998) Effects of monoamine oxidase A inhibition on barrel formation in the mouse somatosensory cortex: determination of a sensitive developmental period. *J Comp Neurol* **393**, 169–184.
94. Persico A. M., Mengual E., Moessner R., et al. (2001) Barrel pattern formation requires serotonin uptake by thalamocortical afferents, and not vesicular monoamine release. *J Neurosci* **21**, 6862–6873.
95. Mazer C., Muneyyirci J., Taheny K., Raio N., Borella A., Whitaker-Azmitia P. (1997) Serotonin depletion during synaptogenesis leads to decreased synaptic density and learning deficits in the adult rat: a possible model of neurodevelopmental disorders with cognitive deficits. *Brain Res* **760**, 68–73.
96. Salichon N., Gaspar P., Upton A. L., et al. (2001) Excessive activation of serotonin (5-HT) 1B receptors disrupts the formation of sensory maps in monoamine oxidase a and 5-ht transporter knock-out mice. *J Neurosci* **21**, 884–896.
97. Salt T. E., Eaton S. A. (1996) Functions of ionotropic and metabotropic glutamate receptors in sensory transmission in the mammalian thalamus. *Prog Neurobiol* **48**, 55–72.
98. Constantine-Paton M., Cline H. T., Debski E. (1990) Patterned activity, synaptic convergence, and the NMDA receptor in developing visual pathways. *Annu Rev Neurosci* **13**, 129–154.
99. Fox K., Daw N. W. (1993) Do NMDA receptors have a critical function in visual cortical plasticity? *Trends Neurosci* **16**, 116–122.
100. Hattori H., Wasterlain C. G. (1990) Excitatory amino acids in the developing brain: ontogeny, plasticity, and excitotoxicity. *Pediatr Neurol* **6**, 219–228.
101. Mcdonald J. W., Johnston M. V. (1990) Physiological and pathophysiological roles of excitatory amino acids during central nervous system development. *Brain Res Brain Res Rev* **15**, 41–70.
102. Pin J. P., Duvoisin R. (1995) The metabotropic glutamate receptors: structure and functions. *Neuropharmacology* **34**, 1–26.
103. Bear M. F., Colman H. (1990) Binocular competition in the control of geniculate cell size depends upon visual cortical N-methyl-D-aspartate receptor activation. *Proc Natl Acad Sci U S A* **87**, 9246–9249.
104. Cline H. T., Constantine-Paton M. (1989) NMDA receptor antagonists disrupt the retinotectal topographic map. *Neuron* **3**, 413–426.
105. Rocha M., Sur M. (1995) Rapid acquisition of dendritic spines by visual thalamic neurons after blockade of N-methyl-D-aspartate receptors. *Proc Natl Acad Sci U S A* **92**, 8026–8030.
106. Datwani A., Iwasato T., Itohara S., Erzurumlu R. S. (2002) NMDA receptor-dependent pattern transfer from afferents to postsynaptic cells and dendritic differentiation in the barrel cortex. *Mol Cell Neurosci* **21**, 477–492.
107. Lee L. J., Iwasato T., Itohara S., Erzurumlu R. S. (2005) Exuberant thalamocortical axon arborization in cortex-specific NMDAR1 knockout mice. *J Comp Neurol* **485**, 280–292.
108. Iwasato T., Datwani A., Wolf A. M., et al. (2000) Cortex-restricted disruption of NMDAR1 impairs neuronal patterns in the barrel cortex. *Nature* **406**, 726–731.
109. Iwasato T., Erzurumlu R. S., Huerta P. T., et al. (1997) NMDA receptor-dependent refinement of somatotopic maps. *Neuron* **19**, 1201–1210.
110. Salt T. E., Turner J. P. (1998a) Modulation of sensory inhibition in the ventrobasal thalamus via activation of group II metabotropic glutamate receptors by 2R,4R-aminopyrrolidine-2,4-dicarboxylate. *Exp Brain Res* **121**, 181–185.
111. Salt T. E., Turner J. P. (1998b) Reduction of sensory and metabotropic glutamate receptor responses in the thalamus by the novel metabotropic glutamate receptor-1-selective antagonist S-2-methyl-4-carboxy-phenylglycine. *Neuroscience* **85**, 655–658.

112. Mateo Z., Porter J. T. (2007) Group II metabotropic glutamate receptors inhibit glutamate release at thalamocortical synapses in the developing somatosensory cortex. *Neuroscience* **146**, 1062–1072.

113. Kossut M., Glazewski S., Siucinska E., Skangiel-Kramska J. (1993) Functional plasticity and neurotransmitter receptor binding in the vibrissal barrel cortex. *Acta Neurobiol Exp* **53**, 161–173.

114. Bortolotto Z. A., Bashir Z. I., Davies C. H., Collingridge G. L. (1994) A molecular switch activated by metabotropic glutamate receptors regulates induction of long-term potentiation. *Nature* **368**, 740–743.

115. Shigemoto R., Abe T., Nomura S., Nakanishi S., Hirano T. (1994) Antibodies inactivating mGluR1 metabotropic glutamate receptor block long-term depression in cultured Purkinje cells. *Neuron* **12**, 1245–1255.

116. O'connor J. J., Rowan M. J., Anwyl R. (1994) Long-lasting enhancement of NMDA receptor-mediated synaptic transmission by metabotropic glutamate receptor activation. *Nature* **367**, 557–559.

117. Dudek S. M. (1996) A discussion of activity-dependent forms of synaptic weakening and their possible role in ocular dominance plasticity. *J Physiol Paris* **90**, 167–170.

118. Gomperts S. N., Carroll R., Malenka R. C., Nicoll R. A. (2000) Distinct roles for ionotropic and metabotropic glutamate receptors in the maturation of excitatory synapses. *J Neurosci* **20**, 2229–2237.

119. Hannan A. J., Blakemore C., Katsnelson A., et al. (2001) PLC-beta1, activated via mGluRs, mediates activity-dependent differentiation in cerebral cortex. *Nat Neurosci* **4**, 282–288.

120. Catania M. V., D'antoni S., Bonaccorso C. M., Aronica E., Bear M. F., Nicoletti F. (2007) Group I metabotropic glutamate receptors: a role in neurodevelopmental disorders? *Mol Neurobiol* **35**, 298–307.

121. Marty S., Carroll P., Cellerino A., et al. (1996) Brain-derived neurotrophic factor promotes the differentiation of various hippocampal nonpyramidal neurons, including Cajal-Retzius cells, in organotypic slice cultures. *J Neurosci* **16**, 675–687.

122. Cellerino A., Maffei L. (1996) The action of neurotrophins in the development and plasticity of the visual cortex. *Prog Neurobiol* **49**, 53–71.

123. Xu B., Gottschalk W., Chow A., et al. (2000) The role of brain-derived neurotrophic factor receptors in the mature hippocampus: modulation of long-term potentiation through a presynaptic mechanism involving TrkB. *J Neurosci* **20**, 6888–6897.

124. Xu B., Zang K., Ruff N. L., et al. (2000) Cortical degeneration in the absence of neurotrophin signaling: dendritic retraction and neuronal loss after removal of the receptor TrkB. *Neuron* **26**, 233–245.

125. Reichardt L. F. (2006) Neurotrophin-regulated signalling pathways. *Philos Trans R Soc Lond B Biol Sci* **361**, 1545–1564.

126. Branchi I., Francia N., Alleva E. (2004) Epigenetic control of neurobehavioural plasticity: the role of neurotrophins. *Behav Pharmacol* **15**, 353–362.

127. Huang E. J., Reichardt L. F. (2003) Trk receptors: roles in neuronal signal transduction. *Annu Rev Biochem* **72**, 609–642.

128. Mattson M. P., Maudsley S., Martin B. (2004) BDNF and 5-HT: a dynamic duo in age-related neuronal plasticity and neurodegenerative disorders. *Trends Neurosci* **27**, 589–594.

129. Lyons W. E., Mamounas L. A., Ricaurte G. A., et al. (1999) Brain-derived neurotrophic factor-deficient mice develop aggressiveness and hyperphagia in conjunction with brain serotonergic abnormalities. *Proc Natl Acad Sci U S A* **96**, 15239–15244.

130. Daws L. C., Munn J. L., Valdez M. F., Frosto-Burke T., Hensler J. G. (2007) Serotonin transporter function, but not expression, is dependent on brain-derived neurotrophic factor (BDNF): in vivo studies in BDNF-deficient mice. *J Neurochem* **101**, 641–651.

131. Grider M. H., Mamounas L. A., Le W., Shine H. D. (2005) In situ expression of brain-derived neurotrophic factor or neurotrophin-3 promotes sprouting of cortical serotonergic axons following a neurotoxic lesion. *J Neurosci Res* **82**, 404–412.

132. Mamounas L. A., Altar C. A., Blue M. E., Kaplan D. R., Tessarollo L., Lyons W. E. (2000) BDNF promotes the regenerative sprouting, but not survival, of injured serotonergic axons in the adult rat brain. *J Neurosci* **20**, 771–782.

133. Zetterstrom T. S., Pei Q., Madhav T. R., Coppell A. L., Lewis L., Grahame-Smith D. G. (1999) Manipulations of brain 5-HT levels affect gene expression for BDNF in rat brain. *Neuropharmacology* **38**, 1063–1073.

134. Angelucci F., Mathe A. A., Aloe L. (2004) Neurotrophic factors and CNS disorders: findings in rodent models of depression and schizophrenia. *Prog Brain Res* **146**, 151–165.

135. Duman R. S. (2004) Role of neurotrophic factors in the etiology and treatment of mood disorders. *Neuromolecular Med* **5**, 11–25.

136. Shoval G., Weizman A. (2005) The possible role of neurotrophins in the pathogenesis and therapy of schizophrenia. *Eur Neuropsychopharmacol* **15**, 319–329.

137. Neumeister A., Yuan P., Young T. A., et al. (2005) Effects of tryptophan depletion on serum levels of brain-derived neurotrophic factor in unmedicated patients with remitted depression and healthy subjects. *Am J Psychiatry* **162**, 805–807.

138. Nelson K. B., Grether J. K., Croen L. A., et al. (2001) Neuropeptides and neurotrophins in neonatal blood of children with autism or mental retardation. *Ann Neurol* **49**, 597–606.

139. Miyazaki K., Narita N., Sakuta R., et al. (2004) Serum neurotrophin concentrations in autism and mental retardation: a pilot study. *Brain Dev* **26**, 292–295.

140. Connolly A. M., Chez M., Streif E. M., et al. (2006) Brain-derived neurotrophic factor and autoantibodies to neural antigens in sera of children with autistic spectrum disorders, Landau-Kleffner syndrome, and epilepsy. *Biol Psychiatry* **59**, 354–363.

141. Mcallister A. K. (2002) Neurotrophins and cortical development. *Results Probl Cell Differ* **39**, 89–112.

142. Galter D., Unsicker K. (2000) Brain-derived neurotrophic factor and trkB are essential for cAMP-mediated induction of the serotonergic neuronal phenotype. *J Neurosci Res* **61**, 295–301.

143. Fryer R. H., Kaplan D. R., Feinstein S. C., Radeke M. J., Grayson D. R., Kromer L. F. (1996) Developmental and mature expression of full-length and truncated TrkB receptors in the rat forebrain. *J Comp Neurol* **374**, 21–40.

144. Kuipers S. D., Bramham C. R. (2006) Brain-derived neurotrophic factor mechanisms and function in adult synaptic plasticity: new insights and implications for therapy. *Curr Opin Drug Discov Devel* **9**, 580–586.

145. Gomes R. A., Hampton C., El-Sabeawy F., Sabo S. L., Mcallister A. K. (2006) The dynamic distribution of TrkB receptors before, during, and after synapse formation between cortical neurons. *J Neurosci* **26**, 11487–11500.

146. Roceri M., Cirulli F., Pessina C., Peretto P., Racagni G., Riva M. A. (2004) Postnatal repeated maternal deprivation produces age-dependent changes of brain-derived neurotrophic factor expression in selected rat brain regions. *Biol Psychiatry* **55**, 708–714.

147. Fumagalli F., Bedogni F., Perez J., Racagni G., Riva M. A. (2004) Corticostriatal brain-derived neurotrophic factor dysregulation in adult rats following prenatal stress. *Eur J Neurosci* **20**, 1348–1354.

148. Garoflos E., Panagiotaropoulos T., Pondiki S., Stamatakis A., Philippidis E., Stylianopoulou F. (2005) Cellular mechanisms underlying the effects of an early experience on cognitive abilities and affective states. *Ann Gen Psychiatry* **4**, 8.

149. Gilmore J. H., Jarskog L. F., Vadlamudi S. (2003) Maternal infection regulates BDNF and NGF expression in fetal and neonatal brain and maternal-fetal unit of the rat. *J Neuroimmunol* **138**, 49–55.

150. Cavus I., Duman R. S. (2003) Influence of estradiol, stress, and 5-HT2A agonist treatment on brain-derived neurotrophic factor expression in female rats. *Biol Psychiatry* **54**, 59–69.

151. Szapacs M. E., Mathews T. A., Tessarollo L., Ernest Lyons W., Mamounas L. A., Andrews A. M. (2004) Exploring the relationship between serotonin and brain-derived

neurotrophic factor: analysis of BDNF protein and extraneuronal 5-HT in mice with reduced serotonin transporter or BDNF expression. *J Neurosci Methods* **140**, 81–92.

152. Vaidya V. A., Terwilliger R. M., Duman R. S. (1999) Role of 5-HT2A receptors in the stress-induced down-regulation of brain- derived neurotrophic factor expression in rat hippocampus. *Neurosci Lett* **262**, 1–4.

153. Croonenberghs J., Wauters A., Devreese K., et al. (2002) Increased serum albumin, gamma globulin, immunoglobulin IgG, and IgG2 and IgG4 in autism. *Psychol Med* **32**, 1457–1463.

154. Malek-Ahmadi P. (2001) Cytokines and etiopathogenesis of pervasive developmental disorders. *Med Hypotheses* **56**, 321–324.

155. Bauer S., Kerr B. J., Patterson P. H. (2007) The neuropoietic cytokine family in development, plasticity, disease and injury. *Nat Rev Neurosci* **8**, 221–232.

156. Pardo C. A., Vargas D. L., Zimmerman A. W. (2005) Immunity, neuroglia and neuroinflammation in autism. *Int Rev Psychiatry* **17**, 485–495.

157. Korvatska E., Van De Water J., Anders T. F., Gershwin M. E. (2002) Genetic and immunologic considerations in autism. *Neurobiol Dis* **9**, 107–125.

158. Licinio J., Alvarado I., Wong M. L. (2002) Autoimmunity in autism. *Mol Psychiatry* **7**, 329.

159. Torrente F., Ashwood P., Day R., et al. (2002) Small intestinal enteropathy with epithelial IgG and complement deposition in children with regressive autism. *Mol Psychiatry* **7**, 375–382, 34.

160. Bailey A., Luthert P., Dean A., et al. (1998) A clinicopathological study of autism. *Brain* **121** (Pt 5), 889–905.

161. Guerin P., Lyon G., Barthelemy C., et al. (1996) Neuropathological study of a case of autistic syndrome with severe mental retardation. *Dev Med Child Neurol* **38**, 203–211.

162. Gupta S., Aggarwal S., Rashanravan B., Lee T. (1998) Th1- and Th2-like cytokines in CD4+ and CD8+ T cells in autism. *J Neuroimmunol* **85**, 106–109.

163. Singh V. K., Warren R., Averett R., Ghaziuddin M. (1997) Circulating autoantibodies to neuronal and glial filament proteins in autism. *Pediatr Neurol* **17**, 88–90.

164. Singh V. K., Lin S. X., Newell E., Nelson C. (2002) Abnormal measles-mumps-rubella antibodies and CNS autoimmunity in children with autism. *J Biomed Sci* **9**, 359–364.

165. Vojdani A., Campbell A. W., Anyanwu E., Kashanian A., Bock K., Vojdani E. (2002) Antibodies to neuron-specific antigens in children with autism: possible cross-reaction with encephalitogenic proteins from milk, Chlamydia pneumoniae and Streptococcus group A. *J Neuroimmunol* **129**, 168–177.

166. Jyonouchi H., Sun S., Le H. (2001) Proinflammatory and regulatory cytokine production associated with innate and adaptive immune responses in children with autism spectrum disorders and developmental regression. *J Neuroimmunol* **120**, 170–179.

167. Molloy C. A., Morrow A. L., Meinzen-Derr J., et al. (2006) Elevated cytokine levels in children with autism spectrum disorder. *J Neuroimmunol* **172**, 198–205.

168. Vargas D. L., Nascimbene C., Krishnan C., Zimmerman A. W., Pardo C. A. (2005) Neuroglial activation and neuroinflammation in the brain of patients with autism. *Ann Neurol* **57**, 67–81.

169. Gandhi R., Hayley S., Gibb J., Merali Z., Anisman H. (2007) Influence of poly I:C on sickness behaviors, plasma cytokines, corticosterone and central monoamine activity: moderation by social stressors. *Brain Behav Immun* **21**, 477–489.

170. Hayley S., Poulter M. O., Merali Z., Anisman H. (2005) The pathogenesis of clinical depression: stressor- and cytokine-induced alterations of neuroplasticity. *Neuroscience* **135**, 659–678.

171. Cloez-Tayarani I., Changeux J. P. (2007) Nicotine and serotonin in immune regulation and inflammatory processes: a perspective. *J Leukoc Biol* **81**, 599–606.

172. Kushnir-Sukhov N. M., Gilfillan A. M., Coleman J. W., et al. (2006) 5-hydroxytryptamine induces mast cell adhesion and migration. *J Immunol* **177**, 6422–6432.

173. Katoh N., Soga F., Nara T., et al. (2006) Effect of serotonin on the differentiation of human monocytes into dendritic cells. *Clin Exp Immunol* **146**, 354–361.
174. Idzko M., Panther E., Stratz C., et al. (2004) The serotoninergic receptors of human dendritic cells: identification and coupling to cytokine release. *J Immunol* **172**, 6011–6019.
175. Tsao C. W., Lin Y. S., Chen C. C., Bai C. H., Wu S. R. (2006) Cytokines and serotonin transporter in patients with major depression. *Prog Neuropsychopharmacol Biol Psychiatry* **30**, 899–905.
176. Maes M., Ombelet W., De Jongh R., Kenis G., Bosmans E. (2001) The inflammatory response following delivery is amplified in women who previously suffered from major depression, suggesting that major depression is accompanied by a sensitization of the inflammatory response system. *J Affect Disord* **63**, 85–92.
177. Russo S., Kema I. P., Fokkema M. R., et al. (2003) Tryptophan as a link between psychopathology and somatic states. *Psychosom Med* **65**, 665–671.
178. Spalletta G., Bossu P., Ciaramella A., Bria P., Caltagirone C., Robinson R. G. (2006) The etiology of poststroke depression: a review of the literature and a new hypothesis involving inflammatory cytokines. *Mol Psychiatry* **11**, 984–991.
179. Capuron L., Miller A. H. (2004) Cytokines and psychopathology: lessons from interferon-alpha. *Biol Psychiatry* **56**, 819–824.
180. Schaefer M., Schwaiger M., Pich M., Lieb K., Heinz A. (2003) Neurotransmitter changes by interferon-alpha and therapeutic implications. *Pharmacopsychiatry* **36** Suppl 3, S203–S206.
181. Sato T., Suzuki E., Yokoyama M., Semba J., Watanabe S., Miyaoka H. (2006) Chronic intraperitoneal injection of interferon-alpha reduces serotonin levels in various regions of rat brain, but does not change levels of serotonin transporter mRNA, nitrite or nitrate. *Psychiatry Clin Neurosci* **60**, 499–506.
182. Kamata M., Higuchi H., Yoshimoto M., Yoshida K., Shimizu T. (2000) Effect of single intracerebroventricular injection of alpha-interferon on monoamine concentrations in the rat brain. *Eur Neuropsychopharmacol* **10**, 129–132.
183. Foley K. F., Pantano C., Ciolino A., Mawe G. M. (2007) IFN-{gamma} and TNF-{alpha} decrease serotonin transporter function and expression in Caco2 cells. *Am J Physiol Gastrointest Liver Physiol* **292**, G779–G784.
184. Myint A. M., Kim Y. K. (2003) Cytokine-serotonin interaction through IDO: a neuro-degeneration hypothesis of depression. *Med Hypotheses* **61**, 519–525.
185. Zhu C. B., Blakely R. D., Hewlett W. A. (2006) The proinflammatory cytokines interleukin-1beta and tumor necrosis factor-alpha activate serotonin transporters. *Neuropsychopharmacology* **31**, 2121–2131.
186. Zhang J., Terreni L., De Simoni M. G., Dunn A. J. (2001) Peripheral interleukin-6 administration increases extracellular concentrations of serotonin and the evoked release of serotonin in the rat striatum. *Neurochem Int* **38**, 303–308.
187. Miyata M., Ito M., Sasajima T., Ohira H., Kasukawa R. (2001) Effect of a serotonin receptor antagonist on interleukin-6-induced pulmonary hypertension in rats. *Chest* **119**, 554–561.
188. Boylan C. B., Blue M. E., Hohmann C. F. (2007) Modeling early cortical serotonergic deficits in autism. *Behav Brain Res* **176**, 94–108.
189. Hohmann C. F., Walker E. M., Boylan C. B., Blue M. E. (2007) Neonatal serotonin depletion alters behavioral responses to spatial change and novelty. *Brain Res* **1139**, 163–177.
190. Hohmann C. F., Richardson C., Pitts E., Berger-Sweeney J. (2000) Neonatal 5,7-DHT lesions cause sex-specific changes in mouse cortical morphogenesis. *Neural Plast* **7**, 213–232.
191. Hardan A. Y., Muddasani S., Vemulapalli M., Keshavan M. S., Minshew N. J. (2006) An MRI study of increased cortical thickness in autism. *Am J Psychiatry* **163**, 1290–1292.
192. Courchesne E. (2002) Abnormal early brain development in autism. *Mol Psychiatry* **7** Suppl 2, S21–S23.

193. Courchesne E., Carper R., Akshoomoff N. (2003) Evidence of brain overgrowth in the first year of life in autism. *JAMA* **290**, 337–344.
194. Courchesne E., Redcay E., Kennedy D. P. (2004) The autistic brain: birth through adulthood. *Curr Opin Neurol* **17**, 489–496.
195. Hazlett H. C., Poe M., Gerig G., et al. (2005) Magnetic resonance imaging and head circumference study of brain size in autism: birth through age 2 years. *Arch Gen Psychiatry* **62**, 1366–1376.
196. Aylward E., Minshew N., Field K., Sparks B., Singh N. (2002) Effects of age on brain volume and head circumference in autism. *Neurology* Jul 23;59(2), 175–183.
197. Redcay E., Courchesne E. (2005) When is the brain enlarged in autism? A meta-analysis of all brain size reports. *Biol Psychiatry* **58**, 1–9.
198. Sacco R., Militerni R., Frolli A., et al. (2007) Clinical, Morphological, and Biochemical Correlates of Head Circumference in Autism. *Biol Psychiatry* **62**, 1038–1047.
199. Dziuk M. A., Gidley Larson J. C., Apostu A., Mahone E. M., Denckla M. B., Mostofsky S. H. (2007) Dyspraxia in autism: association with motor, social, and communicative deficits. *Dev Med Child Neurol* **49**, 734–739.
200. Shemer A. V., Azmitia E. C., Whitaker-Azmitia P. M. (1991) Dose-related effects of prenatal 5-methoxytryptamine (5-MT) on development of serotonin terminal density and behavior. *Brain Res Dev Brain Res* **59**, 59–63.
201. Kahne D., Tudorica A., Borella A., et al. (2002) Behavioral and magnetic resonance spectroscopic studies in the rat hyperserotonemic model of autism. *Physiol Behav* **75**, 403–410.
202. Chugani D. C., Sundram B. S., Behen M., Lee M. L., Moore G. J. (1999) Evidence of altered energy metabolism in autistic children. *Prog Neuropsychopharmacol Biol Psychiatry* **23**, 635–641.
203. Tsujino N., Nakatani Y., Seki Y., et al. (2007) Abnormality of circadian rhythm accompanied by an increase in frontal cortex serotonin in animal model of autism. *Neurosci Res* **57**, 289–295.
204. Schneider T., Przewlocki R. (2005) Behavioral alterations in rats prenatally exposed to valproic acid: animal model of autism. *Neuropsychopharmacology* **30**, 80–89.
205. Markram K., Rinaldi T., Mendola D. L., Sandi C., Markram H. (2007) Abnormal fear conditioning and amygdala processing in an animal model of autism. *Neuropsychopharmacology* **62**, 901–912.
206. Rinaldi T., Kulangara K., Antoniello K., Markram H. (2007) Elevated NMDA receptor levels and enhanced postsynaptic long-term potentiation induced by prenatal exposure to valproic acid. *Proc Natl Acad Sci U S A* **104**, 13501–13506.
207. Rinaldi T., Silberberg G., Markram H. (2007) Hyperconnectivity of local neocortical microcircuitry induced by prenatal exposure to valproic acid. *Cereb Cortex* **18**, 763–770.
208. Narita N., Kato M., Tazoe M., Miyazaki K., Narita M., Okado N. (2002) Increased monoamine concentration in the brain and blood of fetal thalidomide- and valproic acid-exposed rat: putative animal models for autism. *Pediatr Res* **52**, 576–579.
209. Miyazaki K., Narita N., Narita M. (2005) Maternal administration of thalidomide or valproic acid causes abnormal serotonergic neurons in the offspring: implication for pathogenesis of autism. *Int J Dev Neurosci* **23**, 287–297.
210. Rodier P. M., Ingram J. L., Tisdale B., Croog V. J. (1997) Linking etiologies in humans and animal models: studies of autism. *Reprod Toxicol* **11**, 417–422.
211. Arndt T. L., Stodgell C. J., Rodier P. M. (2005) The teratology of autism. *Int J Dev Neurosci* **23**, 189–199.
212. Rodier P. M. (2002) Converging evidence for brain stem injury in autism. *Dev Psychopathol* **14**, 537–557.
213. Ingram J. L., Stodgell C. J., Hyman S. L., Figlewicz D. A., Weitkamp L. R., Rodier P. M. (2000) Discovery of allelic variants of HOXA1 and HOXB1: genetic susceptibility to autism spectrum disorders. *Teratology* **62**, 393–405.

214. Trottier G., Srivastava L., Walker C. D. (1999) Etiology of infantile autism: a review of recent advances in genetic and neurobiological research. *J Psychiatry Neurosci* **24**, 103–115.
215. Ijichi S., Ijichi N. (2002) Minor form of trigonocephaly is an autistic skull shape? A suggestion based on homeobox gene variants and MECP2 mutations. *Med Hypotheses* **58**, 337–339.
216. Rossel M., Capecchi M. R. (1999) Mice mutant for both Hoxa1 and Hoxb1 show extensive remodeling of the hindbrain and defects in craniofacial development. *Development* **126**, 5027–5040.
217. Shuey D. L., Yavarone M., Sadler T. W., Lauder J. M. (1990) Serotonin and morphogenesis in the cultured mouse embryo. *Adv Exp Med Biol* **265**, 205–215.
218. Gharani N., Benayed R., Mancuso V., Brzustowicz L. M., Millonig J. H. (2004) Association of the homeobox transcription factor, ENGRAILED 2, 3, with autism spectrum disorder. *Mol Psychiatry* **9**, 474–484.
219. Benayed R., Gharani N., Rossman I., et al. (2005) Support for the homeobox transcription factor gene ENGRAILED 2 as an autism spectrum disorder susceptibility locus. *Am J Hum Genet* **77**, 851–868.
220. Zhong H., Serajee F. J., Nabi R., Huq A. H. (2003) No association between the EN2 gene and autistic disorder. *J Med Genet* **40**, e4.
221. Cheh M. A., Millonig J. H., Roselli L. M., et al. (2006) En2 knockout mice display neurobehavioral and neurochemical alterations relevant to autism spectrum disorder. *Brain Res* **1116**, 166–176.
222. Kuemerle B., Gulden F., Cherosky N., Williams E., Herrup K. (2007) The mouse Engrailed genes: a window into autism. *Behav Brain Res* **176**, 121–132.
223. Aldridge J. E., Levin E. D., Seidler F. J., Slotkin T. A. (2005) Developmental exposure of rats to chlorpyrifos leads to behavioral alterations in adulthood, involving serotonergic mechanisms and resembling animal models of depression. *Environ Health Perspect* **113**, 527–531.
224. Aldridge J. E., Meyer A., Seidler F. J., Slotkin T. A. (2005) Alterations in central nervous system serotonergic and dopaminergic synaptic activity in adulthood after prenatal or neonatal chlorpyrifos exposure. *Environ Health Perspect* **113**, 1027–1031.
225. Beyrouty P., Stamler C. J., Liu J. N., Loua K. M., Kubow S., Chan H. M. (2006) Effects of prenatal methylmercury exposure on brain monoamine oxidase activity and neurobehaviour of rats. *Neurotoxicol Teratol* **28**, 251–259.
226. Khan I. A., Thomas P. (2004) Aroclor 1254 inhibits tryptophan hydroxylase activity in rat brain. *Arch Toxicol* **78**, 316–320.
227. Mariussen E., Fonnum F. (2001) The effect of polychlorinated biphenyls on the high affinity uptake of the neurotransmitters, dopamine, serotonin, glutamate and GABA, into rat brain synaptosomes. *Toxicology* **159**, 11–21.

Chapter 6
Excitotoxicity in Autism

The Role of Glutamate in Pathogenesis and Treatment

Martin Evers and Eric Hollander

Abstract Autism spectrum disorders are neurodevelopmental disorders characterized by deficits in social skills, communication, and motor function, as well as compulsive and repetitive behaviors and interests. Although these disorders are thought to be of multifactorial origin, with a wide range of genetic and environmental factors implicated, we propose that excitoxicity is the mechanism modulating numerous risk factors. Substantial evidence from a number of sources—including laboratory studies, neuroimaging, postmortem data, and genetic studies—supports a role for excitotoxicity in autism spectrum disorders. These studies often implicate glutamate and glutamatergic dysregulation as the key mechanism driving excitotoxic processes. The relationship of autism spectrum disorders to other diseases in which glutamate plays a critical role gives further support to the glutamatergic theory of autism. If glutamate contributes to the pathology of autism spectrum disorders, it is reasonable to suggest that agents modulating glutamate may have some utility in treatment. To this end, numerous reports have supported roles for medications including memantine, depakote, amantadine, and antipsychotics in the treatment of these disorders. Investigations thus far have consisted mainly of small open-label and uncontrolled studies; larger controlled studies are necessary and are underway.

Keywords Autism · glutamate · neurotoxicity · excitotoxicity · neurodevelopmental

Introduction

Autism spectrum disorders are believed to be of multifactorial origin, with various genetic and environmental factors postulated to play etiologic roles. We theorize that excitotoxicity is an important neurobiological mechanism that modulates these diverse risk factors. The role of excitotoxicity in autism is the focus of this chapter.

M. Evers
Psychiatrist at Northern Westchester Hospital in Mt. Kisco, NY
e-mail: Martin.Evers@mssm.edu

A.W. Zimmerman (ed.), *Autism*, DOI: 10.1007/978-1-60327-489-0_6,
© Humana Press, Totowa, NJ 2008

Glutamate is the primary excitatory central nervous system (CNS) neurotransmitter; it is widely produced in the CNS, and few if any areas of the brain do not receive glutamatergic input [1]. There are multiple glutamate receptors, which may be divided into metabotropic and ionotropic types. Ionotropic receptors may be further divided into three subtypes: N-methyl-D-aspartate (NMDA), α-amino-3-hydroxy-5-methyl-4-isoxazole propionic acid (AMPA), and kainate. The term excitotoxicity was coined by John Olney in a 1969 Science paper [2] discussing brain lesions that resulted from feeding monosodium glutamate to mice. Excitotoxicity is a pathological process by which neurons are damaged or killed by pathological levels or activity of glutamate and other excitotoxins such as NMDA and kainic acid. This occurs when glutamate receptors such as the NMDA and AMPA receptors are overactivated by glutamate and other excitotoxins, leading to excessive calcium influx into the cell. Excessive calcium influx leads to activation of enzymes, including phospholipases, proteases, and endonucleases, which may damage cell structures, including genetic material, membranes, and cytoskeletal components. This excess calcium may also open mitochondrial permeability transition pores, which may cause mitochondria to release more calcium or to release apoptotic proteins. ATP production may also be impaired, damaging electrochemical gradients necessary for proper function of glutamate and other transporters. Extrasynaptic NMDA receptor activation by excess glutamate may also cause loss of mitochondrial membrane potential and apoptosis via activation of a cyclic AMP response element-binding (CREB) protein shutoff. Therefore, excitotoxicity may contribute to, or result from, oxidative stress. The mechanisms by which excitotoxicity leads to specific types of neuronal damage have been elucidated in a number of models.

Excitotoxicity and Autism

Excitotoxicity has been associated with a number of neurodegenerative conditions, including Alzheimer disease, amyotrophic lateral sclerosis, Parkinson disease, and Huntington disease. The exact manner in which excitotoxicity could produce specific autistic phenotypes is a matter of some speculation. Many researchers speculate that glutamatergic overactivity leads to excitotoxicity, causing abnormal neuronal development leading to autism [3]. One hypothesis holds that an imbalance between excitation and inhibition during the development and activation of cortical networks involved in language processing (among other possible functions) leads to deficits in functional connectivity and synchronization of neuronal discharge in the autistic brain [4, 5, 6]. Although it may be coincidence, glutamatergic activity peaks during the second year of life, which coincides with the onset of synaptic pruning and is a common time for symptoms of autism to manifest (although signs of the disorder are often present from the beginning of life) [7].

Evidence for Excitotoxicity in Autism

Laboratory Findings

Studies of plasma levels of glutamate and other amino acids in patients with autism and related disorders have yielded mixed results. A recent study of serum levels of amino acids related to glutamatergic function in 18 adult autism patients and 19 controls found significantly higher levels of glutamate (but not other amino acids) among the individuals with autism [8]. An important additional finding was that glutamate levels correlated positively (but not significantly) with Autism Diagnostic Interview—Revised social scores. A study of 23 patients (both children and adults) with autism or Asperger's disorder and 55 of their relatives found that both patients and their relatives had significantly higher plasma concentrations of glutamate, phenylalanine, lysine, asparagine, tyrosine, and alanine than age-matched controls [9]. A study of plasma amino acid levels in 14 children with autism found higher aspartate and lower glutamine and asparagine levels in the patients than in age-matched controls [10]. By contrast, a comparison of amino acid levels in platelet-rich plasma from 18 drug-naïve autistic children and 14 healthy controls found lower glutamate, aspartate, GABA, and glutamine in the autistic subjects [11]. Thus, results of plasma studies conflict. Interpretation of findings is further made difficult by small sample sizes and differences in characteristics of patient populations (i.e., age and exposure to medications) and study methodologies (i.e., serum vs plasma and time of sample collection). Although the aforementioned studies are of plasma or serum, it has previously been reported that levels of glutamate in human cerebrospinal fluid (CSF) correlate with blood levels [12, 13]. Glutamate has been found to be elevated in the CSF of patients with Rett syndrome compared to patients with autistic disorder and CSF amino acid levels similar to healthy controls [14].

A study of isoprenoid pathway function in autism found elevated levels of digoxin (an endogenous sodium–potassium ATPase inhibitor secreted by the hypothalamus) and reduced RBC membrane sodium–potassium ATPase activity in autism [15]. The authors hypothesized a model for autism as a syndrome of excess secretion of digoxin by the hypothalamus due to upregulation of the isoprenoid pathway. This model proposed that excess digoxin leads to glutamate excitotoxicity via quinolinic acid (an NMDA agonist) and hypomagnesemia induced by membrane sodium–potassium ATPase inhibition. Additionally, altered calcium/magnesium ratios and levels of isoprenoid metabolites could lead to mitochondrial dysfunction and subsequent free radical generation, reduced free radical scavenging, defective apoptosis, and abnormal neuronal function. Abnormal immune activation and autoimmunity is another consequence of this model.

Magnetic Resonance Spectroscopy

Magnetic resonance spectroscopy (MRS) can be used to measure concentrations of brain metabolites; it has been utilized in a limited number of investigations of autism. In one recent study, patients with autism spectrum disorders had significantly higher concentrations of glutamate/glutamine and creatine/phosphocreatine in the amygdala–hippocampal but not parietal regions [16]. A recent MRS study of cortical and cerebellar tissue found significantly lower levels of gray matter N-acetylaspartate and Glx (glutamate + glutamine) in autistic subjects vs controls; these findings affected most cerebral regions and the cerebellum [17]. Other studies have yielded mixed results as to whether glutamate concentration differs among autistic patients and healthy controls [18].

Postmortem Data

Postmortem studies have found some evidence suggesting the involvement of the glutamate system in autism. Two isoforms of glutamic acid decarboxylase (GAD), the enzyme responsible for conversion of glutamate to GABA, were measured in the parietal and cerebellar cortices of individuals with autism and nonautistic controls; one isoform (GAD67) was significantly reduced in the parietal cortex while the other isoform (GAD65) was significantly reduced in the cerebellar cortex of the autistic brains [19]. A more recent study quantifying GAD67 mRNA in the Purkinje cells of adult patients with autism and normal controls found that GAD67 mRNA was reduced by 40% in the autistic subjects [20]. This suggests increased levels of glutamate or glutamate transporter receptor density in the autistic brain. Measurement of mRNA and associated protein levels in the brains of ten patients with autism and 23 matched controls found upregulation of the excitatory amino acid transporter 1 (EAAT1) and glutamate receptor AMPA 1 genes, higher levels of their associated proteins (by Western blotting), and decreased AMPA receptor density in the cerebellum [21]. An autoradiographic study found decreased density of GABA receptors in the hippocampus of autistic subjects [22].

Immune Studies

Abnormal immune and inflammatory processes have been associated with the pathogenesis of autism [23, 24]. It has long been hypothesized that in utero insults such as viral or bacterial infections may lead to autism, schizophrenia, and other neurodevelopmental disorders by triggering maternal and fetal immune activation. In one study, prenatal human influenza viral infection of pregnant mice led to brain region-dependent changes in levels of neuronal nitric oxide synthase (nNOS), a molecule associated with neuronal development and excitotoxicity. Specifically, rostral brain values showed significant increase and then decrease over the first 2 months of life, while middle and caudal brain areas showed reductions in nNOS [25]. Elevated levels of tumor necrosis factor-α (TNF-α) receptor II have been reported in the sera of children with autism spectrum disorders [26]. More recently, elevated levels of TNF-α in CSF were reported in eight boys with autism [27]. Of the eight, four had been treated with immunosuppressants, while four had not; CSF levels of the cytokine were elevated especially in subjects who had not undergone therapy with immunosuppressants.

Relationship to Other Disorders

Further theoretical support for a role of glutamate in autism may be derived from the relationship of autism and seizure disorders. Glutamate is well known to play a role in the pathophysiology of epilepsy and the occurrence of seizure activity, and multiple reports have noted an increased risk of seizure disorders in individuals with autism: approximately one-third of individuals with autism experience clinically apparent seizures [28, 29, 30, 31]. Valproic acid, an anticonvulsant, has neuroprotective effects against glutamate-induced excitotoxicity and has shown benefit in autistic patients with and without clinical seizures [30, 32, 33, 34] (please see section "Implications for Treatment" for further discussion of utility of valproate in the context of autism spectrum disorders). Autism and autistic-type behaviors are also commonly seen in other medical disorders associated with abnormal glutamatergic function, including Fragile X syndrome and tuberous sclerosis [35]; approximately 10% of patients with autism have associated disorders such as fragile X syndrome, tuberous sclerosis, or Rett syndrome.

Genetic Studies

Autism has a strong genetic component. Familial recurrence of autism is 100-fold higher than in the general population, while the concordance rate among monozygotic and dizygotic twins is 70–90% and 0–10%, respectively, for broader autism phenotype [36, 37]. Genetic studies have yielded some evidence implicating genes involved in glutamatergic function. The Autism Genome Project Consortium performed the largest linkage scan to date using 1168 families with at least two affected individuals [38]. Linkage analysis associated chromosome 11p12-p13 with autism, while copy number variation analysis implicated neurexins. Neurexins induce postsynaptic differentiation in contacting dendrites, interacting with neuroligins, which induce presynaptic differentiation in glutamatergic axons. Neurexins are thus important for glutamatergic synaptogenesis. The glutamate receptor ionotropic kainate 2 (*GRIK2*) and glutamate receptor 6 (*GluR6*) genes have been found to be in linkage disequilibrium in autism; an excess of maternal transmission of GRIK2 haplotype was found in the same study [39]. This finding was replicated in a study of Chinese parent–offspring trios [40]. The 6q21 chromosome region containing GRIK2 was identified as a possible autism susceptibility region in a genome scan study.

A mutation in the glutamate receptor gene *GRIK2* on 6q21 is present in 8% of autistic males and 4% of controls [41].

Genetic abnormalities in the 15q11-q13 chromosome region, which contains several GABA type A receptor subunit genes, have been identified in as many as 3% of patients with autism [42]. Within this region, an association of the 155CA-2 marker within GABA receptor subunit B-3 (GABRB3) gene with autism has been reported in two studies [43, 44]; another study has failed to confirm this association [45, 46]. Linkage disequilibrium has been reported for another marker in the GABRB3 region [46]. Another study failed to find linkage disequilibrium for selected SNPs within the *GABRB3* or *GABRA5* gene, but it did find two SNPs within the *GABRG3* gene in disequilibrium [47]. A linkage study of a region containing a number of GABA receptor subunit genes on 15q12 found six markers and several haplotypes across GABRB3, and GABRA5 showed a significant association with autism [48].

Evidence for a susceptibility mutation in linkage disequilibrium with variants in the metabotropic glutamate receptor 8 gene has been found on chromosome 7q [49]. Ramanathan et al. [50] reported a child with autistic disorder and a 19-Mb deletion on 4q32–4q34; among the 33 deleted genes was the *AMPA2* gene that encodes the glutamate receptor GluR2 subunit. An association study found linkage for two single nucleotide polymorphisms (SNPs) of the *SLC25A12* gene, which encodes for the mitochondrial aspartate/glutamate carrier AGC1 [51]; this finding was subsequently confirmed [52]. However, two studies using large sample sizes failed to confirm the association of *SLC25A12* with autism [53, 54]. Another study reported a high frequency of biochemical markers of mitochondrial dysfunction in a large proportion of autistic subjects, while finding no association of variation at the *SLC25A12* gene with these biochemical markers or with a diagnosis of autism [55]. Interestingly, the *SCL25A13* gene, a paralog of SLC25A12 that is likewise a mitochondrial aspartate/glutamate carrier, maps to a region of chromosome 7 that has been linked to autism in multiple studies [56]. *GAD1*, the gene encoding GAD67 (which has been implicated in autism, as discussed above), was not identified as a candidate gene for autism in an association study of genes in its region [57].

Database comparisons of 124 putative candidate genes for autism in an inbred mouse strain evidencing behavioral phenotypes relevant to the core symptoms of autism and a control population found an association of the inbred mouse strain with a polymorphism of the *Kmo* gene [58]. This gene encodes kynurenine 3-hydroxylase, an enzyme that regulates the metabolism of kynurenic acid, a glutamate antagonist with some neuroprotective utility.

A significant proportion (perhaps one-third) of individuals with fragile X syndrome have autism; both full mutation and premutation forms of the fragile X mental retardation 1 (*FMR1*) gene are thought to be associated with autism [59]. Recent research suggests that this may occur via a glutamatergic mechanism. It has recently been proposed that fragile X mental retardation protein (FMRP), an mRNA-binding protein regulating mRNA translation in dendrites downstream of gp1 metabotropic glutamate receptors (mGluRs), regulates the local synthesis of AMPA receptor subunits downstream of mGluR activation [60]. Dysregulation of these subunits could, by altering synaptic transmission and postsynaptic density, impair neuronal plasticity as seen in fragile X syndrome. An FMR1 knockout animal model yielded abnormal axonal branching of Rohon–Beard and trigeminal ganglion neurons as well as defects in the lateral longitudinal fasciculus [61]. The nature of these defects was suggestive of a role for mGluR signaling in neural morphogenesis, thereby supporting the significance of glutamate in the related disorders of fragile X syndrome and autism.

Numerous studies of the glutamate-related genetics of autism have yielded negative results. The Tachykinin 1 (*TAC1*) gene produces substance P and neurokinins, products involved in glutamatergic synaptic transmission and inflammation (which, in turn, may be involved in an excitotoxic pathophysiology of autism). A study of three SNPs of the gene in 170 autistic patients and 214 controls found no association of TAC1 with autism [62]. The Neuronal Pentraxin II (*NPTX2*) gene is associated with neuritic outgrowth and the clustering of synaptic AMPA (i.e., glutamate) receptors; a study of four SNPs in the same population as the previous study found no association of NPTX2 with autism [63].

Implications for Treatment

If excitotoxicity contributes to the pathology of at least a subset of patients with autism spectrum disorders, then glutamatergic drugs may have some utility among this population.

1. Memantine, a noncompetitive NMDA inhibitor with moderate affinity for the receptor, appears to block sustained activation of NMDA receptors by glutamate under pathological conditions, while not interfering with receptor activation under physiological conditions. Memantine has been found to improve learning and memory and to have neuroprotective qualities in animal studies [64, 65, 66]. Memantine has been tested in various neurological conditions and has been approved in the Unites States for Alzheimer's disease. Open-label data has suggested that memantine may have benefits for patients with autistic disorder and other pervasive developmental disorders in various functional domains, including language, attention, motor planning, cognitive function, social reciprocity, social withdrawal, and inattention [67, 68]. Two multicenter trials of memantine in autism are planned—a Cure Autism Now-funded Clinical Trials Network study of the medication's effect on motor planning and expressive language and another multicenter trial of memantine's effect on social and repetitive behavior domains.
2. Amantadine, which has NMDA noncompetitive inhibitor activity at doses used for influenza and Parkinson disease, was found in a double-blind, placebo-controlled trial of autistic children to have only modest efficacy

against irritability and hyperactivity [69]. Significant improvement was noted on the Hyperactivity and Inappropriate Speech subscales of the aberrant behavior checklist (ABC); however, no significant differentiation from placebo occurred in parent measures (per ABC) or clinician-rated global improvement.

3. Lamotrigine, which attenuates some forms of cortical glutamate release via inhibition of sodium, potassium, and calcium channels, was reported to decrease autistic symptoms in eight of thirteen patients given the drug for epilepsy [70]. However, a double-blind, placebo-controlled study of lamotrigine in 28 children with autistic disorder found no difference between lamotrigine and placebo in a number of behaviors and rating scales [71].

4. Sodium valproate (valproic acid, divalproex sodium; which blocks voltage-dependent sodium channels, as well as affects GABAergic enzymes) improved behavior in 10 of 14 children with autism spectrum disorders exhibiting impulsivity/aggression [34]. In an 8-week, double-blind, placebo-controlled trial involving 13 patients with autism spectrum disorder, valproate yielded a significant group difference in improvement in repetitive behaviors as measured by the Children's Yale-Brown Obsessive–Compulsive Scale (C-YBOCS) with a large effect size [33]. In the second phase of this study, treatment with valproate prior to the initiation and standard titration of fluoxetine appeared to prevent symptoms of activation associated with early SSRI treatment [32].

5. D-Cycloserine, a partial NMDA agonist in low doses and antagonist at higher doses, was found in a double–blind, placebo-controlled trial in autistic disorder to significantly improve the CGI and social withdrawal subscale of the ABC [72].

6. Dextromethorphan, an NMDA receptor antagonist, has been reported to improve self-injurious behaviors, tantrums, anxiety, motor planning, socialization, and language in autistic children in a series of case studies and single-subject design studies [73, 74]. A more recent small placebo-controlled study found that, at the group level, dextromethorphan was equivalent to placebo in the treatment of core symptoms and problem behaviors in children with autism [75]. Single-subject analyses did find that subjects with symptoms consistent with attention deficit hyperactivity disorder responded to the medication.

It is possible that one mechanism of action underlying the utility of atypical antipsychotics in autism is the suppression of glutamate release via 5-HT2A antagonism. A study of the effects of four different antipsychotics in mice rendered hypoglutamatergic by chemical treatment (and thereafter displaying behavioral primitivization including defects in habituation and attention and a relative paucity of behaviors) found that risperidone, clozapine, and M100907 (a selective 5-HT2A-receptor antagonist) produced improved intricate patterns of motor activity [76].

Summary

Excitotoxicity is an important aspect of early brain development in autism. We theorize that excitoxicity, primarily driven by dysregulation of glutamatergic function during critical periods of CNS development, is responsible for at least a subset of autism spectrum disorders. As such, this is an important area for further research, with possible future implications for the prevention, diagnosis, and treatment of at least a subset of children with autism spectrum disorders. Evidence from a number of sources, including laboratory, postmortem, genetics, and neuroimaging studies, implicates glutamate in the pathogenesis of autism spectrum disorders. However, at present, significant gaps exist in our knowledge. For instance, while numerous reports support associations of excitotoxic phenomena with autism, understanding of specific mechanisms is generally limited. In addition, investigations of glutamatergic agents in the treatment of autism have yielded somewhat conflicting data regarding specific domains of the disorder improved by these agents, as well as effect sizes. Studies to date have tended to be small in scale and quite variable in terms of research methodologies and the composition and characteristics of study populations. Well-controlled treatment studies, including large-scale and multicenter trials, are needed and are currently underway.

References

1. Carlson, NR (2001). Physiology of behavior (7th edn, pp. 96–129). Boston: Allyn and Bacon.
2. Olney JW. Brain lesions, obesity, and other disturbances in mice treated with monosodium glutamate. Science 1969;164(880):719–21.
3. Bittigau P, Ikonomidou C. Glutamate in neurologic diseases. J Child Neurol 1997;12(8):471–85.
4. Rubenstein JL, Merzenich MM. Model of autism: increased ratio of excitation/inhibition in key neural systems. Genes Brain Behav 2003;2(5):255–67.
5. Belmonte MK, Cook EH Jr, Anderson GM, et al. Autism as a disorder of neural information processing: directions for research and targets for therapy. Mol Psychiatry 2004;9(7):646–63.
6. Polleux F, Lauder JM. Toward a developmental neurobiology of autism. Ment Retard Dev Disabil Res Rev 2004;10(4):303–17.
7. Kornhuber J, Mack-Burkhardt F, Konradi C, Fritze J, Riederer P. Effect of antemortem and postmortem factors on [3H]MK-801 binding in the human brain: transient elevation during early childhood. Life Sci 1989;45(8):745–9.
8. Shinohe A, Hashimoto K, Nakamura K, et al. Increased serum levels of glutamate in adult patients with autism. Prog Neuropsychopharmacol Biol Psychiatry 2006;30(8):1472–7.
9. Aldred S, Moore KM, Fitzgerald M, Waring RH. Plasma amino acid levels in children with autism and their families. J Autism Dev Disord 2003;33(1):93–7.
10. Moreno-Fuenmayor H, Borjas L, Arrieta A, Valera V, Socorro-Candanoza L. Plasma excitatory amino acids in autism. Invest Clin 1996;37(2):113–28.

11. Rolf LH, Haarmann FY, Grotemeyer KH, Kehrer H. Serotonin and amino acid content in platelets of autistic children. Acta Psychiatr Scand 1993;87(5):312–6.
12. McGale EH, Pye IF, Stonier C, Hutchinson EC, Aber GM. Studies of the inter-relationship between cerebrospinal fluid and plasma amino acid concentrations in normal individuals. J Neurochem 1977;29(2):291–7.
13. Alfredsson G, Wiesel FA, Tylec A. Relationships between glutamate and monoamine metabolites in cerebrospinal fluid and serum in healthy volunteers. Biol Psychiatry 1988;23(7):689–97.
14. Hamberger A, Gillberg C, Palm A, Hagberg B. Elevated CSF glutamate in Rett syndrome. Neuropediatrics 1992;23(4):212–3.
15. Kurup RK, Kurup PA. A hypothalamic digoxin-mediated model for autism. Int J Neurosci. 2003;113(11):1537–59.
16. Page LA, Daly E, Schmitz N, et al. In vivo 1H-magnetic resonance spectroscopy study of amygdala-hippocampal and parietal regions in autism. Am J Psychiatry 2006;163(12):2189–92.
17. DeVito TJ, Drost DJ, Neufeld RW, et al. Evidence for cortical dysfunction in autism: a proton magnetic resonance spectroscopic imaging study. Biol Psychiatry 2007;61(4):465–73.
18. Friedman SD, Shaw DW, Artru AA, et al. Regional brain chemical alterations in young children with autism spectrum disorder. Neurology 2003;60(1):100–7.
19. Fatemi SH, Halt AR, Stary JM, Kanodia R, Schulz SC, Realmuto GR. Glutamic acid decarboxylase 65 and 67 kDa proteins are reduced in autistic parietal and cerebellar cortices. Biol Psychiatry 2002;52(8):805–10.
20. Yip J, Soghomonian JJ, Blatt GJ. Decreased GAD67 mRNA levels in cerebellar Purkinje cells in autism: pathophysiological implications. Acta Neuropathol (Berl) 2007;113(5):559–68.
21. Purcell AE, Jeon OH, Zimmerman AW, Blue ME, Pevsner J. Postmortem brain abnormalities of the glutamate neurotransmitter system in autism. Neurology 2001;57(9):1618–28.
22. Blatt GJ, Fitzgerald CM, Guptill JT, Booker AB, Kemper TL, Bauman ML. Density and distribution of hippocampal neurotransmitter receptors in autism: an autoradiographic study. J Autism Dev Disord 2001;31(6):537–43.
23. Cohly HH, Panja A. Immunological findings in autism. Int Rev Neurobiol 2005;71:317–41.
24. Licinio J, Alvarado I, Wong ML. Autoimmunity in autism. Mol Psychiatry 2002;7(4):329.
25. Fatemi SH, Cuadra AE, El-Fakahany EE, Sidwell RW, Thuras P. Prenatal viral infection causes alterations in nNOS expression in developing mouse brains. Neuroreport 2000;11(7):1493–6.
26. Zimmerman AW, Jyonouchi H, Comi AM, et al. Cerebrospinal fluid and serum markers of inflammation in autism. Pediatr Neurol 2005;33(3):195–201.
27. Chez MG, Burton Q, Dowling T, Chang M, Khanna P, Kramer C. Memantine as adjunctive therapy in children diagnosed with autistic spectrum disorders: An observation of initial clinical response and maintenance tolerability. J Child Neurol 2007 May;22(5):574–9.
28. Volkmar FR, Pauls D. Autism. Lancet 2003;362(9390):1133–41.
29. Tuchman R. Autism. Neurol Clin 2003;21(4):915–32, viii.
30. Tuchman R, Rapin I. Epilepsy in autism. Lancet Neurol 2002;1(6):352–8.
31. Hussman JP. Suppressed GABAergic inhibition as a common factor in suspected etiologies of autism. J Autism Dev Disord 2001;31(2):247–8.
32. Anagnostou E, Esposito K, Soorya L, Chaplin W, Wasserman S, Hollander E. Divalproex versus placebo for the prevention of irritability associated with fluoxetine treatment in autism spectrum disorder. J Clin Psychopharmacol 2006;26(4):444–6.
33. Hollander E, Soorya L, Wasserman S, Esposito K, Chaplin W, Anagnostou E. Divalproex sodium vs. placebo in the treatment of repetitive behaviours in autism spectrum disorder. Int J Neuropsychopharmacol 2006;9(2):209–13.

34. Hollander E, Dolgoff-Kaspar R, Cartwright C, Rawitt R, Novotny S. An open trial of divalproex sodium in autism spectrum disorders. J Clin Psychiatry 2001;62(7):530–4.
35. Belmonte MK, Bourgeron T. Fragile X syndrome and autism at the intersection of genetic and neural networks. Nat Neurosci 2006;9(10):1221–5.
36. Veenstra-VanderWeele J, Cook EH Jr. Molecular genetics of autism spectrum disorder. Mol Psychiatry 2004;9(9):819–32.
37. Wassink TH, Brzustowicz LM, Bartlett CW, Szatmari P. The search for autism disease genes. Ment Retard Dev Disabil Res Rev 2004;10(4):272–83.
38. Autism Genome Project Consortium. Mapping autism risk loci using genetic linkage and chromosomal rearrangements. Nat Genet 2007;39(3):319–28.
39. Jamain S, Betancur C, Quach H, et al. Paris Autism Research International Sibpair (PARIS) Study. Linkage and association of the glutamate receptor 6 gene with autism. Mol Psychiatry 2002;7(3):302–10.
40. Shuang M, Liu J, Jia MX, et al. Family-based association study between autism and glutamate receptor 6 gene in Chinese Han trios. Am J Med Genet B Neuropsychiatr Genet 2004;131(1):48–50.
41. Jamain S, Betancur C, Quach H, et al. Paris Autism Research International Sibpair (PARIS) Study. Linkage and association of the glutamate receptor 6 gene with autism. Mol Psychiatry 2002;7(3):302–10.
42. Sutcliffe JS, Nurmi EL, Lombroso PJ. Genetics of childhood disorders: XLVII. Autism, part 6: duplication and inherited susceptibility of chromosome 15q11-q13 genes in autism. J Am Acad Child Adolesc Psychiatry 2003;42(2):253–6.
43. Cook EH Jr, Courchesne RY, Cox NJ, et al. Linkage-disequilibrium mapping of autistic disorder, with 15q11–13 markers. Am J Hum Genet 1998;62(5):1077–83.
44. Buxbaum JD, Silverman JM, Smith CJ, et al. Association between a GABRB3 polymorphism and autism. Mol Psychiatry 2002;7(3):311–16.
45. Maestrini E, Lai C, Marlow A, Matthews N, Wallace S, Bailey A, Cook EH, Weeks DE, Monaco AP. Serotonin transporter (5-HTT) and gamma-aminobutyric acid receptor subunit beta3 (GABRB3) gene polymorphisms are not associated with autism in the IMGSA families. The International Molecular Genetic Study of Autism Consortium. Am J Med Genet 1999;88(5):492–6.
46. Martin ER, Menold MM, Wolpert CM, et al. Analysis of linkage disequilibrium in gamma-aminobutyric acid receptor subunit genes in autistic disorder. Am J Med Genet 2000;96(1):43–8.
47. Menold MM, Shao Y, Wolpert CM, et al. Association analysis of chromosome 15 gabaa receptor subunit genes in autistic disorder. J Neurogenet 2001;15(3–4):245–59.
48. McCauley JL, Olson LM, Delahanty R, et al. A linkage disequilibrium map of the 1-Mb 15q12 GABA(A) receptor subunit cluster and association to autism. Am J Med Genet B Neuropsychiatr Genet 2004;131(1):51–9.
49. Serajee FJ, Zhong H, Nabi R, Huq AH. The metabotropic glutamate receptor 8 gene at 7q31: partial duplication and possible association with autism. J Med Genet 2003;40(4):e42.
50. Ramanathan S, Woodroffe A, Flodman PL, et al. A case of autism with an interstitial deletion on 4q leading to hemizygosity for genes encoding for glutamine and glycine neurotransmitter receptor sub-units (AMPA 2, GLRA3, GLRB) and neuropeptide receptors NPY1R, NPY5R. BMC Med Genet 2004;5:10.
51. Ramoz N, Reichert JG, Smith CJ, et al. Linkage and association of the mitochondrial aspartate/glutamate carrier SLC25A12 gene with autism. Am J Psychiatry 2004 April;161(4):662–9.
52. Segurado R, Conroy J, Meally E, Fitzgerald M, Gill M, Gallagher L. Confirmation of association between autism and the mitochondrial aspartate/glutamate carrier SLC25A12 gene on chromosome 2q31. Am J Psychiatry 2005;162(11):2182–4.
53. Blasi F, Bacchelli E, Carone S, Toma C, Monaco AP, Bailey AJ, Maestrini E; International Molecular Genetic Study of Autism Consortium (IMGSAC). SLC25A12 and

CMYA3 gene variants are not associated with autism in the IMGSAC multiplex family sample. Eur J Hum Genet 2006;14(1):123–6.

54. Rabionet R, McCauley JL, Jaworski JM, Ashley-Koch AE, Martin ER, Sutcliffe JS, Haines JL, DeLong GR, Abramson RK, Wright HH, Cuccaro ML, Gilbert JR, Pericak-Vance MA. Lack of association between autism and SLC25A12. Am J Psychiatry 2006;163(5):929–31.
55. Correia C, Coutinho AM, Diogo L, et al. Brief report: High frequency of bio-chemical markers for mitochondrial dysfunction in autism: no association with the mitochondrial aspartate/glutamate carrier SLC25A12 gene. J Autism Dev Disord 2006;36(8):1137–40.
56. Badner JA, Gershon ES. Regional meta-analysis of published data supports linkage of autism with markers on chromosome 7. Mol Psychiatry 2002;7(1):56–66.
57. Rabionet R, Jaworski JM, Ashley-Koch AE, et al. Analysis of the autism chromosome 2 linkage region: GAD1 and other candidate genes. Neurosci Lett 2004;372(3):209–14.
58. McFarlane HG, Kusek GK, Yang M, Phoenix JL, Bolivar VJ, Crawley JN. Autism-like behavioral phenotypes in BTBR T + tf/J mice. Genes Brain Behav. 2007 Jun 7; [Epub ahead of print]
59. Hagerman RJ, Ono MY, Hagerman PJ. Recent advances in fragile X: A model for autism and neurodegeneration. Curr Opin Psychiatry 2005;18(5):490–6.
60. Muddashetty RS, Kelic S, Gross C, Xu M, Bassell GJ. Dysregulated metabotropic glutamate receptor-dependent translation of AMPA receptor and postsynaptic density-95 mRNAs at synapses in a mouse model of fragile X syndrome. J Neurosci 2007 May 16;27(20):5338–48.
61. Tucker B, Richards RI, Lardelli M. Contribution of mGluR and Fmr1 functional path-ways to neurite morphogenesis, craniofacial development and fragile X syndrome. Hum Mol Genet 2006 Dec 1;15(23):3446–58.
62. Marui T, Funatogawa I, Koishi S, et al. Tachykinin 1 (TAC1) gene SNPs and haplotypes with autism: A case-control study. Brain Dev 2007 Mar 19; [Epub ahead of print]
63. Marui T, Koishi S, Funatogawa I, et al. No association between the neuronal pentraxin II gene polymorphism and autism. Prog Neuropsychopharmacol Biol Psychiatry 2007;31(4):940–3.
64. Barnes CA, Danysz W, Parsons CG. Effects of the uncompetitive NMDA receptor antagonist memantine on hippocampal long-term potentiation, short-term exploratory modulation and spatial memory in awake, freely moving rats. Eur J Neurosci. 1996 Mar;8(3):565–71.
65. Zajaczkowski W, Quack G, Danysz W. Infusion of (+) -MK-801 and memantine—contrasting effects on radial maze learning in rats with entorhinal cortex lesion. Eur J Pharmacol 1996;296(3):239–46.
66. Danysz W, Parsons CG, Quack G. NMDA channel blockers: memantine and amino-aklylcyclohexanes—in vivo characterization. Amino Acids 2000;19(1):167–72.
67. Chez M, Hing P, Chin K, Memon S, Kirschner S. Memantine experience in children and adolescents with autism spectrum disorders. Ann. Neurol. 2004;56:S8.
68. Erickson CA, Posey DJ, Stigler KA, Mullett J, Katschke AR, McDougle CJ. A retro-spective study of memantine in children and adolescents with pervasive developmental disorders. Psychopharmacology (Berl). 2007 Mar;191(1):141–7.
69. King BH, Wright DM, Handen BL, et al. Double-blind, placebo-controlled study of amantadine hydrochloride in the treatment of children with autistic disorder. J Am Acad Child Adolesc Psychiatry 2001;40(6):658–65.
70. Uvebrant P, Bauziene R. Intractable epilepsy in children. The efficacy of lamotrigine treatment, including non-seizure-related benefits. Neuropediatrics 1994;25(6):284–9.
71. Belsito KM, Law PA, Kirk KS, Landa RJ, Zimmerman AW. Lamotrigine therapy for autistic disorder: a randomized, double-blind, placebo-controlled trial. J Autism Dev Disord 2001;31(2):175–81.

72. Posey DJ, Kem DL, Swiezy NB, Sweeten TL, Wiegand RE, McDougle CJ. A pilot study of D-cycloserine in subjects with autistic disorder. Am J Psychiatry 2004;161(11):2115–17.
73. Woodard C, Groden J, Goodwin M, Shanower C, Bianco J. The treatment of the behavioral sequelae of autism with dextromethorphan: A case report. J Autism Dev Disord 2005;35(4):515–18.
74. Welch L, Sovner R. The treatment of a chronic organic mental disorder with dextromethorphan in a man with severe mental retardation. Br J Psychiatry 1992;161:118–20.
75. Woodard C, Groden J, Goodwin M, Bodfish J. A placebo double-blind pilot study of dextromethorphan for problematic behaviors in children with autism. Autism 2007;11(1):29–41.
76. Nilsson M, Waters S, Waters N, Carlsson A, Carlsson ML. A behavioural pattern analysis of hypoglutamatergic mice—effects of four different antipsychotic agents. J Neural Transm 2001;108(10):1181–96.

Chapter 7
Prenatal β₂-Adrenergic Receptor Signaling and Autism:
Dysmaturation and Retained Fetal Function

Susan L. Connors

Abstract The origins of idiopathic autism are prenatal. During fetal life, the β₂-adrenergic receptor (B2AR) is important for growth as well as terminal differentiation of cells. Signaling from this receptor serves different purposes at different times in virtually all tissues during prenatal development, and provides modulation for most organ functions in postnatal life. Because the B2AR is one of the earliest appearing receptors in brain development, interference with it over time during gestation can theoretically affect the development of other neurotransmitter systems, as well as later functioning of the CNS and peripheral organs. Prenatal overstimulation of the B2AR has been linked to autism in dizygotic twins, and a higher prevalence of more active B2AR polymorphisms has been found in autism families.

Animal studies in developmental neurotoxicology show abnormal outcomes for brain and tissue function after prenatal administration of B2AR agonists. These studies have also shown that the fetal B2AR normally does not desensitize, and that several tissues can retain a fetal pattern of signaling after prenatal B2AR overstimulation. This type of dysregulated signaling may also be responsible for the differences in function noted in brain and other tissues of autistic children compared to controls. Results from published studies in many areas of autism research can be related to B2AR second messengers such as cAMP levels or to physiological patterns that are present during fetal life.

Prenatal interference with signaling from the B2AR is not likely to act alone in the development of autism. Downstream pathways stimulated by the B2AR can share components from signaling through other receptors, including those for stress hormones and cytokines. Effects on these shared pathways during gestation may lead to final common mechanisms for the development of autism, and may be a reason that single genes and individual environmental factors have not been identified to explain its causation.

S.L. Connors
Neurology and Developmental Medicine, Kennedy Krieger Institute,
707 North Broadway, Baltimore, MD 21205, USA
e-mail: connorss@kennedykrieger.org

A.W. Zimmerman (ed.), *Autism*, DOI: 10.1007/978-1-60327-489-0_7,
© Humana Press, Totowa, NJ 2008

Keywords β₂-Adrenergic receptor · prenatal · autism · fetal · polymorphisms · Adenylyl cyclase · cAMP

Introduction

Autism is included in a spectrum of neurodevelopmental syndromes with heterogeneous presentation that are currently defined in behavioral terms. The origins of autism are prenatal. Neuroanatomical studies show abnormalities in structures that are well established before birth [1, 2, 3, 4]. Increased levels of neuropeptides have been found at birth in archived blood of children who later developed autism [5]. Both these findings suggest that the underlying alterations in brain development and potential peripheral markers occur long before symptoms become obvious in postnatal life. In fact, maternal risk factors for autism during pregnancy have been published [6, 7]. It is highly likely that the developmental cell programs leading to these disorders are established during gestation.

Nearly every neurotransmitter system, as well as the immune system, has been investigated in autism postmortem brain [8, 9, 10, 11, 12, 13], and data from many studies of peripheral tissues such as the immune, gastrointestinal (GI), and neuroendocrine systems have been published as well [5, 14, 15, 16]. A wide range of results that overlap with normal controls, may change with time, and may or may not correlate with patients' functioning levels has been documented in these studies. Although children with this disorder present with a specific set of core characteristics (Diagnostic and Statistical Manual of Mental Disorders-IV), each individual patient is different, one from another. As well, neuroanatomical studies of the brain show a wide range within each area of abnormality such as the brainstem, cerebral cortex, amygdala, hippocampus, and cerebellum. There have been few consistent or predictive results among these investigations that could apply to all patients with autism, and taken together, the data suggest widespread physiological dysregulation. In addition, the genes involved in autism have been difficult to isolate, though considerable scientific research has been devoted to doing so [17, 18, 19, 20, 21]. When considering the heterogeneity of behavior and genetic findings, as well as dysregulation shown in research data, it is reasonable to conclude that etiologic mechanisms involved in the pathogenesis of autism must occur during gestation and must have the potential to affect and interact with many downstream developmental pathways.

All neurotransmitter systems are important for fetal brain development and interact with each other synchronously to result in normal maturation. A significant abnormality in any of the earliest appearing neurotransmitter systems would impact the development of other systems, could dysregulate development, and lead to a cascade of abnormalities that evolve over time. This chapter will present evidence in support of the theory that during gestation, abnormal signaling in an early appearing transmitter system, that of the catecholamines and especially including the β_2-adrenergic receptor (B2AR), contributes to the etiology of autism, and will relate published data to a model of dysregulated B2AR downstream signaling factors and delayed development of cellular physiology.

The β_2-Adrenergic Receptor

Functions

The catecholamine system is one of the earliest appearing neurotransmitter systems in the human fetal brain [22, 23]. The B2AR is part of this system and is the most studied of the catecholamine cell surface receptors. Cell signaling associated with B2AR stimulation results from the binding of norepinephrine and epinephrine in peripheral tissues and norepinephrine in the CNS as ligands. Although for the great majority of its functions the B2AR couples with the stimulatory G protein, Gs, to activate adenylyl cyclase (AC) and generate cyclic adenosine monophosphate (cAMP), protein kinase A (PKA), and an increase in intracellular calcium levels as second messengers, its activation also stimulates or inhibits MAP kinases to regulate fundamental cell processes such as differentiation, growth, migration, and apoptosis [24, 25]. The cAMP generated influences gene transcription through the cyclic AMP response element (CRE) DNA sequence and its transcription factor or binding protein, CREB (Fig. 7.1). The B2AR is transcribed from its gene on chromosome 5q31–32 as one peptide of 413 amino acid residues [26]. Beta adrenergic receptors (BARs) are widely expressed throughout mammalian fetal tissues including the brain, even in cell types where meager numbers of the receptor will be found in adult life [27]. The B2AR is expressed on mammalian oocytes and preimplantation embryos [28], but whether or not the fetal receptor differs structurally (such as in posttranslational modification) from the mature form is unknown.

Stimulation of the B2AR and the resulting signaling cascades serves different purposes in various tissues and at different times during prenatal development and postnatal life. For example, early in fetal life, B2AR stimulation is coupled to cAMP generation through AC as shown in rat studies [27], and provides signals for growth, but later it promotes differentiation in many tissues and axonal outgrowth in neural cells; in certain tissues, activation can also result in apoptosis or, depending on the degree and duration of stimulation, salvage from apoptosis [27, 29, 30, 31, 32, 33, 34, 35]. In several tissues, stimulation of the B2AR during development causes cells to exit from the cell cycle [36, 37, 38], which is the initial step in the transition from growth by means of cell replication to differentiation and growth as a result of cell enlargement. For that reason, the appearance of the B2AR in various brain regions at different times during development is thought to signal terminal differentiation [39]. Both excesses and decrements in downstream signaling molecules (cAMP)

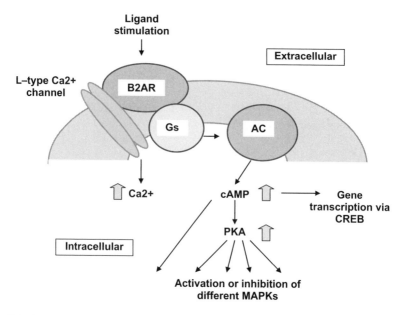

Fig. 7.1 Signaling pathways activated by stimulation of the β₂-adrenergic receptor (B2AR). The B2AR couples with the stimulatory G protein (Gs), which stimulates adenylyl cyclase (AC) to produce cyclic adenosine monophosphate (cAMP) and protein kinase A (PKA) as second messengers. The cAMP and PKA generated activate or inhibit several mitogen-activated protein kinase (MAPK) pathways involved in cell growth, differentiation, and apoptosis. Cyclic AMP influences gene transcription through cAMP response element-binding protein (CREB). An L-type calcium (Ca^{2+}) channel is also part of this complex, and on B2AR stimulation, it is activated so that intracellular calcium levels rise

cause abnormalities in growth cone formation and function in the developing neurons of lower species such as Drosophila [40].

The B2AR system acts as a modulator for cellular signaling in postnatal life. Stimulation of the B2AR facilitates long-term GABAergic transmission to Purkinje cells in the cerebellum [41, 42] and modulates L-type calcium channels in dendritic spines of pyramidal hippocampal neurons in the rat [43]. Presynaptic BAR stimulation results in long lasting increases in synaptic transmission in rodent amygdala [44], and beta adrenergic activity is essential for enabling glucocorticoid modulation of memory consolidation in the human amygdala [45]. β₂-Adrenergic receptor signaling activates rodent astrocytes, provides neuroprotection [46], participates in the regulation of a number of cytokines produced by microglia in different situations [47, 48], and increases HLA-DR expression in glioblastoma cells [49]. Signaling through the receptor has many effects on the circulating immune system, such as regulating the amount of IgG1 antibody produced by B lymphocytes in the peripheral blood [50, 51] and regulating immune responses [52, 53]. B2AR second messengers and signaling pathways are also involved in numerous functions and responses in the GI tract and its immune system, an organ system that, in addition to the brain, has been the subject of many investigations in autism research [15, 54, 55].

Regulation of β_2-Adrenergic Receptor Signaling

Postnatal signaling by the B2AR is regulated by desensitization (decreased signaling), which can be homologous (involving just the B2AR) or heterologous (involving other receptors that share the same signaling pathway). Homologous desensitization involves two distinct processes: uncoupling of the B2AR from its ability to activate the Gs and, with prolonged receptor stimulation, downregulation (decreased numbers of receptors on the cell surface). The primary mechanism for homologous desensitization involves phosphorylation of the receptor, followed by binding of arrestins to the phosphorylated receptor, which uncouples it from its Gs, ending signal generation. Downregulation is accomplished through events shared with desensitization (phosphorylation and association with arrestins), followed by endocytosis and internalization of receptors, and finally their degradation in lysosomes [56]. Decreasing receptor synthesis and increasing the rate of degradation can also contribute to downregulation. Heterologous desensitization, which also occurs with prolonged B2AR stimulation, involves phosphorylation and uncoupling of other receptors that act through Gs, or loss of function of Gs and AC itself. Together, both desensitization and downregulation terminate cell signaling in the face of excessive input, an essential homeostatic mechanism designed to protect the cell.

Polymorphisms of the β_2-Adrenergic Receptor

Polymorphisms (single nucleotide substitutions) of the B2AR gene exist in human populations, and three of these code for changes in the amino acid sequence of the receptor that have physiological significance for receptor signaling: glycine at codon 16 (Gly16), glutamic acid at codon 27 (Glu27), and isoleucine at codon 164 (Ile164). Although the Gly16 and Glu27 polymorphisms are associated with enhanced signaling through the receptor, the Ile164 polymorphism results in reduced affinity for ligand binding and lower levels of second messenger formation [57]. Ligand stimulation of Gly16 and Glu27 receptors in vivo results in decreased desensitization and downregulation compared to the wild-type variants Arg16 and Gln27 [58, 59].

Polymorphisms of the B2AR gene have been associated with susceptibility to and prognosis in several disease states, including outcome in congestive heart failure [60, 61], medication response in asthma [62], obesity [63], type 2 diabetes [64], Graves' disease [65], myasthenia gravis [66], rheumatoid arthritis [67], and psychological coping [68]. Because specific combinations of genetic polymorphisms can change the physiology of receptor function and contribute to predispositions for diverse disease states, and because the B2AR is important for normal brain and organ development, it is probable that certain polymorphisms that increase or decrease signaling could become genetic risk factors during gestation for neurodevelopmental disorders, in a similar way as those linked to disease in peripheral organs.

Animal Studies

The functional characteristics of the B2AR have been extensively investigated in the developing rat by Slotkin's group at Duke University. Studies have shown that protective regulatory mechanisms for B2AR signaling are not intrinsic properties of cells, but are acquired during ontogenesis. In fact, the arrival of innervation in target tissues provides a timing signal for the development of receptor desensitization [69]. Fetal and newborn tissues not only are resistant to BAR desensitization but actually show the opposite: agonist stimulation of the fetal receptor enhances net physiological responses instead of producing desensitization, as in adult tissues [70].

Work in rodents has clarified the mechanisms that underlie fetal sensitization of continued B2AR signaling, and although the earliest studies utilized neonatal cardiac tissue, further research resulted in similar findings in the central nervous system in several mammalian species [71, 72, 73, 74]. Changes in signal transduction after overstimulation of the fetal B2AR depend primarily on changes in receptor coupling and response elements downstream from the receptor, rather than on receptor numbers [75]. Enhanced fetal responses involve increased expression of membrane-associated Gs (which is stimulatory for AC), decreased expression of Gi (which is inhibitory for AC), increased concentration of a more active splice variant of the alpha subunit of Gs, and elevated expression of AC molecules [76, 77, 78]. In addition, the expression of muscarinic type 2 cholinergic receptors (m2AChR) that couple with Gi, as well as their ability to inhibit AC, is decreased, at least in the heart [79]. These differences in fetal tissue promote signaling through AC and decrease its inhibition, resulting in increased production of second messengers such as cAMP and PKA. It is also important to note that sensitization in the fetus is "heterologous," meaning that downstream signaling generated from activation of other receptors that, like the B2AR, utilize Gs and AC (such as the glucagon and β_1-adrenergic receptors) is enhanced as well [70] (Fig. 7.2).

The signaling changes described above in rodent studies, resulting from overstimulation of the B2AR, do not occur uniformly throughout all brain regions at all ages. As in many other studies involving manipulation of gestational cell signaling [80, 81, 82], these responses depend upon the region investigated, gender, and maturational stage at exposure, and they change with age.

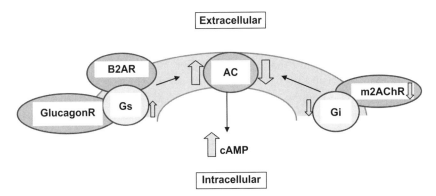

Fig. 7.2 Cellular mechanisms for enhanced adenylyl cyclase (AC) signaling in the fetus. Stimulatory G protein (Gs) function/expression is enhanced, as is the expression of AC molecules, both of which increase AC function. Sensitization is heterologous: signaling from stimulation of other receptors that couple with Gs (such as glucagon) is enhanced as well. In addition, the expression/function of the inhibitory G protein (Gi) is reduced, as is its ability to inhibit AC. The expression of at least one receptor that couples with Gi, the muscarinic cholinergic receptor (m2AChR), is reduced (in cardiac tissue). Block arrows indicate the direction of expression or function

Maturational stage is the predominant factor determining the net signaling response to B2AR agonist exposure during brain development. For example, fetal exposure to terbutaline, a selective B2AR agonist, during the developmental period equivalent to the early-to-mid second trimester of human pregnancy (GD17–20) [83], results in nongender-dependent, enhanced AC responses in whole brain during the immediate period after treatment, compared to controls [71]. Later administration of terbutaline to neonatal rats (PN 2–5), equivalent to the late second and early third trimester in human pregnancy, produces similar changes, but only in specific regions that follow a maturational timetable of susceptibility [71]. After this neonatal exposure schedule, by PN 45 (the end of adolescence in the rat), significant increases in AC responses in males and reductions in females are found in the cerebellum, the last brain structure to develop [71]. Other areas such as the brainstem and striatum show decreases in both genders. Thus, in adolescence, components of the pattern seen with fetal (GD17–20) administration of terbutaline, specifically enhanced Gs and AC signaling, persist into the postnatal period in the developing rat according to a regional pattern that reflects the timetable for maturation of each brain area. Other regions at different times show no effect or decreased AC signaling. By adulthood (PN 60), decrements in AC signaling were found in the cerebral cortex, an area that had shown no net changes in adolescence [75]. To date, AC responses after prenatal or neonatal exposure to B2AR overstimulation have not been measured in the rat brain at a time equivalent in human development to young childhood, when behavioral symptoms of neurodevelopmental disorders often emerge.

Excessive B2AR signaling during PN 2–5 in the rat results in abnormalities in AC function that continue in adolescence and adulthood but with regional and quantitative (over- or undersignaling) differences from those found at the outset. Thus, it is probable that overstimulation of the B2AR or factors activating similar mechanisms alters the "program" for development of cell signaling through G proteins and AC. Because AC and cAMP are involved in countless cell processes that include gene expression through cAMP response element-binding (CREB) protein and neuronal function, abnormalities in AC signaling, such as those described, likely lead to alterations in neuronal cell differentiation, cytoarchitecture, and synaptic signaling.

The fetus has little or no protective mechanism to decrease effects from prolonged B2AR signaling, and exposures during pregnancy that increase B2AR signaling, or overstimulate the receptor, could have widespread effects, the severity of which may depend upon the dose, timing, and duration of interference in the specific brain regions and organs affected. Although decreases in receptor binding can occur in some regions of the CNS and in peripheral tissues of the fetus with excessive B2AR stimulation, it is downstream signaling pathways that are upregulated and provide increased responses [70, 71]. These functional changes occur without differences in form; terbutaline treatment in rats does not affect brain or body weight or rate of growth, characteristics that may be analogous to the situation in autism.

What is the Relationship to Autism?

Prenatal overstimulation of the B2AR, in combination with its more active polymorphisms, likely contributes to the etiology and pathogenesis of autism. Indeed, these two factors have been linked in human studies to this disorder. Exposure for 2 weeks or longer to terbutaline, a selective B2AR agonist that was originally developed for use in asthma and that has been used extensively to arrest or prevent preterm labor [84], has been linked to concordance for autism spectrum disorders (ASDs) in dizygotic twins (relative risk 4.4 in male twins with no family history of ASDs) [85]. This study supports earlier work showing poor cognitive and abnormal psychiatric outcomes in children exposed to B2AR agonists for preterm labor [86, 87]. Terbutaline crosses the placenta and blood–brain barrier and stimulates B2ARs in all tissues of the fetus [27, 88, 89]. In addition, an increased prevalence of the B2AR polymorphisms Glu27 and Gly16 has been found in dizygotic twin sets compared to the general population [85], and the Glu27 homozygous variant has been linked to an increased risk for autism in parent–child trios from the Autism Genetic Resource Exchange (AGRE) population [90, 91].

Because signaling through the B2AR contributes to the shift from neural cell proliferation to differentiation, and because the B2AR is part of one of the earliest appearing chemical transmitter systems in brain and tissue development

[23, 92], interference with its function over time during gestation can affect developmental programming and can influence the maturation of other neurotransmitter systems, as well as the later functioning of the CNS and peripheral organs. Rodent studies, in addition to the work cited above, have shown that prenatal overstimulation of the B2AR by administration of terbutaline, on PN 2–5 and 11–14, results in several neuroanatomical abnormalities in the CNS that are analogous to those noted in postmortem autism brain, such as loss of cerebellar Purkinje cells, smaller cells in the sensory cortex, and neuroimmune activation [2, 14, 93, 94]. In addition, prenatal administration of terbutaline has not resulted in abnormalities in form (birth weight and rate of growth were unaffected), but in abnormal function (measured by receptor signaling in membrane preparations) of lungs, liver, heart, and kidneys in the developing rat [72, 79, 95]. Administration of this drug to neonatal rats has also resulted in microglial activation in brain areas that correlate with neuroinflammation in postmortem autism brain, and in juvenile rats at PN 30, it has led to later emerging hyperactivity and auditory sensitivity in preliminary behavioral studies [14, 96].

These results all point to a likely scenario for neurodevelopmental changes, resulting in the pattern seen in autism. Stimulation of the B2AR can cause a premature exit from the cell cycle, a mechanism by which the receptor's signaling decreases cellular proliferation in favor of differentiation [36, 37, 38]. In humans, if excessive BAR stimulation is inappropriately timed and occurs in neural pathways that have not yet completed full innervation of their target tissues, abnormal connections would be formed, and, just as important, the tissues awaiting final innervation and synapse formation might remain in a response state similar to that seen in fetal life, with enhanced AC signaling and decreased inhibition of that enzyme (Fig. 7.3). The net effect of cellular responses at first would be excitatory in pathways that depend upon AC signaling. Later, areas of increased or decreased AC signaling would become region-, gender- and age-dependent, similar to those differences found in the terbutaline animal model. More importantly, patterns of signaling abnormalities would differ among individual patients, since they would depend upon the maturational stage of the CNS at the time of exposure to factors that could overstimulate the B2AR. Later responses and adaptations to the environment that could affect programming of other neurotransmitter systems would be influenced by these early signaling abnormalities, further adding to heterogeneity and disordered maturation.

With this model in mind, the behavioral disorder called autism can be viewed as a biological one marked by dysregulation involving abnormalities in AC function and cAMP formation. Signaling through AC leads to transduction signals shared by numerous neuronal and hormonal pathways. Cyclic AMP is a ubiquitous molecule that not only provides direct signaling within a cell, but also effects gene transcription through the generation of CREB protein. Abnormalities in the AC system, then, could be considered an epigenetic factor that would influence gene expression and developmental trajectory. Autism

Fig. 7.3 Theoretical mechanisms by which overstimulation of the β₂-adrenergic receptor (B2AR) may cause miswiring, with premature "maturation" in upstream tissues and persistent fetal modes of functioning in downstream tissues. B2AR overstimulation can cause cells to prematurely exit from the cell cycle, decreasing proliferation in favor of differentiation and causing synapse formation at inappropriate targets. The downstream tissues that are targets for innervation under normal conditions continue to show enhanced fetal response patterns (failure to desensitize) since innervation provides a cellular signal for the development of desensitization

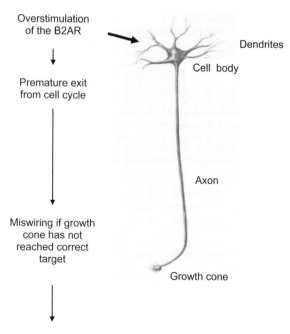

Overstimulation of the B2AR

↓

Premature exit from cell cycle

↓

Miswiring if growth cone has not reached correct target

↓

Downstream tissue retains fetal responses due to lack of adequate innervation

Dendrites

Cell body

Axon

Growth cone

may also be a disorder reflecting fetal physiology of the catecholamine system, due to the possibility that the B2AR and its signaling molecules can effect an exit from the cell cycle during the process of innervation. Upstream tissues would then be inappropriately "mature," while downstream elements might retain fetal responses, and inappropriate, misplaced synapses would create abnormal "wiring" of the CNS. Overall development and functioning would certainly be disordered, as they are in autism.

Relating the Model to Autism Research

Many findings from autism research reflect increased or deceased AC signaling, when parameters being studied are influenced by AC. Alternatively, the findings may reflect a fetal pattern of functioning due to dysregulated AC signaling over time, with some brain regions and other tissues exhibiting sensitized, enhanced responses. Results from these investigations then take on new significance. This section will correlate some of the more recent findings (or those with the greatest impact) in autism literature with these possibilities.

Neuroglial Activation

It is unknown whether the neuroglial activation noted in postmortem autism brain [13] is detrimental, reparative, developmental, or a mixture of all three. Many of the cytokines reported as elevated in cerebrospinal fluid (CSF) and brain tissue by Vargas et al. [13], such as MCP-1 and IL-6, act as growth and differentiation factors during gestation [97, 98], and expression of these two molecules is influenced by cAMP, the former being inhibited by it and the latter being upregulated [99, 100]. When measured by HLA-DR staining, activated and more numerous microglia (the immune cells of the brain) were a major finding in the work by Vargas and colleagues. Microglial HLA-DR expression is present in the human fetal brain from the second trimester and is involved in normal development [101]. Transient overexpression of activated microglia occurs normally in the cerebral white matter of the human fetus [102]; thus the finding of increased expression of HLA-DR in autism brain may reflect dysmaturation. HLA-DR expression in vitro is increased in glioblastoma cells in response to cAMP [103], even though B2AR stimulation (and thus increased cAMP) inhibits proliferation of microglia from adult rat brain in vitro [104]. Increased numbers of activated microglia may also be a reflection of low AC responses in autism, part of dysregulated AC; norepinephrine (the ligand for the B2AR) depletion in newborn rats results in activated microglia in the cerebellum [105]. Activated microglia may reflect, in part, a developmentally delayed process as well, since in the rodent cerebellum, microglia promote the death of developing Purkinje cells [106], lower numbers of which have been repeatedly noted in cerebellar tissue from autism brain [1, 107, 108].

Increased Cerebral White Matter

Increased white matter on magnetic resonance imaging (MRI) has been documented in children with autism [109]. Some of this increase has been attributed to activated microglia noted in postmortem autism brain [13], since microglia promote myelin formation in cocultures with oligodendrocytes from developing rats [110]. This finding of increased white matter can also be related to enhanced B2AR stimulation and increased cAMP levels, since both processes induce expression of myelin basic protein in immature rodent oligodendrocytes [111, 112]. At a developmental stage immediately preceding the beginning of the active period of myelin synthesis in the rat, the cAMP-dependent pathway that leads to myelin production is stimulated only by B2AR agonists. Thus, delayed or disordered oligodendrocyte maturation may be responsible for the findings by Herbert et al. [109].

Insulin-like Growth Factor-1

Insulin-like growth factor-1 (IGF-1) is a neurotrophic factor that is important in early brain development and axonal assembly at the growth cone [113, 114].

Low levels of IGF-1 in CSF of autistic children [115] may be related to enhanced cAMP signaling, since this molecule inhibits expression of IGF-1 in cultured cells [116] (see Chapter 20 by Riikonen).

Glial Fibrillary Acidic Protein, Bcl-2, and GAD67

Glial fibrillary acidic protein (GFAP) is an astrocytic marker. Astrocytes play important roles in neuronal function, synaptic plasticity, and detoxification [117, 118]. Elevated levels of GFAP in postmortem autism brain [119] correlate with astrogliosis noted by Vargas et al. [13]. They also may reflect increased B2AR signaling, since BAR stimulation increases the expression of GFAP in astrocytes [120]. In autism brain tissue, decreased expression of Bcl-2, a marker for apoptosis, and GAD67, which catalyzes the conversion of glutamate to GABA [121, 122], can also be related to cAMP signaling. Increased levels of cAMP cause reductions in the expression of GAD67 in C6 glioma cells [123], and levels of Bcl-2 are directly correlated with those of cAMP [124, 125].

Epilepsy

Up to 40% of children with autism develop epileptic seizures, the majority of which have their onset by adolescence [126]. The catecholamines, specifically norepinephrine, have long been known to have anticonvulsant effects in the CNS in animal studies [127, 128]. Work done with the rodent model of prenatal overstimulation of the B2AR described previously resulted in diverse areas of over- and undersignaling through AC that change with age. By adulthood, decrements in cortical AC signaling are apparent in rats after B2AR over-stimulation by terbutaline during early development [75]. It is possible that decreased AC signaling may contribute to a propensity to develop seizures in patients with autism as they grow older.

Sulfation, Methylation, and Oxidative Stress

Abnormal sulfation, as it relates to glutathione synthesis and methylation, has been investigated in autism. Lower plasma levels of methionine, S-adenosylmethionine (SAM), homocysteine, cystathionine, cysteine, and thus, total glutathione have been found in children with autism compared to controls [129] (see Chapter 10) (Fig. 7.4). Glutathione is a three amino acid molecule that provides the major defense against reactive oxygen species. These values, along with higher levels of S-adenosylhomocysteine (SAH), adenosine, and oxidized glutathione, may certainly reflect increased oxidative stress in the peripheral circulation and could reflect impaired methylation capacity. However, these results may also be consistent with additional abnormalities and

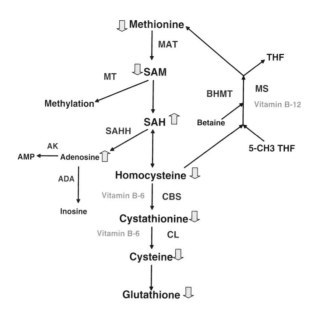

Fig. 7.4 The methionine cycle and generation of glutathione. Homocysteine is remethylated by either of two enzymatic reactions that are folate-vitamin B12-dependent (via methionine synthase, MS) or -independent (via betaine homocysteine methyltransferase, BHMT). After a series of steps involving methyl group transfer and donation for methyltransferase (MT) reactions, methionine metabolites are converted back to homocysteine, which may be metabolized to cysteine through two enzymatic steps involving cystathionine β-synthase (CBS) and cystathionine lyase (CL). Cysteine is incorporated into the tripeptide glutathione, a major defense against reactive oxygen species. Block arrows indicate the direction of plasma levels reported in autistic children by James et al. [129] Other abbreviations: MAT, methionine adenosyltransferase; SAM, *S*-adenosylmethionine; SAH, *S*-adenosylhomocysteine; SAHH, SAH hydrolase; AK, adenosine kinase; ADA, adenosine deaminase; THF, tetrahydrofolate; 5CH3THF, 5-methyl THF

fluctuations in cAMP levels as well as dysmaturation. The expression of the enzymes cystathionine β-synthase (CBS) and cystathionine lyase (CL), which catalyze the conversion of homocysteine to cystathionine and cystathionine to cysteine, respectively, are lower in fetal than in postnatal rodent brain and liver, and their expression is cAMP-dependent in human and rat fetal liver [130, 131]. High levels of SAH and low levels of homocysteine would, at first glance, appear to be consistent with decreased *S*-adenosylhomocysteine hydrolase (SAHH) activity due to high adenosine levels, since increases in adenosine can reduce this enzyme's activity. However, cAMP competes with adenosine to inhibit this enzyme as well, and fluctuations in cAMP (higher levels) could result in similar findings [132, 133]. Finally, increased plasma levels of adenosine found in autistic children [129], and hypothesized to be due to inhibition of adenosine kinase (AK) by oxidative stress, may also reflect low levels of expression of the enzyme, as is found in the fetus and newborn compared to older infants and children [134].

Although oxidative stress is most likely involved in the metabolic functioning of individuals with autism [129, 135, 136, 137], the origin of these findings may not be obvious at first glance. Low activities of antioxidant enzymes such as superoxide dismutase, glutathione peroxidase, and catalase have also been documented in red blood cells from autistic patients [138, 139, 140], as have high levels of nitric oxide [139]. These data make sense when related to fetal functioning since there is a developmental lag in the appearance of superoxide dismutase compared to other antioxidant enzymes in the brain [141], and both glutathione peroxidase and catalase are lower in fetal than in postnatal fibroblasts [142]. In addition, nitric oxide is formed in response to B2AR stimulation in some tissues, such as endothelial cells and platelets [143, 144].

Overstimulation of the B2AR by terbutaline in the fetal rat brain and during the first and second postnatal weeks (equivalent to the second and third trimesters of human gestation) results in an increase in markers of lipid peroxidation compared to nonexposed rats [145], the hallmark of oxidative stress. Similar findings (though using different markers of lipid peroxidation) have been documented in postmortem autism brain [146].

Porphyrins

Increased urinary coproporphyrins have been proposed as a sign of heavy metal toxicity, specifically due to mercury, in children with autism. Geier and Geier [147] and Nataf et al. [148] suggest that high levels of coproporphyrin in both studies, and precoproporphyrin in the latter, are due to heavy metal inhibition of two downstream enzymes in the heme pathway, uroporphyrinogen decarboxylase and coproporphyrinogen oxidase. The inhibitory actions of heavy metals on the functions of these enzymes have been demonstrated in previous literature [149, 150]. An alternate explanation, however, could include a contribution from dysregulated levels of cAMP, since raising cellular levels of this molecule in rodent hepatocyte cultures causes accumulation of coproporphyrins [151].

It is important to note that data from many studies in autism, including those of urinary porphyrins, sulfation, and methylation, are reflections of peripheral organ function, not that of the brain. Although Nataf et al. [148] found that heavy metal chelation tended to normalize the urinary porphyrin levels on paper, it is unknown whether chelation could change the CNS level of mercury. Interestingly, though heavy metal sequestration is a function of the metal carrier metallothionein (MT), and if future research documents heavy metal sequestration in the autistic brain, liver, and bone marrow, this could be a sign of increased levels of MT. Cyclic AMP increases MT levels [152] and induces tissue-specific redistribution of heavy metals for sequestration [153]. The expression of MT in human liver is higher in the fetus than in children over 6 months of age [154, 155]. Thus, increased levels of MT could be responsible for sequestration of heavy metals such as cadmium and mercury, as a result of higher levels of cAMP or a fetal pattern of physiological functioning.

Plasma and Whole Blood Serotonin

Although whole blood serotonin (5HT) levels are increased in some patients with autism [156, 157], low levels of plasma 5HT that are available for receptor binding have been documented in autistic adult patients compared to normal controls [158], in a study using the platelet and its 5HT transporter (5HTT) as a reflection of serotonin's actions at the synapse. Elevated whole blood serotonin and low-plasma 5HT could be the result of (1) increased 5HT production, or (2) increased expression of the transporter on the platelet surface, or both, among other reasons. Circulating 5HT is produced in, and released from, enterochromaffin cells in the bowel, in a process regulated by neurotransmitter systems that include BARs [159], and is then incorporated into the platelet by the 5HTT. If signaling elements downstream from BARs, such as AC and cAMP, are dysregulated or enhanced, then release of 5HT from the bowel may be disordered or increased. Notably, Vered et al. [160] found a dysregulated response of 5HT levels in platelet-poor plasma from autistic adults after a carbohydrate-rich meal: an initial increase followed by a deficit compared to age-matched normal controls.

Ninety-nine percent of circulating serotonin is contained in platelets as a result of the function of the 5HT transporter [161]. The promoter region of the human serotonin transporter gene contains a CRE motif, and the promoter is inducible by cAMP signal transduction pathways [162, 163]. In addition, an increased density of platelet 5HTT expression has been documented in autism, using paroxetine binding as a specific label [164]. This could be the result of dysregulated AC functioning (leading to enhanced cAMP production) that may have been part of cell programming in bone marrow megakaryocytes before platelet fragmentation. This scenario would be expected after prenatal overstimulation of the B2AR. Because of this, increased transport of 5HT into the platelet could occur, partly due to increased production of 5HT in the bowel as above, and partly because of the increased expression of the 5HTT, both resulting from increased BAR activation. This may therefore contribute to increased platelet levels of serotonin and decreased levels of 5HT in the plasma of autistic patients.

Oxytocin

Oxytocin (OT) is a neuropeptide that mediates complex social and emotional behaviors [165]. Several studies have proposed that excess OT or OT abnormalities play a role in the etiology of autism [166, 167, 168]. Higher levels of the unprocessed, C-peptide-extended prohormone, OT-X, compared to the completely processed OT have been found in plasma from autistic children compared to age-matched normal controls [169]. A higher ratio of unprocessed: processed OT is similar to that seen in studies of fetal animals [170, 171]. Increased levels of OT-X in autism may therefore reflect the persistence of fetal physiology associated with B2AR dysregulation.

The Gastrointestinal Tract

The GI tract has been a subject of great research interest in autism. Findings of reflux esophagitis and disaccharidase deficiency [55] can be correlated with fluctuations or dysregulation in cAMP and fetal/neonatal functioning. Gastroesophageal reflux is common in premature human infants and neonates, and experimentally increased cAMP levels relax the lower esophageal sphincter in animal models [172, 173]. Disaccharidase expression in the human fetus is relatively low during midgestation and increases toward term [174]. Lactase levels in preterm infants are very low, especially in those younger than 37-weeks gestation [175], making it understandable that these enzymes might be deficient in children with autism, if their intestinal physiology is developmentally delayed and similar to that found in the fetus.

Abnormal motility contributes to diarrhea and constipation, symptoms often reported by caregivers of young children with autism [176, 177]. Peristaltic contractions of the GI tract begin in the stomach and are regulated by "pacemaker" cells, the interstitial cells of Cajal (ICC) [178, 179]. The ICC produce "slow waves," or the basic electrical rhythm of the smooth muscle layers, upon which electrical action potentials must be superimposed in order to generate true contractions. Human ICC are present from the second trimester of gestation, though they are not well-networked, even at term [180]. If ICC were delayed in their development in autism, peristalsis may be abnormal as a result.

Increased intestinal permeability to lactulose has been documented in 43% of a small group ($N = 21$) of children with high-functioning autism compared to controls [54]. D'Eufemia and colleagues point out that because lactulose is absorbed through the paracellular pathway, its increased absorption may be evidence of damage to intercellular tight junctions, as may be seen with inflammation. However, none of the patients in this study had GI symptoms, such as one might expect from the inflammation documented in later investigations [55, 181], suggesting an alternative explanation for increased permeability. In relating this research to delayed physiology or dysregulated AC signaling, it is known that intestinal permeability to sugars is high in neonates [182], and that intracellular cAMP levels are related to the integrity of intracellular tight junctions [183, 184].

Inflammation in the GI tract has been well-documented in autism, though its impact on the course of the disorder is unknown. The findings in the bowel [185] and stomach [186] in autistic children can be explained only by the interaction of many factors since the inflammatory process involves the immune and GI systems, both of which are constantly in a state of stimulus–response reactivity. Explanations for this process in relationship to the B2AR theory are beyond the scope of this chapter. A predisposition to inflammation, however, may be inherent in bowel functions of autistic children, as it is in the premature intestine; stimulated fetal enterocytes produce more interleukin-8 (IL-8), a proinflammatory cytokine, than those of older children [187], and transcription of IL-8 is increased by stimulation of B2ARs [188]. DeFelice et al. [189] found no increase in the basal production of IL-8 from endoscopic intestinal biopsies in patients with pervasive developmental disorders, but the measurements were performed with unstimulated enterocytes.

The Immune System

Dysregulated AC cell programming and signaling, along with delayed physiology in the bowel and its immune system, would have far-reaching consequences, because these two organ systems interact continuously to respond to exposures from the external "environment" of the gut lumen. In this situation, an immature, disordered immune system would attempt to respond to stimuli along with an equally dysregulated system of intestinal responses, the results of which could become virtually impossible to predict as static measures, and would change over time. It has become apparent that dysregulated immune responses characterize the inflammation documented in the bowel in autism, as well as the circulating immune system [15], at times reflecting tendencies to T helper type 2 (TH2) responses [181, 190], both TH1 and TH2 activation [191], innate-type immune responses [192], and, less often, normal results [193]. Although predicting one consistent "snapshot" of immune functioning is unattainable given this theory, and indeed, seems impossible in general in autism research, many parameters in the circulating immune system relate to fetal/neonatal functioning or fluctuations in B2AR or AC signaling, and several will be explained here.

Low levels of expression of the TH2 cytokine IL-2 on T lymphocytes [194], "incomplete activation" of T cells, or increased DR + T cells without a corresponding increase in IL-2 [195], and low intracellular levels of IL-2 and IFN-γ [196] may be explained by enhanced AC signaling, since cAMP inhibits IL-2 production and expression on T helper cells [197], and B2AR stimulation inhibits the production of IFN-γ by T lymphocytes [198]. In addition, cord blood lymphocytes express lower levels of IL-2 receptors [199] and produce lower amounts of IL-2 [200] and IFN-γ than lymphocytes from older children [201], supporting a relationship between these parameters in autism and delayed physiological functioning.

Investigations of serum antibodies in children with autism have revealed many targets for binding in rodent and human CNS tissue [202, 203, 204, 205, 206, 207, 208, 209], suggesting that this disorder may be an autoimmune disease.

However, none of these sera was tested against autism brain. In the more recent work mentioned above, serum antibodies of autistic patients were produced at higher levels than in controls. Interestingly, stimulation of the B2AR regulates the production of immunoglobulin G1 (IgG1) and IgE from B lymphocytes, as well as the number of B cells secreting IgM [210, 211, 212], so that an alternate explanation for these findings might relate them to fluctuations in signaling downstream from the B2AR. In addition, immature CD5 + (or B1) B cells may contribute to increased antibody production in children with autism. These lymphocytes are the predominant B cell during fetal and early neonatal life, and are reactive against autoantigens [213]. They generally produce low-affinity auto- and polyreactive antibodies (usually IgM), including those directed against single-stranded DNA [214, 215], as has been found in children with ASDs and Landau–Kleffner syndrome [204]. Higher levels of antibody binding to proteins of specific molecular weight may be the result of these B-cell effects, which in turn may have been "programmed" by prenatal B2AR overstimulation.

If the immune system in young children with autism is physiologically delayed, the T-cell receptor (TCR) may be immature as well. The CDR3 region of the TCR provides fine antigen recognition and is shorter in length during fetal life compared to that in neonates [216]. The enzyme necessary for lengthening the CDR3 region (terminal deoxynucleotidal transferase) is ontogenically expressed [217] and is cAMP-dependent [218]. In turn, deficient fine antigen recognition may contribute to the formation of polyreactive autoantibodies in autism.

The Autonomic Nervous System

Both published research [219, 220, 221, 222] and anecdotal reports (urinary retention, sluggish pupillary response to light, and clammy extremities) have noted differences in autonomic nervous system function in patients with autism compared to controls. Ming et al. [223] measured baseline cardiovascular autonomic function in children with autism (ages 4–14) and age-matched healthy controls, using a device that can simultaneously measure heart rate and blood pressure, derive a continuous index of cardiac vagal tone (CVT), and monitor cardiac sensitivity to baroreflex (CSB), in real-time. Both CVT and CSB parameters reflect parasympathetic activity. The CVT and CSB were significantly lower in association with a significant elevation in heart rate, mean arterial blood pressure, and diastolic blood pressure in children with autism compared to controls. Levels of CVT and CSB were lower in autistic children with symptoms of autonomic dysfunction compared to those without, and were not related to age. These low levels suggest impaired cardiac parasympathetic activity, with unrestrained and perhaps hyperactivity of the sympathetic nervous system.

The findings of Ming et al. [223] can be correlated with results from overstimulation of BARs in fetal (GD17–20) or newborn rats (PN 2–5; equivalent to the second trimester of human gestation). Garofolo et al. [79] administered isoproterenol, a combined β_1 and β_2 agonist, and terbutaline, a selective B2AR agonist, to rats during these two time periods. Both treatments caused downregulation of m2 type acetylcholine receptors (m2AChRs) as well as a loss of the ability of m2AChR stimulation to inhibit AC in the heart, which could reduce effects from parasympathetic input. Premature exposure of the developing heart to BAR agonists also promotes the function and sensitization of signaling through AC [70], enhancing sympathetic input. These effects may parallel changes found in sympathetic and parasympathetic tone in autistic children [223].

Differences in Signal Transduction

If autism is a biological disorder characterized, in part, by dysregulated AC functioning that leads to disordered development, then markers of increased AC signaling should be expressed at higher levels in some tissues in patients with autism compared to those in controls. Examples are the enzymes involved in the cAMP-dependent second messenger pathways such as PKA and cAMP. Chauhan et al. [224] have studied signal transduction in children with autism from the AGRE database and found increased expression of PKA in membrane preparations of lymphoblasts from the autistic children compared to normal sibling controls. This finding lends support to the B2AR/AC theory. Further studies investigating levels of cAMP and AC in lymphoblasts are underway by this group. It will be interesting and important to study the expression of other molecules that could be predicted from Theodore Slotkin's previously described rat model, such as a decrement in Gi proteins and deficits in their ability to inhibit AC [77, 78]. This would provide support for a hypothetical physiological signaling imbalance due to prenatal overstimulation of the B2AR.

Different Mechanisms, Similar Result

No single prenatal insult, such as overstimulation of the B2AR during critical periods of brain development, can be responsible in all cases for a complex neurodevelopmental disorder with heterogeneous presentations such as autism. Prenatal B2AR overstimulation in the previously described rat model leads to enhanced signaling through AC early in development, and to gender-dependent dysregulation (enhancement and deficit) of AC signaling as development progresses, in various regions of the brain. It is possible that this sort of downstream signaling abnormality that has the potential to affect numerous processes and cell programming through CREB protein and other components is a "final common pathway" in the development of autism. Similar findings have been documented in animal models that have been used to investigate mechanisms of insults different from β_2 receptor overstimulation during development.

Several of these mechanisms have been linked to autism in human studies, such as maternal stress [225], or have been theoretically associated in animal models such as the "stress" of infections during pregnancy [226, 227] (see Patterson chapter). Stress itself elevates levels of epinephrine and norepinephrine (B2AR ligands) in the pregnant mother, but these do not cross the placenta effectively; Moreover, stress also causes increases in endogenous levels of glucocorticoids, which do cross to the fetus. Glucocorticoid excess in maternal stress can result from psychological, infectious, or physical causes. Prenatal administration of the glucocorticoid, dexamethasone, to rats results in early, enhanced AC activity in the developing brain and liver [228] that persists, at least in the forebrain, as increased activity of the AC catalytic subunit in postnatal life [229]. After pregnant guinea pigs are exposed to synthetic glucocorticoids, the offspring exhibits alterations in pituitary–adrenal function and responses to endogenous glucocorticoids throughout life, as well as differences in activity levels compared to controls [230, 231]. Effects of maternal adversity (as opposed to exposure to synthetic glucocorticoids) have been studied in the guinea pig as well. Male offspring of stressed dams develops altered basal plasma cortisol concentrations and exhibits increased anxiety compared to controls [232]. Interestingly, the male offspring of glucocorticoid-exposed dams developed elevated plasma testosterone levels [230], which could be analogous to the elevated plasma testosterone levels reported in autistic children [233].

Subtoxic maternal exposure to organophosphate pesticides, which represent half of all the insecticide use in the world, may be another factor in the development of autism. The most used organophosphate, chlorpyrifos, though prohibited now from use in the home in the United States, is still widely applied in the chemical preparation of homesites and throughout agriculture, and thus nearly the entire population is exposed to this agent. In a rodent model, prenatal exposure to this pesticide results in enhanced AC signaling in the brain that is region- and sex-dependent and persists into adulthood [82, 234], results that converge with the outcomes after terbutaline treatment. It is possible that the study of such prenatal disturbances that result in dysregulated AC function will lead further to defining some of the mechanisms involved in the etiology of autism.

Prenatal β_2-Adrenergic Receptor Overstimulation and Sensitization to Other Factors

It may not be that one process alone, namely B2AR overstimulation during critical periods for prenatal brain development, consistently in all cases leads to a complex neurodevelopmental disorder with heterogeneous presentation such as autism. The presence of polymorphisms of the B2AR that amplify this tendency could further increase the chances that the disorder might develop in an individual child. It is also possible that prenatal B2AR overstimulation sensitizes the CNS and peripheral tissues to other, later factors that could affect development. For example, in rats, B2AR overstimulation by terbutaline during the neurodevelopmental equivalent of the second trimester in humans results in dysregulated AC signaling as described previously. However, when these same neonatal rats are later exposed at PN 11–14 to the organophosphate insecticide, chlorpyrifos, larger alterations in more widespread areas of the developing brain are found than are seen with either agent alone [75]. A good deal of attention is understandably paid to genetic etiologies of differential susceptibility to environmental toxicants, but it is equally plausible that nongenetic factors such as prior chemical or drug exposure in mothers, children, and fetuses may also define subpopulations that are especially vulnerable to alterations in developmental trajectories.

Effects on Other Neurotransmitter Systems

During postnatal life, B2AR signaling interacts with and affects all other neurotransmitter systems in various tissues. The same may be true for brain development during fetal life. Prenatal B2AR overstimulation not only leads to dysregulation of development in the catecholamine/AC signaling system, but also influences the development of other neurotransmitter systems, such as serotonin (5HT). Overstimulation of the B2AR during PN 2–5 in the rat elicits global increases in 5HT receptors and the 5HT transporter in brain regions possessing 5HT cell bodies (midbrain and brainstem) as well as in the hippocampus [235, 236]. Slotkin and Seidler [236] found that the changes in the 5HT system were demonstrable after this exposure, as late as adolescence. Interestingly, lower total serotonin content was noted in the brains of treated rats compared to controls [236]. This may be analogous to lower levels of brain serotonin synthesis over time in autistic patients compared to controls [237] (see Blue chapter 5).

The Role of Maternal Factors

In considering prenatal factors that may result in the dysregulation of AC functioning and disordered development, one must take into account that effects on the fetus are the result of, and influenced by, the physiology of the pregnant mother and her adaptive responses to the environment. It may be that downstream cell signaling that results from factors such as inflammatory (cytokine) responses in the mother [227], maternal antibodies [238, 239], and stress [225] shares components that are also elicited by overstimulation of the B2AR. Additional factors may worsen the tendency for abnormal developmental trajectories, such as low maternal plasma serotonin [240] and genetic polymophisms that could increase or decrease cell responses. Because abnormalities in development induced by the pregnancy environment can become transgenerational [241], studies of mothers and grandmothers of children with autism may be necessary in the future.

Prenatal β_2-Adrenergic Receptor Understimulation

The opposite case, i.e., blocking BAR function or adrenergic denervation during early development, also produces unique patterns of receptor responses in animal studies. In comparison with prenatal B2AR overstimulation, which would lead to dysregulation of AC signaling, after adrenergic denervation with 6-hydroxydopamine, developing rats exhibit a delay in the onset of desensitization in the heart (PN 25 instead of PN 15) and lack of any desensitization in the liver, even at PN 25 [242]. The resulting signal enhancement continues as an immature pattern of sensitization for a prolonged period compared to controls [242]. Effects can be permanent: receptors in some tissues never acquire some of their essential responses [243]. Denervation in this context, then, shifts a developmental change (one from sensitization to desensitization) toward an older age and creates a physiological developmental delay. This scenario of understimulation may relate to maternal antibodies that can potentially block receptors [238, 239] and viral infections such as influenza in the pregnant mother [227] or postnatal herpes [244], both of which decrease AC signaling and lower cAMP production [245, 246].

Speiser et al. [247] treated pregnant rats during GD8–22 with the BAR antagonists propranolol (a β_1 and β_2 receptor blocker) and atenolol (a specific β_1 receptor blocker). The offspring of rats treated with propranolol demonstrated increased motor activity and poor performance in the active avoidance test, whereas the offspring exposed to atenolol during gestation exhibited no behavioral changes, showing that blocking B2ARs during brain development can be linked to behavioral abnormalities in animals. Blocking BARs in rats during an age equivalent to the second trimester has not been studied as extensively as has B2AR overstimulation in the same period, but there are two studies pointing toward alterations in neural cell development that may be relevant to neurodevelopmental disorders such as autism. In newborn rats, destruction of presynaptic noradrenergic terminals with the neurotoxin 6-hydroxydopamine results in blunting of the development of the ability of B2ARs to elicit cellular responses [248], again showing the importance of a critical period in which the appropriate exposure of receptors to the natural neurotransmitter ligand "programs" the development of cell responses. Similarly, chronic administration of propranolol to pregnant rats throughout gestation delays the development of B2AR coupling to cellular responses [248]. In the long run, interference with B2AR transmission during pregnancy produces a permanent change in the responsiveness to noradrenergic input in the brain [249]. Since these mechanisms result in abnormalities in AC signaling, they all are potentially important in autism pathogenesis.

Conclusion

Historically, there seem to be several stages in unraveling the etiology of a neurodevelopmental disorder. After recognition that a specific disorder exists (for example, Rett syndrome was first described in 1966), a prolonged period ensues, marked by a plethora of research studies that describe the physiology of the disorder, and proposes mechanisms of pathogenesis as well as candidate genes involved in it. Finally, the cause is discovered and research into treatment begins (the abnormality in the MECP2 gene was discovered in 1999). At this time in history, autism research is still in the descriptive phase of this process, but with cohesive theories on which to base research efforts, it will move from description to discover causes. Autism is clearly heterogeneous in its clinical presentation, with biological research results that show widespread dysregulation. The theory of prenatal B2AR signaling outlined in this chapter, relating environmental influences with genetic predispositions to dysregulation in an enzyme system that is ubiquitous and influences a multitude of developmental pathways, accounts for multiple disparate findings and disordered development. This type of approach will help to unravel the etiologies of autism.

Acknowledgments I thank Theodore A. Slotkin of Duke University Medical Center for his decades of excellent research on the β_2-adrenergic receptor, without which this chapter could not have been written, and for his thoughtful review and suggestions for this manuscript, and Dorothy E. Crowell of Housatonic, Massachusetts, a mother of autistic triplets, who first proposed a connection between β_2-adrenergic receptor and autism.

References

1. Bailey A, Luthert P, Dean A, Harding B, Janota I, Montgomery M, Rutter M, Lantos P (1998) A clinicopathological study of autism. *Brain* 121(Pt 5): 889–905.
2. Casanova MF, van Kooten IA, Switala AE, van Engeland H, Heinsen H, Steinbusch HW, Hof PR, Trippe J, Stone J, Schmitz C (2006) Minicolumnar abnormalities in autism. *Acta Neuropathol* 112: 287–303.
3. Bauman ML, Kemper TL (2005) Neuroanatomic observations of the brain in autism: a review and future directions. *Int J Dev Neurosci* 23: 183–187.
4. Rodier (2000) The early origins of autism. *Sci Am* 282: 56–63.
5. Nelson KB, Grether JK, Croen LA, Dambrosia JM, Dickens BF, Jelliffe LL, Hansen RL, Phillips TM (2001) Neuropeptides and neurotrophins in neonatal blood of children with autism or mental retardation. *Ann Neurol* 49: 597–606.
6. Leonard H, de Klerk N, Bourke J, Bower C (2006) Maternal health in pregnancy and intellectual disability in the offspring: a population-based study. *Ann Epidemiol* 16: 448–454.

7. Robinson PD, Schutz CK, Macciardi F, White BN, Holden JJA (2001) Genetically determined low maternal serum dopamine β-hydroxylase levels and the etiology of autism spectrum disorders. *Am J Med Gen* 100: 30–36.

8. Chugani DC, Muzik O, Behen M, Rothermel R, Janisse JJ, Lee J, Chugani HT (1999) Developmental changes in brain serotonin synthesis capacity in autistic and nonautistic children. *Ann Neurol* 45: 287–295.

9. Blatt GJ, Fitzgerald CM, Guptill JT, Booker AB, Kemper TL, Bauman ML (2001) Density and distribution of hippocampal neurotransmitter receptors in autism: an autoradiographic study. *J Autism Dev Disord* 31: 537–543.

10. Martin-Ruiz CM, Lee M, Perry RH, Baumann M, Court JA, Perry EJ (2004) Molecular analysis of nicotinic receptor expression in autism. *Brain Res Mol Brain Res* 123: 81–90.

11. Purcell AE, Jeon OH, Zimmerman AW, Blue ME, Pevsner J (2001) Postmortem brain abnormalities of the glutamate neurotransmitter system in autism. *Neurology* 57:1618–1628.

12. Zimmerman AW, Jyonouchi H, Comi AM, Connors SL, Milstien S, Varsou A, Heyes MP (2005) Cerebrospinal fluid and serum markers of inflammation in autism. *Pediatr Neurol* 33:195–201.

13. Vargas DL, Nascimbene C, Krishnan C, Zimmerman AW, Pardo CA (2005). Neuroglial activation and neuroinflammation in the brain of patients with autism. *Ann Neurol* 57: 67–81.

14. Ashwood P, Wills S, Van de Water J (2006) The immune response in autism: a new frontier for autism research. *J Leukoc Biol* 80: 1–15.

15. Erickson CA, Stigler KA, Corkins MR, Posey DJ, Fitzgerald JF, McDougle CJ (2005) Gastrointestinal factors in autistic disorder: a critical review. *J Autism Dev Disord* 35: 713–727.

16. Modahl C, Green L, Fein D, Morris M, Waterhouse L, Feinstein C, Levin H (1998) Plasma oxytocin levels in autistic children. *Biol Psychiatry* 43: 270–277.

17. Ylisaukko-oja T, Alarcon M, Cantor RM, Auranen M, Vanhala R, Kempas E, von Wendt L, Jarvela I, Geschwind DH, Peltonen L (2006) Search for autism loci by combined analysis of Autism Genetic Resource Exchange and Finnish families. *Ann Neurol* 59:145–155.

18. Sykes NH, Lamb JA (2007) Autism: the quest for the genes. *Expert Rev Mol Med* 9: 1–15.

19. Campbell DB, Sutcliffe JS, Ebert PJ, Militemi R, Bravaccio C, Trillo S, Elia M, Schneider C, Melmed R, Sacco R, Persico AM, Levitt P (2006) A genetic variant that disrupts MET transcription is associated with autism. *Proc Natl Acad Sci USA* 103: 16834–16839.

20. Swackhammer R, Tatum OL, (2006) Survey of candidate genes for autism susceptibility. *J Assoc Genet Technol* 33: 8–16.

21. Stone JL, Merriman B, Cantor RM, Geschwind DH, Nelson SF (2007) High density SNP association study of a major autism linkage region on chromosome 17. *Hum Mol Genet* 16: 704–715.

22. Zimmerman AW, Connors SL, Pardo-Villamizar CA (2006): Neuroimmunology and neurotransmitters in autism; in Tuchman R, Rapin I (eds): *Autism: A Neurological Disorder of Early Brain Development*. London, MacKeith Press, pp. 141–159.

23. Sundström E, Kölare S, Souverbie F, Samuelsson E-B, Pschera H, Lunell N-O, Seiger Å (1993) Neurochemical differentiation of human bulbospinal monoaminergic neurons during the first trimester. *Brain Res Dev Brain Res* 75: 1–12.

24. Crespo P, Cachero TG, Xu N, Gutkind JS (1995) Dual effect of beta-adrenergic receptors on mitogen-activated protein kinase. Evidence for a beta gamma-dependent activation and a G alpha s-cAMP-mediated inhibition. *J Biol Chem* 270: 25259–25265.

25. Schmitt JM, Stork PJ (2000) Beta 2-adrenergic receptor activates extracellular signal-related kinases (ERKs) via the small G protein rap1 and the serine/threonine kinase B-Raf. *J Biol Chem* 275: 25342–25350.

26. Kobilka BK, Dixon RA, Frielle T, Dohlman HG, Bolanowski MA, Sigal IS, Yang-Feng TL, Francke U, Caron MG, Lefkowitz RJ (1987) cDNA for the human beta 2-adrenergic

receptor: a protein with multiple membrane-spanning domains and encoded by a gene whose chromosomal location is shared with that of the receptor for platelet-derived growth factor. *Proc Natl Acad Sci USA* 84: 46–50.

27. Slotkin TA, Lau C, Seidler FJ (1994) Beta-adrenergic receptor overexpression in the fetal rat: distribution, receptor subtypes, and coupling to adenylate cyclase activity via G-proteins. *Toxicol Appl Pharmacol* 129: 223–234.

28. Cikos S, Veselá J, Il'ková G, Rehák P, Czikková S, Koppel J (2005) Expression of beta adrenergic receptors in mouse oocytes and preimplantation embryos. *Mol Reprod Dev* 71: 145–153.

29. Lai LP, Mitchell J (2008) Beta(2)-adrenergic receptors expressed on murine chondrocytes stimulate cellular growth and inhibit the expression of Indian hedgehog and collagen type X. *J Cell Biochem* 104: 545–553.

30. Duncan CP, Seidler FJ, Lappi SE, Slotkin TA (1990) Dual control of DNA synthesis by α- and β-adrenergic mechanisms in normoxic and hypoxic neonatal rat brain. *Brain Res Dev Brain Res* 55: 29–33.

31. Kwon JH, Eves EM, Farrell S, Segovia J, Tobin AJ, Wainer BH, Downen M (1996) Beta-adrenergic receptor activation promotes process outgrowth in an embryonic rat basal forebrain cell line and in primary neurons. *Eur J Neurosci* 8: 2042–2055.

32. Gu C, Ma YC, Benjamin J, Littman D, Chao MV, Huang XY (2000) Apoptotic signaling through the beta-adrenergic receptor. A new Gs effector pathway. *J Biol Chem* 275: 20726–20733.

33. Gharami K, Das S (2000) Thyroid hormone-induced morphological differentiation and maturation of astrocytes are mediated through the beta-adrenergic receptor. *J Neurochem* 75: 1962–1969.

34. Burniston JG, Tan LB, Goldspink DF (2005) Beta2-Adrenergic receptor stimulation in vivo induces apoptosis in the rat heart and soleus muscle. *J Appl Physiol* 98: 1379–1386.

35. Zhu WZ, Zheng M, Koch WJ, Lefkowitz RJ, Kobilka BK, Xiao RP (2001) Dual modulation of cell survival and cell death by beta(2)-adrenergic signaling in adult mouse cardiac myocytes. *Proc Natl Acad Sci USA* 98: 1607–1612.

36. Stewart AG, Harris T, Fernandes DJ, Schachte LC, Koutsoubos V, Guida E, Ravenhall CE, Vadiveloo P, Wilson JW (1999) Beta2-adrenergic receptor agonists and cAMP arrest human cultured airway smooth muscle cells in the G(1) phase of the cell cycle: role of proteasome degradation of cyclin D1. *Mol Pharmacol* 56: 1079–1086.

37. Sewing A, Bürger C, Brüsselbach S, Schalk C, Lucibello FC, Müller R (1993) Human cyclin D1 encodes a labile nuclear protein whose synthesis is directly induced by growth factors and suppressed by cyclic AMP. *J Cell Sci* 104 (Pt 2): 545–555.

38. Cocks BG, Vairo G, Bodrug SE, Hamilton JA (1992) Suppression of growth factor-induced CYL1 cyclin gene expression by antiproliferative agents. *J Biol Chem* 267: 12307–12310.

39. Lidow MS, Rakic P (1994) Unique profiles of the alpha 1-, alpha 2-, and beta-adrenergic receptors in the developing cortical plate and transient embryonic zones of the rhesus monkey. *J Neurosci* 14: 4064–4078.

40. Kim YT, Wu CF (1996) Reduced growth cone motility in cultured neurons from Drosophila memory mutants with a defective cAMP cascade. *J Neurosci* 16: 5593–5602.

41. Saitow F, Satake S, Yamada J, Konishi S (2000) Beta-adrenergic receptor-mediated presynaptic facilitation of inhibitory GABAergic transmission at cerebellar interneuron-Purkinje cell synapses. *J Neurophysiol* 84: 2016–2025.

42. Saitow F, Suzuki H, Konishi S (2005) Beta-Adrenoceptor-mediated long-term up-regulation of the release machinery at rat cerebellar GABAergic synapses. *J Physiol* 565 (Pt 2): 487–502.

43. Hoogland TM, Saggau P (2004) Facilitation of L-type Ca^{2+} channels in dendritic spines by activation of beta2 adrenergic receptors. *J Neurosci* 24: 8416–8427.

44. Wang SJ, Cheng LL, Gean PW (1999) Cross-modulation of synaptic plasticity by beta-adrenergic and 5-HT1A receptors in the rat basolateral amygdala. *J Neurosci* 19: 570–577.

45. Roozendaal B, Nguyen BT, Power AE, McGaugh JL (1999) Basolateral amygdala noradrenergic influence enables enhancement of memory consolidation induced by hippocampal glucocorticoid receptor activation. *Proc Natl Acad Sci USA* 96: 11642–11647.

46. Junker V, Becker A, Hühne R, Zembatov M, Ravati A, Culmsee C, Krieglstein J (2002) Stimulation of beta-adrenoceptors activates astrocytes and provides neuroprotection. *Eur J Pharmacol* 446: 25–36.

47. Tomozawa Y, Yabuuchi K, Inoue T, Satoh M (1995) Participation of cAMP and cAMP-dependent protein kinase in beta-adrenoceptor-mediated interleukin-1 beta mRNA induction in cultured microglia. *Neurosci Res* 22: 399–409.

48. Prinz M, Hausler KG, Kettenmann H, Hanisch U (2001) beta-adrenergic receptor stimulation selectively inhibits IL-12p40 release in microglia. *Brain Res* 899: 264–270.

49. Basta PV, Moore TL, Yokota S, Ting JP (1989) A beta-adrenergic agonist modulates DR alpha gene transcription via enhanced cAMP levels in a glioblastoma multiforme line. *J Immunol* 142: 2895–2901.

50. Podojil JR, Sanders VM (2003) Selective regulation of mature IgG1 transcription by CD86 and beta 2-adrenergic receptor stimulation. *J Immunol* 170: 5143–5151.

51. Kasprowicz DJ, Kohm AP, Berton MT, Chruscinski AJ, Sharpe A, Sanders VM (2000) Stimulation of the B cell receptor, CD86 (B7–2), and the beta 2-adrenergic receptor intrinsically modulates the level of IgG1 and IgE produced per B cell. *J Immunol* 165: 680–690.

52. Nance DM, Sanders VM (2007) Autonomic innervation and regulation of the immune system (1987–2007). *Brain Behav Immun* 21: 736–745.

53. Takamoto T, Hori Y, Koga Y, Toshima H, Hara A, Yokoyama MM (1991) Norepinephrine inhibits human natural killer cell activity in vitro. *Int J Neurosci* 58: 127–131.

54. D'Eufemia P, Celli M, Finocchiaro R, Pacifico L, Viozzi L, Zaccagnini M, Cardi E, Giardini O (1996) Abnormal intestinal permeability in children with autism. Acta Paediatric 85: 1076–1079.

55. Horvath K, Papadimitriou JC, Rabsztyn A, Drachenberg C, Tildon JT (1999) Gastrointestinal abnormalities in children with autistic disorder. *J Pediatr* 135: 559–563.

56. Kohout TA, Lefkowitz RJ (2003) Regulation of G protein-coupled receptor kinases and arrestins during receptor desensitization. *Mol Pharmacol* 63: 9–18.

57. Green SA, Rathz DA, Schuster AJ, Liggett SB (2001) The Ile 164 beta(2)-adrenoceptor polymorphism alters salmeterol exosite binding and conventional agonist coupling to G(s). *Eur J Pharmacol* 421: 141–147.

58. Dishy V, Sofowora GG, Xie HG, Kim RB, Byrne DW, Stein CM, Wood AJ (2001) The effect of common polymorphisms of the beta2-adrenergic receptor on agonist-mediated vascular desensitization. *N Engl J Med* 345: 1030–1035.

59. Cockcroft JR, Gazis AG, Cross DJ, Wheatley A, Dewar J, Hall IP, Noon JP (2000) Beta(2)-adrenoceptor polymorphism determines vascular reactivity in humans. *Hypertension* 36: 371–375.

60. Liggett SB, Wagoner LE, Craft LL, Hornung RW, Hoit BD, McIntosh TC, Walsh RA (1998) The Ile 164 beta2-adrenergic receptor polymorphism adversely affects the outcome of congestive heart failure. *J Clin Invest* 102: 1534–1539.

61. Wolk R, Snyder EM, Somers VK, Turner ST, Olson LJ, Johnson BD (2007) Arginine 16 glycine beta2-adrenoceptor polymorphism and cardiovascular structure and function in patients with heart failure. *J Am Soc Echocardiogr* 20: 290–297.

62. Lee DK, Currie GP, Hall IP, Lima JJ, Lipworth BJ (2004) The arginine-16 beta2-adrenoceptor polymorphism predisposes to bronchoprotective subsensitivity in patients treated with formoterol and salmeterol. *Br J Clin Pharmacol* 57: 68–75.

63. Kawaguchi H, Masuo K, Katsuya T, Sugimoto K, Rakugi H, Ogihara T, Tuck ML (2006) Beta2- and beta3-Adrenoceptor polymorphisms relate to subsequent weight gain and blood pressure elevation in obese nornotensive individuals. *Hypertens Res* 29: 951–959.

64. Pinelli M, Giacchetti M, Acquaviva F, Cocozza S, Donnarumma G, Lapice E, Riccardi G, Romano G, Vaccaro O, Monticelli A (2006) Beta2-adrenergic receptor and UCP3 variants modulate the relationship between age and type 2 diabetes mellitus. *BMC Med Genet* 7: 85.

65. Jazdzewski K , Bednarczuk T, Stepnowska M, Liyanarachchi S, Suchecka-Rachon K, Limon J, Narkiewicz K (2007) Beta-2-adrenergic receptor gene polymorphism confers susceptibility to Graves' disease. *Int J Mol Med* 19: 181–186.

66. Xu BY, Huang D, Pirskanen R, Lefvert AK (2000) Beta2-adrenergic receptor gene polymorphisms in myasthenia gravis (MG). *Clin Exp Immunol* 119: 156–160.

67. Xu B, Arlehag L, Rantapää-Dahlquist SB, Lefvert AK (2004) Beta2-adrenergic receptor gene single-nucleotide polymorphisms are associated with rheumatoid arthritis in northern Sweden. *Scand J Rheumatol* 33: 395–398.

68. Busjahn A, Freier K, Faulhaber HD, Li GH, Rosenthal M, Jordan J, Hoehe MR, Timmermann B, Luft FC (2002) Beta-2 adrenergic receptor gene variations and coping styles in twins. *Biol Psychol* 61: 97–109.

69. Slotkin TA, Saleh JL, Zhang J, Seidler FJ (1996) Ontogeny of beta-adrenoceptor/adenylyl cyclase desensitization mechanisms: the role of neonatal innervation. *Brain Res* 742: 317–328.

70. Slotkin TA, Auman JT, Seidler FJ (2003) Ontogenesis of β-adrenoceptor signaling: Implications for perinatal physiology and for fetal effects of tocolytic drugs. *J Pharmacol Exp Ther* 306: 1–7.

71. Slotkin TA, Tate CA, Cousins MM, Seidler FJ (2001) Beta-Adrenoceptor signaling in the developing brain: sensitization or desensitization in response to terbutaline. *Brain Res Dev Brain Res* 131: 113–125.

72. Auman JT, Seidler FJ, Tate CA, Slotkin TA (2001) Beta-adrenoceptor-mediated cell signaling in the neonatal heart and liver: responses to terbutaline. *Am J Physiol Regul Integr Comp Physiol* 281: R1895–R1901.

73. Stein HM, Oyama K, Sapien R, Chappell BA, Padbury JF (1992) Prolonged beta-agonist infusion does not induce desensitization or down-regulation of beta-adrenergic receptors in newborn sheep. *Pediatr Res* 31: 462–467.

74. Sun LS (1999) Regulation of myocardial beta-adrenergic receptor function in adult and neonatal rabbits. *Biol Neonate* 76: 181–192.

75. Meyer A, Seidler FJ, Aldridge JE, Slotkin TA (2005) Developmental exposure to terbutaline alters cell signaling in mature rat brain regions and augments the effects of subsequent neonatal exposure to the organophosphorus insecticide chlorpyrifos. *Toxicol Appl Pharmacol* 203: 154–166.

76. Zeiders JL, Seidler FJ, Iaccarino G, Koch WJ, Slotkin TA (1999) Ontogeny of cardiac beta-adrenoceptor desensitization mechanisms: agonist treatment enhances receptor/G-protein transduction rather than eliciting uncoupling. *J Mol Cell Cardiol* 31: 413–423.

77. Zeiders JL, Seidler FJ, Slotkin TA (2000) Ontogeny of G-protein expression: control by beta-adrenoceptors. *Brain Res Dev Brain Res* 120: 125–134.

78. Auman JT, Seidler FJ, Slotkin TA (2002) Beta-adrenoceptor control of G protein function in the neonate: determinant of desensitization or sensitization. *Am J Physiol Regul Integr Comp Physiol* 283: R1236–R1244.

79. Garofolo MC, Seidler FJ, Auman JT, Slotkin TA (2002) Beta-Adrenergic modulation of muscarinic cholinergic receptor expression and function in developing heart. *Am J Phys Reg Integr Comp Physiolo* 282: R1356–R1363.

80. Owen D, Matthews SG (2007) Repeated maternal glucocorticoid treatment affects activity and hippocampal NMDA receptor expression in juvenile guinea pigs. *J Physiol* 578: 249–257.

81. Stanwood GD, Levitt P (2007) Prenatal exposure to cocaine produces unique developmental and long-term adaptive changes in dopamine D1 receptor activity and subcellular distribution. *J Neurosci* 27: 152–157.

82. Meyer A, Seidler FJ, Aldridge JE, Tate CA, Cousins MM, Slotkin TA (2004) Critical periods for chlorpyrifos-induced developmental neurotoxicity: alterations in adenylyl cyclase signaling in adult rat brain regions after gestational or neonatal exposure. *Environ Health Perspect* 112: 295–301.
83. Rice D, Barone S Jr (2000) Critical periods of vulnerability for the developing nervous system: evidence from humans and animal models. *Environ Health Perspect* 108(Suppl 3): 511–533.
84. Lam F, Elliott J, Jones JS, Katz M, Knuppel RA, Morrison J, Newman R, Phelan J, Willcourt R (1998) Clinical issues surrounding the use of terbutaline sulfate for preterm labor. *Obstet Gynecol Surv* 53(11Suppl): S85–S95.
85. Connors SL, Crowell DE, Eberhart CG, Copeland J, Newschaffer CJ, Spence SJ, Zimmerman AW (2005) Beta2-adrenergic receptor activation and genetic polymorphisms in autism: data from dizygotic twins. *J Child Neurol* 20: 876–884.
86. Hadders-Algra M, Touwen BCL, Huisjes HJ (1986) Long-term follow-up of children prenatally exposed to ritodrine. *Br J Obstet Gynaecol* 93: 156–161.
87. Pitzer M, Schmidt MH, Esser G, Laucht M (2001) Child development after maternal tocolysis with b-sympathomimetic drugs. *Child Psychiatry Hum Dev* 31: 165–182.
88. Bergman B, Bokstrom H, Borga O, Enk L, Hedner T, Wangberg B (1984) Transfer of terbutaline across the human placenta in late pregnancy. *Eur J Respir Dis Suppl* 134: 81–86.
89. Hsu CH, Robinson CP, Basmadjian GP (1994) Tissue distribution of 3H-terbutaline in rabbits. *Life Sci* 54: 1465–1469.
90. Geschwind DH, Sowinski J, Lord C, Iversen P, Shestack J, Jones P, Ducat L, Spence SJ; AGRE Steering Committee (2001) The autism genetic resource exchange: a resource for the study of autism and related neuropsychiatric conditions. *Am J Hum Genet* 69: 463–466.
91. Cheslack-Postava K, Fallin MD, Avramopoulos D, Connors SL, Zimmerman AW, Eberhart CG, Newschaffer CJ (2007) Beta2-Adrenergic receptor gene variants and risk for autism in the AGRE cohort. *Mol Psychiatry* 12: 283–291.
92. Fujinaga M, Scott JC (1997) Gene expression of catecholamine synthesizing enzymes and β adrenoceptor subtypes during rat embryogenesis. *Neurosci Lett* 231: 108–112.
93. Bauman M, Kemper TL (1985) Histoanatomic observations of the brain in early infantile autism. *Neurology* 35: 866–874.
94. Rhodes MC, Seidler FJ, Abdel-Rahman A, Tate CA, Nyska A, Rincavage HL, Slotkin TA (2004) Terbutaline is a neurotoxicant: effects on neuroproteins and morphology in cerebellum, hippocampus, and somatosensory cortex. *J Pharmacol Exp Ther* 308: 529–537.
95. Kudlacz EM, Navarro HA, Kavlock RJ, Slotkin TA (1990) Regulation of postnatal beta-adrenergic receptor/adenylate cyclase development by prenatal agonist stimulation and steroids: alterations in rat kidney and lung after exposure to terbutaline or dexamethasone. *J Dev Physiol* 14: 273–281.
96. Zerrate MC, Pletnikov M, Connors SL, Vargas DL, Seidler FJ, Zimmerman AW, Slotkin TA, Pardo CA (2007) Neuroinflammation and behavioral abnormalities after neonatal terbutaline treatment in rats: implications for autism. *J Pharmacol Exp Ther* 322: 16–22.
97. Meng SZ, Oka A, Takashima S (1999) Developmental expression of monocyte chemoattractant protein-1 in the human cerebellum and brainstem. *Brain Dev* 21: 30–35.
98. Dame JB, Juul SE (2000) The distribution of receptors for the pro-inflammatory cytokines interleukin (IL)-6 and IL-8 in the developing human fetus. *Early Hum Dev* 58: 25–39.
99. Nishio Y, Kashiwagi A, Takahara N, Hidaka H, Kikkawa R (1997) Cilostazol, a cAMP phosphodiesterase inhibitor, attenuates the production of monocyte chemoattractant protein-1 in response to tumor necrosis factor-alpha in vascular endothelial cells. *Horm Metab Res* 29: 491–495.

100. Hannila SS, Filbin MT (2008) The role of cyclic AMP signaling in promoting axonal regeneration after spinal cord injury. *Exp Neurol* 209: 321–332.

101. Wierzba-Bobrowicz T, Kosno-Kruszewska E Gwiazda E, Lechowicz W (2000) Major Histocompatibility complex class II (MHC II) expression during the development of human fetal cerebral occipital lobe, cerebellum, and hematopoietic organs. *Folia Neuropathol* 38: 11–117.

102. Billiards SS, Haynes RL, Folkerth RD, Trachtenberg FL, Liu LG, Volpe JJ, Kinney HC (2006) Development of microglia in the cerebral white matter of the human fetus and infant. *J Comp Neurol* 497: 199–208.

103. Basta PV, Moore TL, Yokota S, Ting JP (1989) Beta-adrenergic agonist modulates DR alpha gene transcription via enhanced cAMP levels in a glioblastoma multiforme line. *J Immunol* 142: 895–2901.

104. Fujita H, Tanaka J, Maeda N, Sakanaka M (1998) Adrenergic agonists suppress the proliferation of microglia through beta 2-adrenergic receptor. *Neurosci Lett* 242: 37–40.

105. Podkletnova I, Rothstein JD, Helén P, Alho H (2001) Microglial response to the neurotoxicity of 6-hydroxydopamine in neonatal rat cerebellum. *Int J Dev Neurosci* 19: 47–52.

106. Marin-Teva JL, Dusart I, Colin C, Gervais A, van Rooijen N, Mallat M (2004) Microglia promote the death of developing Purkinje cells. *Neuron* 41: 535–547.

107. Kemper TL, Bauman ML (2002) Neuropathology of infantile autism. *Mol Psychiatry* 7(Suppl 2): S12–S13.

108. Ritvo ER, Freeman BJ, Scheibel AB, Duong T, Robinson H, Guthrie D, Ritvo A (1986) Lower Purkinje cell counts in the cerebella of four autistic subjects: initial findings of the UCLA-NSAC Autopsy Research Report. *Am J Psychiatry* 143: 862–866.

109. Herbert MR, Ziegler DA, Makris N, Filipek PA, Kemper TL, Normandin JJ, Sanders HA, Kennedy DN, Caviness VS Jr (2004) Localization of white matter volume increase in autism and developmental language disorder. *Ann Neurol* 55: 530–540.

110. Hamilton SP, Rome LH (1994) Stimulation of in vitro myelin synthesis by microglia. *Glia* 11: 326–335.

111. Sato-Bigbee C, Pal S, Chu AK (1999) Different neuroligands and signal transduction pathways stimulate CREB phosphorylation at specific developmental stages along oligodendrocyte differentiation. *J Neurochem* 72: 139–147.

112. Afshari FS, Chu AK, Sato-Bigbee C (2001) Effect of cyclic AMP on the expression of myelin basic protein species and myelin proteolipid protein in committed oligodendrocytes: differential involvement of the transcription factor CREB. *J Neurosci Res* 66:37–45.

113. Pfenninger KH, Laurino L, Peretti D, Wang X, Rosso S, Morfini G, Cáceres A, Quiroga S (2003) Regulation of membrane expansion at the nerve growth cone. *J Cell Sci* 116(Pt 7):1209–1217.

114. Shambaugh G 3rd, Glick R, Radosevich J, Unterman T (1993) Insulin-like growth factor-I and binding protein-1 can modulate fetal brain cell growth during maternal starvation. *Ann N Y Acad Sci* 692:270–272.

115. Riikonen R, Makkonen I, Vanhala R, Turpeinen U, Kuikka J, Kokki H (2006) Cerebrospinal fluid insulin-like growth factors IGF-1 and IGF-2 in infantile autism. *Dev Med Child Neurol* 48:751–755.

116. Wang L, Adamo ML (2001) Cyclic adenosine 3′,5′-monophosphate inhibits insulin-like growth factor 1 gene expression in rat glioma cell lines: evidence for regulation of transcription and messenger ribonucleic acid stability. *Endocrinology* 142: 3041–3050.

117. Fields RD, Stevens-Graham B (2002) New insights into neuron–glia communication. *Science* 298:556–562.

118. Nedergaard M, Takano T, Hansen AJ (2002) Beyond the role of glutamate as a neurotransmitter. *Nat Rev Neurosci* 3: 748–755.

119. Laurence JA, Fatemi SH (2005) Glial fibrillary acidic protein is elevated in superior frontal, parietal and cerebellar cortices of autistic subjects. *Cerebellum* 4:206–210.
120. Hodges-Savola C, Rogers SD, Ghilardi JR, Timm DR, Mantyh PW (1996) Beta-adrenergic receptors regulate astrogliosis and cell proliferation in the central nervous system in vivo. *Glia* 17: 52–62.
121. Fatemi SH, Halt AR, Stary JM, Realmuto GM, Jalali-Mousavi M (2001) Reduction in anti-apoptotic protein Bcl-2 in autistic cerebellum. *Neuroreport* 12:929–933.
122. Fatemi SH, Halt AR, Stary JM, Kanodia R, Schultz SC, Realmuto GR (2002) Glutamic acid decarboxylase 65 and 67 kDa proteins are reduced in autistic parietal and cerebellar cortices. *Biol Psychiatry* 52:805–810.
123. Salero-Coca E, Vergara P, Segovia J (1995) Intracellular increases of cAMP induce opposite effects in glutamic acid decarboxylase (GAD 67) distribution and glial fibrillary acidic protein immunoreactivities in C6 cells. *Neurosci Lett* 191: 9–12.
124. Kitagawa K (2007) CREB and cAMP response element-mediated gene expression in the ischemic brain. *FEBS J* 274: 3210–3217.
125. Freeland K, Boxer LM, Latchman DS (2001) The cyclic AMP response element in the Bcl-2 promoter confers inducibility by hypoxia in neuronal cells. *Brain Res Mol Brain Res* 92: 98–106.
126. Tuchman R, Rapin I (2002) Epilepsy in autism. *Lancet Neurol* 1: 352–358.
127. Szot P, Weinshenker D, White SS, Robbins CA, Rust NC, Schwartzkroin PA, Palmiter RD (1999) Norepinephrine-deficient mice have increased susceptibility to seizure-inducing stimuli. *J Neurosci* 19: 10985–10992.
128. Weinshenker D, Szot P, Miller NS, Rust NC, Hohmann JG, Pyati U, White SS, Palmiter RD (2001) Genetic comparison of seizure control by norepinephrine and neuropeptide Y. *J Neurosci* 21: 7764–7769.
129. James SJ, Cutler P, Melnyk S, Jernigan S, Janak L, Gaylor DW, Neubrander JA (2004) Metabolic biomarkers of increased oxidative stress and impaired methylation capacity in children with autism. *Am J Clin Nutr* 80:1611–1617.
130. Enokido Y, Suzuki E, Iwasawa K, Namekata K, Okazawa H, Kimura H (2005) Cystathione beta-synthase, a key enzyme for homocysteine metabolism, is preferentially expressed in the radial glia/astrocyte lineage of developing mouse CNS. *FASEB J* 19: 1854–1856.
131. Heinonen K, Räihä NC (1974) Induction of cystathionase in human foetal liver. *Biochem J* 144: 607–609.
132. Kloor D, Danielyan L, Osswald H (2002) Characterization of the cAMP binding site of purified S-adenosyl-homocysteine hydrolase from bovine kidney. *Biochem Pharmacol* 64: 1201–1206.
133. de la Haba G, Agostini S, Bozzi A, Merta A, Unson C, Cantoni GL (1986) S-Adeno-sylhomocysteinase: mechanism of reversible and irreversible inactivation by ATP, cAMP, and 2′-deoxyadenosine. *Biochemistry* 25: 8337–8342.
134. Renouf JA, Thong YH, Chalmers AH (1987) Activities of purine metabolising enzymes in lymphocytes of neonates and young children: correlates with immune function. *Immunol Lett* 15:161–166.
135. Paşca SP, Nemeş B, Vlase L, Gagyi CE, Dronca E, Miu AC, Dronca M (2006) High levels of homocysteine and low serum paraoxonase 1 arylesterase activity in children with autism. *Life Sci* 78:2244–2248.
136. Ming X, Stein TP, Brimacombe M, Johnson WG, Lambert GH, Wagner GC (2005) Increased excretion of a lipid peroxidation biomarker in autism. *Prostaglandins Leukot Essent Fatty Acids* 73: 379–384.
137. Chauhan A, Chauhan V, Brown WT, Cohen I (2004) Oxidative stress in autism: increased lipid peroxidation and reduced serum levels of ceruloplasmin and transferrin – the antioxidant proteins. *Life Sci* 75: 2539–2549.

138. Yorbik O, Sayal A, Akay C, Akbiyik DL, Sohmen T (2002) Investigation of antioxidant enzymes in children with autistic disorder. *Prostaglandins Leukot Essent Fatty Acids* 67: 341–343.
139. Söğüt S, Zoroğlu SS, Ozyurt H, Yilmaz HR, Ozuğurlu F, Sivasli E, Yetkin O, Yanik M, Tutkun H, Sava HA, Tarakçioğlu M, Akyol O (2003) Changes in nitric oxide levels and antioxidant enzyme activities may have a role in the pathophysiological mechanisms involved in autism. *Clin Chim Acta* 331: 111–117.
140. Zoroglu SS, Armutcu F, Ozen S, Gurel A, Sivasli E, Yetkin O, Meram I (2004) Increased oxidative stress and altered activities of erythrocyte free radical scavenging enzymes in autism. *Eur Arch Psychiatry Clin Neurosci* 254: 14314–14317.
141. Folkerth RD, Haynes RL, Borenstein NS, Belliveau RA, Trachtenberg F, Rosenberg PA, Volpe JJ, Kinney HC (2004) Developmental lag in superoxide dismutases relative to other antioxidant enzymes in premyelinated human telencephalic white matter. *J Neuropathol Exp Neurol* 63: 990–999.
142. Keogh BP, Allen RG, Pignolo R, Horton J, Tresini M, Cristofalo VJ (1996) Expression of hydrogen peroxide and glutathione metabolizing enzymes in human skin fibroblasts derived from donors of different ages. *J Cell Physiol* 167: 512–522.
143. Queen LR, Xu B, Horinouchi K, Fisher I, Ferro A (2000) Beta(2)-adrenoceptors activate nitric oxide synthase in human platelets. *Circ Res* 87: 39–44.
144. Queen LR, Ji Y, Xu B, Young L, Yao K, Wyatt AW, Rowlands DJ, Siow RC, Mann GE, Ferro A (2006) Mechanisms underlying beta2-adrenoceptor-mediated nitric oxide generation by human umbilical vein endothelial cells. *J Physiol* 576(Pt 2): 585–594.
145. Slotkin TA, Oliver CA, Seidler FJ (2005) Critical periods for the role of oxidative stress in the developmental neurotoxicity of chlorpyrifos and terbutaline, alone or in combination. *Brain Res Dev Brain Res* 157: 172–180.
146. Vargas D, Bandaru V, Zerrate MC, Zimmerman AW, Haughey N, Pardo CA (2006) Oxidative stress in brain tissues from autistic patients: increased concentration of isoprostanes. IMFAR, June 1–3, PS2.6 Montreal, Canada.
147. Geier DA, Geier MR (2006) A clinical trial of combined anti-androgen and anti-heavy metal therapy in autistic disorders. *Neuro Endocrinol Lett* 27: 833–838.
148. Nataf R, Skorupka C, Amet L, Lam A, Springbett A, Lathe R (2006) Porphyrinuria in childhood autistic disorder: implications for environmental toxicity. *Toxicol Appl Pharmacol* 214: 99–108.
149. Woods JS, Kardish RM (1983) Developmental aspects of hepatic heme biosynthetic capability and hematotoxicity-II. Studies on uroporphyrinogen decarboxylase. *Biochem Pharmacol* 32: 73–78.
150. Woods JS, Echeverria D, Heyer NJ, Simmonds PL, Wilkerson J, Farin FM (2005) The association between genetic polymorphisms of coproporphyrinogen oxidase and an atypical porphyrinogenic response to mercury exposure in humans. *Toxicol Appl Pharmacol* 206: 113–120.
151. De Matteis F, Harvey C, (2005) Inducing coproporphyria in rat hepatocyte cultures using cyclic AMP and cyclic AMP-releasing agents. *Arch Toxicol* 79: 381–389.
152. Nebes VL, DeFranco D, Morris SM Jr (1988) Cyclic AMP induces metallothionein gene expression in rat hepatocytes but not in rat kidney. *Biochem J* 255: 741–743.
153. Dunn MA, Cousins RJ (1989) Kinetics of zinc metabolism in the rat: effect of dibutyryl cAMP. *Am J Physiol* 256: E420–E430.
154. Fuller CE, Elmes ME, Jasani B (1990) Age-related changes in metallothionein, copper, copper-associated protein, and lipofuscin in human liver: a histochemical and immunohistochemical study. *J Pathol* 161: 167–172.
155. Zlotkin SH, Cherian MG (1988) Hepatic metallothionein as a source of zinc and cysteine during the first year of life. *Pediatr Res* 24: 326–329.

156. Leventhal BL, Cook EH Jr, Morford M, Ravitz A, Freedman DX (1990) Relationships of whole blood serotonin and plasma norepinephrine within families. *J Autism Dev Disord* 20: 499–511.

157. Anderson GM, Freedman DX, Cohen DJ, Volkmar FR, Hoder EL, McPhedran P, Minderaa RB, Hansen CR, Young JG (1987) Whole blood serotonin in autistic and normal subjects. *J Child Psychol Psychiatry* 28: 885–900.

158. Spivak B, Golubchik P, Mozes T, Vered Y, Nechmad A, Weizman A, Strous RD (2004) Low platelet-poor plasma levels of serotonin in adult autistic patients. *Neuropsychobiology* 50: 157–160.

159. Racke K, Reimann H, Kilbinger H (1996) Regulation of 5-HT release from enterochromaffin cells. *Behav Brain Res* 73: 83–87.

160. Vered Y, Golubchik P, Mozes T, Strous R, Nechmad A, Mester R, Weizman A, Spivak B (2003) The platelet-poor plasma 5-HT response to carbohydrate rich meal administration in adult autistic patients compared with normal controls. *Hum Psychopharmacol* 18: 395–399.

161. Stahl SM (1977) The human platelet. A diagnostic and research tool for the study of biogenic amines in psychiatry and neurologic disorders. *Arch Gen Psychiatry* 34: 509–516.

162. Heils A, Teufel A, Petri S, Seemann M, Bengel D, Balling U, Riederer P, Lesch KP (1995) Functional promoter and polyadenylation site mapping of the human serotonin (5-HT) transporter gene. *J Neural Transm Gen Sect* 102: 247–254.

163. Heils A, Mössner R, Lesch KP (1997) The human serotonin transporter gene polymorphism – basic research and clinical implications. *J Neural Transm* 104:1005–1014.

164. Marazziti D, Muratori F, Cesari A, Masala I, Baroni S, Giannaccini G, Dell'OssoL, Cosenza A, Pfanner P, Cassano GB (2000) Increased density of the platelet serotonin transporter in autism. *Pharmacopsychiatry* 33: 165–168.

165. Kirsch P, Esslinger C, Chen Q, Mier D, Lis S, Siddhanti S, Gruppe H, Mattay VS, Gallhofer B, Meyer-Lindenberg A (2005) Oxytocin modulates neural circuitry for social cognition and fear in humans. *J Neurosci* 25: 11489–11493.

166. Wahl RU (2004) Could oxytocin administration during labor contribute to autism and related behavioral disorders? – A look at the literature. *Med Hypotheses* 63: 456–460.

167. Hollander E, Novotny S, Hanratty M, Yaffe R, DeCaria CM, Aronowitz BR, Mosovich S (2003) Oxytocin infusion reduces repetitive behaviors in adults with autistic and Asperger's disorders. *Neuropsychopharmacology* 28: 193–198.

168. Modahl C, Green L, Fein D, Morris M, Waterhouse L, Feinstein C, Levin H (1998) Plasma oxytocin levels in autistic children. *Biol Psychiatry* 43: 270–277.

169. Green L, Fein D, Modahl C, Feinstein C, Waterhouse L, Morris M (2001) Oxytocin and autistic disorder: alterations in peptide forms. *Biol Psychiatry* 50: 609–613.

170. Morris M, Castro M, Rose JC (1992) Alterations in oxytocin prohormone processing during early development in the fetal sheep. *Am J Physiol* 263: R738–R740.

171. Altstein M, Gainer H (1988) Differential biosynthesis and posttranslational processing of vasopressin and oxytocin in rat brain during embryonic and postnatal development. *J Neurosci* 8: 3967–3977.

172. Barnette MS, Grous M, Torphy TJ, Ormsbee HS 3rd (1990) Activation of cyclic AMP-dependent protein kinase during canine lower esophageal shincter relaxation. *J Pharmacol Exp Ther* 252: 1160–1166.

173. Hagen BM, Bayguinov O, Sanders KM (2006) VIP and PACAP regulate localized Ca^{2+} transients via cAMP-dependent mechanism. *Am J Physiol Cell Physiol* 291: C375–C385.

174. Antonowicz I, Lebenthal E (1977) Developmental pattern of small intestinal enterokinase and disaccharidase activities in the human fetus. *Gastroenterology* 72: 1299–1303.

175. Mobassaleh M, Montgomery RK, Biller JA, Grand RJ (1985) Development of carbohydrate absorption in the fetus and neonate. *Pediatrics* 75: 160–166.

176. Afzal N, Murch S, Thirrupathy K, Berger L, Fagbemi A, Heuschkel R (2003) Constipation with acquired megarectum in children with autism. *Pediatrics* 112: 939–942.

177. Molloy CA, Manning-Courtney P (2003) Prevalence of chronic gastrointestinal symptoms in children with autism and autistic spectrum disorders. *Autism* 7: 165–671.
178. Liu LW, Thuneberg L, Huizinga JD (1994) Selective lesioning of interstitial cells of Cajal by methylene blue and light leads to loss of slow waves. *Am J Physiol* 266(3 Pt 1): G485–G496.
179. Thuneberg L (1982) Interstitial cells of Cajal: intestinal pacemaker cells? *Adv Anat Embryol Cell Biol* 71: 1–130.
180. Wester T, Eriksson L, Olsson Y, Olsen L (1999) Interstitial cells of Cajal in the human fetal small intestine as shown by c-kit immunohistochemistry. *Gut* 44: 65–71.
181. Furlano RI, Anthony A, Day R, Brown A, McGarvey L, Thomson MA, Davies SE, Berelowitz M, Forbes A, Wakefield AJ, Walker-Smith JA, Murch SH (2001) Colonic CD8 and γδ T-cell infiltration with epithelial damage in children with autism. *J Pediatr* 138: 366–372.
182. Catassi C, Bonucci A, Coppa GV, Carlucci A, Giorgi PL (1995) Intestinal permeability changes during the first month: effect of natural versus artificial feeding. *J Pediatr Gastroenterol Nutr* 21: 383–386.
183. Ballard ST, Hunter JH, Taylor AE (1995) Regulation of tight junction permeability during nutrient absorption across the intestinal epithelium. *Annu Rev Nutr* 15: 35–55.
184. Lawrence DW, Comerford KM, Colgan SP (2002) Role of VASP in reestablishment of epithelial tight junction assembly after Ca^{2+} switch. *Am J Physiol Cell Physiol* 282: C1235–C1245.
185. Torrente F, Ashwood P, Day R, Machado N, Furlano RI, Anthony A, Davies SE, Wakefield AJ, Thomson MA, Walker-Smith JA, Murch SH (2002) Small intestinal enteropathy with epithelial IgG and complement deposition in children with regressive autism. *Mol Psychiatry* 7: 375–382.
186. Torrente F, Anthony A, Heuschkel RB, Thomson MA, Ashwood P, Murch SH (2004) Focal-enhanced gastritis in regressive autism with features distinct from Crohn's and Helicobacter pylori gastritis. *Am J Gastroenterol* 99: 598–605.
187. Nanthakumar NN, Fusunyan RD, Sanderson I, Walker WA (2000) Inflammation in the developing human intestine: A possible pathophysiologic contribution to necrotizing enterocolitis. *Proc Natl Acad Sci USA* 97: 6043–6048.
188. Kavelaars A, van de Pol M, Zijlstra J, Heijnen CJ (1997) Beta 2-adrenergic activation enhances interleukin-8 production by human monocytes. *J Neuroimmunol* 77: 211–216.
189. DeFelice ML, Ruchelli ED, Markowitz JE, Strogatz M, Reddy KP, Kadivar K, Mulberg AE, Brown KA (2003) Intestinal cytokines in children with pervasive developmental disorders. *Am J Gastroenterol* 98: 1777–1782.
190. Gupta S, Aggarwal S, Heads C (1996) Brief report: Dysregulated immune system in children with autism: beneficial effects of intravenous immune globulin on autistic characteristics. *J Autism Dev Disord* 26: 439–452.
191. Molloy CA, Morrow AL, Meinzen-Derr J, Schleifer K, Dienger K, Manning-Courtney P, Altaye M, Wills-Karp M (2006) Elevated cytokine levels in children with autism spectrum disorder. *J Neuroimmunol* 172: 198–205.
192. Jyonouchi H, Sun S, Itokazu N (2002) Innate immunity associated with inflammatory responses and cytokine production against common dietary proteins in patients with autism spectrum disorder. *Neuropsychobiology* 46: 76–84.
193. Stern L, Francoeur MJ, Primeau MN, Sommerville W, Fombonne E, Mazer BD (2005) Immune function in autistic children. *Ann Allergy Asthma Immunol* 95: 558–565.
194. Denney DR, Frei BW, Gaffney GR (1996) Lymphocyte subsets and interleukin-2 receptors in autistic children. *J Autism Dev Disord* 26: 87–97.
195. Warren RP, Yonk J, Burger RW, Odell D, Warren WL (1995) DR-positive T cells in autism: association with decreased plasma levels of the complement C4B protein. *Neuropsychobiology* 31: 53–57.
196. Gupta S, Aggarwal S, Rashanravan B, Lee T (1998) Th1- and Th2-like cytokines in CD4+ and CD8+ T cells in autism. *J Neuroimmunol* 85: 106–109.

197. Anastassiou ED, Paliogianni F, Balow JP, Yamada H, Boumpas DT (1992) Prostaglandin E2 and other cyclic AMP-elevating agents modulate IL-2 and IL-2R alpha gene expression at multiple levels. *J Immunol* 148: 2845–2852.

198. Sanders VM (1998) The role of norepinephrine and beta-2-adrenergic receptor stimulation in the modulation of Th1, Th2, and B lymphocyte function. *Adv Exp Med Biol* 437: 269–278.

199. Zola H, Ridings J, Elliott S, Nobbs S, Weedon H, Wheatland L, Haslam R, Robertson D, Macardle PJ (1998) Interleukin 2 receptor regulation and IL-2 function in the human infant. *Hum Immunol* 59: 615–624.

200. Macardle PJ, Wheatland L, Zola H (1999) Analysis of the cord blood T lymphocyte response to superantigen. *Hum Immunol* 60: 127–139.

201. Gupta AK, Rusterholz C, Holzgreve W, Hahn S (2005) Constant IFN gamma mRNA to protein ratios in cord and adult blood T cells suggests regulation of IFNgamma expression in cord blood T cells occurs at the transcriptional level. *Clin Exp Immunol* 140: 282–288.

202. Singh VK, Warren R, Averett R, Ghaziuddin M (1997) Circulating autoantibodies to neuronal and glial filament proteins in autism. *Pediatr Neurol* 17: 88–90.

203. Plioplys AV, Greaves A, Yoshida W (1989) Anti-CNS antibodies in childhood neurologic diseases. *Neuropediatrics* 20: 93–102.

204. Connolly AM, Chez MG, Pestronk A, Arnold ST, Mehta S, Deuel RK (1999) Serum autoantibodies to brain in Landau-Kleffner variant, autism, and other neurologic disorders. *J Pediatr* 134: 607–613.

205. Connolly AM, Chez M, Streif EM, Keeling RM, Golumbek PT, Kwon JM, Riviello JJ, Robinson RG, Neuman RJ, Deuel RM (2006) Brain-derived neurotrophic factor and autoantibodies to neural antigens in sera of children with autistic spectrum disorders, Landau-Kleffner syndrome, and epilepsy. *Biol Psychiatry* 59: 354–363.

206. Croonenberghs J, Wauters A, Devreese K, Verkerk R, Scharpe S, Bosmans E, Egyed B, Deboutte D, Maes M (2002) Increased serum albumin, gamma globulin, immunoglobulin IgG, and IgG2 and IgG4 in autism. *Psychol Med* 32: 1457–1463.

207. Trajkovski V, Ajdinski L, Spiroski M (2004) Plasma concentration of immunoglobulin classes and subclasses in children with autism in the Republic of Macedonia: retrospective study. *Croat Med J* 45: 746–749.

208. Singer HS, Morris CM, Williams PN, Yoon DY, Hong JJ, Zimmerman AW (2006) Antibrain antibodies in children with autism and their unaffected siblings. *Neuroimmunol* 178: 149–155.

209. Cabanlit M, Wills S, Goines P, Ashwood P, Van de Water J (2007) Brain-specific autoantibodies in the plasma of subjects with autistic spectrum disorder. *Ann N Y Acad Sci* 1107: 92–103.

210. Sanders VM, Powell-Oliver FE (1992) Beta 2-adrenoceptor stimulation increases the number of antigen-specific precursor B lymphocytes that differentiate into IgM-secreting cells without affecting burst size. *J Immunol* 148: 1822–1828.

211. Podojil JR, Sanders VM (2003) Selective regulation of mature IgG1 transcription by CD86 and beta 2-adrenergic receptor stimulation. *J Immunol* 170: 5143–5151.

212. Pongratz G, McAlees JW, Conrad DH, Erbe RS, Haas KM, Sanders VM (2006) The level of IgE produced by a B cell is regulated by norepinephrine in a p38 MAPK- and CD23-dependent manner. *J Immunol* 177: 2926–2938.

213. Hannet I, Erkeller-Yuksel F, Lydyard P, Deneys V, DeBruyère M (1992) Developmental and maturational changes in human blood lymphocyte subpopulations. *Immunol Today* 13: 215–218.

214. Chen ZJ, Wheeler CJ, Shi W, Wu AJ, Yarboro CH, Gallagher M, Notkins AL (1998) Polyreactive antigen-binding B cells are the predominant cell type in the newborn B cell repertoire. *Eur J Immunol* 28: 989–994.

215. Lydyard PM, Lamour A, MacKenzie LE, Jamin C, Mageed RA, Youinou P (1993) CD5+ B cells and the immune system. *Immunol Lett* 38: 159–166.

216. Schelonka RL, Raaphorst FM, Infante D, Kraig E, Teale JM, Infante AJ (1998) T cell receptor repertoire diversity and clonal expansion in human neonates. *Pediatr Res* 43: 396–402.
217. Deibel MR Jr, Riley LK, Coleman MS, Cibull ML, Fuller SA, Todd E (1983) Expression of terminal deoxynucleotidyl transferase in human thymus during ontogeny and development. *J Immunol* 131: 195–200.
218. Siden EJ, Gifford A, Baltimore D (1985) Cyclic AMP induces terminal deoxynucleotidyl transferase in immature B cell leukemia lines. *J Immunol* 135:1518–1522.
219. Palkovitz RJ, Wiesenfeld AR (1980) Differential autonomic responses of autistic and normal children. *J Autism Dev Disord* 10: 347–360.
220. Barry RJ, James AL (1988) Coding of stimulus parameters in autistic, retarded, and normal children: evidence for a two-factor theory of autism. *Int J Psychophysiol* 6: 139–149.
221. Thirumalai SS, Shubin RA, Robinson R (2002) Rapid eye movement sleep behavior disorder in children with autism. *J Child Neurol* 17: 173–178.
222. Wakefield AJ, Anthony A, Murch SH, Thomson M, Montgomery SM, Davies S, O'Leary JJ, Phil D, Berelowitz M, Walker-Smith JA (2000) Enterocolitis in children with developmental disorders. *Am J Gastroenterol* 95: 2285–2295.
223. Ming X, Julu POO, Brimacombe M, Connor S, Daniels ML (2005) Reduced cardiac parasympathetic activity in children with autism. *Brain Dev* 27: 509–516.
224. Chauhan V, Chauhan A, Cohen I, Sheikh A (2007) Abnormalities in signal transduction in autism. IMFAR, May 3–5 PS 6.40 Seattle, WA.
225. Beversdorf DQ, Manning SE, Hillier A, Anderson SL, Nordgren RE, Walters SE, Nagaraja HN, Cooley WC, Gaelic SE, Bauman ML (2005) Timing of prenatal stressors and autism. *J Autism Dev Disord* 35: 471–478.
226. Patterson PH (2002) Maternal infection: window on neuroimmune interactions in fetal brain development and mental illness. *Curr Opin Neurobiol* 12: 115–118.
227. Shi L, Fatemi SH, Sidwell RW, Patterson PH (2003) Maternal influenza infection causes marked behavioral and pharmacological changes in the offspring. *J Neurosci* 23: 297–302.
228. Slotkin TA, Lau C, McCook EC, Lappi SE, Seidler FJ (1994) Glucocorticoids enhance intracellular signaling via adenylate cyclase at three distinct loci in the fetus: a mechanism for heterologous teratogenic sensitization? *Toxicol Appl Pharmacol* 127: 64–75.
229. Slotkin TA, McCook EC, Seidler FJ (1993) Glucocorticoids regulate the development of intracellular signaling: enhanced forebrain adenylate cyclase catalytic subunit activity after fetal dexamethasone exposure. *Brain Res Bull* 32: 359–364.
230. Liu L, Li A, Matthews SG (2001) Maternal glucocorticoid treatment programs HPA regulation in adult offspring: sex-specific effects. *Am J Physiol Endocrinol Metab* 280: E729–E739.
231. Owen D, Matthews SG (2007) Prenatal glucocorticoid exposure alters hypothalamic-pituitary-adrenal function in juvenile guinea pigs. *J Neuroendocrinol* 19: 172–180.
232. Kapoor A, Matthews SG (2005) Short periods of prenatal stress affect growth, behaviour and hypothalamo–pituitary–adrenal axis activity in male guinea pig offspring. *J Physiol* 566(Pt 3): 967–977.
233. Geier DA, Geier MR (2006) A clinical and laboratory evaluation of methionine cycle-transsulfuration and androgen pathway markers in children with autistic disorders. *Horm Res* 66: 182–188.
234. Meyer A, Seidler FJ, Cousins MM, Slotkin TA (2003) Developmental neurotoxicity elicited by gestational exposure to chlorpyrifos: when is adenylyl cyclase a target? *Environ Health Perspect* 111: 1871–1876.
235. Aldridge JE, Meyer A, Seidler FJ, Slotkin TA (2005) Developmental exposure to terbutaline and chlorpyrifos: pharmacotherapy of preterm labor and an environmental

neurotoxicant converge on serotonergic systems in neonatal rat brain regions. *Toxicol Appl Pharmacol* 203: 132–144.

236. Slotkin TA, Seidler FJ (2007) Developmental exposure to terbutaline and chlorpyrifos, separately or sequentially, elicits presynaptic serotonergic hyperactivity in juvenile and adolescent rats. *Brain Res Bull* 73: 301–309.

237. Chugani DC, Muzik O, Behen M, Rothermel R, Janisse JJ, Lee J, Chugani HT (1999) Developmental changes in brain serotonin synthesis capacity in autistic and nonautistic children. *Ann Neurol* 45: 287–295.

238. Zimmerman AW, Connors SL, Matteson KJ, Lee LC, Singer HS, Castaneda JA, Pearce DA (2007) Maternal antibrain antibodies in autism. *Brain Behav Immun* 21: 351–357.

239. Singer HS, Morris CM, Gause CD, Gillin PK, Crawford S, Zimmerman AW (2008) Antibodies against fetal brain in sera of mothers with autistic children. *J Neuroimmunol* 194: 165–172.

240. Connors SL, Matteson KJ, Sega GA, Lozzio CB, Carroll RC, Zimmerman AW (2006) Plasma serotonin in autism. *Pediatr Neurol* 35:182–186.

241. Kapoor A, Petropoulos S, Matthews SG (2008) Fetal programming of hypothalamic-pituitary-adrenal (HPA) axis function and behavior by synthetic glucocorticoids. *Brain Res Rev* 57: 586–595.

242. Slotkin TA, Saleh JL, Zhang J, Seidler FJ (1996) Ontogeny of beta-adrenoceptor/ adenylyl cyclase desensitization mechanisms: the role of neonatal innervation. *Brain Res* 742: 317–328.

243. Hou QC, Seidler FJ, Slotkin TA (1989) Development of the linkage of beta-adrenergic receptors to cardiac hypertrophy and heart rate control: neonatal sympathectomy with 6-hydroxydopamine. *J Dev Physiol* 11: 305–311.

244. DeLong GR, Bean SC, Brown FR 3rd (1981) Acquired reversible autistic syndrome in acute encephalopathic illness in children. *Arch Neurol 38*: 191–194.245.

245. Lee TP (1980) Virus exposure diminishes beta-adrenergic response in human leukocytes. *Res Commun Chem Path Pharmacol* 30: 469–476.

246. Stanwick TL, Anderson RW, Nahmias AJ (1977) Interaction between cyclic nucleotides and herpes simplex viruses: productive infection. *Infect Immun* 18: 342–347.

247. Speiser Z, Gordon I, Rehavi M, Gitter S (1991) Behavioral and biochemical studies in rats following prenatal treatment with beta-adrenoceptor antagonists. *Eur J Pharmacol* 195: 75–83.

248. Wagner JP, Seidler FJ, Slotkin TA (1991) Presynaptic input regulates development of β-adrenergic control of rat brain ornithine decarboxylase: effects of 6-hydroxydopamine or propranolol.*Brain Res. Bull* 26:885–890.

249. Kudlacz EM, Spencer JR, Slotkin TA (1991) Postnatal alterations in β-adrenergic receptor binding in rat brain regions after continuous prenatal exposure to propranolol *via* maternal infusion. *Res Commun Chem Pathol Pharmacol* 71:153–61.

Part III
Endocrinology, Growth, and Metabolism

Chapter 8
A Role for Fetal Testosterone in Human Sex Differences

Implications for Understanding Autism

Bonnie Auyeung and Simon Baron-Cohen

Abstract Autism spectrum conditions (ASCs) may be an extreme manifestation of specific male-typical characteristics. Evidence for this theory is provided by the empathizing–systemizing (E–S) theory of sex-typical behavior, which suggests ASCs as an extreme form of the male brain (EMB). In this chapter, we review the evidence supporting EMB theory and examine the effect of hormones on the development of sex differences related to ASCs. An important candidate mechanism for the development of sex-typical behavior is the effect of fetal testosterone (fT) during pregnancy. Evidence that elevated levels of fT may be a risk factor for ASC is also discussed. Many neurodevelopmental conditions occur in males more often than females, including autism, dyslexia, attention-deficit hyperactivity disorder (ADHD), and early onset persistent antisocial behavior [1]). Autism in particular has been described as an extreme manifestation of some sexually dimorphic traits or an "extreme male brain" [2]. In this chapter, we review the reasons why this condition in particular has been viewed in this light and the evidence related to it.

Keywords Fetal testosterone · sex differences · autism

The Extreme Male Brain Theory of Autism

Autism, high-functioning autism, Asperger syndrome, and pervasive developmental disorder (not otherwise specified, PDD/NOS) are thought to lie on the same continuum and can be referred to as autism spectrum

B. Auyeung
Autism Research Centre, Department of Psychiatry, University of Cambridge,
Douglas House, 18B Trumpington Rd, Cambridge, CB2 8AH, UK
e-mail: ba251@cam.ac.uk

A.W. Zimmerman (ed.), *Autism*, DOI: 10.1007/978-1-60327-489-0_8,
© Humana Press, Totowa, NJ 2008

conditions (ASCs)[1]. These conditions are characterized by impairments in reciprocal social interaction, in verbal and nonverbal communication, alongside strongly repetitive behaviors and unusually narrow interests [3]. Recent epidemiological studies have shown that as many as 1% of people could have an ASC [4]. The incidence of ASC is strongly biased toward males [5, 6, 7] with a male: female ratio of 4:1 for classic autism [8] and as high as 8:1 for Asperger syndrome [9]. The cause of the observed sex difference in ASC remains a topic of debate. It is possible that males have a lower threshold for expressing the condition [10]. ASCs have a strong neurobiological and genetic component [11]; however, the specific factors (hormonal, genetic, or environmental) that are responsible for the higher male incidence in ASC are still unclear.

The extreme male brain (EMB) theory of autism is an extension of the empathizing–systemizing (E–S) theory of typical sex differences [2, 12] which proposes that females on average have a stronger drive to empathize (to identify another person's emotions and thoughts and to respond to these with an appropriate emotion) while males tend to have a stronger drive to systemize (to analyze or construct rule-based systems, whether mechanical, abstract, or another type) [12]. The empathizing quotient (EQ) [13] and systemizing quotient (SQ) [14] were developed to measure these dimensions in an individual. Using the difference between a person's EQ and SQ, individual "brain types" can be calculated [15, 16], where individuals who are equal in their E and S are said to have a balanced (B) brain type (E = S). The type S (S > E) brain type is more common in males while the type E (E > S) is more common in females [16]. Extreme types are also found [16], and the majority (61.6%) of adults with ASC fall in the extreme S (S >> E), compared to 1% of typical females [16] (Fig. 8.1).

Experimental evidence at the psychological level relevant to the EMB theory of autism includes the following:

Individuals with ASC score higher on the SQ, an instrument on which typical males score higher than typical females in both adults [14, 16] and children [17]. Individuals with ASC are superior to controls on the embedded figures task (EFT), a task on which typical males perform better than typical females [18, 19]. The EFT requires good attention to detail, a prerequisite of systemizing. Individuals with ASC have also been found to have either intact or superior functioning on tests of intuitive physics [20, 21], a domain which shows a sex difference in favor of males in adulthood [21]. Sex differences have been found on the block design subscale of the WISC-R intelligence test,

---

[1] The American Psychiatric Association uses the term ASD for autism spectrum disorders. We prefer the use of the term ASC as those at the higher functioning end of the autistic spectrum who do not necessarily see themselves as having a "disorder," and the profile of strengths and difficulties in ASC can be conceptualized as atypical but not necessarily disordered. ASC remains a medical diagnosis, hence the use of the term "condition," which signals that such individuals need support. We feel that the use of the term ASCs is more respectful to differences; recognizes that the profile in question does not fit a simple "disease" model but includes areas of strength (e.g., in attention to detail) as well as areas of difficulty; and does not identify the individual purely in terms of the latter.

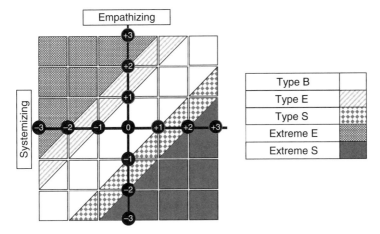

Fig. 8.1 The Empathizing–Systemizing Model of Typical Sex Differences. The main brain types are illustrated on axes of Empathizing (E) and Systemizing (S) dimensions (numbers represent standard deviations from the mean). Balanced brain (Type B); female brain (Type E), male brain (Type S); the extreme Types E and S lie at the outer borders. According to the 'extreme male brain' theory of autism, people with ASC will generally fall in the Extreme Type S region. Modified from Ref. [153]

with typical males performing better than females [22], and children with autism demonstrating superior functioning on this test [23, 24, 25].

Studies using the EQ [13] report that individuals with ASC score lower on both the adult and the child versions than the control groups [13, 17], where typical females score higher than typical males [13, 17]. Individuals with ASC are also impaired on certain measures where women tend to score higher than men. For example, individuals with ASC score lower than control males in the "Reading the Mind in the Eyes" task (considered to be an advanced test of empathizing) [26], the Social Stories Questionnaire [21], which involves recognition of complex emotions from videos of facial expressions or audios of vocalizations [27], and on the Friendship and Relationship Questionnaire (which tests the importance of emotional intimacy and sharing in relationships).

The Childhood Autism Spectrum Test (CAST) [28, 29], (formerly known as the childhood Asperger syndrome test, renamed because it can be used for all subgroups on the autistic spectrum [30]) is a parent-report measure developed to screen for Asperger syndrome and ASCs in a typical population on which [28] boys score higher than girls [31], and children with ASC score higher than typically developing children [29]. Another measure is the Autism Spectrum Quotient (AQ), which was developed to help quantify the number of autistic traits an individual displays [32], and individuals with Asperger syndrome or high-functioning autism score higher than those without a diagnosis [32]. Among controls, males again score higher than females [32], and these results are consistent in adults, adolescents, and children [32, 33, 34] as well as cross-culturally [35, 36, 37, 38]. Furthermore, similar results have been found using the Social Responsiveness Scale (SRS), a 65-item rating scale designed to measure the severity of autistic symptoms, demonstrating that individuals with an ASC diagnosis score higher than typical males, who in turn score higher than typical females [39].

In addition to the evidence at the psychological level, it has been suggested that characteristics of neurodevelopment in autism, such as larger overall brain volumes and greater growth of the amygdala during childhood, may also represent an exaggeration of typical sex differences in brain development [40]. Studies using fMRI indicate that typical females show increased activity in the extrastriate cortex during the Embedded Figures Test and increased activity bilaterally in the inferior frontal cortex during the "Reading the Mind in the Eyes" test. Parents of children with ASC also tend to show hyper-masculinization of brain activity [41], suggesting that hyper-masculinization may be part of the broader autism phenotype.

It remains important to identify the biological mechanisms that cause such sexual dimorphism. One study has shown sexual dimorphism in looking preferences in 102 newborn infants who were approximately 37 hours old. Boys were found to exhibit a preference for mobiles while girls tended to prefer looking at faces [42]. Although these simple experiments with stimuli are not an indication of ASC, these early sex differences suggest a biological sex difference in behavior, as there had been no opportunity for postnatal influence of social or cultural factors. One possible biological mechanism is the effect of prenatal exposure to hormones, in particular the androgen testosterone [43].

Hormones and Sexual Differentiation

Hormones are essential to reproduction, growth and development, maintenance of the internal environment and the production, use and storage of energy [44]. There are marked physical and behavioral consequences of exposure to hormones throughout life. Prenatally, the presence or absence of specific hormones (or their receptors) is known to be essential to the sexual differentiation of the fetus. In addition to stimulating development of physical characteristics such as genitalia [45, 46, 47, 48, 49], there is increasing evidence that prenatal hormones have a substantial effect on gender-typical aspects of behavior [48, 50]. If this is observed to be the case, then the occurrence of these hormones prenatally may have a substantial bearing on the development of an extreme male profile responsible for autistic traits.

The links between hormone levels, physical development, and behavior are complex and not yet fully understood, particularly in terms of effects on early development. Hormone levels can be measured at particular points in time but levels could vary on a daily basis [51], and prenatal measurements are very difficult (and potentially dangerous) to obtain for research purposes alone. Furthermore, correlations with behavioral measurements are always complicated by the need to determine the presence of a particular trait, without artificially inducing the behavior or creating bias in the result. A useful way of controlling some of these variables is to examine results from animals, where it has been possible to directly manipulate and monitor the levels of hormones throughout pregnancy and to control for environmental effects. As a result, we often look to confirm effects measured in animals with similar measurements in humans. Even then, the correlation between animals and humans is not always clear cut, with the potential for quite different mechanisms.

Though genetic sex is determined at conception, it is the gonadal hormones (i.e., androgens, estrogens, and progestins [52]) that are responsible for differentiation of the male and female phenotypes in the developing human fetus [45, 46, 47, 48, 49]. Androgens such as testosterone and dihydrotestosterone (DHT—a hormone formed from testosterone) are of particular interest to the study of male-typical behavior because when these androgens and the appropriate receptors are present, the male genital phenotype will develop. If androgens (or their receptors) are not present, then the female genital phenotype will develop (such as in female fetuses or males with Androgen insensitivity syndrome) [53, 54, 55, 56]. Another hormone that forms from prenatal testosterone is the estrogen hormone estradiol, which has been observed to promote male-typical behavior in rats and other rodents [57]. The relative contributions of DHT and estradiol to development of male-typical human behaviors are less certain.

Behavioral studies in nonhuman mammals have shown that the same prenatal hormones that are involved in sexual differentiation of the body are also involved in sexual differentiation of behavior [58, 59]. In animals, higher doses of hormones have been seen to masculinize behavior more than lower doses, though the effect of concentration is not uniform for different behaviors [59]. Effects are also likely to be nonlinear and include both lower and upper threshold values, beyond which changes in concentration have no effect [50]. Interaction between hormones may also be important, as described above [59].

Atypical Fetal Hormone Environments

In humans, the manipulation or even direct measurement of hormone levels is considered unethical because of the potential dangers involved. However, some information is available from specific abnormalities which occur naturally. Such abnormalities can lead to considerable difficulties for the individual and fortunately such instances are rare. However, some studies have obtained sufficient participation to render useful information about how abnormal environments influence behavior. A detailed review of many of the studies surrounding these conditions has been provided elsewhere [43, 44, 48, 50], so we focus our discussion here on findings relevant to characteristics of ASC.

Congenital Adrenal Hyperplasia

Congenital adrenal hyperplasia (CAH) is a genetic disorder affecting both sexes that causes excess adrenal androgen production beginning prenatally [60]. CAH affects both males and females but is most clearly observed in females because of their typically low androgen levels. Female fetuses with CAH have similar androgen levels to those found in typical males [48]. Behavioral studies of females with CAH show a more masculinized profile compared to unaffected female siblings or matched controls.

In terms of specific behaviors, girls with CAH show masculinization of ability in activities typically dominated by males. These include spatial orientation, visualization, targeting, personality, cognitive abilities, and sexuality [61, 62, 63]. Females with CAH may also be more likely to be left handed [64] and are more interested in male-typical activities and less interested in female-typical activities throughout life [65, 66, 67, 68, 69].

Studies relating CAH and autism are limited. Since the condition typically introduces masculinization, effects are more apparent in girls than boys. For these girls, behavior tends to become more aligned with expectations of behavior from typical males and few cases of ASC with CAH are reported. Results from one study of girls with CAH suggest that they exhibit more autistic traits, measured by the AQ, compared to their unaffected sisters [70]. Individuals with CAH also demonstrate higher levels of language and learning difficulties compared to unaffected family members [63], as do people with ASCs. Whilst CAH provides an interesting window on additional androgen exposure, the relatively rare occurrence of CAH in conjunction with ASCs makes it difficult to obtain large enough sample sizes for generalization to the wider population. In addition, some researchers have suggested that CAH-related disease characteristics, rather than prenatal androgen exposure, could be responsible for atypical cognitive profiles [71, 72].

Complete Androgen Insensitivity Syndrome

Complete androgen insensitivity syndrome (CAIS) occurs when there is a complete deficiency of androgen receptors and is more common in males, with incidence between 1 in 60,000 and 1 in 20,000 births. At birth, genetic male infants with CAIS are phenotypically female despite an XY (male typical) complement and are usually raised as girls with no knowledge of the underlying disorder. Although breasts develop, diagnosis usually takes place when menarche fails to occur [52, 73].

Investigation of behavior such as gender identity, sexual orientation, gender role behavior in childhood and adulthood, personality traits that show sex differences, and hand preferences have suggested that males with this condition do not significantly differ from same-sex controls [72, 74]. However, other results suggests that individuals with CAIS tend to show feminized performance on tests of visuo-spatial ability [75]. If replicable, this finding lends support to the notion that androgens enhance male-typical behaviors. Specific evidence for ASCs is not available due to the low incidence of this condition.

Idiopathic Hypogonadotropic Hypogonadism

Idiopathic hypogonadotropic hypogonadism (IHH) occurs when an individual's gonads lack sufficient stimulation to produce normal levels of hormones, and the disorder can occur congenitally or after puberty. These individuals have normal male genitalia at birth, so it can be assumed that their prenatal testosterone levels were normal [44]. Men with IHH perform worse on the Embedded Figures Test, the space relations subtests from the differential aptitude tests and the block design subtest of the Weschler Adult Intelligence Scale, when compared with normal males and males with acquired hypogonadotropic hypogonadism after puberty [76]. However, another study found that males with IHH do not show deficits on the same scale [77]. More research needs to be conducted to resolve these findings and relate the effects to ASC.

Hormonal Effects: Indirect Studies in Typical Populations

There is a steady body of evidence that indicates that fetal hormone levels influence certain physical characteristics that can be observed after birth. These "proxy" measurements have been used to indicate the levels of prenatal androgen expose and have been examined extensively in relation to behavioral traits. Several reviews of these measurements exist [49, 50], and we focus the discussion here on studies related to behaviors associated with ASC.

Digit Ratio (2D:4D)

The ratio between the length of the 2nd and 4th digit (2D:4D) has been found to be sexually dimorphic, being lower in males than in females. 2D:4D ratio is thought to be fixed by week 14 of fetal life, and it has been hypothesized that it might reflect fetal exposure to prenatal sex hormones in early gestation [78].

Measurements indicate an association between fT levels and 2D:4D ratio for the right hand after controlling for sex [79]. For subjects with CAH, females show lower (more masculinized) 2D:4D on the right hand compared to unaffected females, and men with CAH have lower 2D:4D on the left hand compared to unaffected males [80]. Results in this sample are consistent with the notion that prenatal androgen exposure masculinizes 2D:4D ratio. This measure has been widely used as a proxy for prenatal testosterone exposure due to the ease and simplicity of measurement. However, it is likely that 2D:4D ratio is affected by multiple factors [50].

The findings in studies with 2D:4D ratio tend to support the suggestion that higher f T levels are a risk factor for ASC. Lower (i.e., hyper-masculinized) digit ratios have been found in children with autism compared to typically developing children, and this was also found in the siblings and parents of children with autism, suggesting genetically based elevated f T levels in autism [81, 82].

Dermatoglyphics

Dermatoglyphics, or fingerprints, have also been used as a proxy measure for prenatal exposure to testosterone. The number of dermal ridges is thought to be fixed by about the 4th month of gestation [83]. Researchers have used total finger ridge count and asymmetry between left and right hands. Sex differences have been observed in ridge count with males exhibiting more ridges in total than females. Sex differences have also been observed in asymmetry although both sexes have more ridges on the right hand than on the left hand (R > L). The left greater than right (L > R) pattern is more common in females than in males [49].

Studies examining total ridge count in adults and children have shown that for both men and women who exhibit L > R, performance was better for tasks that show a female superiority such as verbal fluency and perceptual speed [84, 85, 86]. The opposite pattern was found for those exhibiting the R > L pattern, who demonstrated better performance for tasks that show a male superiority [84, 85, 86].

Data from dermatoglyphic patterns and their relation to autism are limited and conflicting. In one study 78 children with autism were compared to the same number of matched controls [87]. Analysis of ridge patterns and ridge counts resulted in significant differences between the children with and without autism. Children with autism typically exhibited lower ridge count and less distinct fingerprint features [87]. However, a smaller comparison of children with autism, learning difficulties (then called "retardation"), and typical children found no significant differences for ridge counts [88]. It was argued that dermatoglyphics may be ineffective in delineating autism from other typical populations [88].

As with the 2D:4D ratio, studies using dermatoglyphics may be useful, but more evidence is needed to establish whether there is a link between dermatoglyphics and prenatal hormone exposure. In addition, further studies are needed to understand the potential links with ASCs. The few studies of dermatoglyphics in ASCs are quite old, and in more recent decades, diagnostic clinics have become more alert to detecting autism in higher functioning individuals (such as those with Asperger syndrome), and it would be of interest to repeat these early studies with the range of subgroups on the autistic spectrum.

Lateralization

It has been proposed that some observable sex differences in human behavior and cognition may be accounted for by differences in cerebral lateralization [89]. In addition to research investigating functional asymmetries in the brain, body asymmetries (other than fingerprint asymmetries) have been associated with prenatal sex hormones [49].

Fetal testosterone (fT) has been implicated in left-handedness and asymmetrical lateralization [90, 91, 92, 93]. Left-handedness and ambidexterity are more common in typical males [94] as well as in individuals with autism [95]. In addition, the typical male brain is heavier than the female brain [96], a difference that may in part be due to early fT exposure [48].

Pubertal Onset

Pubertal onset has been used to investigate variations in hormones. Females typically enter puberty earlier than males [50]. Research examining the physical indicators of hormone exposure and autism has found that a subset of male adolescents with autism show hyper-androgeny, or elevated levels of androgens, and precocious puberty [97]. These findings suggest that individuals with autism have atypical hormonal activity around this time. Other research has also shown that androgen-related medical conditions such as polycystic ovary syndrome (PCOS), ovarian growths, and hirsutism [98] occur with elevated rates in both women with Asperger syndrome and in mothers of children with autism [98]. Delayed menarche has also been observed in females with Asperger syndrome [98, 99]. These may reflect early abnormalities in level of fT, though this would require testing in a longitudinal study.

Co-Twin Sex

Other indirect studies of the relationship between prenatal hormones and behavior come from studies examining the effects of having an opposite or same-sex twin. Nonhuman studies examining the effects of animal position in the uterus have suggested that the sex of littermates can affect the development of sex-typical behaviors [100]. For rodents, masculinization of females was seen to occur when they were between two males in the uterus. For multiple littermates, the blood supply is channeled between fetuses, and in another study it was found that females developed more male typical traits if they were "downstream" of their male littermates [48].

For human twins, it is thought that females adjacent to a male will demonstrate masculinized behavior as a result of testosterone from the male [101, 102, 103]. There is also some evidence that human males with an opposite-sex twin exhibit feminized gender-role behavior [104]. However, most studies have not observed feminization [105, 106, 107, 108, 109]. Other investigations of gender-typical play have also failed to find opposite-sex twin effects [108, 110]. Such findings in humans are difficult to interpret because these findings may be a result of being reared with a female, rather than an effect of hormonal exposure during gestation [50].

It is widely accepted that genes play a role in the etiology of autism. In the absence of any known gene or genes, the main support for this is derived from family and twin studies. Two recent studies [111, 112] suggest that the twinning process itself may be an important risk factor in the development of autism. Both studies compared the number of affected twin pairs among affected sibling pairs to expected values in two separate samples and reported a significant excess of twin pairs. However, data from other studies do not support twinning as a substantial risk factor in the etiology of autism [113, 114, 115]. The high proportion of twins found in affected-sib-pair studies could be explained by the high ratio of concordance rates in monozygotic (MZ) twins versus siblings [114]. Researchers have suggested that environmental factors associated with various demographic characteristics such as sex, multiple births, maternal age, and education may interact with genetic vulnerability to increase the risk of autism [113] but no firm conclusions can be drawn at the present time.

Hormonal Effects: Measurements of Fetal Testosterone

Whilst many convenient methods have been recommended, the ability to infer prenatal hormone exposure through abnormal environments or proxy measures has obvious limitations. Although evidence for theories surrounding the influence of androgens can be obtained, there is (as yet) little direct support for these predictors as a way of studying prenatal hormone influence.

Ideally, we would like to make direct measurements of testosterone at regular intervals throughout gestation and into postnatal life. Whilst some indication of fetal exposure to androgens might be gained from maternal samples, there is little evidence to suggest that these correlate well with the fetal environment that is protected by the placenta [48].

The timing of hormonal effects is also crucial when studying lasting effects on development. There are thought to be two general types of hormonal effects: organizational and activational [116]. Organizational effects are most likely to occur during early development when most neural structures are becoming established and produce permanent changes in the brain [116], whereas activational effects are short term and are dependent on current hormone levels. Since ASC are typically persistent with an early onset, any hormonal influence on the development of ASC is likely to be organizational in nature.

It is widely thought that organizational effects are maximal during sensitive periods, which are hypothetical windows of time in which a tissue can be formed

[48]. Outside the sensitive period, the effect of the hormone will be limited, protecting the animal from disruptive influences. This means, for example, that circulating sex hormones necessary for adult sexual functioning does not cause unwanted alterations to tissues even though the same hormones might have been essential to the initial development of those tissues. Different behaviors may also have different sensitive periods for development [117]. The importance of sensitivity to organizational effects was seen by Goy et al., who showed that androgens masculinize different behaviors at different times during gestation for rhesus macaques [117].

For typical human males, there is believed to be a surge in f T at around 8–24 weeks of gestation [43, 48, 57], with a decline to barely detectable levels from the end of this period until birth. As a result, any effects of f T on development are most likely to be determined in this period. For typical human females, levels are typically very low throughout pregnancy and childhood [48].

In addition to the fetal surge, two other periods of elevated testosterone have been observed in typical males. The first takes place shortly after birth and lasts for approximately 3–4 months [118], after which levels return to very low levels until puberty. Results show that neonatal testosterone is important for genital development [119], but the evidence for its role in behavioral development is unclear. Early pubertal effects are the first visible effects of rising androgen levels in childhood and occur in both boys and girls. Due to the early onset of ASC, the pubertal surge in testosterone is of little interest in determining etiology of these conditions. Few studies have been conducted on the effects of neonatal testosterone; however, there is an increasing body of evidence that suggests that prenatal androgens may be involved in determining sexually dimorphic traits. In the remainder of this chapter, we discuss direct measurements of testosterone and our ability to correlate this with the development of ASC.

Maternal Sampling During Pregnancy

Various studies have measured testosterone levels in maternal blood during pregnancy [120, 121, 122]. One study found that androgen exposure in the second trimester was positively associated with male-typical behavior in adult females [121]. Similar findings in another study revealed that higher levels of testosterone in mothers were associated with masculinized gender-role behavior in 3.5-year-old girls, but not boys. These findings may be a result of a genetic predisposition for women with high testosterone levels to pass these genes on to their daughters [120]. Another possibility is that raised maternal testosterone levels promote more male-typical behavior in girls [120]. No study to date has used maternal testosterone levels to investigate the development of autistic traits, and it would be interesting to examine whether maternal testosterone during pregnancy is related to f T or future development of autistic traits.

Samples from the Umbilical Cord

A series of studies have examined relationships between umbilical cord (perinatal) hormones and later behavior such as temperament and mood. High perinatal testosterone and estradiol levels were significantly related to low timidity in boys [123, 124, 125]. In girls, no relationships were observed. Other studies of umbilical cord hormones have shown inconsistent results [126, 127, 128].

An important factor to consider when using umbilical samples is that fT levels are typically at very low levels from about week 24 of gestation, whereas the neonatal peak has not yet appeared. In addition, the cord contains blood from the mother as well as the fetus, and hormone levels may vary due to labor itself [124].

Amniotic Fluid

One of the most promising methods for obtaining information about the fetal exposure to androgens appears to be the direct sampling of fT levels in amniotic fluid, obtained from routine diagnostic amniocentesis. This is performed for clinical reasons to detect genetic abnormalities in the fetus. As a result, it is typically performed in a relatively narrow time window that is thought to coincide with the peak in fT for male fetuses. This peak is also apparent in amniotic fluid, and several studies have documented a large sex difference in amniotic androgens [129, 130, 131, 132, 133]. There are significant risks associated with the procedure itself, so that it cannot be performed solely for research. However, the process itself does not appear to have any negative effects for later development [131].

The origins of androgens in amniotic fluid are not fully understood, but the main source seems to be the fetus itself [50]. Hormones enter the amniotic fluid in two ways: via diffusion through the fetal skin in early pregnancy and via fetal urine in later pregnancy [131, 134]. Given the risk entailed in obtaining blood from the fetus, there are very limited data directly comparing testosterone in amniotic fluid to that in fetal blood. Androgens in amniotic fluid are unrelated to androgens measured in maternal blood in the same period, as shown in studies in early and mid-gestation [51, 135]. Based on these findings, testosterone obtained in amniotic fluid appears to be a good reflection of the levels in the fetus and represents an alternative to direct assay of the more risky process of collecting fetal serum [50].

Finegan et al. [136] conducted the first study that explored the relationship between prenatal hormone levels in amniotic fluid and later behavior on a broad range of cognitive functions at age 4. The findings are difficult to interpret since the authors used measures that did not show sex differences. However, the same children were followed up at 7 years of age, and associations between spatial ability and fT were examined [137]. A significant positive association between f T levels and faster performance on a mental rotation task was observed in a small subgroup of girls, but not boys. At 10 years of age, prenatal testosterone levels were found to relate to handedness and dichotic listening tasks [138], and the results were interpreted as providing support for the hypothesis that higher levels of prenatal sex hormones are related to lateralization in boys and girls [139].

Cambridge Fetal Testosterone Project

The Cambridge Fetal Testosterone Project is an ongoing longitudinal study investigating the relationship between f T levels and the development of behaviors relating to ASC [43, 140]. Mothers of participating children had all undergone amniocentesis for clinical reasons between 1996 and 2001 and gave birth to healthy singleton infants. To date, these children have been tested postnatally at 12 months, 18 months, 24 months, 4 years, and 6–8 years of age.

Fetal Testosterone and Eye Contact at 12 Months

The first study aimed to measure f T and estradiol levels in relation to eye contact for a sample of 70 typically developing, 12-month old children [141]. Reduced eye contact is a characteristic common in children with autism [141, 142]. Frequency and duration of eye contact were measured using videotaped sessions. Sex differences were found, with girls making significantly more eye contact than boys. The amount of eye contact varied quadratically with f T levels when the sexes were combined. Within the sexes, a relationship was only found for boys [141]. No relationships were observed between the outcome and estradiol levels. Results were taken to indicate that f T may play a role in shaping the neural mechanisms underlying social development [141].

Fetal Testosterone and Vocabulary at 18 and 24 Months

Another study (of 87 children) focused on the relationship between vocabulary size in relation to fT and estradiol levels from amniocentesis. In some subgroups within ASCs, such as classic autism, vocabulary development is also delayed [143]. Vocabulary size was measured using the Communicative Development Inventory that is a self-administered checklist of words for parents to complete [144]. Girls were found to have a significantly larger vocabulary than boys at both time points [145]. Results showed that levels of fT inversely predicted the rate of vocabulary development in typically developing children between the ages of 18 and 24 months [145]. Within sex analyses showed no significant relationships in boys or girls, which the authors believe may have been due to the relatively small sample sizes. No relationships between estradiol and vocabulary size were found. Despite the lack of significant results within sex, the significant findings in the combined sample suggest that fT may be involved in shaping the neural mechanisms underlying communicative development [145].

Fetal Testosterone and Empathy at Age 4

These children were next followed-up at 4 years of age. Thirty-eight children completed a "moving geometric shapes" task where they were asked to describe cartoons with two moving triangles whose interaction with each other suggested social relationships and psychological motivations [146]. Sex differences were observed with girls using more mental and affective state terms to describe the cartoons compared to boys; however, no relationships between fT levels and mental or affective state terms were observed. Girls were found to use more intentional propositions than males, and a negative relationship between fT levels and frequency of intentional propositions was observed when the sexes were combined and in boys. Boys used more neutral propositions than females, and fT was related with the frequency of neutral propositions when the sexes were combined. However, no significant relationships were observed when boys and girls were examined separately. In addition, no relationships with estradiol were observed. These results are consistent with the EMB theory since other studies have found that individuals with ASC perform lower than typical males on a similar moving geometric shapes task [147].

Fetal Testosterone, Restricted Interests, and Social Relationships at Age 4

A second follow-up at 4 years of age in this same cohort of children utilized a measure called the Children's Communication Checklist [148]. The quality of social relationships subscale demonstrated higher fT levels to be associated with poorer quality of social relationships for both sexes combined but not individually. A limitation to within-sex analysis was the sample size ($n = 58$).

fT levels were also associated with more narrow interests when the sexes were combined and in boys only [149]. Sex differences are reported, with males scoring higher (i.e., having more narrow interests) than females [149]. Individuals with ASC demonstrate more restricted interests as well as difficulties with social relationships [149].

Fetal Testosterone, Systemizing, Empathizing at Ages 6–8

In 2004, the Cambridge fT project sample size was increased by recruiting more mothers who had undergone amniocentesis during the same period. The parents of these children ($n = 204$) were asked to complete children's versions of the Systemizing Quotient (SQ-C) and Empathy Quotient (EQ-C).

In systemizing, boys scored higher than girls on the SQ-C, and levels of fT positively predicted SQ-C scores in boys and girls individually [150]. Sex differences were observed in EQ-C scores, with girls scoring higher than boys.

A significant negative correlation between fT levels and EQ-C was observed when the sexes were combined and within boys.

A regression analysis was used to identify the main contributions to EQ-C. Whilst the main effect of sex was found, there was no main effect of fT. However, the effect of fT cannot be disregarded, since sex and fT are strongly correlated [151]. A subset of these children ($n = 78$) were also invited to participate in further cognitive tests, and the children's version of the Reading the Mind in the Eyes task (Eyes-C) was administered. No significant differences were found between sexes though a significant relationship between fT levels and Eyes-C was observed for both boys and girls [151].

Behaviors Associated with ASC

In the study outlined above, a series of measurements go some way toward experimentally linking direct measurements of fT and behaviors associated with ASC. At ages 12 months, 18 months, 24 months, and 4 years, behavioral traits associated with empathizing appear to be linked to lower levels of fT. At 6–8 years of age, SQ and EQs both appear to show sexual dimorphism – consistent with the E–S theory. In addition, fT was positively correlated with systemizing for both boys and girls.

The lack of ability to empathize and drive to systemize appear to be characteristic of ASC [2]. Whilst these behaviors do not confirm a clear link between fT and ASC, the results are broadly consistent with a role for fT in shaping sexually dimorphic behavior.

Fetal Testosterone and Autistic Traits

In light of some of the above results, a more direct approach of evaluating the links between autistic traits and fT was implemented. In this study, effects of fT were directly evaluated against autistic traits as measured by the CAST [28, 29] and the Child Autism Spectrum Quotient (AQ-C) [152]. The CAST was used because it has shown good test-retest reliability, good positive predictive value (50%), and high specificity (97%) and sensitivity (100%) for ASCs [29]. The AQ-C has also shown good test-retest reliability, high sensitivity (95%), and high specificity (95%) [34].

fT levels were positively associated with higher scores (indicating greater number of autistic traits) on the CAST as well as on the AQ-C. For the AQ-C, this relationship was seen within sex as well as when the sexes were combined, suggesting that this is an effect of fT rather than an effect of sex. The relationship between CAST scores and fT was also seen within males, but not within females [152]. These findings, from two measures of autistic traits, are consistent with the notion that higher levels of fT may be associated with the development of autistic traits.

Limitations of Amniocentesis Methods

Research suggests that amniotic fluid provides the best, direct measurement of fetal hormones compared to maternal serum and is therefore probably the best choice for studying the behavioral effects of variations in prenatal androgen exposure [51, 136]. Using this method, research has shown that f T levels are significantly associated with behaviors associated with ASC, providing strong evidence for a role for prenatal hormones on typical development.

A drawback of amniocentesis is that it can only be conducted for purposes of diagnosing fetal anomalies. This means that the samples studied are selected in several ways that may influence the generalizability of the results. In addition, in these amniotic fluid studies, total extractable (or free) testosterone is utilized. However, free testosterone may not be directly responsible for the interactions which masculinize behavior [48]. A more detailed understanding of the chemistry of masculinization would be useful in extrapolating the effects of free testosterone in the development of conditions such as ASC. Against these limitations should be weighed the obvious strengths of amniocentesis, which mainly concern its timing and measurement of the fetal environment whilst avoiding unnecessary additional risk.

Conclusions

ASCs are characterized by social impairments, restricted and repetitive interests accompanied by language delay. ASC are believed to lie on a spectrum, reflecting the range of individual ability in each of these areas.

Many of the behaviors that are characteristic of ASC have also been linked to extremes of certain male-typical behaviors. Evidence includes superior performance on a range of tasks where male individuals typically outperform females but increased impairment compared to typical males on tasks with female superiority. Additional evidence linking ASC to an extreme form of the male brain comes from measurements of physical characteristics where males and females typically differ.

Physical sexual differentiation is largely attributed to the gonadal hormones, and in particular testosterone and its derivatives. Animal studies have suggested that the same hormones might also control the development of sex-typical behaviors. In humans, the direct manipulation of hormones in early development is not possible. Studies of atypical hormone environments yield some information about the role of testosterone in behavioral development, but sample sizes are very limited, particularly for individuals who also have an ASC.

More information about the hormones affecting ASC can be extrapolated from indicators of prenatal hormone levels such as finger length ratio and lateralization. These measurements are typically easy to obtain but their links to hormone levels are not clear.

Direct study of the effects of testosterone is difficult because levels rise and fall in the fetal environment over the course of gestation. In addition, maternal levels are not representative of fetal levels. The optimal way to directly measure fT appears to be via amniotic fluid obtained during clinical amniocentesis. This method is not ideal because it limits the sample available. However, some results have been obtained linking behavior and fT levels in a group of typically developing children in the UK.

In this sample, findings suggest that behaviors known to be affected by ASC also tend to be related to elevated fT levels. These behaviors include measures of social and communicative development, empathy and systemizing. In addition, a more recent study using the CAST and the Autism Spectrum Quotient-Children's Version (AQ-C) has demonstrated a relationship between the number of autistic traits a child exhibits and the fT levels measured via amniocentesis.

The findings presented in this chapter lend support to the "Extreme Male Brain" theory of ASC and its link to fT. However, a proper evaluation of this theory will require testing not just for associations between fT and autistic traits, but between fT and clinically diagnosed ASC. The latter will require much larger samples than have previously been available. Nevertheless, exposure to elevated levels of fT is indicated to be a possible factor in the development of ASCs, a hypothesis that needs more definitive testing.

Acknowledgments B.A. was supported by a scholarship from Trinity College and S.B.C. by grants from the Nancy Lurie Marks Family Foundation and the MRC during the period of this work. We are grateful to Emma Chapman, Svetlana Lutchmaya, Malcolm Bang, and Rebecca Knickmeyer for the valuable discussions.

References

1. Rutter M, Caspi A, Moffitt TE. Using sex differences in psychopathology to study causal mechanisms: Unifying issues and research strategies. *Journal of Child Psychology and Psychiatry* 2003;44(8):1092–1015.
2. Baron-Cohen S. The extreme male brain theory of autism. *Trends in Cognitive Sciences* 2002;6(6):248–254.
3. APA. *DSM-IV Diagnostic and Statistical Manual of Mental Disorders*, 4th ed. Washington DC: American Psychiatric Association, 1994.
4. Baird G, Simonoff E, Pickles A, et al. Prevalence of disorders of the autism spectrum in a population cohort of children in South Thames: The special needs and autism project (SNAP). *Lancet* 2006;368(9531):210–215.
5. Fombonne E. The changing epidemiology of autism. *Journal of Applied Research in Intellectual Disabilities* 2005;18(4):281–294.
6. Tidmarsh L, Volkmar FR. Diagnosis and epidemiology of autism spectrum disorders. *Canadian Journal of Psychiatry* 2003;48(8):517–525.
7. Bryson SE, Smith IM. Epidemiology of autism: Prevalence, associated characteristics, and implications for research and service delivery. *Mental Retardation and Developmental Disabilities Research Reviews*, Special Issue: Autism 1998;4(2):97–103.

8. Chakrabarti S, Fombonne E. Pervasive developmental disorders in preschool children: Confirmation of high prevalence. *American Journal of Psychiatry* 2005;162(6): 1133–1141.
9. Scott F, Baron-Cohen S, Bolton P, Brayne C. Prevalence of autism spectrum conditions in children aged 5–11 years in Cambridgeshire, UK. *Autism* 2002;6(3):231–237.
10. Kraemer S. The fragile male. *British Medical Journal* 2000;321(7276):1609–1612.
11. Stodgell CJ, Ingram JI, Hyman SL. The role of candidate genes in unraveling the genetics of autism. *International Review of Research in Mental Retardation* 2001;23:57–81.
12. Baron-Cohen S. *The Essential Difference: Men, Women and the Extreme Male Brain.* London: Penguin, 2003.
13. Baron-Cohen S, Wheelwright S. The empathy quotient: an investigation of adults with asperger syndrome or high functioning autism, and normal sex differences. *Journal of Autism and Developmental Disorders* 2004;34(2):163–175.
14. Baron-Cohen S, Richler J, Bisarya D, Gurunathan N, Wheelwright S. The systemising quotient (SQ): An investigation of adults with asperger syndrome or high functioning autism and normal sex differences. *Philosophical Transactions of the Royal Society* 2003;358:361–374.
15. Goldenfeld N, Baron-Cohen S, Wheelwright S. Empathizing and systemizing in males, females and autism. *International Journal of Clinical Neuropsychology* 2005; 2:338–345.
16. Wheelwright S, Baron-Cohen S, Goldenfeld N, et al. Predicting Autism Spectrum Quotient (AQ) from the Systemizing Quotient-Revised (SQ-R) and Empathy Quotient (EQ). *Brain Research* 2006;1079(1):47–56.
17. Auyeung B, Baron-Cohen S, Wheelwright S, Samarawickrema N, Atkinson M. The Children's Empathy Quotient (EQ-C) and Systemizing Quotient (SQ-C): A study of sex differences in typical development and of autism spectrum conditions.
18. Shah A, Frith U. An islet of ability in autistic children: A research note. *Journal of Child Psychology and Psychiatry* 1983;24 (4):613–620.
19. Jolliffe T, Baron-Cohen S. Are people with autism and Asperger syndrome faster than normal on the Embedded Figures Test? *Journal of Child Psychology and Psychiatry* 1997;38(5):527–534.
20. Baron-Cohen S, Wheelwright S, Spong A, Scahill L, Lawson J. Are intuitive physics and intuitive psychology independent? A test with children with Asperger Syndrome. *Journal of Developmental and Learning Disorders* 2001;5:47–78.
21. Lawson J, Baron-Cohen S, Wheelwright S. Empathising and systemising in adults with and without Asperger syndrome. *Journal of Autism and Developmental Disorders* 2004;34(3):301–310.
22. Lynn R, Raine A, Venables PH, Mednick SA, Irwing P. Sex differences on the WISC-R in Mauritius. *Intelligence* 2005;33:527–533.
23. Allen MH, Lincoln AJ, Kaufman AS. Sequential and simultaneous processing abilities of high-functioning autistic and language-impaired children. *Journal of Autism and Developmental Disorders* 1991;21(4):483–502.
24. Lincoln AJ, Courchesne E, Kilman BA, Elmasian R, Allen M. A study of intellectual abilities in high-functioning people with autism. *Journal of Autism and Developmental Disorders* 1988;18(4):505–524.
25. Shah A, Frith C. Why do autistic individuals show superior performance on the block design task? *Journal of Child Psychology and Psychiatry* 1993;34:1351–1364.
26. Baron-Cohen S, Jolliffe T, Mortimore C, Robertson M. Another advanced test of theory of mind: Evidence from very high functioning adults with autism or Asperger Syndrome. *Journal of Child Psychology and Psychiatry* 1997;38(7):813–822.
27. Golan O, Baron-Cohen S, Hill J. The Cambridge Mindreading (CAM) face-voice battery: Testing complex emotion recognition in adults with and without Asperger syndrome. *Journal of Autism and Developmental Disorders* 2006;36(2):169–183.

28. Scott FJ, Baron-Cohen S, Bolton P, Brayne C. The CAST (Childhood Asperger Syndrome Test): Preliminary development of a UK screen for mainstream primary-school-age children. *Autism* 2002;6(1):9–13.
29. Williams J, Scott F, Stott C, et al. The CAST (Childhood Asperger Syndrome Test): Test accuracy. *Autism* 2005;9(1):45–68.
30. Baron-Cohen S, Scott FJ, Allison C, et al. Estimating autism spectrum prevalence in the population: A school based study from the UK. Submitted.
31. Williams J, Scott F, Stott C, et al. Sex differences in social and communication skills in primary school aged children.
32. Baron-Cohen S, Wheelwright S, Skinner R, Martin J, Clubley E. The Autism Spectrum Quotient (AQ): Evidence from Asperger syndrome/high functioning autism, males and females, scientists and mathematicians. *Journal of Autism and Developmental Disorders* 2001;31:5–17.
33. Baron-Cohen S, Hoekstra R, Knickmeyer R, Wheelwright S. The Autism-Spectrum Quotient (AQ) – Adolescent version. *Journal of Autism and Developmental Disorders* 2006;36(3):343–350.
34. Auyeung B, Baron-Cohen S, Wheelwright S, Allison C. Development of the Autism Spectrum Quotient – Children's Version (AQ-Child). *Journal of Autism and Developmental Disorders*, in press.
35. Wakabayashi A, Tojo Y, Baron-Cohen S, Wheelwright S. The Autism-Spectrum Quotient (AQ) Japanese version: Evidence from high-functioning clinical group and normal adults. *Japanese Journal of Psychology* 2004;75(1):78–84.
36. Wakabayashi A, Baron-Cohen S, Uchiyama T, et al. The Autism-Spectrum Quotient (AQ) children's version in Japan: A cross-cultural comparison. *Journal of Autism and Developmental Disorders* 2006;36:263–270.
37. Wakabayashi A, Baron-Cohen S, Wheelwright S, Tojo Y. The Autism-Spectrum Quotient (AQ) in Japan: A cross-cultural comparison. *Journal of Autism and Developmental Disorders* 2006;36(2):263–270.
38. Hoekstra R, Bartels M, Cath DC, Boomsma DI. Factor structure of the broader autism phenotype and its diagnostic validity: A study using the Dutch translation of the autism-spectrum quotient (AQ). *Journal of Autism and Developmental Disorders*, in press.
39. Constantino JN, Todd RD. Autistic traits in the general population. *Archives of General Psychiatry* 2003;60:524–530.
40. Baron-Cohen S, Knickmeyer R, Belmonte MK. Sex differences in the brain: Implications for explaining autism. *Science* 2005;310:819–823.
41. Baron-Cohen S, Ring H, Chitnis X, et al. fMRI of parents of children with Asperger Syndrome: a pilot study. *Brain Cognition* 2006;61(1):122–130.
42. Connellan J, Baron-Cohen S, Wheelwright S, Batki A, Ahluwalia J. Sex differences in human neonatal social perception. *Infant Behavior & Development* 2000;23(1):113–118.
43. Baron-Cohen S, Lutchmaya S, Knickmeyer R. *Prenatal Testosterone in Mind.* Cambridge, Massachusetts: The MIT Press, 2004.
44. Knickmeyer RC, Baron-Cohen S. Fetal testosterone and sex differences in typical social development and in Autism. *Journal of Child Neurology* 2006;21(10):825–845.
45. Novy M, Resko J (eds). *Fetal Endocrinology.* New York, New York: Academic Press, 1981.
46. Fuchs F, Klopper A. *Endocrinology of Pregnancy.* Philadelphia: Harper & Row, 1983.
47. Tulchinsky D, Little AB. *Maternal-Fetal Endocrinology,* 2nd ed. Philadelphia, London: W.B. Saunders, 1994.
48. Hines M. *Brain Gender.* New York, New York: Oxford University Press, Inc., 2004.
49. Kimura D. *Sex and Cognition.* Cambridge, MA: The MIT Press, 1999.
50. Cohen-Bendahan CC, van de Beek C, Berenbaum SA. Prenatal sex hormone effects on child and adult sex-typed behavior: Methods and findings. *Neuroscience and Biobehavioral Reviews* 2005;29(2):353–384.

51. Van de Beek C, Thijssen JHH, Cohen-Kettenis PT, Van Goozen SH, Buitelaar JK. Relationships between sex hormones assessed in amniotic fluid, and maternal and umbilical cord blood: What is the best source of information to investigate the effects of fetal hormonal exposure? *Hormones and Behavior* 2004;46:663–669.
52. Larsen PR, Kronenberg HM, Melmed S, Polonsky KS (eds). *Williams Textbook of Endocrinology,* 10th ed. Philadelphia: Saunders, 2002.
53. Jost A. The role of foetal hormones in prenatal development. *Harvey Lectures* 1961; 55:201–226.
54. Jost A. Hormonal factors in the sex differentiation of the mammalian foetus. *Philosophical Transactions of the Royal Society of London: B Biological Sciences* 1970; 259(828):119–130.
55. Jost A. A new look at the mechanism controlling sexual differentiation in mammals. *John Hopkins Medical Journal* 1972;130:38–53.
56. George FW, Wilson JD. Embryology of the Genital Tract. In: Walsh PC, Retik AB, Stamey TA, (eds) *Campbell's Urology,* 6th ed., Philadelphia: WB Saunders, 1992: 1496–1508.
57. Collaer ML, Hines M. Human behavioural sex differences: A role for gonadal hormones during early development? *Psychological Bulletin* 1995;118:55–107.
58. Breedlove SM. Sexual dimorphism in the vertebrate nervous system. *The Journal of Neuroscience* 1992;12:4133–4142.
59. Goy RW, McEwen BS. *Sexual Differentiation of the Brain.* Cambridge, MA: The MIT Press, 1980.
60. New MI. Diagnosis and management of congenital adrenal hyperplasia. *Annual Review of Medicine* 1998;49:311–328.
61. Hampson E, Rovet JF, Altmann D. Spatial reasoning in children with congenital adrenal hyperplasia due to 21-hydroxylase deficiency. *Developmental Neuropsychology* 1998; 14(2):299–320.
62. Hines M, Fane BA, Pasterski VL, Matthews GA, Conway GS, Brook C. Spatial abilities following prenatal androgen abnormality: Targeting and mental rotations performance in individuals with congenital adrenal hyperplasia. *Psychoneuroendocrinology* 2003; 28:1010–26.
63. Resnick SM, Berenbaum SA, Gottesman II, Bouchard TJ. Early hormonal influences on cognitive functioning in congenital adrenal hyperplasia. *Developmental Psychology* 1986;22(2):191–198.
64. Nass R, Baker S, Speiser P, et al. Hormones and handedness: left-hand bias in female adrenal hyperplasia patients. *Neurology* 1987;37:711–715.
65. Berenbaum SA, Hines M. Early androgens are related to childhood sex-typed toy preferences. *Psychological Science* 1992;3:203–206.
66. Berenbaum SA, Snyder E. Early hormonal influences on childhood sex-typed activity and playmate preferences: Implications for the development of sexual orientation. *Developmental Psychology* 1995;31:31–42.
67. Berenbaum SA. Effects of early androgens on sex-typed activities and interests in adolescents with congenital adrenal hyperplasia. *Hormones and Behavior* 1999;35(1):102–110.
68. Ehrhardt AA, Baker SW. Fetal androgens, human central nervous system differentiation, and behavior sex differences. In: Friedman RC, Richart RR, Vande Wiele RL, (eds) *Sex Differences in Behavior.* New York: Wiley, 1974:33–51.
69. Hines M, Brook C, Conway GS. Androgen and psychosexual development: Core gender identity, sexual orientation and recalled childhood gender role behavior in women and men with congenital adrenal hyperplasia (CAH). *Journal of Sex Research* 2004; 41(1):75–81.
70. Knickmeyer RC, Fane BA, Mathews G, et al. Autistic traits in people with congenital adrenal hyperplasia: A test of the fetal testosterone theory of autism. *Hormones and Behavior* 2006;50(1):148–153.

71. Fausto-Sterling A. *Myths of Gender*. New York: Basic Books, 1992.
72. Quadagno DM, Briscoe R, Quadagno JS. Effects of perinatal gonadal hormones on selected nonsexual behavior patterns: A critical assessment of the nonhuman and human literature. *Psychological Bulletin* 1977;84:62–80.
73. Nordenstrom A, Servin A, Bohlin G, Larsson A, Wedell A. Sex-typed toy play behavior correlates with the degree of prenatal androgen exposure assessed by the CYP21 genotype in girls with congenital adrenal hyperplasia. *The Journal of Clinical Endocrinology and Metabolism* 2002;87(11):5119–5124.
74. Hines M, Ahmed SF, Hughes IA. Psychological outcomes and gender-related development in complete androgen insensitivity syndrome. *Archives of Sexual Behavior* 2003;32(2):93–101.
75. Money J, Schwartz M, Lewis V. Adult erotosexual status and fetal hormonal masculinization and demasculinization: 46 XX congenital adrenal hyperplasia and 46 XY androgen insensitivity syndrome compared. *Psychoneuroendocrinology* 1984;9:405–414.
76. Hier DB, Crowley WF. Spatial ability in androgen-deficient men. *New England Journal of Medicine* 1982;306:1202–1205.
77. Cappa SF, Guariglia C, Papagno C, et al. Patterns of lateralization and peformance levels for verbal and spatial tasks in congenital androgen deficiency. *Behavioural Brain Research* 1988;31:177–183.
78. Manning JT, Bundred PE, Flanagan BF. The ratio of 2nd to 4th digit length: A proxy for transactivation activity of the androgen receptor gene? *Medical Hypotheses* 2002;59(3):334–336.
79. Lutchmaya S, Baron-Cohen S, Raggatt P, Knickmeyer R, Manning JT. 2nd to 4th digit ratios, fetal testosterone and estradiol. *Early Human Development* 2004;77:23–28.
80. Brown WM, Hines M, Fane B, Breedlove SM. Masculinized finger length patterns in human males and females with congenital adrenal hyperplasia. *Hormones and Behavior* 2002;42:380–386.
81. Manning JT, Baron-Cohen S, Wheelwright S, Sanders G. The 2nd to 4th digit ratio and autism. *Developmental Medicine & Child Neurology* 2001;43(3):160–164.
82. Milne E, White S, Campbell R, Swettenham J, Hansen P, Ramus F. Motion and form coherence detection in autistic spectrum disorder: Relationship to motor control and 2:4 digit ratio. *Journal of Autism and Developmental Disorders* 2006;36(2):225–237.
83. Holt SB. *The Genetics of Dermal Ridges*. Springfield, IL: Charles C. Thomas, 1968.
84. Kimura D, Carson MW. Dermatoglyphic asymmetry: Relation to sex, handedness and cognitive pattern. *Personality and Individual Differences* 1995;19(4):471–478.
85. Kimura D, Clarke PG. Cognitive pattern and dermatoglyphic asymmetry. *Personality and Individual Differences* 2001;30(4):579–586.
86. Sanders G, Waters F. Fingerprint asymmetry predicts within sex differences in the performance of sexually dimorphic tasks. *Personality and Individual Differences* 2001;31(7):1181–1191.
87. Walker HA. A dermatoglyphic study of autistic patients. *Journal of Autism and Childhood Schizophrenia* 1977;7(1):11–21.
88. Hartin PJ, Barry RJ. A comparative dermatoglyphic study of autistic, retarded, and normal children. *Journal of Autism and Developmental Disorders* 1979;9(3):233–246.
89. Hines M, Shipley C. Prenatal exposure to diethylstilbestrol (DES) and the development of sexually dimorphic cognitive abilities and cerebral lateralization. *Developmental Psychology* 1984;20(1):81–94.
90. Fein D, Waterhouse L, Lucci D, Pennington B, Humes M. Handedness and cognitve functions in pervasive developmental disorders. *Journal of Autism and Developmental Disorders* 1985;15:323–333.
91. McManus IC, Murray B, Doyle K, Baron-Cohen S. Handedness in childhood autism shows a dissociation of skill and preference. *Cortex* 1992;28(3):373–381.

92. Satz P, Soper H, Orsini D, Henry R, Zvi J. Handedness subtypes in autism. *Psychiatric Annals* 1985;15:447–451.
93. Soper H, Satz P, Orsini D, Henry R, Zvi J, Schulman M. Handedness patterns in autism suggest subtypes. *Journal of Autism and Developmental Disorders* 1986;16:155–167.
94. Peters M. Sex differences in human brain size and the general meaning of differences in brain size. *Canadian Journal of Psychology* 1991;45(4):507–522.
95. Gillberg C. Autistic children's hand preferences: Results from an epidemiological study of infantile autism. *Psychiatry Research* 1983;10(1):21–30.
96. Harden AY, Minshew NJ, Mallikarjuhn M, Keshavan MS. Brain volume in autism. *Journal of Child Neurology* 2001;16(6):421–424.
97. Tordjman S, Ferrari P, Sulmont V, Duyme M, Roubertoux P. Androgenic activity in autism. *American Journal of Psychiatry* 1997;154(11):1626–1627.
98. Ingudomnukul E, Baron-Cohen S, Wheelwright S, Knickmeyer R. Elevated rates of testosterone-related disorders in women with autism spectrum conditions. *Hormones and Behavior* 2007;51(5):597–604.
99. Knickmeyer RC, Wheelwright S, Hoekstra R, Baron-Cohen S. Age of menarche in females with autism spectrum conditions. *Developmental Medicine and Child Neurology* 2006;48(12):1007–1008.
100. Clark MM, Galef BG. Effects of intraurine position on the behavior and genital morphology of litter-bearing rodents. *Developmental Neurology* 1998;14:197–211.
101. Fels E, Bosch LR. Effect of prenatal administration of testosterone on ovarian function in rats. *American Journal of Obstetrics and Gynecology* 1971;111(7):964–969.
102. Meisel RL, Ward IL. Fetal female rats are masculinized by male littermates located caudally in the uterus. *Science* 1981;213:239–242.
103. Even MD, Dhar MG, vom Saal FS. Transport of steroids between fetuses via amniotic fluid in relation to the intrauterine position phenomenon in rats. *Journal of Reproduction and Fertility* 1992;96(2):709–716.
104. Elizabeth PH, Green R. Childhood sex-role behaviors: Similarities and differences in twins. *Acta Geneticae Medicae et Gemellologiae (Roma)* 1984;33(2):173–179.
105. Miller EM, Martin N. Analysis of the effect of hormones on opposite-sex twin attitudes. *Acta Geneticae Medicae et Gemellologiae (Roma)* 1995;44(1):41–52.
106. Resnick SM, Gottesman II, McGue M. Sensation seeking in opposite-sex twins: an effect of prenatal hormones? *Behavior Genetics* 1993;23(4):323–329.
107. Cole-Harding S, Morstad AL, Wilson JR. Spatial ability in members of opposite-sex twin pairs (abstract). *Behavior Genetics* 1988;18:710.
108. Rodgers CS, Fagot BI, Winebarger A. Gender-typed toy play in dizygotic twin pairs: A test of hormone transfer theory. *Sex Roles* 1998;39 (3–4):173–184.
109. Elkadi S, Nicholls ME, Clode D. Handedness in opposite and same-sex dizygotic twins: Testing and testosterone hypothesis. *Neuroreport* 1999;10(2):333–336.
110. Henderson BA, Berenbaum SA. Sex-typed play in opposite-sex twins. *Developmental Psychobiology* 1997;31:115–123.
111. Betancur C, Leboyer M, Gillberg C. Increased rate of twins among affected sibling pairs with autism. *American Journal of Human Genetics* 2002;70(5):1381–1383.
112. Greenberg DA, Hodge SE, Sowinski J, Nicoll D. Excess of twins among affected sibling pairs with autism: implications for the etiology of autism. *American Journal of Human Genetics* 2001;69(5):1062–1067.
113. Croen LA, Grether JK, Selvin S. Descriptive epidemiology of autism in a California population: Who is at risk? *Journal of Autism and Developmental Disorders* 2002;32(3):217–224.
114. Hallmayer J, Glasson EJ, Bower C, et al. On the twin risk in autism. *American Journal of Human Genetics* 2002;71(4):941–946.
115. Hultman CM, Sparen P, Cnattingius S. Perinatal risk factors for infantile autism. *Epidemiology* 2002;13(4):417–423.

116. Phoenix CH, Goy RW, Gerall AA, Young WC. Organizing action of prenatally administered testosterone propionate on the tissues mediating mating behavior in the female guinea pig. *Endocrinology* 1959;65:369–382.
117. Goy RW, Bercovitch FB, McBrair MC. Behavioral masculinization is independent of genital masculinization in prenatally androgenized female rhesus macaques. *Hormones and Behavior* 1988;22:552–571.
118. Smail PJ, Reyes FI, Winter JSD, Faiman C. The Fetal Hormonal Environment and its Effect on the Morphogenesis of the Genital System. In: Kogan SJ, Hafez ESE, (eds) *Pediatric Andrology*. Boston: Martinus Nijhoff, 1981:9–19.
119. Brown GR, Nevison CM, Fraser HM, Dixson AF. Manipulation of postnatal testosterone levels affect phallic and clitoral development in infant rhesus monkeys. *International Journal of Andrology* 1999;22:119–128.
120. Hines M, Golombok S, Rust J, Johnston KJ, Golding J. Testosterone during pregnancy and gender role behavior of preschool children: A longitudinal, population study. *Child Development* 2002;73(6):1678–1687.
121. Udry JR, Morris NM, Kovenock J. Androgen effects on women's gendered behaviour. *Journal of Biosocial Science* 1995;27:359–368.
122. Udry JR. Biological limits of gender construction. *American Sociological Review* 2000;65:443–457.
123. Jacklin CN, Maccoby EE, Doering CH. Neonatal sex-steroid hormones and timidity in 6-18-month-old boys and girls. *Developmental Psychobiology* 1983;16(3):163–168.
124. Jacklin CN, Wilcox KT, Maccoby EE. Neonatal sex-steroid hormones and cognitive abilities at six years. *Developmental Psychobiology* 1988;21(6):567–574.
125. Marcus J, Maccoby EE, Jacklin CN, Doering CH. Individual differences in mood in early childhood: their relation to gender and neonatal sex steroids. *Developmental Psychobiology* 1985;18(4):327–340.
126. Abramovich DR, Rowe P. Foetal plasma testosterone levels at mid-pregnancy and at term: Relationship to foetal sex. *Journal of Endocrinology* 1973;56(3):621–622.
127. Forest MG, Sizonenko PC, Cathiard AM, Bertrand J. Hypophyso-gonadal function in humans during the first year of life: I. Evidence for testicular activity in early infancy. *The Journal of Clinical Investigation* 1974;53:819–828.
128. Pang S, Levine LS, Chow DM, Faiman C, New MI. Serum androgen concentrations in neonates and young infants with congenital adrenal hyperplasia due to 21-hydroxylase deficiency. *Clinical Endocrinology* 1979;11:575–584.
129. Dawood MY, Saxena BB. Testosterone and dihydrotestosterone in maternal and cord blood and in amniotic fluid. *American Journal of Obstetrics and Gynecology* 1977; 129:37–42.
130. Finegan J, Bartleman B, Wong PY. A window for the study of prenatal sex hormone influences on postnatal development. *Journal of Genetic Psychology* 1989;150(1): 101–112.
131. Judd HL, Robinson JD, Young PE, Jones OW. Amniotic fluid testosterone levels in midpregnancy. *Obstetrics and Gynecolology* 1976;48(6):690–692.
132. Nagamani M, McDonough PG, Ellegood JO, Mahesh VB. Maternal and amniotic fluid steroids throughout human pregnancy. *American Journal of Obstetrics and Gynecology* 1979;134(6):674–680.
133. Robinson J, Judd H, Young P, Jones D, Yen S. Amniotic fluid androgens and estrogens in midgestation. *Journal of Clinical Endocrinology* 1977;45:755–761.
134. Schindler AE. Hormones in human amniotic fluid. *Monographs on Endocrinology* 1982;21:1–158.
135. Rodeck CH, Gill D, Rosenberg DA, Collins WP. Testosterone levels in midtrimester maternal and fetal plasma and amniotic fluid. *Prenatal Diagnosis* 1985;5(3):175–181.
136. Finegan JK, Niccols GA, Sitarenios G. Relations between prenatal testosterone levels and cognitive abilities at 4 years. *Developmental Psychology* 1992;28(6):1075–1089.

137. Grimshaw GM, Sitarenios G, Finegan JK. Mental rotation at 7 years: Relations with prenatal testosterone levels and spatial play experiences. *Brain and Cognition* 1995;29(1):85–100.
138. Grimshaw GM, Bryden MP, Finegan JK. Relations between prenatal testosterone and cerebral lateralization in children. *Neuropsychology* 1995;9(1):68–79.
139. Witelson SF. Sex-differences in neuroanatomical changes with aging. *New England Journal of Medical* 1991;325(3):211–212.
140. Knickmeyer RC, Baron-Cohen S. Fetal testosterone and sex differences. *Early Human Development* 2006;82(12):755–760.
141. Lutchmaya S, Baron-Cohen S, Raggatt P. Foetal testosterone and eye contact in 12 month old infants. *Infant Behavior and Development* 2002;25:327–335.
142. Swettenham J, Baron-Cohen S, Charman T, et al. The frequency and distribution of spontaneous attention shifts between social and non-social stimuli in autistic, typically developing, and non-autistic developmentally delayed infants. *Journal of Child Psychology and Psychiatry* 1998;9:747–753.
143. Rutter M. Diagnosis and Definition. In: Schopler IMRE, (ed.) *Autism: A Reappraisal of Concepts and Treatment*. New York: Plenum Press, 1978:1–26.
144. Hamilton A, Plunkett K, Shafer G. Infant vocabulary development assessed with a British Communicative Inventory: Lower scores in the UK than the USA. *Journal of Child Language* 2000;27(3):689–705.
145. Lutchmaya S, Baron-Cohen S, Raggatt P. Foetal testosterone and vocabulary size in 18- and 24-month-old infants. *Infant Behavior & Development* 2002;24(4):418–424.
146. Knickmeyer R, Baron-Cohen S, Raggatt P, Taylor K, Hackett G. Fetal testosterone and empathy. *Hormones and Behavior* 2006;49:282–292.
147. Klin A. Attributing social meaning to ambiguous visual stimuli in higher-functioning Autism and Asperger syndrome: the Social Attribution Task. *Journal of Child Psychology and Psychiatry* 2000;7:831–846.
148. Bishop DVM. Development of the children's communication checklist (CCC): A method for assessing qualitative aspects of communicative impairment in children. *Journal of Child Psychology and Psychiatry* 1998;6:879–891.
149. Knickmeyer R, Baron-Cohen S, Raggatt P, Taylor K. Foetal testosterone, social relationships, and restricted interests in children. *Journal of Child Psychology and Psychiatry* 2005;46(2):198–210.
150. Auyeung B, Baron-Cohen S, Chapman E, Knickmeyer R, Taylor K, Hackett G. Foetal testosterone and the Child Systemizing Quotient. *European Journal of Endocrinology* 2006;155(Suppl. 1):S23–S30.
151. Chapman E, Baron-Cohen S, Auyeung B, Knickmeyer R, Taylor K, Hackett G. Fetal testosterone and empathy: Evidence from the Empathy Quotient (EQ) and the 'Reading the Mind in the Eyes' test. *Social Neuroscience* 2006;1:135–148.
152. Auyeung B, Baron-Cohen S, Chapman E, Knickmeyer R, Taylor K, Hackett G. Foetal testosterone and autistic traits. *British Journal of Psychology*, in press.
153. Baron-Cohen S, Wheelwright S, Griffin R, Lawson J, Hill J. The Exact Mind: Empathising and Systemising in Autism Spectrum Conditions. In: Goswami U, (ed.) *Handbook of Cognitive Development*. Blackwell, 2002.

Chapter 9
Interaction between Genetic Vulnerability and Neurosteroids in Purkinje cells as a Possible Neurobiological Mechanism in Autism Spectrum Disorders

Flavio Keller, Roger Panteri, and Filippo Biamonte

Abstract Autism has a strong genetic basis, but other risk factors, acting during brain development, probably contribute to the clinical manifestations of the disease. Male sex is an important risk factor in autism, as shown by the much higher frequency of autism in males. Some authors have proposed the theory of the "autistic brain" as an extreme form of "male brain." This theory is supported by the existence of sexual dimorphisms between male and female brains, by the effects of estrogens on several neurotransmitter systems, and by observations that prenatal androgens produce sex differences in brain and behavior. The investigation of sex-related differences in brain structures that are implicated in autism thus appears to be an extremely interesting research avenue.

A major challenge, however, is to develop animal models where the interaction between genetic predisposition and non-genetic risk factors in autism can be tested. The heterozygous *reeler* mouse brain, which has reduced levels of the protein Reelin, reproduces some of the developmental abnormalities observed in the autistic brain, such as loss of Purkinje cells (PCs). It has been demonstrated that male (but not female) heterozygous *reeler* mice show a loss of PCs in the cerebellum, thus suggesting that decreased function of Reelin in the male gender provides a selective vulnerability to PC loss. Females may be protected against genetic vulnerability because of their high estrogen/androgen balance during brain development.

Thus, our efforts are directed toward understanding how male and female sex steroids and Reelin interact during cerebellar development and how they impinge on the regulation of PC generation and survival. This research may define novel targets for therapies to prevent PC loss, with the hypothesis that drugs that prevent or reduce these developmental abnormalities in the animal model may also be useful to prevent or ameliorate autistic behaviors in humans.

Keywords Autism · sex · genes · neurosteroids · purkinje cells

F. Keller
Laboratory of Developmental Neuroscience, Università Campus Bio-Medico, Rome, Italy
e-mail: f.keller@unicampus.it

A.W. Zimmerman (ed.), *Autism*, DOI: 10.1007/978-1-60327-489-0_9,
© Humana Press, Totowa, NJ 2008

Introduction

Autism spectrum disorders (ASD) are a behaviorally defined group of disorders characterized by impaired social skills, delayed and impaired language, and restricted areas of interest. Additional features may include poor eye contact, repetitive and stereotypic behaviors, sensory modulation dysfunction, and varying levels of cognition and motor disturbances. Males are affected more frequently than females by a ratio of 4:1 for the syndrome as a whole. Twin studies support a strong genetic component. The high concordance rate for autism found in monozygotic twins, together with the increased male prevalence, strongly suggests that autism has a genetic component that may be modulated by sex hormones, acting during critical periods of brain development. From a genetic point of view, this would be equivalent to a modulation of the penetrance of autism-causing mutations by sex hormones. Variable penetrance of genetic mutations in ASD is currently hypothesized to explain phenotype variability among individuals carrying the same mutation, e.g., in the case of neuroligin mutations [1]. According to this hypothesis, sex hormones acting on the developing brain may either mask or enhance genetic vulnerability for autism, depending on the underlying genetic vulnerability, the type of hormone, the cellular target, and the critical time window in which the mutation exerts its effects. In other words, gonadal hormones could lead to gender differences in the expression of genetic vulnerability, through convergent effects of genetic mutations and gonadal hormones on the same neural systems. Unfortunately, although there is a substantial body of evidence about hormonally induced sexual dimorphisms related to reproductive behavior, evidence about sexual dimorphisms of neural systems related to cognitive, emotional, and social functions (i.e., those systems that are probably impaired in ASD) is much more limited. Importantly, greater sexual dimorphisms of adult brain volumes were observed in areas of the human brain that are characterized, according to the animal literature, by developmentally high levels of androgen and estrogen receptors (ER) during critical periods of brain development, in comparison with other brain regions [2].

An early attempt to establish a link between ASD and altered sex hormone levels is the Geschwind–Behan–Galaburda (GBG) hypothesis [3], that posits a key role for testosterone in the development of cognitive problems in males, through slowing down of the growth of the left cerebral hemisphere during fetal life. It is interesting that, in their original paper, the authors point out that "*The association of sinistrality, learning disorders, and immune disorders suggests the possibility of a common origin that may help in accounting for certain findings in autism, one of the forms of developmental cognitive disorder*"[3]. Early experimental attempts to prove the validity of the GBG model failed. The GBG model is presently not thought to be valid in its original formulation [4, 5], but there may be a grain of truth in it. Indeed, a recent version of the GBG hypothesis, which is substantiated by converging psychological evidence, is the extreme male brain theory of ASD [[6]; see also the chapter by Auyeung and Baron-Cohen in this book (Chapter 8)].

An assessment of the role of testosterone on brain development is complicated by the fact that two prominent testosterone peaks are observed in male individuals, across mammalian species. The first peak occurs during early gestation (around the 13th–15th gestational week in humans) when the male testes start to secrete testosterone. The second peak occurs around birth and is observed in many mammalian species, including rats, mice, and humans [7]. Furthermore, one should not forget that androgen precursors are converted into estrogens in the brain, by the cytochrome P450 aromatase enzyme, implying that estradiol, not testosterone, is the true "masculinizing hormone" for many brain regions. One of the regions examined in neural sexual differentiation studies, the sexually dimorphic nucleus of the medial preoptic area (SDN-POA), which is involved in many behaviors including masculine sexual and social behavior, was found to be five to six times larger in volume in male than in female rats [8]. Brief exposure of newborn female rats to very high levels of estradiol masculinizes the volume of the sexually dimorphic nucleus of the preoptic area by reducing apoptotic cell death [9]. Because of the perinatal testosterone surge, the developing male brain is exposed to high levels of estrogens, whereas the female brain is exposed to high estrogen levels only after puberty. Therefore, it could be that estrogen, rather than testosterone, is the critical hormone to understand the skewed sex ratio in ASD.

Another important issue for ASD concerns the receptors and signaling pathways through which sex hormones sculpt brain circuits. For example, estrogens are well known to exert their effects through two different ER, ER-α and ER-β. ER-α is thought to be the fundamental receptor mediating estrogen action on reproductive organs and behavior, whereas ER-β is thought to mediate at least some of the effects of estrogens on behaviors that are not directly associated with reproduction, such as locomotor activity, arousal, fear responses, anxiety, and learning and memory processes [10]. Indeed, ER-β appears to be the principal ER expressed in brain areas such as the cerebral cortex, the hippocampus, and cerebellum [11]. That ER-α and ER-β have different functions during brain development is suggested by the fact that ER-β knockout mice show a deficit of neurons in the neocortex, starting from late embryonic life, whereas ER-α knockouts show normal neuronal numbers [12].

In our opinion, ASD research is in urgent need of animal models where interactions between genetic background and gender/gonadal hormones can be reliably investigated and the results translated back to the human context. Ideally, these models should recapitulate, at least in part, the anatomical, biochemical, and behavioral alterations observed in ASD. Mouse models showing gender-dependent effects of genetic mutations, associated with Purkinje cell (PC) loss, include the heterozygous *reeler* and *staggerer* mutants [13, 14]. Also relevant for ASD is the gender-dependent rescue of the *Engrailed-1* knockout phenotype by *Engrailed-2* ([15], see also Chapter 1 by Ian T. Rossman and E. DiCicco-Bloom) In these three models, the male gender is more severely affected. Another potentially interesting model is the *FoxP2*-knockout mouse, in which the disruption of a single copy of the *FoxP2* gene leads to significant alteration in ultrasound vocalization by pups in response to maternal separation. Clearly, this is a behaviorally interesting model of early communication deficits in ASD. Cerebellar abnormalities were also observed in the *Foxp2* mutant, in particular PCs were abnormally placed in the heterozygous mice [16]. To our knowledge, it is not known whether there is a PC loss in the *FoxP2* knockout, nor whether the penetrance of the *Foxp2* mutation is gender-dependent. In summary, it is intriguing that all four mutations mentioned above affect the development of the cerebellum, a brain region that many investigators consider critical for ASD, and that at least three of the four mutations are expressed in a sexually dimorphic manner.

A recent neuroanatomical study on monozygotic twins by Kates et al. [17] confirms that the cerebellum could be an important mediator between genetic predisposition and behavioral phenotype in autism. Furthermore, this same study suggests that, although cerebellar morphometry is under a high level of genetic control in *typically developing* monozygotic twins, there is more variability in cerebellar morphometry in *autistic* monozygotic twins, implying that this variability may be mediated by non-genetic factors. As Kates et al. suggest, "... it is possible that the genes that underlie the phenotype for autism affect the cerebellum in a way that increases its susceptibility to environmental risk factors." [17]

This review focuses on the interaction between the *reeler* mutation and the perinatal levels of sex hormones in PCs as a possible biological model of ASD. We will begin with a brief review of the organizing role of sex hormones in the brain and its mechanisms.

Organizing Role of Sex Hormones in Neuronal Growth and Survival

Hormones influence growth, differentiation, maturation, and function of a wide variety of reproductive and non-reproductive tissues, including those of the nervous system. Organizational effects of sex steroids in the brain occur during fetal development in man and other long gestation species, and perinatally in short gestation species such as the rat and the mouse.

In particular, the endogenous gonadal steroid 17β-estradiol (estrogen) plays an important role in the development, maturation, and function of central nervous system (CNS) tissue. The actions of estrogen at target tissues can be divided into long-term "genomic" actions that are mediated by the binding to the intracellular ER-α and ER-β, and rapid membrane actions that modulate a diverse array of intracellular signal transduction cascades. ER-α and ER-β are members of the steroid/thyroid superfamily of transcription-factor receptors and are transiently expressed at high levels in the developing rodent cerebral cortex [18, 19, 20, 21, 22]. Each receptor binds estradiol with high affinity and interacts with estrogen-responsive elements (ERE) to influence transcription of responsive genes [23, 24]. Numerous studies have also demonstrated that estradiol can rapidly influence cellular physiology in many different cell types of reproductive and non-reproductive tissues through the activation of a diverse array of intracellular signaling mechanisms. These rapid mechanisms probably involve the binding of estradiol at membrane-associated ERs that are closely related to the classical intracellular receptors [25]. Thus, the total physiological effects of estradiol may result both from rapid non-genomic mechanisms and from longer-duration genomic mechanisms.

In the brain, estradiol is well known as a fundamental regulator of physiological and neuroendocrine functions related to sexual behavior. Along with this role, estradiol also plays a significant role during the normal development and genderization of the mammalian CNS, and it exerts important neurotrophic and neuroprotective functions in the brain [26, 27]. Apart from the actions important for normal neuronal development and function, such as the modulation of neuronal excitability, relevant to this review are the actions of estradiol shown to influence pathways resulting in increased neuronal survival. For example, in cultured cortical neurons, estradiol can rapidly activate the phosphatidylinositol-3-kinase [28] and MAPK signaling pathways [29, 30], and initiate mechanisms that increase cell survival. Interestingly, in addition to cortical neurons, estradiol can rapidly activate MAPK signaling in developing cerebellar granule cells [31].

There are substantial reciprocal interactions between estrogen-signaling and growth factor-mediated signaling pathways [32, 33, 34, 35]. Neurotrophins [nerve growth factor (NGF), brain-derived neurotrophic factor (BDNF), neurotrophin-3 (NT-3), and neurotrophin-4 (NT-4)] regulate survival and differentiation in multiple neural systems through two receptor classes [36]. The first, the neurotrophin-selective tyrosine kinases (Trks), mediate neuroprotective mechanisms. A second receptor-subtype, p75NTR, binds all neurotrophins and modulates their interaction with Trks, but can also induce death [37, 38]. Estrogen downregulates p75NTR mRNA while upregulating trkA mRNA [39], suggesting that estrogen may attenuate death and promote survival.

The mechanisms through which estradiol modulates neuronal survival partly depend on the activation of growth factor-dependent MAPK signaling in developing brain cells [40]. In neurons, the ERK1/2 MAPK-signaling pathway is activated both by the binding of neurotrophins at their cognate receptor Trks and by the interaction of estradiol with ER-α [29, 30].

In addition to the interactions with neurotrophins, estradiol interacts in the brain with the signaling mechanisms of insulin-like growth factor-1 (IGF-1) [41]. IGF-1 and its receptors, for example, are expressed by PCs in the cerebellum [42, 43], and IGF-1 is a trophic factor for PCs and rescues PC loss in some experimental models [44]. Thus, there appears to be a direct interaction between components of the neurotrophin IGF-1, and estradiol-mediated signaling pathways in several neuronal populations to regulate cell survival.

Estradiol also appears to exhibit opposing, region-specific regulation of death and proliferation. Thus, estrogen stimulates neurogenesis in the avian song control nuclei [45], while blocking proliferation of neuroblastoma cells [46] and transformed hypothalamic cells [47], and mediates opposing region-specific effects on neuronal death in the hypothalamus and preoptic area [47, 9]. Estrogen regulation of death and proliferation may therefore be context- and region-specific.

Sex Steroids Regulate Neuronal Death and Survival

In correlation with the survival-promoting effects of estradiol, differences in cell number are among the most common types of sex difference reported. These differences may arise as the result of differential neurogenesis, migration, or cell death in males and females. Currently, there is weak evidence linking the hormonal control of neurogenesis or migration to sexual dimorphisms seen in the adult brain [48, 49]. In contrast, several well-studied sex differences are attributed to the hormonal control of cell survival. Sex steroids appear to influence cell-survival mechanisms in several neural regions in rodents, such as the spinal nucleus of the bulbocavernosus (SNB), the central portion of the medial preoptic nucleus (MPNc; a subportion of the sexually dimorphic nucleus of the preoptic area), the principal nucleus of the bed nucleus of the stria terminalis (BNSTp), and the anteroventral periventricular nucleus (AVPV). Testosterone decreases cell death in the SNB, MPNc, and BNSTp, while increasing cell death in AVPV [50, 51, 52, 53].

Testosterone or its metabolites may regulate sexual differentiation of cell number in several systems by acting through classical intracellular estrogen/androgen receptors [54, 55]. One example is given by estrogens that control cell number in the AVPV and MPNc, and ER are expressed by neurons in these nuclei during the cell death period [56]. Thus, it is possible that aromatized products of testosterone act directly on cells in AVPV and the MPNc to control their survival. Another possibility is that hormones act on afferents or targets of AVPV or MPNc neurons or on glial cells within the region.

In the SNB, hormones have indirect effects in the control of neuron number. Testosterone acts through androgen receptors to rescue SNB cells from death [54], yet the motoneurons themselves do not express androgen receptors during the perinatal cell death period [57]. In fact, the rescue of SNB motoneurons is mediated by androgens acting at the target muscles in the system to produce trophic factors that retrogradely influence the survival of innervating SNB motoneurons [58]. Interestingly, SNB motoneurons express abundant ciliary neurotrophic factor-α (CNTFR-α) receptors during perinatal life, and the normal sex difference in the SNB is absent in knockout mice lacking the CNTFR-α [59, 60].

All this evidence supports the view that pathways dependent upon gonadal hormones and trophic factors closely interact to influence neuronal survival. The convergence of these signaling pathways is fundamental in instructing cells to decide whether to undergo a program of apoptotic cell death.

Many aspects of the molecular bases of apoptosis have been elucidated over the past several years, and proteins of the Bcl-2 family have emerged as crucial regulators of death in many cell types [61]. In mice over-expressing the survival-promoting protein Bcl-2, the sex differences in neuron number in the SNB and AVPV are reduced [62], and in mice knockout for the proapoptotic protein Bax the sex differences in overall cell number are eliminated [63, 64]. Steroid hormones may thus normally control cell death in the SNB and AVPV through a Bcl-2 dependent pathway. Interestingly, steroid hormones have been shown to regulate the expression of Bcl-2 family members in neural tissue [65, 66].

Although gonadal steroid hormones may regulate the expression of Bcl-2 and/or Bax to control neuronal cell death, other possibilities exist such as differences in hormone receptor subtypes or the expression of genes involved in the steroidogenic pathway, which may be investigated through the use of microarray analysis in combination with laser microdissection technology in the regions that show sex differences in cell number, to identify key genes that control sexually dimorphic cell death.

The Loss of Cerebellar Purkinje Cells in ASD

Several neuroanatomic studies have reported abnormalities of the cerebellum in ASD. One of the most commonly reported microscopic findings in this disorder has been the presence of reduced numbers of Purkinje Cells (PC) primarily in the posterior inferior regions of the cerebellar hemispheres [67, 68, 69, 70, 71]. Although the cerebellar vermis has also been reported to show reduced numbers of PCs in some cases [68, 70, 71] and selective vermal hypoplasia has been reported with magnetic resonance imaging [72], a detailed microscopic analysis of PC number in all lobules of the vermis by Bauman and Kemper [69] has found no reduction in cell number in this cerebellar region.

The PCs are inhibitory neurons that carry the only output of the cerebellar cortex; they influence, through deep cerebellar nuclei and vestibular nuclei, many motor and non-motor regions of the cerebral cortex and spinal cord, thus affecting almost all aspects of behavior. The cerebellum, besides its role in the execution of motor programs, participates also in non-motor functions such as emotional behavior and fear learning. For example, it has been shown that a fear-learning paradigm, provided by a neutral stimulus paired with a painful one, can induce a freezing response in the animal, that correlates to an increase in the strength of the parallel fiber-PC synapse [73].

Although a reduction of PCs has been reported in the majority of cases, it is not yet understood how this PC loss comes about. In principle, reduced numbers of PCs may result from a failure of neurogenesis, from migration into their usual cerebellar location, or from loss at a later time. Because no corresponding loss of neurons has been observed in the inferior olivary nucleus, which innervates PC through the climbing fibers, it has been hypothesized that the PC loss takes place during fetal life, possibly before the 28th–30th week of gestation [74]. A critical issue is to know whether there is also similar loss of granule cells. Because PCs stimulate granule cell proliferation, a loss of PCs before peak neurogenesis of granule cells is expected to lead to reduced granule cell numbers. Unfortunately, there are no reports of granule cell numbers in autistic cerebella. This may be due to the technical difficulty of obtaining cell counts from such a large cell population.

In the rodent cerebellum, PC neurogenesis takes place between E13 and E15. Purkinje neurons complete their migration and maturation just after birth [75]. Experimental evidence in rats shows that PCs undergo a critical period of vulnerability during the first postnatal week (between P0 and P7), in which they undergo massive cell death after axotomy, whereas their sensitivity to axotomy is much lower in earlier and later developmental periods [76]. This vulnerable period in the first postnatal week is characterized by intense synaptogenesis and dendritic remodeling of PCs. It is not known whether the enhanced PC vulnerability during the first postnatal week in rodents is selective for axotomy, or reflects a more general type of vulnerability to insults, including genetic insults. It is perhaps not a fortuitous coincidence that, according to published tables relating CNS development across mammals [77], the first postnatal week in rats corresponds approximately to the 13th–18th gestational week in humans, i.e., slightly earlier than the period to which PC loss in ASD has been traced.

Many mouse mutants characterized by a cerebellar phenotype show decreased numbers of PC. The timing of onset of PC deficiency shows substantial differences among mutants (see Table 9.1). At one extreme, *Engrailed-2* mutants are characterized by deficient neurogenesis of PCs and other neurons of the olivocerebellar circuit [78]; at the other extreme, mutants such as *weaver, nervous*, and *Purkinje cell degeneration* are born with a normal complement of PCs, and PC begin to die postnatally. The PC deficit in male heterozygous *reeler* mice is already observable in the second postnatal week (see below).

The different characteristics of PC loss in different mutants should be taken into account in the discussion of mouse models of cerebellar abnormalities in ASD. On the basis of published literature, we have summarized the characteristics of the PC loss observed in different cerebellar mouse mutants in Table 9.1.

Table 9.1 Mouse mutations leading to Purkinje cell loss

Mutant mouse strain	gender	Total number of PCs in control mice	Time of observed onset of PC loss	% PC loss relative to control mice	References
Reeler homozygous (RL)	Males (m)	~177,000	??	50%	[125]
Reeler heterozygous (HZ)	m. females (f)	~200,000	P12–P16	20% (only in m)	[13, 113]
Staggerer (Rora^{sg}/Rora^{sg})	m, f	~182,000 (m)	1–3 months (m, hz)	10% at 3 months (m, hz)	[14]
		~179,000 (f)	9–13 months (f, hz)	25% at13–24 months (m, f hz)	[126]
				75% for adult homozygous	
Engrailed-2 (En2) null	–		?? (reduction of cerebellar size from E15)	30–40% in adults	[78]
En2 overexpression transgenic (L7-En2)	–		P9	41% in (vermis)	[127]
				40% (hemispheres), in adults	
Lurcher (Lc)	–	~177,000	P8–P10	10% at 26 days	[128]
				~100% at 90 days	
Hyperspiny Purkinje cells (Hpc)	–	~216,000	Adults	27% at 91 days	[129]
Purkinje cell degeneration (Pcd) null	–		P15–P18	25–50% at P22–P24	[130]
Pcd heterozygous	m, f	~175,000	Adults	~95% at P30–P40	[131]
				18% at 17 months	
Nervous (Nr) null	–		P23	85–90% between 4th and 7th week	[132]
Nr heterozygous	m, f		Adults	No loss	[131]
Ataxia and male sterility (AMS)	–		P21	99% at 6 weeks	[133]
Weaver (wv)	–	~183,000	P5	14% (heterozygous), 28% (null mutant) at 3–9 months	[134]
Leaner (Cacna1a la/la)	–		P25	50% at P60	[135]
Toppler	–		P21	30% at P30	[136]
				90% at 6 months	

Estrogens Influence Cerebellar PC Growth, Maturation, and Survival

The cerebellar cortex is a region that has received much attention for investigating the actions of sex steroids both on neuronal growth and maturation and on neuronal survival. In particular, the PC is known to be a major site of neurosteroid synthesis in the brain [79]. Neurosteroids are a class of steroids that are synthesized de novo from cholesterol in the central and peripheral nervous systems of vertebrates through mechanisms at least partly independent of peripheral steroidogenic glands [79, 80].

The first step in the synthesis of steroid hormones in PCs is the conversion of cholesterol to pregnenolone, a 3βhydroxy-Δ steroid and a main precursor of steroid hormones, catalyzed by the cytochrome P450 side-chain cleavage enzyme (P450scc). Human P450scc is encoded by a single gene on chromosome 15, the *CYP11A1* gene [81]. P450scc mRNA and protein are present in the placenta, the primitive gut, and the brain [82, 83, 84]. The presence of CYP11A1 mRNA in human brain tissue provides evidence that pregnenolone can be produced in the CNS.

Once pregnenolone is synthesized, it can be converted to progesterone through 3β-hydroxysteroid dehydrogenase-isomerase (3β-HSD) or it can be 17-hydroxylated by 17α-hydroxylase/c17,20 lyase (P450c17), a microsomal P450, to produce dehydroepiandrosterone (DHEA). There is a substrate specificity for the 17 sterhydroxylase reaction, with some species preferring Δ^5 steroids (i.e., pregnenolone) and others preferring Δ^4 steroids (i.e., progesterone) [85]. DHEA can then be converted to androgens, through a reaction mediated by tissue-specific 17β-HSDs, to form androstenediol, and then to testosterone, through 3β-HSDs. The cytochrome P450 aromatase (P450arom) is the product of the CYP19 gene, which has been cloned and sequenced [86, 87]. The aromatase pathway then converts testosterone into estradiol and androstenedione into estrone, whereas 5α-reductase converts testosterone to DHT [88].

The PC possesses several kinds of steroidogenic enzymes, such as cytochrome P450scc and 3β-HSD, and actively produces progesterone during neonatal life, when cerebellar neuronal circuit formation occurs [89, 90, 91]. The PC also expresses the enzyme that mediates DHT formation, 5α-reductase [92] as well as the key enzyme of estrogen formation, P450arom, and appears to actively produce estradiol in the neonate [93]. From these recent studies it emerges that the PC, which is known to play an important role in the process of cerebellar memory and learning, represents an excellent cellular model to study the organizing actions of neurosteroids.

In the rodent cerebellum, PCs complete their migration and maturation just after birth [75] when cerebellar levels of estrogen are high [93]. Thus, we may hypothesize that estrogen synthesized de novo in the developing PCs may be critically involved in the cerebellar cortical formation during neonatal life. Furthermore, it has been demonstrated that estrogen can influence neuronal function in the cerebellum, such as modulating the response of cerebellar PCs to the excitatory neurotransmitter glutamate [94, 95], Interestingly, neonatal and adult rodent PCs express ER-β [96], suggesting that estradiol acts on PCs through ER-β to exert its cellular effects during cerebellar development. Consistent with this, treatment of cerebellar cultures with exogenous estradiol promotes the dendritic growth and spine formation of PCs in a dose-dependent manner, with active doses being in the range of physiological levels of estradiol measured in the cerebellum [93]. Similar morphological effects were also obtained by the in vivo treatment with estradiol, whereas these effects were reversed by treatment with the ER antagonist tamoxifen.

Thus, both these in vitro and in vivo studies suggest that a high level of cerebellar estradiol during neonatal life is essential for the promotion of dendritic growth of PCs. In addition, the in vitro treatment with estradiol also increased the density of dendritic spines on PCs. The authors considered the changes in dendritic spines as possibly reflecting changes in the number of synapses.

To understand the mode of action of estradiol, the identification of ERs and the analysis of their distribution in neonatal cerebellum is essential. It has been reported that the neonatal rat PCs express ER-β [96]. Consequently, estradiol may act directly on the PC through ER-β-mediated mechanisms, to promote neuronal growth and spine formation in PCs during cerebellar cortical formation. One finding that supports this hypothesis is the reversal of the effects of estrogen on PC dendritic growth by the ER modulator tamoxifen [93]. It is known that tamoxifen binds to ER-α and ER-β and exerts ligand activation properties opposite to those mediated by estrogen [97]. The absence of almost all the dendritic spine-like structures in PCs after the treatment with tamoxifen suggests that estradiol acts on PCs through these receptors. However, estradiol might act on PCs through non-nuclear ERs. In fact, it has been shown that locally applied estrogens and antiestrogens can rapidly potentiate glutamate-evoked excitation of PCs, whereas prior administration of the anti-estrogen tamoxifen did not prevent the estrogen effects [94, 95], whereas other investigators have suggested that the effect of estradiol on hippocampal CA1 pyramidal cell dendrite spine density requires the activation of NMDA receptors [98].

As discussed previously, brain trophic factors may interact with signaling pathways induced by neurosteroids; thus, neurotrophic factors represent putative modulators of PC dendrite and spine development. It has been reported that neurotrophic factors, such as BDNF or NT-3, are highly expressed in the developing cerebellum and are critical for proper development of PCs and granule cells [99, 100]. Thus, the effects of estrogens on PC maturation and development may well depend upon the timely release and action of specific neurotrophic factors.

The Male Heterozygous *Reeler* Mouse: A Possible Heuristic Model of Cerebellar Pathology in ASD

Several years ago, Mariani and colleagues noticed that adult male hetero-zygous *reeler* (RLHZ) or *staggerer* mice have reduced numbers of PCs, in comparison with female heterozygous and with wild-type (WT) mice of either sex [13, 14]. At 1 month of age, the number of PCs was not different in WT and heterozygous *staggerer* males, the loss of PCs began between 1 and 3 months and was aggravated regularly up to 13 months. On the other side, male RLHZ mice are haploinsufficient for Reelin, a gene which is involved in regulating neuron migration and positioning during early cerebral cortex formation [101], and which has been linked to ASD in a number of studies [102, 103, 104, 105]. Reelin, which is produced by marginal zone Cajal–Retzius cells and by cerebellar granule cells, binds to the very low-density lipoprotein receptor (VLDLR) and apolipoprotein E receptor 2 (ApoER2) to induce tyrosine phosphorylation of the adaptor protein Dab1 [106, 107]. Both VLDLR and ApoER2 are expressed by rodent and human PCs [108, 109]. In addition, PCs of neonatal and adult rodents express ER-β [96, 110, 111] and androgen receptors [112], thus suggesting estrogens and androgens, as well as Reelin, act on PCs to exert some physiological effects during cerebellar development. This suggested to us that there may be an interaction between reduced Reelin levels in the cerebellum and male sex hormones in the regulation of PC survival. We therefore decided to follow up Mariani's findings. Data from our laboratory have confirmed the PC loss in male RLHZ mice, showing moreover that the PC loss is already detectable in the early postnatal period, as early as postnatal days 14–16, and perhaps even earlier [113], thus indicating an interaction between male gender and decreased Reelin function during early postnatal development of the cerebel-lum. This observation is interesting as a biological model of an interaction between genetic vulnerability and male gender.

Because estrogens synthesized in the brain by PCs promote dendritic growth, spinogenesis, and synaptogenesis in the PCs through their respective receptors, these factors may be part of an epigenetic program that fine tunes the expression of genes involved in the terminal differentiation of PCs and guides the proper integration of these cells into the functional circuitry of the cerebellum [114]. In our investigations, we have also found that 17β-estradiol, injected into the cisterna magna of newborn mice just after birth (on postnatal day 4), increases the number of PCs in male RL HZ mice, but does not affect PC number in female WT or RLHZ mice. Conversely, treatment with tamox-ifen, a potent ER modulator, and ICI 182,780, a pure and potent ER antago-nist, dramatically decreases PC numbers both in female WT and in female RLHZ mice, whereas PCs in WT or RLHZ males are not affected. Estradiol thus appears to exert a neuroprotective action on PCs during the period of maximal vulnerability of PCs [76], and we hypothesize that male RLHZ mice are more susceptible to PC loss because of a decreased availability of estrogens. Alternatively, male RLHZ mice may lose PC cells because of testosterone excess.

The hypothesis of an imbalance of the estrogen/androgen ratio in the male RLHZ cerebellum is supported by our quantitative determinations of neuroactive steroids in the cerebella of WT and RLHZ mice of both sexes, obtained through mass spectrometry. These data indicate that in the cerebellum of male RLHZ mice just after birth there are higher levels of testosterone compared with that of the other groups [113].

Our investigations thus strongly suggest that Reelin interacts with estrogens during mouse cerebellar development. In male RLHZ mice, where we observe an excess of testosterone and a parallel reduction of estradiol, the haploinsufficiency of Reelin leads to reduced PC numbers, whereas in female RLHZ mice, where the estrogen/androgen ratio is high, PCs may be protected from the Reelin deficit by high estrogens. A cartoon showing possible mechanisms of convergence of the Reelin and the steroid-signaling pathways in PCs, based on available literature, is shown in Fig. 9.1.

One fundamental question faced in the investigation of the mechanisms regulating PC development in *reeler* mice is whether PCs are generated in normal numbers, correctly placed within the cerebellar cortex, and then later lost, or whether they are never correctly placed or perhaps even never generated in the cerebellar primordium. Some authors have reported evidence for naturally occurring PC death during development, reporting that the loss of PCs in certain mouse strains may be attributed to increased apoptosis of the progenitor cells in the embryonic cerebellum [115, 116]. Interestingly, because the PC loss specifically affects male RLHZ mice, which show higher testosterone levels than their WT and RLHZ female littermates, it is possible that testosterone negatively regulates Reelin expression [117], thereby leading to reduced PC

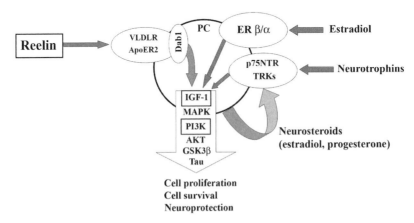

Fig. 9.1 Convergence of Reelin and neurosteroid signaling pathways in PCs. Reelin (see text), PIK3 [137] and IGF-1 [138] have been implicated in autism

neurogenesis. However, at the present time we cannot tell whether the selective reduction of PCs we observe in the young postnatal male RLHZ cerebellum is due to increased neuronal apoptosis or rather reduced neurogenesis and proliferation of PCs during embryogenesis.

It has been proposed that PCs regulate the process of proliferation of granule cell precursors [118]. Indeed, when PCs are not located immediately under the external granular layer (EGL), like in the homozygous *reeler* mutant, or when these cells are abnormal and in very low numbers (like in homozygous *staggerer*), the amount of granule cells generated is extremely reduced [119]. PCs express *Sonic hedgehog* [120], a secreted factor that plays a role in cell proliferation and cell fate determination [121], and dividing cells in the EGL express *Patched* (Ptc) and *Gli1*, two target genes of Shh [120]. Antibody-blocking of Shh action results in a marked reduction of granule cell production. In contrast, treatment with Shh prevents the EGL precursors to exit from their cycle and prolongs their proliferation [120, 122]. Therefore, because PCs regulate proliferation of granule cell precursors through secretion of Shh, a reduced number of generated PCs (or correctly placed PCs, or correctly functioning PCs, respectively) leads to a smaller number of granule cells. However, a decrease in the number of PCs after generation of the granule cells does not result in a decline in the number of granule cells.

Keeping in mind this model, we are currently evaluating whether the reduction of PCs in the young male RLHZ cerebellum is accompanied by a reduction of granule cell numbers. Through this analysis we intend to distinguish between two alternative hypotheses: (1) the generation–migration–loss hypothesis, according to which male and female WT/RLHZ generate the same numbers of PCs, which are later lost after migration selectively in the male RLHZ cerebellum, or (2) the incorrect generation/misplacement hypothesis, according to which PCs are never generated, or are lost before they enter the migration phase.

Interestingly, several years ago, Vogel and colleagues found in knockout mouse mutants of the pro-apoptotic protein Bax that the numbers of PCs increased by over 30%, but the numbers of granule cells were unchanged compared with controls [123]. The presumed rescue of PCs from cell death in the absence of Bax activity on the one hand supports the conclusions of previous studies on Bcl-2-overexpressing transgenics that PCs undergo a period of programmed cell death [124], but on the other hand, because of the lack of an effect on granule cell numbers in the absence of Bax expression and in the presence of increased PCs, raises questions about the role of Bax in granule cell death and the mechanisms by which PCs regulate granule cell numbers. One hypothesis, advanced by the authors, is that the excess PCs in the Bax knockout mutants are rescued from cell death that occurs after the period of granule cell generation. This hypothesis is also consistent with our observations: in male RLHZ mice, we are able to rescue the PC loss by treatment with estradiol during early postnatal life (postnatal day 4), thus suggesting that the period of naturally occurring PC death, at least in these mice, extends beyond the embryonic stages of cerebellar development.

Future Research Avenues

The basic idea underyling this review is that research into the neurobiological underpinnings of sexual dimorphisms in brain areas that are not directly related to reproductive function will yield important insights into the neurobiology of ASD. One interesting brain area, in this respect, appears to be the cerebellum.

Several mutations have been described in mice, which lead to impaired cerebellar development and/or neuronal degeneration in the cerebellum. Mice homozygous for the recessive mutations *staggerer* and *reeler* exhibit a marked ataxia associated with an atrophic cerebellum and a very severe loss of PCs during the first postnatal weeks. Instead, mice heterozygous for these mutations show a progressive and age-related loss of PCs, which is gender-dependent. The heterozygous *reeler* mouse therefore offers an interesting model to test the interaction between sex, age, and genetic background on the development and maintenance of cerebellar neuronal populations. One question that remains to be investigated in the RLHZ mouse is whether there is a topographic pattern of selectivity to the loss of PCs. Several studies have shown that, in normal mice, PCs in the cerebellar vermis and hemispheres are distributed according to precise neurogenetic gradients and exhibit a differential vulnerability in several mutant strains. To resolve this issue, we are currently quantifying PC loss in the individual cerebellar lobules, both in mice and in human autistic cerebellar specimens (in collaboration with Dr. M.L. Bauman and colleagues at Boston University). Our aim is to understand whether the PC loss in RLHZ mice is more pronounced in the lateral hemispheres than in the cerebellar vermis, as repeatedly observed in humans.

The reason why the *reeler* allele exerts its deleterious effect in young males, while sparing females is not yet clear, although our results tentatively suggest that an imbalance in the estrogen/androgen ratio in the male gender leads to a diminished neuroprotective function in RLHZ males. It would be worthwhile to search in humans for any evidence of hormonal imbalance, altered expression of sex-steroid-synthesizing enzymes, or sex-steroid receptors, during fetal/early postnatal development in autistic patients. Moreover, we should encourage research in humans aimed at defining the timing and regional selectivity of PC loss in autistic cerebella, and at evaluating whether the pattern of PC loss is more evident and severe in males than in females.

Furthermore, it will be particularly important to find out whether the estrogen action on PC is mediated by ER-α or ER-β, this issue having consequences for future therapeutic strategies.

One interesting research avenue is the use of microarray analysis in combination with laser microdissection technology in the cell populations that show sex differences in cell number, to identify key genes that are involved in the pathways that control sexually dimorphic cell loss. These molecular pathways, possibly involved in mediating the action of sex steroids on PCs, could represent candidate signaling pathways in ASD and lead to potentially new therapeutic targets.

Research on gender differences in brain aging has shown a greater decline in males than in females in various structures, such as the MPNc and the hippocampal formation, in rodents. In humans, several age-related cortical atrophies in the frontal and temporal lobes appear to be significantly greater in elderly men than women. Our studies on the RLHZ mouse, together with data from other groups, supports the idea that a third parameter, the genetic background, can interact with age and gender to influence the maintenance of specific neuronal populations.

Our research and that of other groups on the hormonal regulation of cell survival in the CNS is currently proceeding to understand the basic disease mechanisms, involving the interactions between the Reelin and the estrogen signaling pathways, which govern PC generation and survival, and possibly define novel pharmacological targets for therapies aimed at preventing PC loss, through the modulation of Reelin and estrogen signaling.

Acknowledgments The authors thank Mitchell Glickstein, Luis Miguel Garcia-Segura, Piergiorgio Strata, and Catherina Becker for helpful discussions and criticisms on the manuscript. Work by the authors has been supported by grants from *Autism Speaks* and from the *Fondation Jerôme Lejeune*.

References

1. Yan J, Oliveira G, Coutinho A, Yang C, Feng J, Katz C, Sram J, Bockholt A, Jones IR, Craddock N, Cook EH, Vicente A, Sommer SS. Analysis of the neuroligin 3 and 4 genes in autism and other neuropsychiatric patients. *Mol Psychiatry* 2005; 10: 329–335.
2. Goldstein JM, Seidman LJ, Horton NJ, Makris N, Kennedy DN, Caviness VS, Faraone SV, Tsuang MT. Normal sexual dimorphism of the adult human brain assessed by in vivo magnetic resonance imaging. *Cereb Cortex* 2001; 11: 490–497.
3. Geschwind N, Behan P. Left-handedness: Association with immune disease, migraine, and developmental learning disorder. *Proc Natl Acad Sci USA* 1982; 79: 5097–5100.
4. Berenbaum SA, Denburg SD. Evaluating the empirical support for the role of testosterone in the Geschwind-Behan-Galaburda model of cerebral lateralization: Commentary on Bryden, McMANUS, and Bulman-Fleming. *Brain Cogn* 1995; 27: 79–83.
5. Mathews GA, Fane BA, Pasterski VL, Conway GS, Brook C, Hines M. Androgenic influences on neural asymmetry: Handedness and language lateralization in individuals with congenital adrenal hyperplasia. *Psychoneuroendocrinology* 2004; 29: 810–822.
6. Baron-Cohen S. The extreme male brain theory of autism. *Trends Cogn Sci* 2002; 8: 248–254.
7. Corbier P, Edwards DA, Roffi J. The neonatal testosterone surge: A comparative study. *Arch Int Physiol Biochim Biophys* 1992; 100: 127–131.
8. Gorski RA, Gordon JH, Shryne JE, Southam AM. Evidence for a morphological sex difference within the medial preoptic area of the rat brain. *Brain Res* 1978; 148: 333–346.
9. Arai Y, Sekine Y, Murakami S. Estrogen and apoptosis in the developing sexually dimorphic preoptic area in female rats. *Neurosci Res* 1996; 25: 403–407.
10. Krezel W, Dupont S, Krust A, Chambon P, Chapman PF. Increased anxiety and synaptic plasticity in estrogen receptor β-deficient mice. *Proc Natl Acad Sci USA* 2001; 102: 12278–12282.

11. Bodo C, Rissman EF. New roles for estrogen receptor beta in behavior and neuroendo-crinology. *Front Neuroendocrinol* 2006; 27: 217–232.

12. Wang L, Andersson S, Warner M, Gustafsson JA. Estrogen receptor (ER)β knockout mice reveal a role for ERβ in migration of cortical neurons in the developing brain. *Proc Natl Acad Sci USA* 2003; 100: 703–708.

13. Hadj-Sahraoui N, Frederic F, Delhaye-Bouchaud N, Mariani J. Gender effect on Purkinje cell loss in the cerebellum of the heterozygous reeler mouse. *J Neurogenet* 1996; 11: 45–58.

14. Doulazmi M, Frederic F, Lemaigre-Dubreuil Y, Hadj-Sahraoui N, Delhaye-Bouchaud N, Mariani J. Cerebellar Purkinje cell loss during life span of the heterozygous staggerer mouse (Rora(+)/Rora(sg)) is gender-related. *J Comp Neurol* 1999; 411: 267–273.

15. Kuemerle B, Gulden F, Cherosky N, Williams E, Herrup K. The mouse *Engrailed* genes: A window into autism. *Behav Brain Res* 2007; 176: 121–132.

16. Shu W, Cho JY, Jiang Y, Zhang M, Weisz D, Elder GA, Schmeidler J, De Gasperi R, Sosa MA, Rabidou D, Santucci AC, Perl D, Morrisey E, Buxbaum JD. Altered ultrasonic vocalization in mice with a disruption in the Foxp2 gene. *Proc Natl Acad Sci USA* 2005; 102: 9643–9648.

17. Kates WR, Burnette CP, Eliez S Strunge LA, Kaplan D, Landa R, Reiss AL, Pearlson GD. Neuroanatomic variation in monozygotic twin pairs discordant for the narrow phenotype for autism. *Am J Psychiatry* 2004; 161: 539–546.

18. Friedman WJ, McEwen BS, Toran-Allerand CD, Gerlach JL. Perinatal development of hypothalamic and cortical estrogen receptors in mouse brain: Methodological aspects. *Brain Res Dev Brain Res* 1983; 11: 19–27.

19. Gerlach JL, McEwen BS, Toran-Allerand CD, Friedman WJ. Perinatal development of estrogen receptors in mouse brain assessed by radioautography, nuclear isolation and receptor assay. *Brain Res Dev Brain Res* 1983; 11: 7–18.

20. Miranda R, Toran-Allerand CD. Developmental regulation of estrogen receptor mRNA in the rat cerebral cortex: A non-isotopic *in situ* hybridization study. *Cereb Cortex* 1992; 2: 1–15.

21. Shughrue P, Scrimo P, Lane M, Askew R, Merchenthaler I. The distribution of estrogen receptor-β mRNA in forebrain regions of the estrogen receptor-a knockout mouse. *Endocrinology* 1997a; 138: 5649–5652.

22. Shughrue PJ, Lane MV, Merchenthaler I. Comparative distribution of estrogen receptor-α and -β mRNA in the rat central nervous system. *J Comp Neurol* 1997b; 388: 507–525.

23. Kuiper GG, Carlsson B, Grandien K, Enmark E, Haggblad J, Nilsson S, Gustafsson JA. Comparison of the ligand binding specificity and transcript tissue distribution of estrogen receptors alpha and beta. *Endocrinology* 1997; 138: 863–870.

24. Paech K, Webb P, Kuiper GG, Nilsson S, Gustafsson J, Kushner PJ, Scanlan TS. Differential ligand activation of estrogen receptors ERalpha and ERbeta at AP1 sites. *Science* 1997; 277: 1508–1510.

25. Levin ER. Cellular functions of the plasma membrane estrogen receptor. *Trends Endocrinol Metab* 1999; 10: 374–377.

26. Toran-Allerand CD, Singh M, Setalo GJ. Novel mechanisms of estrogen action in the brain: New players in an old story. *Front Neuroendocrinol* 1999; 20: 97–121.

27. Belcher SM, Zsarnovszky A. Estrogenic actions in the brain: Estrogen, phytoestrogens, and rapid intracellular signaling mechanisms. *J Pharmacol Exp Ther* 2001; 299: 408–414.

28. Honda K, Sawada H, Kihara T, Urushitani M, Nakamizo T, Akaike A, Shimohama S. Phosphatidylinositol 3-kinase mediates neuroprotection by estrogen in cultured cortical neurons. *J Neurosci Res* 2000; 60: 321–327.

29. Singer CA, Figueroa-Masot XA, Batchelor RH, Dorsa DM. The mitogen activated protein kinase pathway mediates estrogen neuroprotection after glutamate toxicity in primary cortical neurons. *J Neurosci* 1999; 19: 2455–2463.

30. Singh M, Setalo G Jr, Guan XP, Warren M, Toran-Allerand CD. Estrogen induced activation of mitogen-activated protein kinase in cerebral cortical explants: Convergence of estrogen and neurotrophin signaling pathways. *J Neurosci* 1999; 19: 1179–1188.

31. Belcher SM, Le HH, Spurling L, Wong JK. Rapid estrogenic regulation of extracellular signal-regulated kinase 1/2 signaling in cerebellar granule cells involves a G protein- and protein kinase A-dependent mechanism and intracellular activation of protein phosphatase 2A. *Endocrinology* 2005; 146: 5397–5406.

32. Singh M, Meyer E, Simpkins J. The effect of ovariectomy and estradiol replacement on brain derived neurotrophic factor messenger ribonucleic acid expression in cortical and hippocampal brain regions of female Sprague-Dawley rats. *Endocrinology* 1995; 136: 2320–2324.

33. Sohrabji F, Miranda R, Toran-Allerand D. Identification of a putative estrogen response element in the gene encoding brain-derived neurotrophic factor. *Proc Natl Acad Sci USA* 1995; 92: 11110–11114.

34. McMillan P, Singer C, Dorsa D. The effects of ovariectomy and estrogen replacement on trkA and cholineacetyltransferase mRNA expression in the basal forebrain of the adult female sprague-dawley rat. *J Neurosci* 1996; 16: 1860–1865.

35. Miranda R, Sohrabji F, Singh M, Toran-Allerand CD. Nerve growth factor (NGF) regulation of estrogen receptors in explant cultures of the developing forebrain. *J Neurobiol* 1996; 31: 77–87.

36. Chao MV, Hempstead B. p75 and Trk: A two-receptor system. *Trends Neurosci* 1995; 18: 321–326.

37. Rabizadeh S, Oh J, Zhong L, Yang J, Bitler C, Butcher L, Bredesen D. Induction of apoptosis by the low-affinity NGF receptor. *Science* 1993; 261: 345–348.

38. Majdan M, Lachance C, Gloster A, Aloyz R, Zeindler C, Bamji SX, Bhakar A, Belliveau D, Fawcett J, Miller FD, Baker PA. Transgenic mice expressing the intracellular domain of p75 neurotrophin receptor undergo neuronal apoptosis. *J Neurosci* 1997; 17: 6988–6998.

39. Sohrabji F, Miranda R, Toran-Allerand CD. Estrogen differentially regulates estrogen and nerve growth factor receptor mRNAs in adult sensory neurons. *J Neurosci* 1994b; 14: 459–471.

40. Toran-Allerand CD. Mechanisms of estrogen action during neural development: Mediation by interactions with the neurotrophins and their receptors? *J Steroid Biochem Mol Biol* 1996; 56: 169–178.

41. Mendez P, Wandosell F, Garcia-Segura LM. Cross-talk between estrogen receptors and insulin-like growth factor-I receptor in the brain: Cellular and molecular mechanisms. *Front Neuroendocrinol* 2006; 27: 391–403.

42. Aguado F, Sanchez-Franco F, Cacidedo L, Fernandez T, Rodrigo J, Martinez-Murillo R. Subcellular localization of insulin-like growth factor I (IGF-I) in purkinje cells of the adult rat: An immunocytochemical study. *Neurosci Lett* 1992; 135: 171–174.

43. Bondy C, Werner H, Roberts CT Jr, LeRoith D. Cellular pattern of type-I insulin-like growth factor receptor gene expression during maturation of the rat brain: Comparison with insulin-like growth factors I and II. *Neuroscience* 1992; 46: 909–923.

44. Fukudome Y, Tabata T, Miyoshi T, Haruki S, Araishi K, Sawada S, Kano M. Insulin-like growth factor-I as a promoting factor for cerebellar purkinje cell development. *Eur J Neurosci* 2003; 17: 2006–2016.

45. Nordeen EJ, Nordeen KW. Estrogen stimulates the incorporation of new neurons into the avian song nuclei during adolescence. *Dev Brain Res* 1989; 49: 27–32.

46. Ma ZQ, Spreafico E, Pollio G, Santagati S, Conti E, Cattaneo E, Maggi A Activated estrogen receptor mediates growth arrest and differentiation of a neuroblastoma cell line. *Proc Natl Acad Sci USA* 1993; 90: 3740–3744.

47. Rasmussen J, Torres-Aleman I, McLusky N, Naftolin F, Robbins R. The effects of estradiol on the growth patterns of estrogen receptor-positive hypothalamic cell lines. *Endocrinology* 1990; 126: 235–240.

48. Henderson RG, Brown AE, Tobet SA. Sex differences in cell migration in the preoptic area/anterior hypothalamus of mice. *J Neurobiol* 1999; 41: 252–266.

49. Orikasa C, Kondo Y, Hayashi S, McEwen B, Sakuma Y. Sexually dimorphic expression of estrogen receptor β in the anteroventral periventricular nucleus of the rat preoptic area. Implication in luteinizing hormone surge. *Proc Natl Acad Sci USA* 2002; 99: 3306–3311.

50. Nordeen EJ, Nordeen KW, Sengelaub DR, Arnold AP. Androgens prevent normally occurring cell death in a sexually dimorphic spinal nucleus. *Science* 1985; 229: 671–673.

51. Murakami S, Arai Y. Neuronal death in the developing sexually dimorphic periventricular nucleus of the preoptic area in the female rat effect of neonatal androgen treatment. *Neurosci Lett* 1989; 102: 185–190.

52. Davis EC, Popper P, Gorski RA. The role of apoptosis in sexual differentiation of the rat sexually dimorphic nucleus of the preoptic area. *Brain Res* 1996; 734: 10–18.

53. Chung WC, Swaab DF, De Vries GJ. Apoptosis during sexual differentiation of the bed nucleus of the stria terminalis in the rat brain. *J Neurobiol* 2000; 43: 234–243.

54. Breedlove SM, Arnold AP. Sexually dimorphic motor nucleus in the rat lumbar spinal cord response to adult hormone manipulation, absence in androgen-insensitive rats. *Brain Res* 1981; 225: 297–307.

55. Simerly RB, Zee MC, Pendleton JW, Lubahn DB, Korach KS. Estrogen receptor-dependent sexual differentiation of dopaminergic neurons in the preoptic region of the mouse. *Proc Natl Acad Sci USA* 1997; 94: 14077–14082.

56. DonCarlos LL, Handa RJ. Developmental profile of estrogen receptor mRNA in the preoptic area of male and female neonatal rats. *Dev Brain Res* 1994; 79: 283–289.

57. Jordan CL, Breedlove SM, Arnold AP. Ontogeny of steroid accumulation in spinal lumbar motoneurons of the rat implications for androgen's site of action during synapse elimination. *J Comp Neurol* 1991; 313: 441–448.

58. Freeman LM, Watson NV, Breedlove SM. Androgen spares androgen-insensitive motoneurons from apoptosis in the spinal nucleus of the bulbocavernosus in rats. *Horm Behav* 1996; 30: 424–433.

59. Forger NG, Howell ML, Bengston L, MacKenzie L, DeChiara TM, Yancopoulos GD. Sexual dimorphism in the spinal cord is absent in mice lacking the ciliary neurotrophic factor receptor. *J Neurosci* 1997; 17: 9605–9612.

60. Varela CR, Bengston L, Xu J, MacLennan AJ, Forger NG. Additive effects of ciliary neurotrophic factor and testosterone on motoneuron survival; differential effects on motoneuron size and muscle morphology. *Exp Neurol* 2000; 165: 384–393.

61. Merry DE, Korsmeyer SJ. Bcl-2 gene family in the nervous system. *Annu Rev Neurosci* 1997; 20: 245–267.

62. Zup SL, Carrier H, Waters E, Tabor A, Bengston L, Rosen G, Simerly R, Forger NG. Overexpression of Bcl-2 reduces sex differences in neuron number in the brain and spinal cord. *J Neurosci* 2003; 23: 2357–2362.

63. Forger NG, Rosen GJ, Waters EM, Jacob D, Simerly RB, de Vries GJ. Deletion of *Bax* eliminates sex differences in the mouse forebrain. *Proc Natl Acad Sci USA* 2004; 101: 13666–13671.

64. Jacob DA, Bengston CL, Forger NG. Effects of *Bax* gene deletion on muscle and motoneuron degeneration in a sexually dimorphic neuromuscular system. *J Neurosci* 2005; 25: 5638–5644.

65. Garcia-Segura LM, Cardona-Gomez P, Naftolin F, Chowen JA. Estradiol upregulates Bcl-2 expression in adult brain neurons. *Neuroreport* 1998; 9: 593–597.

66. Dubal DB, Shughrue PJ, Wilson ME, Merchenthaler I, Wise PM. Estradiol modulates bcl-2 in cerebral ischemia a potential role for estrogen receptors. *J Neurosci* 1999; 19: 6385–6393.

67. Bauman ML, Kemper TL. Histoanatomic observations of the brain in early infantile autism. *Neurology* 1985; 35: 866–874.

68. Ritvo ER, Freeman BJ, Scheibel AB, Duong T, Robinson H, Guthrie D, Ritvo A. Lower Purkinje cell counts in the cerebella of four autistic subjects: Initial findings of the UCLA-NSAC autopsy research report. *Am J Psychiatry* 1986; 146: 862–866.

69. Kemper TL, Bauman ML. Neuropathology of infantile autism. *J Neuropathol Exp Neurol* 1998; 57: 645–652.
70. Bailey A, Luthert P, Dean A, Harding B, Janota I, Montgomery M, Rutter M, Lantos P. A clinicopathological study of autism. *Brain* 1998; 121: 889–905.
71. Lee M, Martin-Ruiz C, Graham A, Court J, Jaros E, Perry R, Iverson P, Bauman M, Perry E. Nicotinic receptor abnormalities in the cerebellar cortex in autism. *Brain* 2002; 15: 1483–1495.
72. Courchesne E, Yeung-Courchesne R, Press GA, Hesselink JR, Jernigan TL. Hypoplasia of the cerebellar vermal lobules VI and VII in autism. *N Engl J Med* 1988; 318: 1349–1354.
73. Sacchetti B, Scelfo B, Tempia F, Strata P. Long-term synaptic changes induced in the cerebellar cortex by fear conditioning. *Neuron* 2004; 42: 973–982.
74. Bauman ML, Kemper TL. Neuroanatomic observations of the brain in autism: A review and future directions *Int J Dev Neurosci* 2005; 23: 183–187.
75. Altman J. Postnatal development of the cerebellar cortex in the rat. II. Phases in the maturation of Purkinje cells and of the molecular layer. *J Comp Neurol* 1972; 145: 399–464.
76. Dusart I, Airaksinen MS, Sotelo C. Purkinje cell survival and axonal regeneration are age dependent: An in vitro study. *J Neurosci* 1997; 17: 3710–3726.
77. Clancy B, Darlington RB, Finlay BL. Translating developmental time across mammalian species. *Neuroscience* 2001; 105: 7–17.
78. Kuemerle B, Zanjani H, Joyner A, Herrup K. Pattern deformities and cell loss in Engrailed-2 mutant mice suggest two separate patterning events during cerebellar development. *J Neurosci* 1997; 17: 7881–7889.
79. Tsutsui K, Ukena K, Usui M, Sakamoto H, Takase M. Novel brain function: Biosynthesis and actions of neurosteroids in neurons. *Neurosci Res* 2000; 36: 261–273.
80. Compagnone NA, Mellon SH. Neurosteroids: Biosynthesis and function of these novel neuromodulators. *Front Neuroendocrinol* 2000; 21: 1–56.
81. Chung BC, Matteson KJ, Voutilainen R, Mohandas TK, Miller WL. Human cholesterol side-chain cleavage enzyme, P450SCC: cDNA cloning, assignment of the gene to chromosome 15, and expression in the placenta. *Proc Natl Acad Sci USA* 1986; 83: 8962–8966.
82. Simpson ER, MacDonald PC. Endocrine physiology of the placenta. *Annu Rev Physiol* 1981; 43: 163–188.
83. Keeney DS, Ikeda Y, Waterman MR, Parker KL. Cholesterol side-chain cleavage cytochrome P450 gene expression in the primitive gut of the mouse embryo does not require steroidogenic factor 1. *Mol Endocrinol* 1995; 9: 1091–1098.
84. Mellon S, Deschepper CF. Neurosteroid biosynthesis: Genes for adrenal steroidogenic enzymes are expressed in the brain. *Brain Res* 1993; 629: 283–292.
85. Miller W, Tyrell J. (1994) The adrenal cortex, in: Felig P, Baxter J, Frohman L. (Eds.), *Endocrinology & Metabolism*, McGraw Hill, New York, pp. 555–711.
86. Corbin CJ, Graham-Lorence S, McPhaul M, Mendelson CR, Simpson ER. Isolation of a full-length cDNA insert encoding human aromatase system cytochrome P-450 and its expression in nonsteroidogenic cells. *Proc Natl Acad Sci USA* 1988; 85: 8948–8952.
87. Harada N. Cloning of a complete cDNA encoding human aromatase: Immunochemical identification and sequence analysis. *Biochem Biophys Res Commun* 1988; 156: 725–732.
88. Mellon SH. Neurosteroids: Biochemistry, modes of action, and clinical relevance. *J Clin Endocrinol Metab* 1994; 78: 1003–1008.
89. Ukena K, Usui M, Kohchi C, Tsutsui K. Cytochrome p450 side-chain cleavage enzyme in the cerebellar Purkinje neuron and its neonatal change in rats. *Endocrinology* 1998; 139: 137–147.
90. Ukena K, Kohchi C, Tsutsui K. Expression and activity of 3β-hydroxysteroid dehydrogenase/Δ^5–Δ^4 isomerase in the rat Purkinje neuron during neonatal life. *Endocrinology* 1999; 140: 805–813.

91. Lavaque E, Mayen A, Azcoitia I, Tena-Sempere M, Garcia-Segura LM. Sex differences, developmental changes, response to injury and cAMP regulation of the mRNA levels of steroidogenic acute regulatory protein, cytochrome p450scc, and aromatase in the olivocerebellar system. *J Neurobiol* 2006; 66: 308–318.

92. Agís-Balboa RC, Pinna G, Zhubi A, Maloku E, Veldic M, Costa E, Guidotti A. Characterization of brain neurons that express enzymes mediating neurosteroid biosynthesis. *Proc Natl Acad Sci USA* 2006; 103: 14602–14607.

93. Sakamoto H, Mezaki Y, Shikimi H, Ukena K, Tsutsui K. Dendritic growth and spine formation in response to estrogen in the developing Purkinje cell. *Endocrinology* 2003; 144: 4466–4477.

94. Smith SS, Waterhouse BD, Woodward DJ. Locally applied estrogens potentiate glutamate-evoked excitation of cerebellar Purkinje cells. *Brain Res* 1988; 475: 272–282.

95. Smith SS. Estrogen administration increases neuronal responses to excitatory amino acids as a long-term effect. *Brain Res* 1989; 503: 354–357.

96. Price RH, Handa RJ. Expression of estrogen receptor-beta protein and mRNA in the cerebellum of the rat. *Neurosci Lett* 2000; 288: 115–118.

97. Webb P, Lopez GN, Uht RM, Kushner PJ. Tamoxifen activation of the estrogen receptor/AP-1 pathway: Potential origin for the cell-specific estrogen-like effects of antiestrogens. *Mol Endocrinol* 1995; 9: 443–456.

98. Woolley CS, McEwen BS. Estradiol regulates hippocampal dendritic spine density via an *N* -methyl-d-aspartate receptor-dependent mechanism. *J Neurosci* 1994; 14: 7680–7687.

99. Schwartz PM, Borghesani PR, Levy RL, Pomeroy SL, Segal RA. Abnormal cerebellar development and foliation in BDNF –/– mice reveals a role for neurotrophins in cns patterning. *Neuron* 1997; 19: 269–281.

100. Bates B, Rios M, Trumpp A, Chen C, Fan G, Bishop JM, Jaenisch R. Neurotrophin-3 is required for proper cerebellar development. *Nat Neurosci* 1999; 2: 115–117.

101. D'Arcangelo G, Miao GG, Chen SC, Soares HD, Morgan JI, Curran T. A protein related to extracellular matrix proteins deleted in the mouse mutant reeler. *Nature* 1995; 374, 719–723.

102. Persico AM, D'Agruma L, Maiorano N, Totaro A, Militerni R, Bravaccio C, Wassink TH, Schneider C, Melmed R, Trillo S, Montecchi F, Palermo M, Pascucci T, Puglisi-Allegra S, Reichelt KL, Conciatori M, Marino R, Quattrocchi CC, Baldi A, Zelante L, Gasparini P, Keller F. Collaborative linkage study of autism. *Mol Psychiatry* 2001; 6: 150–159.

103. Lugli G, Krueger JM, Davis JM, Persico AM, Keller F, Smalheiser NR. Methodological factors influencing measurement and processing of plasma reelin in humans. *BMC Biochem* 2003; 4: 9.

104. Skaar DA, Shao Y, Haines JL, Stenger JE, Jaworski J, Martin ER, DeLong GR, Moore JH, McCauley JL, Sutcliffe JS, Ashley-Koch AE, Cuccaro ML, Folstein SE, Gilbert JR, Pericak-Vance MA. Analysis of the RELN gene as a genetic risk factor for autism. *Mol Psychiatry* 2005; 10: 563–571.

105. Serajee FJ, Zhong H, Mahbubul Huq AH. Association of Reelin gene polymorphisms with autism. *Genomics* 2006; 87: 75–83.

106. D'Arcangelo G, Homayouni R, Keshvara L, Rice DS, Sheldon M, Curran T. Reelin is a ligand for lipoprotein receptors. *Neuron* 1999; 24: 471–479.

107. Hiesberger T, Trommsdorff M, Howell BW, Goffinet A, Mumby MC, Cooper JA, Herz J. Direct binding of reelin to VLDL receptor and ApoE receptor 2 induces tyrosine phosphorylation of disabled -1 and modulates Tau phosphorylation. *Neuron* 1999; 24: 481–489.

108. Kim DH, Iijima H, Goto K, Sakai J., Ishii H, Kim HJ, Suzuki H, Kondo H, Saeki S, Yamamoto T. Human apolipoprotein E receptor 2. A novel lipoprotein receptor of the low density lipoprotein receptor family predominantly expressed in brain. *J Biol Chem* 1996; 271: 8373–8380.

109. Perez-Garcia CG, Tissir F, Goffinet AM, Meyer G. Reelin receptors in developing laminated brain structures of mouse and human. *Eur J Neurosci* 2004; 20: 2827–2832.
110. Jakab RL, Wong JK, Belcher SM. Estrogen receptor beta immunoreactivity in differentiating cells of the developing rat cerebellum. *J Comp Neurol* 2001; 430: 396–409.
111. Mitra SW, Hoskin E, Yudkovitz J, Pear L, Wilkinson HA, Hayashi S, Pfaff DW, Ogawa S, Rohrer SP, Schaeffer JM, McEwen BS, Alves S.E. Immunolocalization of estrogen receptor beta in the mouse brain: Comparison with estrogen receptor alpha. *Endocrinology* 2003; 144: 2055–2067.
112. Simerly RB, Chang C, Muramatsu M, Swanson LW. Distribution of androgen and estrogen receptor mRNA-containing cells in the rat brain: An in situ hybridization study. *J Comp Neurol* 1990; 294: 76–95.
113. Biamonte F, Assenza G, Marino R, Caruso D, Crotti S, Melcangi RC, Cesa R, Strata P, Keller F. Interaction between estrogens and reelin in Purkinje cell development. Program 322.18/C5 Abstract Viewer/Itinerary Planner. Atlanta, Georgia: Society for Neuroscience, Online 2006.
114. Hatten ME, Heintz N. Mechanisms of neural patterning and specification in the developing cerebellum. *Annu Rev Neurosci* 1995; 18: 385–408.
115. Fritzsch B. Observations on degenerative changes of Purkinje cells during early development in mice and in normal and otocyst-deprived chickens. *Anat Embryol* 1979; 158: 95–102.
116. Sawada K, Sakata-Haga H, Jeong YG, Azad MA, Ohkita S, Fukui Y. Purkinje cell loss in the cerebellum of ataxic mutant mouse, dilute-lethal: A fractionator study. *Congenit Anom* (Kyoto) 2004; 44: 189–195.
117. Absil P, Pinxten R, Balthazart J, Eens M. Effects of testosterone on Reelin expression in the brain of male European starlings. *Cell Tissue Res* 2003; 312: 81–93.
118. Mariani J, Crepel F, Mikoshiba K, Changeux JP, Sotelo C. Anatomical, physiological and biochemical studies of the cerebellum from Reeler mutant mouse. *Philos Trans R Soc Lond B Biol Sci* 1977; 281: 1–28.
119. Sotelo C. Cellular and genetic regulation of the development of the cerebellar system. *Prog Neurobiol* 2004; 72: 295–339.
120. Wallace VA. Purkinje-cell-derived sonic hedgehog regulates granule neuron precursor cell proliferation in the developing mouse cerebellum. *Curr Biol* 1999; 9: 445–448.
121. Hammerschmidt M, Brook A, McMahon AP. The world according to hedgehog. *Trends Genet* 1997; 13: 14–21.
122. Wechsler-Reya RJ, Scott MP. Control of neuronal precursor proliferation in the cerebellum by Sonic Hedgehog. Neuron 1999; 22: 103–114.
123. Fan H, Favero M, Vogel MW. Elimination of Bax expression in mice increases cerebellar Purkinje cell numbers but not the number of granule cells. *J Comp Neurol* 2001; 436: 82–91.
124. Zanjani HS, Vogel MW, Delhaye-Bouchaud N, Martinou JC, Mariani J. Increased cerebellar Purkinje cell numbers in mice overexpressing a human *bcl* -2 transgene. *J Comp Neurol* 1996; 374: 332–341.
125. Heckroth JA, Goldowitz D, Eisenman LM. Purkinje cell reduction in the reeler mutant mouse: A quantitative immunohistochemical study. *J Comp Neurol* 1989; 279: 546–555.
126. Herrup K, Mullen RJ. Role of the Staggerer gene in determining Purkinje cell number in the cerebellar cortex of mouse chimeras. *Brain Res* 1981; 227: 475–485.
127. Baader SL, Sanlioglu S, Berrebi AS, Parker-Thornburg J, Oberdick J. Ectopic overexpression of engrailed-2 in cerebellar Purkinje cells causes restricted cell loss and retarded external germinal layer development at lobule junctions. *J Neurosci* 1998; 18: 1763–1773.
128. Caddy KW, Biscoe TJ. Structural and quantitative studies on the normal C3H and Lurcher mutant mouse. *Philos Trans R Soc Lond B Biol Sci* 1979; 287: 167–201.

129. Frederic F, Hainaut F, Thomasset M, Guenet JL, Delhaye-Bouchaud N, Mariani J. Cell counts of Purkinje and inferior olivary neurons in the 'Hyperspiny Purkinje Cells' mutant mouse. *Eur J Neurosci* 1992; 4: 127–135.
130. Mullen RJ, Eicher EM, Sidman RL. Purkinje cell degeneration, a new neurological mutation in the mouse. *Proc Natl Acad Sci USA* 1976; 73: 208–212.
131. Doulazmi M, Hadj-Sahraoui N, Frederic F, Mariani J. Diminishing Purkinje cell populations in the cerebella of aging heterozygous Purkinje cell degeneration but not heterozygous nervous mice. *J Neurogenet* 2002; 16: 111–123.
132. Sidman RL, Green MC. "Nervous," a new mutant mouse with cerebellar disease. *Coll Int Centre Natl Recherche Sci* 1970; 924: 69–79.
133. Harada T, Pineda LL, Nakano A, Omura K, Zhou L, Iijima M, Yamasaki Y, Yokoyama M. Ataxia and male sterility (AMS) mouse. A new genetic variant exhibiting degeneration and loss of cerebellar Purkinje cells and spermatic cells. *Pathol int Jun* 2003; 53: 382–389.
134. Blatt GJ, Eisenman LM. A qualitative and quantitative light microscopic study of the inferior olivary complex of normal, reeler, and weaver mutant mice. *J Comp Neurol* 1985; 232: 117–128.
135. Zanjani H, Herrup K, Mariani J. Cell number in the inferior olive of nervous and leaner mutant mice. *J Neurogenet* 2004; 18: 327–339.
136. Duchala CS, Shick HE, Garcia J, Deweese DM, Sun X, Stewart VJ, Macklin WB. The toppler mouse: A novel mutant exhibiting loss of Purkinje cells. *J Comp Neurol* 2004; 476: 113–129.
137. Serajee FJ, Nabi R, Zhong H, Mahbubul Huq AH. Association of INPP1, PIK3CG, and TSC2 gene variants with autistic disorder: Implications for phosphatidylinositol signalling in autism. *J Med Genet* 2003; 40: e119.
138. Riikonen R, Makkonen I, Vanhala R, Turpeinen U, Kuikka J, Kokki H. Cerebrospinal fluid insulin-like growth factors IGF-1 and IGF-2 in infantile autism. *Dev Med Child Neurol* 2006; 48: 751–755.

Chapter 10
Insulin-Like Growth Factors

Neurobiological Regulators of Brain Growth in Autism?

Raili Riikonen

Abstract In autism, disruption of normal neurobiological mechanisms is found, but it is not known which specific developmentally important molecules might be involved in this disorder. Increased cerebral volume or brain weight is found across studies in autism. Pathological brain growth and premature developmental arrest are suggested to be restricted to the first years of life. We found a correlation between insulin-like growth factor-1 (IGF-1) concentrations and head growth in children with autism but not in the controls. At an early stage of infantile autism, the cerebrospinal fluid (CSF) concentration of IGF-1 was lower than in the comparison group, but not in older children. This suggests a disruption of normal neurobiological mechanisms at an early age. Furthermore, we also found normal CSF nerve growth factor (NGF) in autism, but low NGF and normal IGF-1 in Rett syndrome (RS). There is evidence that IGF-1 is important for cerebellar development, and low CSF IGF-1 concentrations may lead to cerebellar abnormalities. In autism, almost all neuropathological studies have reported decreased numbers of Purkinje cells in the cerebellum. NGF is important for cholinergic neurons of the forebrain, and the cholinergic system is affected in RS. Therefore, autism and RS could be distinguished by their different levels of the two growth factors. This is in agreement with the different morphological and neurochemical findings in the two syndromes. In autism, there seems to be a disruption of normal neurobiological mechanisms due to "premature growth without guidance." The data suggest that the IGF system may play an important role in the pathophysiology of autism. Autism is a behaviorally defined condition in which social interaction and reciprocal communication are disturbed. Aberrant behavioral expression in children with autism is usually noticed by the parents, typically between 12 and 24 months of age. The diagnosis is based on clinical criteria. The etiopathogenesis is unknown. Disruption of normal neurobiological mechanisms has been found, but it is not known what specific developmentally important molecules might be involved in this disorder. Nelson et al. 2001 [1] have suggested that in autism there is a relationship between abnormal concentrations of growth factors and

R. Riikonen
Children's Hospital, University of Kuopio, P.O. Box 1627, FIN 70211 Kuopio, Finland
e-mail: raili.riikonen@kolumbus.fi

A.W. Zimmerman (ed.), *Autism*, DOI: 10.1007/978-1-60327-489-0_10,
© Humana Press, Totowa, NJ 2008

abnormal patterns of brain growth. Knowledge of the important roles of neurotrophic factors in relation to the development of the brain is growing.

Keywords Insulin-like growth factors · brain growth · autism

What are Neurotrophic Factors?

Neurotrophic factors are important for neuronal growth, differentiation and
survival of neurons, and synaptic formation during the development of the
infant brain [2]. The most important neurotrophic factors are nerve growth
factor (NGF), brain derived neurotrophic factor (BDNF), glial cell line derived
neurotrophic factor (GDNF), neurotrophin-3/4 (NT-3/4), and insulin-like
growth factor-1 (IGF-1).

The patterns of expression of neurotrophic factors and their action in the
CNS are quite distinct. Neurotrophic factors have specific receptors and are
located on specific chromosomes.

The action of neurotrophic factors is most important in the developing brain.
Early brain development is very active: the number of brain cells doubles from the
30th gestational week to the age of 6 months and trebles by the age of 1 year. During
early development, an excess of motor and sensory neurons are produced. Only
those neurons that have made tight contacts with their target tissues receive sufficient amounts of neurotrophic factors necessary to continue to develop further. The
other neurons are eliminated by programmed cell death (apoptosis). Withdrawal of
neurotrophic factors will lead to apoptosis at early ages, but not later. It is important
to note that different neurotrophins are expressed at distinct developmental stages
at the neuromuscular junction, brain, and CSF [3, 4, 5, 6]. Neurotrophins are
involved in the selection of optimal neuronal connections. Roles for neurotrophic
factors in various neurological disorders have been suggested [7, 8].

Insulin-Like Growth Factors

IGF-1 and IGF-2 are members of the insulin-gene family which stimulate
cellular proliferation and differentiation during embryonic and postnatal development. IGF-1 and IGF-2 are bound to six different binding proteins. These
binding proteins are thought to have roles in the modulation of IGF action and
targeting of IGFs to certain tissues [2, 9]. Both IGF-1 and IGF-2 exert their
mitogenic effects via the membrane-bound tyrosine kinase IGF-1 receptor [10].
IGF-1 and IGF-2 are widely expressed in the central nervous system. However,
in contrast to IGF-1, IGF-2 is most highly expressed in non-neuronal tissues of
the nervous system: in the choroid plexus, leptomeninges, microvasculature,
myelin sheaths, and CSF [11]. The highest expression of IGF-2 is during fetal
and early postnatal development of the human brain [12]. In contrast, IGF-1

mRNA is widely expressed in foetal brain tissue, with more discrete expression postnatally in sites showing ongoing growth and differentiation.

We wanted to see whether (1) growth factors IGF-1 and IGF-2 in CSF correlate to age and head growth in patients with autism or whether (2) CSF NGF or IGF-1 could differentiate two syndromes: infantile autism and Rett syndrome (RS). At the onset, these two conditions overlap widely with respect to emotional, perceptual, and motor control.

Insulin-Like Growth Factors and Head Growth

There is growing evidence that the IGF system is important in relation to cell survival and growth in the developing brain [9]. It has been suggested that IGF-1 is important in the process of neuronal differentiation, maturation of dendrites, and synaptogenesis. It is thought that IGF-1 is important in the developing brain, not only as a trophic factor but also as a mediator of the action of other neuro-trophic factors [13]. IGF-1 has a specific role on axonal growth and myelination and pro-oligodendrocyte survival by inhibiting apoptosis and increasing levels of myelin basic protein mRNA and proteolipid protein [14].

Brain growth is extremely sensitive to IGF-1 levels. In animals, IGF-1 significantly affects the brain size in the developing nervous system [13]. Mice with null mutations of IGF-1 or IGF-1 receptor have small brains [15]. Conversely, IGF-1 overexpression increases brain growth, with increased widths of myelinated axons and increased width of myelin sheaths. Schoenle et al. 1986 [16] found elevated concentrations of IGF-2 in an infant with macrocephaly in CSF and also in the brain tissue. The concentration of IGF-2 in the frontal cortex was more than five times higher than in the control specimen.

Insulin-Like Growth Factors and Specific Brain Cells

In the CNS, certain distinct cell types are sensitive to IGF-1 deficiency, including cerebellar Purkinje cells and granulosa cells, oligodendrocytes, and motoneurons, whereas other cells are independent of the presence of IGF-1 [2]. Both neurons and glial cells have been observed to synthetise IGF-1 mRNA [11, 17] and cause immature glial precursors to develop into oligodendrocytes [14]. By acting as a mitogen, IGF-1 increases both the number of oligodendrocytes and the amount of axonal myelin they produce.

Oligodendrocytes myelinate axons and constitute the vast majority of cells in the white matter. IGF-1 may be related to degeneration of myelin or cerebellar granule neurons. There is also evidence that IGF-1 is important for cerebellar development [18, 19, 20]. In autism, almost all neuropathological studies have reported decreased numbers of Purkinje cells in the cerebellum. It has also been suggested that the IGF system might be involved in progressive cerebellar diseases [18, 21, 22].

Autism and Head Growth

Increased cerebral volume or brain weight is also found across studies in autism although not consistently [23, 24, 25, 26, 27, 28]. Head size was increased in about a quarter of individuals in the study by Davincovitch et al. [27].

Age

The age when macrocephaly develops in autism has until now been unknown. However, recent data indicate that head circumference is normal or even somewhat small at birth, dramatically increases within the first year of life, but then plateaues so that by adulthood the majority of cases are within the normal range [28]. The pathological growth and arrest has already passed by the typical age of clinical diagnosis (i.e., 3–4 years).

In autism, the postnatal increase of head growth has been attributed to overgrowth of white matter [26]. Herbert et al. [26] suggest that the enlargement of the brain develops postnatally and is caused by a temporally modulated process. The distribution of volume changes suggests a process that alters some non-axonal component of white matter, possibly myelin itself. Recently, localization of this white matter enlargement was studied volumetrically and found to be localized to the radiate white matter in all lobes [26]. In the study of Courchesne et al. [24], hyperplasia was present not only in cerebral and cerebellar white matter but also in the grey matter in young children with autism. These patterns of enlargement are consistent with postnatal increases reported in autism. Although the causes of aberrant head growth in autism are unknown, there seems to be a disruption of normal neurobiological mechanisms or "premature growth without guidance" [24]. Early brain growth abnormalities are deleterious because they occur during a critical time in human development.

Neuropathology

In autism, many studies show hippocampal or cerebellar developmental abnormalities. Evidence for dysgenetic brain development is thought to begin before 30 weeks of gestation, perhaps through a signaling abnormality of some neurotrophic factor or aberrant programmed cell death [29]. Decreased numbers of Purkinje cells in the cerebellum have been reported in almost all postmortem histologic examinations [30, 31] and are considered to be of prenatal origin. Postmortem studies also support the hypothesis that there is a combination of both abnormal excess and reduction of cellularity at the microstructural level [32, 33, 34].

Studies on Autism and Neural Growth Factors

Nelson et al. 2001 [1] found overexpression of certain neuropeptides and neuro-trophins in peripheral blood drawn in the first days of life in children with autism or mental retardation but not in cerebral palsy or in comparison without autism.

The Finnish Study

CSF IGF-1 and IGF-2 in Autism

We measured IGF-1 and IGF-2 from cerebrospinal fluid (CSF) by radio-immunoassay in 25 children with autism (median age 5 years 5 months; range 1 year 11 months to 15 years 10 months; and 16 age-matched controls [35]). The diagnosis of autism was based on the Diagnostic and Statistical Manual of Mental Disorders, 3rd edition (DSM-III, American Psychiatric Association 1980), criteria. Patients were admitted to the Hospital for Children and Adolescents, Helsinki, and Kuopio University Hospital, Kuopio, Finland. All the patients underwent extensive studies that included physical examination (with special attention to signs of neurocutaneous disorders), neuroimaging, chromosomal analysis, electroencephalography (EEG) for subclinical seizures, and tests of blood and urine for metabolic disorders. The autism was idiopathic. The mean intelligence quotient (IQ) was 84 (SD 19, range 31–109) measured by the Leiter test (median IQ 93). We have been able to use "healthy" controls. The children were admitted for surgery on the lower part of the body to be performed under spinal anesthesia. None of the comparison group had any signs of malnutrition, hepatic failure, chronic inflammation, diabetes, or hypothyroidism, conditions that might have influenced IGF concentrations.

Serum and CSF samples were frozen and stored at $-70°C$ until analysis. The concentrations of CSF IGF-1 and IGF-2 were determined by means of radio-immunoassay (Mediagnost GmbH, Reutlingen, Germany). Lumbar puncture was performed in the patients with autism when the patients were lightly sedated. The Ethical Committees of the Hospital for Children and Adolescents, Helsinki, and the Kuopio University Hospital, Kuopio, approved this study. Written, informed consent for participation and publication was obtained from the parents of the patients and of the comparison group.

In the Finnish study CSF IGF-1 concentrations of children with autism under 5 years of age were lower than their age-matched comparisons ($p = 0.014$) (Fig. 10.1). In children of 5 years and older, there was no difference. There was no difference between the two groups in the CSF-IGF-2 concentrations. Head circumference is a useful index of brain size and growth, particularly in children. Weight, height, and head circumference were measured at regular intervals from birth until the point of study. We did not find any macrocephaly at the point of the study. The pathological brain growth is largely restricted to the first years of life before the typical age of clinical identification [28] as also seen in our series. However, we found a correlation

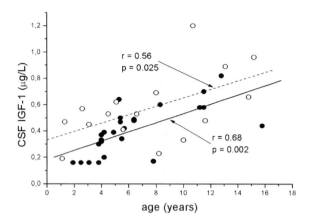

Fig. 10.1 Concentrations of cerebrospinal fluid insulin-like growth factor-1 (CSF IGF-1) and age in children with autism (•; $n = 25$) and their age-matched comparisons (○; $n = 16$). There was a significant correlation between CSF IGF-1 concentrations and age both in children with autism ($p = 0.002$) and in their age-matched comparisons ($r = 0.68$, $p = 0.025$), but IGF-1 concentration was significantly lower in the group with autism than in the comparison group ($r = 0.58$, $p = 0.02$). Published with permission from Riikonen et al. *Dev Med Child Neurol* 2006 [35]

between head growth (head circumference) and IGF-1 concentrations in children with autism (at the mean age of 5.5 years) but not in the comparison group (Fig. 10.2).

   Low concentrations of IGF-1 during the critical period of rapid brain growth in autism might be insufficient for survival of normal Purkinje cells in the cerebellum. Planimetric magnetic resonance imaging in autism has suggested

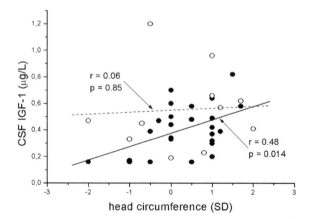

Fig. 10.2 Concentrations of cerebrospinal fluid insulin-like growth factor-1 (CSF IGF-1) with head circumference (SD) in children with autism (•; $n = 14$) and their age-matched comparisons (○; $n = 16$). There was a significant correlation between IGF-1 and head circumference in children with autism ($r = 0.48$, $p = 0.0014$) but not in the comparison group ($r = 0.06$, $p = 0.85$). Published with permission from Riikonen et al. *Dev Med Child Neurol* 2006 [35]

developmental abnormalities in the cerebellum, particularly in vermal regions [24]. Furthermore, the cerebellum is involved in shifting attention [36, 37], which in turn underlines its importance in social and cognitive development.

Serotonin and IGF

There are disturbances of the serotonergic system in autism. Low serotonin synthesis capacity has been shown on PET scans in young patients with autism [38]. We have also shown disturbed serotonin transporter (SERT) binding capacity in the medial frontal cortex and midbrain in autism [39]. The connection between growth factors and serotonin has recently been examined. Some antidepressant drugs, already in clinical use, increase the synthesis of neurotrophic factors [40]. Activation of serotonin receptors that stimulate the adenylyl cyclase pathways positively regulates IGF-1 expression in cultured craniofacial mesenchymal cells [41].

Serotonin reuptake inhibitors have been proved to palliate symptoms in autism, and SERT has been the exact site of action of these drugs increasing serotonin availability at synapses [42, 43]. We hypothesize that an early decrease in serotonergic innervation could lead to low brain concentrations of IGFs, which are critical for Purkinje cell development.

Autism and Rett Syndrome

RS is believed to be a neurodevelopmental rather than a neurodegenerative disease because of a lack of neurodegenerative changes in postmortem brains. "The major result of mutations in the MECP2 gene is to cause age-related disruption of synaptic proliferation and pruning in the first decade" [44]. At their onset, both RS and autism overlap widely. In their very early stages, clinical differentiation of RS from infantile autism was not always easy before the MECP2 gene mutation was found in RS. However, there are a number of differences in the two syndromes (Table 10.1). In our earlier study, patients with

Table 10.1 Differences between Rett syndrome and infantile autism

Head growth
  Microcephaly in Rett syndrome [51, 52]
  Normal or macrocephaly in autism [28]
Neurons
  Cholinergic neurons affected in Rett syndrome [53]
  Serotoninergic neurons affected in autism [38, 54, 55, 56]
Neuropathology
  Frontal cortex pathology in Rett syndrome [51, 52, 57, 53, 58, 59]
  Cerebellum and hippocampus pathology in autism [30, 31, 60]
Onset
  Postnatal onset in Rett syndrome [44]
  Early prenatal origin in autism: beginning at the time of neural-tube closure [29].

Table 10.2 Basal forebrain in Rett syndrome

Reduced volume [60, 58]
Reduced cholinergic neurons [51]
Reduced choline acetyl transferase [53]
Reduced NGF concentrations postmortem [57]
Hypometabolism [58, 59]

RS had low levels of CSF NGF [45], whereas patients with autism had normal levels [46]. Conversely, patients with RS had normal levels of CSF IGF-1 compared with controls [47] and patients with autism had low levels [48, 49].

In RS, the forebrain is more severely affected than the other cortical areas (Table 10.2). Therefore, low CSF NGF is consistent with the evidence for loss of basal forebrain cholinergic neurons in RS.

Our findings of mainly normal CSF NGF in autism and low to negligible values in RS, along with decreased IGF-1 in autism and normal in RS (Fig. 10.3),

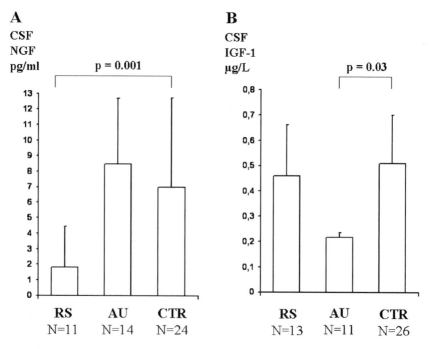

Fig. 10.3 Cerebrospinal fluid nerve growth factor (CSF NGF) and insulin-like growth factor (IGF-1) levels in patients with Rett syndrome (RS), patients with autism (AU), and controls (CTR). (**A**) Patients with RS have low to negligible levels of CSF NGF ($p = .001$). CSF NGF levels of patients with autism did not differ from the controls. (**B**) Patients with autism had low levels of CSF IGF-1 compared with controls ($p = .03$), whereas CSF IGF-1 levels of patients with RS did not differ from the controls. Published with permission from Riikonen R. *J Child Neurol* 2003 [49]

are in agreement with the different morphological and neurochemical findings in the two syndromes (Table 10.2).

Serum IGF-1

We found no significant correlation between serum and CSF IGF-1 in RS. This may indicate an independent role of IGF system in the central nervous system, making serum IGF-1 measurement unreliable as an indicator of disturbed function in the central nervous system. In autism, the serum concentrations also did not differ from the controls. Mills et al. 2007 [50] found elevated serum levels of growth-related hormones (including IGF-1) in 71 children aged 4–8 years with autism and autism spectrum disorder . This is not in discrepancy with our findings of low CSF IGF-1 in younger children and normal in older children. They found a significant correlation between serum IGF-1 levels and head circumference in autistic children but not in the controls. This correlation was also seen between CSF IGF-1 levels and head circumference in our series [35].

Conclusion

In autism, many studies have shown defects in cerebellar development. Low CSF IGF-1 concentrations in children with autism may lead to cerebellar abnormalities. Pathological brain growth and premature arrest are suggested to be restricted to the first years of life. Our study dealt with older children. We expected a correlation between head growth (head circumference) and IGFs, and we found such a correlation in children with autism but not in the comparison group. We speculate that this may partly explain abnormal brain size in autism. In theory, IGF-1 administration might increase the brain IGF-1 concentration and thus be of therapeutic importance. However, the therapy should be given early to support the survival of Purkinje cells. Further research focussing on the early process of brain pathology will likely be critical to elucidate the etiology of autism

Acknowledgments This study received financial support from the EVO fund of the Hospital District of Northern Savo, Kuopio, Finland and the Jonty Foundation, NJ, USA. There exists no conflict of interests related to this work.

References

1. Nelson K, Grether J, Groen L, Dambrosia J, Dickens B, Jelliffe L, Hansen R, Phillips T. Neuropeptides and neurotrophins in neonatal blood of children with autism and mental retardation. *Ann Neurol* 2001; 49: 597–606.

2. Leventhal P, Russel J, Feldman E. IGFs and the nervous system. In: Rosenfeld R, Roberts C eds: *The IGF System: Molecular Biology, Physiology, and Clinical Application.* Totawa, NJ: Humana Press, 1999: 425–455.
3. Mozell R, McMorris F. Insulin-like growth factor I stimulates oligodendrocyte development and myelination in rat brain aggregate culture. *J Neurosci* 1999; 30: 382–390.
4. Laudiero L, Aloe L, Levi-Montalcini R, Buttnelli C, Schiffre D, Offesen S, Otten U. Multiple sclerosis patients express increased levels of beta-nerve growth factor in the cerebrospinal fluid. *Neurosci Lett* 1992; 147; 9–12.
5. Xie K, Wang T, Olafsson P, Mizuno K, Lu B. Activity-dependent expression of NT-3 muscle cells in culture: Implications in the development of neuromuscular junctions. *J Neurosci* 1997; 17; 2947–2958.
6. Torres-Aleman I, Villaba M, Nieto-Bona M. Insulin-like growth factor-1 modulation of cerebellar cell populations is developmentally stage-dependent and mediated by specific intracellular pathways. *Neuroscience* 1998; 83: 321–334.
7. Yuen E, Mobley W. Therapeutic potential of neurotrophic factors for neurological disorders. *Ann Neurol* 1996; 40: 346–354.
8. Dore S, Kar S, Quirion R. Rediscovering an old friend, IGF-1: Potential use in the treatment of neurodegenerative diseases. *Trends Neurosci* 1997; 20; 326–331.
9. Russo V, Gluckman E, Feldman L, Werther G. The insulin-like growth factor system and its pleiotropic functions in brain. *Endocr. Rev.* 2005; 26: 916–943.
10. DMello S, Borodezt K, Soltoff S. Insulin-like growth factor and potassium depolarisation maintain neuronal survival by distinct pathways: Possible involvement of Pl 3-kinase in IGF-1 signalling. *J Neurosci* 1997; 17: 1548–1560.
11. Bondy C, Werner H, Roberts C, LeRoith D. Cellular patterns of type-1 insulin-like growth factor receptor gene expression during maturation of the rat brain: Comparison of the insulin-like growth factors I and II. *Neuroscience* 1992; 46: 909–923.
12. McKelvie P, Rosen K, Kinney H, Villa-Komaroff L. Insulin-like growth factor II expression in the developing brain. *J Neuropathol Exp Neurol* 1992; 51: 464–471.
13. Han V. Is the central nervous system a target for growth hormone and insulin-like growth factors? *Acta Paediatr* 1995; 411 (Suppl):3–8.
14. Barres B, Hart I, Coles H. et al. Cell death and control of cell survival in oligodendrocyte lineage. *Cell* 1992; 70: 31–42.
15. Beck K, Powell-Braxton L, Widmer H, Valverd J, Hefti F. Igf1 gene disruption results in reduced brain size, CNS hypomyelination, and loss of hippocampal granule and striatal parvalbumin-containing neurons. *Neuron* 1995; 14: 717–730.
16. Schoenle E, Haselbacher G, Briner J, Janzer R, Gammeltoff S, Humbel R, Prader A. Elevated concentration of IGF II in brain tissue from an infant with macrocephaly. *J Pediats* 1986; 108: 737–740.
17. Rotwein P, Burges S, Milbrandt J, Krause J. Differential expression of insulin-like growth factor genes in rat central nervous system. *Proc Natl Acad Sci USA* 1988; 85: 265–269.
18. Torres-Aleman I, Barrios W, Liedo A, Bericiano J. The insulin-like growth factor I system in cerebellar degeneration. *Ann Neurol* 1996; 39: 335–342.
19. Werther G, Russo V, Baker N, Buler G. The role of insulin-like growth factor system in developing brain. *Horm Res* 1998; 49: 37–40.
20. Fukudome Y, Tabata T, Miyoshi T et al. Insulin-like growth factor-1 as a promoting factor for cerebellar Purkinje cell development. *Eur J Neurosci* 2003; 17: 2006–2016.
21. Riikonen R, Somer M, Turpeinen U. Low insulin-like growth factor (IGF-1) in cerebrospinal fluid of children with progressive encephalopathy, hypsarrhythmia, and optic atrophy (PEHO) syndrome and cerebellar degeneration. *Epilepsia* 1999; 40: 1642–1648.
22. Riikonen R, Vanhanen S-L, Tyynelä J, et al. CSF insulin-like growth factor-1 in infantile neuronal ceroid lipofuscinosis. *Neurology* 2000; 54: 1828–1832.
23. Piven J, Saliba K, Bailey J, Arndt S. An MRI study of autism. The cerebellum revisted. *Neurology* 1997; 49: 546–551.

24. Courchesne E, Karns C, Davis H, Ziccardi R, Carper R, Tigue Z, Chisum H, Moses P, Pierce K, Lord C, Lincoln A, Pizzo S, Schreibman L, Haas R, Akshoomoff N, Courchesne R. Unusual brain growth patterns in early life in patients with autistic disorder. An MRI Study. *Neurology* 2001; 57: 243–254.

25. Sparks B, Friedman S, Shaw D, Aylward E, Echelard D, Artru A, Maravilla K, Giedd J, Munson J, Dawson G, Dager S. Brain structural abnormalities in young children with autism spectrum disorder. *Neurology* 2002; 59: 184–192.

26. Herbert M, Ziegler A, Makris N, Filipek P, Filipek P, Kemper T, Normandin J, Sanders H, Kennedy D, Caviness V. Localization of white matter volume increase in autism and developmental language disorder. *Ann Neurol* 2004; 55: 530–540.

27. Davincovitch M, Patterson B, Gartside P. Head circumference measurements in children with autism. *J Child Neurol* 1996; 11: 389–393.

28. Redcay E, Courchesene E. When is the brain enlarged in autism? A meta-analysis of all brain size reports. *Biol Psychaitry* 2005; 58: 1–9.

29. Rodier P, Ingram J, Tisdale B et al. Embryological origin for autism; developmental anomalies of the cranial nerve motor nuclei. *J Comp Neurol* 1996; 370: 247–261.

30. Bauman M, Kemper T. Histoanatomic observations of the brain in early infantile autism. *Neurology* 1985; 35: 866–874.

31. Rapin I, Katzman R. Neurobiology of autism. *Ann Neurol* 1998; 43: 7–14.

32. Casanova M, Buxhoeveden D, Switala A, Roy E. Minicolumnar pathology in autism. *Neurology* 2002; 58: 428–432.

33. Courchesne E, Pierce K. Brain overgrowth in autism during a critical time in development: Implications for frontal pyramidal neuron and interneuron development and connectivity. *Int J Dev Neurosci* 2005; 23: 153–170.

34. Vargas D, Nascimbene C, Krishnan C, Zimmerman A, Pardo C. Neuroglial activation and neuroinflammation in the brain of patients with autism. *Ann Neurol* 2005; 57: 67–81.

35. Riikonen R, Makkonen I, Turpeinen U, Kuikka J, Kokki H. Cerebrospinal fluid insulin-like growth factors IGF-1 and IGF-2 in infantile autism. *Dev Med Child Neurol* 2006; 48: 751–755.

36. Townsend J, Courchesne E, Covington J, Westerfield M, Harris N, Lyden P, Lowren T, Press G. Spatial attention deficits in patients with acquired or developmental cerebellar abnormality. *J Neurosci* 1999; 19: 5632–5563.

37. Allen G, Muller R, Courchesne E. Cerebellar function in autism: functional magnetic resonance image activation during a simple motor task. *Biol Psychiatry* 2004; 56: 269–278.

38. Chugani D, Muzik O, Behen M, Rothermel R, Janisse J, Lee J, Chugani H. Developmental changes in brain serotonin synthesis capacity in autistic and nonautistic children. *Ann Neurol* 1999; 45: 287–295.

39. Makkonen I, Riikonen R, Kokki H, Airaksinen M, Kuikka J. Serotonin and dopamine transporter binding in children and adolescents with autism. *Dev Med Child Neurol* 2008 August, in press.

40. Castren E. Neurotrophic effects of antidepressant drugs. *Curr Opin Pharmacol* 2004; 4: 58–64.

41. Lambert H, Weiss E, Lauder J. Activation of 5-HT receptors that stimulate the adenylyl cyclase pathway positively regulates IGF-1 in cultured craniofacial mesenchymal cells. *Dev Neurosci* 2001; 23: 70–77.

42. DeLong G, Rich C, Burch S. Fluoxetine response in children with autistic spectrum disorders: Correlation with familial major affective disorder and intellectual achievement. *Dev Med Child Neurol* 2002; 44: 652–659.

43. Hollander E, Phillips A, Chaplin W, Zagursky K, Novotny S, Wasserman S, Iyengar R. A placebo controlled crossover trial of liquid fluoxetine on repetitive behaviors in childhood and adolescent autism. *Neuropsychopharmacology* 2005; 30; 582–589.

44. Johnston M, Jeon O, Pevsner J, Blue M, Naidu S. Neurobiology of Rett syndrome: A genetic disorder of synapse development. *Brain Dev* 2001; 23 (Suppl):1S206–213.

45. Lappalainen R, Lindholm D, Riikonen R. Low levels of nerve growth factor in cerebrospinal fluid of children with Rett syndrome. *J Child Neurol* 1996; 11: 296–300.
46. Riikonen R, Vanhala R. Levels of cerebrospinal fluid nerve-growth factor differ in infantile autism and Rett syndrome. *Dev Med Child Neurol* 1999; 41: 148–152.
47. Vanhala R, Turpeinen U, Riikonen R. Insulin-like growth factor-1 in cerebrospinal fluid and serum in Rett syndrome. *J Child Neurol* 2000; 15: 797–802.
48. Vanhala R, Turpeinen U, Riikonen R. Low levels of insulin-like growth factor-1 in cerebrospinal fluid in children with autism. *Dev Med Child Neurol* 2001; 43: 614–616.
49. Riikonen R. Neurotrophic factors in the pathogenesis of Rett syndrome. *J Child Neurol* 2003; 18: 693–697.
50. Mills L, Hediger M, Molloy C, Chrousos G, Manning-Courtney P, Yu K, Brasington M, England L. Elevated levels of growth-related hormones in autism and autism spectrum disorders. *Clin Endocrinol* 2007; 67: 230–237.
51. Armstrong D, Dunn J, Antalffy B, Trivedi R. Selective dendritic alterations in the cortex of Rett syndrome. *J Neuropathol Exp Neurol* 1995; 54; 195–201.
52. Armstrong D, Dunn J, Schultz R et al. Organ growth in Rett syndrome; a post-mortem examination and analysis. *Paediatr Neurol* 1999; 20: 125–129.
53. Cook E, Courchesne R, Lord C et al. Evidence of linkage between the serotonin transporter and autistic disorder. *Mol Psychiatry* 1997; 2: 247–250.
54. Lipani J, Battacharjee M, Corey D, Lee D. Reduced nerve growth factor in Rett syndrome D postmortem brain tissue. *J Neuropathol Exp Neurol* 2000; 59; 889–895.
55. Wenk G, Naidu S, Casanova M, Kitt C, Moser H. Altered neurochemical markers in Rett's syndrome. *Neurology* 1991; 41: 1753–1756.
56. Coleman P, Romano J, Lapham I, Simon W. Cell counts in cerebral cortex in autistic patient. *J Autism Dev Disord* 1985; 15: 245–255.
57. Anderson G, Freedman D, Cohen D et al. Whole blood serotonin in autistic and normal subjects. *J Child Psychol Psychiatry* 1987; 28: 885–900.
58. Lappalainen R, Liewenthal K, Sainio K, Riikonen R. Brain perfusion SPECT and EEG findings in Rett syndrome. *Acta Paediatr Scand* 1997; 95: 44–50.
59. Raymond G, Bauman M, Kemper T. Hippocampus in autism: A Golgi analysis. *Acta Neuropathol (Berl)* 1996; 91: 117–119.
60. Uvebrandt P, Bjure J, Sixt R et al. Regional cerebral blood flow: SPECT as a tool for localization of brain dysfunction. In: Hagberg B, (ed.) *Rett Syndrome – Clinical and Biological Aspects*. London: Mac Keith Press, 1993; 80–85.

Chapter 11
Oxidative Stress and the Metabolic Pathology of Autism

S. Jill James

Abstract Chronic metabolic imbalance in the cellular microenvironment is often a primary factor in the development of complex disease. An integrated metabolic profile reflects the combined influence of genetic, epigenetic, and environmental factors that affect the candidate pathway of interest. In this way, the metabolic phenotype of an individual reflects the combined influence of both endogenous and exogenous factors on genotype and provides a window through which the cumulative impact of genes and environment may be viewed. Although both genetic and environmental factors appear to be necessary, in the majority of cases neither is independently sufficient for the autistic phenotype. A metabolic endophenotype provides an intermediate biomarker that is influenced by both genes and environment and can offer insights into relevant candidate genes and pathways.

Moreover, chronic or systemic metabolic imbalance can leave a metabolic footprint that can be followed analytically to gain mechanistic insights into the pathophysiology and pathogenesis of autism and thereby open new windows for therapeutic intervention. Escalating evidence suggests that many autistic children may be under chronic oxidative stress. The scientific question posed in this chapter is whether the autism phenotype reflects multiple and variable susceptibility alleles that converge to create a fragile, environmentally sensitive homeostasis with diminished ability to control and resolve pro-oxidant exposures.

Keywords Autism · metabolic · oxidative stress · glutathione · gene-environment · redox

S.J. James
Arkansas Children's Hospital Research Institute, 1120 Marshall St., Slot 512-40B,
Little Rock, AR 72202, USA
e-mail: jamesjill@uams.edu

A.W. Zimmerman (ed.), *Autism*, DOI: 10.1007/978-1-60327-489-0_11,

Introduction

Research into the metabolic phenotype of autism has been relatively less explored compared with broad-scale genomic and proteomic approaches, although metabolic abnormalities have been implicated in the pathogenesis of many other neurologic disorders [1, 2, 3, 4, 5]. We have used a targeted approach to autism "metabolomics" by focusing on the dynamics of an integrated metabolic pathway that is important for the regulation of normal redox homeostasis and cellular methylation. In a recent case–control study, we reported that the metabolic profile of children diagnosed with an autism spectrum disorder was severely abnormal relative to that of unaffected control children [6, 7]. Briefly, the mean ratio of plasma S-adenosylmethionine (SAM) to S-adenosylhomocysteine (SAM/SAH ratio), an index of methylation capacity, was significantly reduced and the mean level of glutathione (GSH), the major intracellular antioxidant, was also significantly decreased. The oxidized disulfide form of glutathione (GSSG) was significantly increased resulting in a twofold reduction in the GSH/GSSG redox ratio. Several metabolic precursors for glutathione synthesis were also lower in the autistic children suggesting that GSH synthesis may be inadequate. A summary of these results is presented in Table 11.1. These new findings are of clinical concern because they indicate a significant decrease in cellular methylation capacity ($\downarrow$SAM/SAH) and in antioxidant/detoxification capacity ($\downarrow$GSH/GSSG), and an increase in oxidative stress ($\uparrow$GSSG). Associated with the abnormal metabolic phenotype, preliminary evidence suggests that many autistic children exhibit an increased frequency of several genetic polymorphisms that negatively affect the flux through these metabolic pathways [7].

A diagram of the three interconnected pathways of folate, methionine, and glutathione metabolism found to be abnormal in many autistic children is

Table 11.1 Transmethylation and transsulfuration metabolites in autistic cases and controls

	Control* ($n = 73$)	Autistic* ($n = 80$)	p value
Methionine (µmol/L)	28.0 ± 6.5	20.6 ± 5.2	<0.0001
SAM (nmol/L)	93.8 ± 18	84.3 ± 11	<0.0001
SAH (nmol/L)	18.8 ± 4.5	23.3 ± 7.9	<0.0001
SAM/SAH ratio	5.5 ± 2.8	4.0 ± 1.7	<0.0001
Homocysteine (µmol/L)	6.0 ± 1.3	5.7 ± 1.2	0.03
Cysteine (µmol/L)	207 ± 22	165 ± 14	<0.0001
Total GSH (µmol/L)	7.53 ± 1.7	5.1 ± 1.2	<0.0001
Free GSH (µmol/L)	2.2 ± 0.9	1.4 ± 0.5	<0.0001
GSSG (µmol/L)	0.24 ± 0.1	0.40 ± 0.2	<0.0001
Total GSH/GSSG ratio	28.2 ± 7.0	14.7 ± 6.2	<0.0001
Free GSH/GSSG ratio	7.9 ± 3.5	4.9 ± 2.2	<0.0001

SAM: S-adenosylmethionine; SAH: S-adenosylhomocysteine; GSH: glutathione; GSSG: glutathione disulfide.
*Means $\pm$ SD.

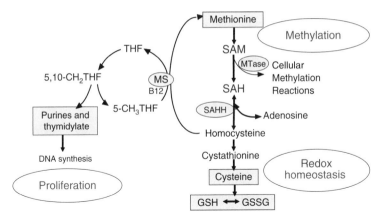

Fig. 11.1 A diagram of tetrahydrofolate (THF)-dependent methionine transmethylation and glutathione synthesis is presented in Fig. 11.1. The methionine cycle (transmethylation) involves the regeneration of methionine from homocysteine by the B12-dependent transfer of a methyl group from 5-methyl-tetrahydrofolate (5-CH$_3$THF) through the methionine synthase (MS) reaction [138]. Methionine is then activated to S-adenosylmethionine (SAM), the methyl donor for multiple cellular methyltransferase (MTase) reactions and the methylation of essential molecules such as DNA, RNA, proteins, phospholipids, creatine, and neurotransmittors [139]. The transfer of the methyl group from SAM results in the demethylated product S-adenosylhomocysteine (SAH). The reversible hydrolysis of SAH to homocysteine and adenosine by the SAH hydrolase (SAHH) reaction completes the methionine cycle. Homocysteine can then be either remethylated to methionine or irreversibly removed from the methionine cycle by cystathionine beta synthase (CBS). This is a one-way reaction that permanently removes homocysteine from the methionine cycle and initiates the transsulfuration pathway for the synthesis of cysteine and glutathione as indicated in Fig. 11.1 [140]. Glutathione is shown in its active reduced form (GSH) and its inactive oxidized disulfide form (GSSG). Glutathione is present in millimolar concentrations inside the cell and is the major determinant of intracellular redox homeostasis. Cell functions affected by perturbations in these interwoven pathways include proliferation (e.g., immune function, DNA synthesis, and repair), essential methylation (e.g., DNA, RNA, protein, phospholipid, neurotransmittors, creatine), and redox homeostasis (e.g., cell signaling, detoxification, stress response, cell cycle progression, and apoptosis)

presented in Fig. 11.1 with a detailed metabolic description. Note that methionine metabolism is directly connected to glutathione synthesis through homocysteine and cysteine. Methionine is necessary for the synthesis of SAM, the major methyl donor for all cellular methylation reactions. It is also the major precursor for cysteine, the rate-limiting amino acid for glutathione synthesis. Methionine levels can be negatively affected by genetic and environmental factors that reduce folate availability and/or oxidatively inhibit the methionine synthase enzyme [8]. Because these three metabolic pathways are mutually interdependent, genetic or environmental perturbation of folate or methionine metabolism will indirectly impact glutathione synthesis, and conversely, alterations in glutathione synthesis will alter flux through pathways of folate and methionine metabolism [9]. As diagrammed in Fig. 11.1, this interdependency translates into broader impact on (1) DNA synthesis/repair and proliferation; (2) cellular methylation including DNA, RNA, proteins, phospholipids and neurotransmitters; and (3) glutathione redox homeostasis and the essential reducing environment inside the cell.

Oxidative Stress

Oxidative stress occurs when cellular antioxidant defense mechanisms fail to counterbalance and control endogenous reactive oxygen and nitrogen species (ROS/RNS) generated from normal oxidative metabolism or from pro-oxidant environmental exposures. The potential role of oxidative stress in the pathogenesis of autism has not been extensively investigated, although the developing brain is highly vulnerable to oxidative damage [10, 11]. The brain utilizes more than 20% of oxygen consumed by the body yet comprises only 2% of body weight [12]. High energy demands from oxidative metabolism plus a high concentration of polyunsaturated fatty acids and relatively low antioxidant enzyme activity is thought to render the brain more vulnerable to oxidative insult than most organs [12, 13, 14].

Recent reviews of the literature lend support to the hypothesis that oxidative stress may be a contributing factor to the pathology of autism [15, 16]. In addition to low glutathione and GSH/GSSG redox ratio [6, 7], higher plasma and red blood cell levels of pro-oxidant nitric oxide have been documented in autistic children [17, 18]. Several independent reports have documented that the antioxidant enzymes, glutathione peroxidase and superoxide dismutase (SOD1), are lower in autistic children relative to controls [17, 19]. In addition, evidence consistent with chronic brain inflammation and increased lipid peroxidation in autistic individuals has been recently reported [20, 21, 22, 23]. However, because these biomarkers of oxidative stress were documented in children who already had autism, it is not possible to discern whether oxidative stress contributes to the cause or is simply a consequence of having autism. Prospective studies in high-risk children will be required to answer this provocative question.

Glutathione and Redox Homeostasis

Glutathione is a tripeptide of cysteine, glycine, and glutamate that is synthesized de novo in every cell of the body and serves as the major intracellular redox buffer. It is the sulfhydryl (SH) group of the cysteine moiety that provides the reducing equivalents for all glutathione functions. An essential reducing environment is maintained inside the cell by the high ratio of reduced GSH (mM) to the oxidized disulfide GSSG (μM) [24]. The GSH/GSSG redox equilibrium regulates a pleiotropic range of functions that include free radical scavenger and intracellular redox homeostasis [25], maintenance of protein redox conformation and regulation of redox sensitive enzyme activity [26], cell membrane integrity and signal transduction [27, 28], transcription factor binding and gene expression [29], phase II detoxification [30], and regulation of proliferation, differentiation, and apoptosis [31, 32].

Under normal physiologic conditions, the GSH/GSSG redox ratio is maintained by the enzyme glutathione reductase that converts GSSG back to GSH. However, excessive oxidative stress that exceeds the capacity of glutathione reductase will induce GSSG export to the plasma in an attempt to regain intracellular redox homeostasis. This loss of glutathione effectively increases the requirement for cysteine, the rate-limiting amino acid for de novo glutathione synthesis [24]. In approximately 60% of autistic children, the observed decrease in plasma GSH and cysteine and the increase in GSSG provide evidence of chronic intracellular oxidative stress [7]. An inherent or genetic deficiency in GSH synthesis in autistic children could provide a biologic explanation for glutathione-related pathology such as gastrointestinal (GI) dysfunction [33], impaired detoxification [34], and microglial activation [21, 35] that are associated with both oxidative stress and autism.

It is important to distinguish acute reversible redox signaling from chronic irreversible redox imbalance and oxidative damage. Because most studies focus on the damaging effects of oxidative stress, it is not commonly appreciated that subtle changes in redox status are essential for multiple mechanisms of redox-mediated cell signaling [36, 37]. These include signals for cell cycle regulation and differentiation [38], inhibition and activation of redox sensitive enzymes [39], transcription factor binding and gene expression [29], activation of innate immunity and the acute inflammatory response [40, 41]. As diagrammed in Fig. 11.2a, these reversible subtle shifts in redox equilibrium (designated by bidirectional arrows) act as a rheostat to maintain intracellular homeostasis and also comprise an integrated cell response to environmental stressors. As shown in Fig. 11.2b, unopposed oxygen- and nitrogen-free radicals that exceed the redox buffering capacity of glutathione can disrupt protein structure and function [42], promote membrane lipid peroxidation [43], oxidative glutamate neurotoxicity [44, 45] and nuclear and mitochondrial DNA damage leading to apoptotic cell death [43, 46]. During prenatal and postnatal development, these aberrations would promote dysregulated redox enzyme activity, unresolved excitation and inflammation, faulty membrane signal transduction, and disrupted signals for proliferation, differentiation, and apoptosis. Several neuropsychiatric disorders including schizophrenia, bipolar disorder, Parkinson's disease, and amphetamine abuse have been associated with oxidative stress [47, 48, 49, 50, 51]. In addition, chronic oxidative stress and glutathione depletion have been implicated in the etiology of birth defects, aging, and complex diseases including cardiovascular disease, cancer, and autoimmune disease [52, 53, 54].

A

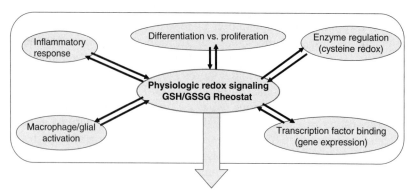

Functional Metabolism Viability

B

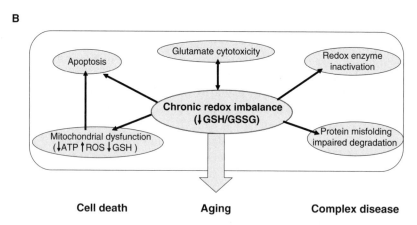

Fig. 11.2 Subtle transient and reversible glutathione-dependent redox signaling pathways **(a)** are essential for cell viability whereas chronic severe and irreversible pathways associated with glutathione depletion lead to a self-perpetuating cycle of oxidative stress, damage, and death **(b)**

Hypothesis and Evidence Linking Oxidative Stress and Autism

Despite major progress in identifying chromosomal regions associated with autism [55], the transition from chromosome to candidate gene has proven to be enormously complex. Similar to the poorly defined role of environmental factors, no single gene has been reproducibly identified as a major risk factor for autism. The current genetic model predicts that 10–100 small-effect genes may interact for the phenotypic expression of the autism phenotype [55, 56].

A polygenic model with interactions between multiple genes and multiple environmental factors further complicates and confounds the search for candidate genes. Moreover, if an environmental trigger is a requisite factor, the same genetic profile could be present in unaffected individuals who did not receive the environmental insult during critical developmental windows.

Whole genome screens, cytogenetic analyses, and linkage disequilibrium studies are widely utilized in the search for candidate genes. This approach contrasts with a hypothesis-driven search that predicts gene involvement based on metabolic evidence indicating that specific gene products are altered. Even small variations in gene expression and enzyme activity, if expressed chronically, could have a significant impact on downstream metabolic dynamics. By definition, a biologic hypothesis must demonstrate applicability to the many disparate conditions and etiologies associated with autism as well as common clinical symptoms such as immunological, GI, and neurologic dysfunction. The hypothesis that oxidative stress and impaired methylation capacity may contribute, in part, to the development of the autistic phenotype is supported by our metabolic evidence and the extensive literature review of risk factors and co-morbid conditions outlined in the sections below.

Single Gene Disorders

Several widely disparate single gene disorders including tuberous sclerosis (TS), phenylketonuria (PKU), Rett and Fragile X syndromes, have autism as a co-morbid feature. Notably, each of these disorders exhibit direct or indirect evidence of oxidative stress and/or abnormal DNA methylation. For example, TS is an autosomal disorder that is characterized by multiple benign brain tumors. Although the prevalence of autism in TS is approximately 100 times greater than in the general population [57], only a subset of TS children develop autism. Among TS monozygotic twins who are discordant for autism, the autistic twin uniquely exhibits intractable early-onset seizures [58, 59]. Because chronic seizure activity severely depletes brain glutathione [60, 61], the autistic twin is thereby exposed to greater neuronal oxidative stress [60, 62]. Autism is also more prevalent among singleton TS patients with seizures [63]; thus, seizure-related oxidative stress may increase autism risk among children with TS. Similarly, the incidence of epileptic seizures is approximately 30% in "ideopathic" autism and increases with clinical severity [64].

Autism prevalence is also increased among patients with Rett and Fragile X syndromes, and both disorders exhibit abnormal DNA methylation function as well as seizure-related oxidative stress that increases with severity of disease [65, 66, 67, 68]. In Rett syndrome, DNA methylation is intact but the mutation in the methyl-binding protein, MeCP2, prevents it from binding to DNA methyl groups. If the methyl groups occur in gene promoter regions, the failure of MeCP2 to bind results in over-expression of genes that are normally silenced by the MeCP2 enzyme complex [69]. In Fragile X, an abnormal dinucleotide repeat expansion of methyl-binding sites in DNA promotes silencing of the *FMR-1* gene leading to disease expression [70]. Both syndromes are associated with seizure-related oxidative stress [68, 71].

Approximately 10% of children with Down syndrome are autistic [72]. Overexpression of SOD1 resulting from three copies of chromosome 21 leads to hydrogen peroxide-induced oxidative stress and glutathione depletion [73, 74, 75]. We and others have shown that these children also have abnormal DNA methylation presumably because of the presence of DNA methyltransferase 3 L on chromosome 21 [5, 76]. Finally, the frequency of autism, glutathione depletion, and oxidative stress is increased in patients and animal models of PKU, and the associated oxidative stress has been shown to be reversible with antioxidant supplementation [77, 78, 79, 80].

Taken together, considerable evidence exists to suggest that oxidative stress and abnormal DNA methylation may be a common theme among these apparently diverse genetic disorders. As shown in Fig. 11.1, cellular methylation and glutathione synthesis are metabolically linked within the same metabolic pathway. Interestingly, functional aberrations in both are often associated with the pathogenesis of other neurologic disorders and complex diseases [5, 81, 82].

Prenatal and Peri-Natal Environmental Exposures

Prenatal exposures to valproate [83], thalidomide [84], and ethylmercury preservative [85, 86] have been implicated as possible risk factors for autism. In experimental studies, prenatal valproic acid (VPA) given during gestation has been shown to reduce fetal methionine and glutathione levels and to induce DNA hypomethylation [87, 88]. VPA treatment of rats increased oxidative stress as evidenced by elevated isoprostane levels whereas glutathione treatment conferred protection against oxidative mitochondrial damage induced by VPA [89]. Valproate has been recently shown to be an inhibitor of histone deacetylase [90]. The failure to deacetylate histones in a timely manner promotes a more open and vulnerable conformation of DNA resulting in the over expression of normally silenced genes and an increased vulnerability to oxidative damage [91, 92].

Maternal thalidomide exposure at a developmental stage just before limb development is one of the established risk factors for autism [84]. Consistent with our hypothesis, the mechanism of thalidomide toxicity has been recently shown to involve free radical generation, glutathione depletion, and modulation of redox sensitive gene expression [93, 94]. Oxidative stress during critical developmental windows could permanently disrupt normal differentiation, proliferation, and cell ontogeny.

Ethylmercury in vaccines given to infants and pregnant mothers is a controversial risk factor that may have biologic plausibility [86]. Mercury is a developmental neurotoxin and has been shown to induce oxygen-free radicals, glutathione depletion, and apoptosis in vivo and in cultured human brain cells and lymphocytes [95, 96, 97]. Cell death associated with mercury is mediated by mitochondrial glutathione depletion and is prevented by pretreatment with glutathione [95, 97]. Thus, oxidative stress and glutathione depletion has been shown to occur with several environmental exposures associated with autism.

Gender Disparity in Autism

Male gender is the strongest most consistent risk factor for autism, yet the basis for the striking gender disparity is unknown. One plausible explanation is that females are protected against factors that increase risk of autism. Consistent with this possibility, estrogen is well known to be a neuroprotectant and has potent antioxidant activity associated with up-regulation of relevant genes [98, 99, 100, 101, 102]. Furthermore, estrogen has been shown to increase plasma and mitochondrial GSH levels in a dose-dependent manner by up-regulating the activity of enzymes involved in the synthesis and maintenance of GSH [103, 104, 105]. Male infants have lower GSH levels and glutathione reductase activity than female infants and are well known to be more vulnerable to oxidative insult [106, 107, 108]. Until menopause, females have lower homocysteine levels than males because of faster turnover of the methionine cycle and increased synthesis of the major methyl donor, SAM [109, 110]. Thus, a preponderance of evidence exists to suggest that both cellular methylation capacity and glutathione antioxidant activity are higher in females than in males. If our hypothesis is correct, males with lower estrogen levels and consequently lower methylation potential and glutathione-mediated antioxidant/detoxification capacity would be inherently more vulnerable to oxidative stress and methylation dysregulation and thereby may be more vulnerable to the development of autism.

Clinical Pathology of Autism and Oxidative Stress

If increased vulnerability to oxidative stress is a constitutive feature of autism phenotype, then autistic children should exhibit systemic evidence of oxidative stress. Multiple clinical studies have documented a high prevalence of GI inflammation and increased mucosal permeability in both upper and lower intestine in autistic children [33, 111, 112]. Biliary glutathione protects the mucosa from oxidative injury and is essential for normal GI function and membrane integrity [113]. Furthermore, impaired intestinal glutathione synthesis has been shown to result in severe mucosal degeneration and inflammation [33, 114]. A recent report documented the presence of chronic inflammation in the autistic brain that appears to be mediated by innate microglial activation and proinflammatory cytokines [20]. Decreased glutathione levels promote inflammation, and chronic inflammation further depletes glutathione, thereby promoting a vicious cycle that could both initiate and exacerbate GI and central nervous system (CNS) inflammation associated with autism.

Finally, viral infections have also been implicated as a pathogenic mechanism for autism [115, 116]. Viral infection and replication induces glutathione depletion and oxidative stress in host cells [117]. Influenza and RNA virus infections are effectively suppressed in a dose–response manner by exogenous glutathione treatment whereas chronic viral infections are promoted by low glutathione status [117, 118].

Summary

Although the evidence described above for a relationship between autism and oxidative stress is clearly associative, it strengthens the novel hypothesis that oxidative stress and abnormal methylation may be contributing factors that link the diverse metabolic and genetic disorders associated with autism. The hypothesis predicts that autistic children will be at the low end of the population distribution of methionine, SAM/SAH, cysteine, and glutathione. It would also predict that a genetic predisposition to impaired methylation capacity and glutathione synthesis is necessary but not sufficient to promote the autism phenotype and that other endogenous and/or exogenous promoters interact to expose the genetic liability.

Gene–Environment Interactions Affecting Glutathione and Redox Imbalance

The concept of a gene–environment interaction implies that the relative impact of an environmental exposure will vary depending on the genetic background of an individual, and that in the absence of the environmental insult, there would be no pathology. The frequent phenotypic discordance among monozygotic twins provides the best supportive evidence for the involvement of environmental and/or epigenetic factors (DNA/histone methylation) in the development of autism. Using strict diagnostic criteria, only 60% concordance for severe autism was reported whereas concordance for milder forms of autism was 90% [119]. Other reports have documented phenotypic discordance in cerebellar gray and white matter volumes among monozygotic twins [120]. DNA and histone methylation are reversible epigenetic modifications that permit the sequential activation and repression of gene expression during the developmental program and during the response to environmental stressors. Thus, it is probable that epigenetic phenomena that do not involve a change in DNA sequence also contribute to the discordance between monozygotic twins. The individual vulnerability and phenotypic heterogeneity within the autism spectrum most likely reflect variable gene penetrance and epigenetic factors interacting with severity of environmental exposure and developmental timing.

The developmental neurotoxicity of commonly encountered environmental toxicants is a major public health concern and challenge because their identification and regulation could lead to preventive measures. Grandjean and Landrigan elegantly raised this issue in a recent review and provided evidence that the multitude of untested and unregulated chemicals present in drinking water, soil, and air could be contributing to a "silent pandemic" of neurodevelopmental disorders in modern society [121]. Designated "safe limits" of environmental pollutants often underestimate the toxic threshold because they are based on independent evaluation of each chemical in isolation and the evaluations are generally conducted in adult subjects. Because children are commonly exposed to complex mixtures of chemicals present in the environment, it is important to emphasize that *sub*toxic doses of individual chemicals can reach a toxic threshold when these exposures are combined. Additive interaction at subtoxic doses obviates published "safe limits" of exposure. Moreover, children have a lower threshold of neurotoxicity compared with adults because their nervous systems are developmentally immature and especially vulnerable. For example, the developmental trajectory of CNS progenitor cells can be derailed by environmentally relevant pro-oxidant exposures [122]. Because normal neurodevelopment requires rigid control over a precisely defined series of events, environmental exposures that interfere with this developmental program can disrupt normal cell ontogeny and brain development.

Pro-Oxidant Environmental Exposures

A wide variety of chemically and structurally diverse environmental toxicants share a common mechanism of toxicity by inducing oxidative stress and/or depleting glutathione [122]. A literature search reveals that many ubiquitous and environmentally relevant metals (cadmium, arsenic, lead, nickel, and cobalt) are toxic pro-oxidants. In addition, commonly encountered chemical solvents such as alcohol, benzene, and chlorinated organocompounds also induce oxidative stress. Furthermore, the toxicity of several industrial chemicals including polychlorinated biphenyls and many herbicides and pesticides are associated with ROS generation and glutathione depletion. Although these disparate environmental toxicants have little in common chemically or structurally, the shared mechanism of toxicity is provocative and suggests that combined exposures to these environmentally relevant compounds is likely to be additive. Because glutathione provides the major pathway for detoxification and elimination of pro-oxidant toxins, the low levels of glutathione in many autistic children would render them less able to detoxify these

Fig. 11.3 The cytotoxicity of pro-oxidant exposure depends on the glutathione-dependent redox reserve. A robust GSH/GSSG ratio will buffer a toxic insult so that the toxic threshold is not reached whereas a low GSH/GSSG ratio with limited reserve creates a fragile homeostasis that will precipitate toxicity at an equivalent dose

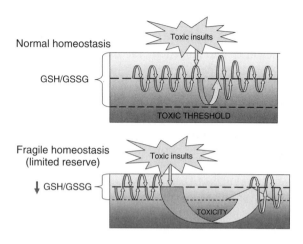

exposures. As diagrammed in Fig. 11.3, robust glutathione reserves create a buffer against environmental insults such that the threshold for toxicity is never reached. On the contrary, a low GSH/GSSG redox ratio creates a fragile homeostasis that is more easily disrupted by environmental toxins, i.e., the toxic threshold will be reached at a lower dose or reached earlier at an equivalent dose. By definition, the redox ratio will be more severely affected by small increases in GSSG when GSH levels are low and will be less affected when GSH levels are high. The low GSH observed in many autistic children is consistent with a fragile redox homeostasis and increased sensitivity to environmental toxicants.

Recent genetic analysis of autism family trios (mother, father, child) suggests the majority of idiopathic autism may be the result of spontaneous mutations that have a high penetrance in males and weak penetrance in females [123]. This genetic model of autism proposes that most cases of autism are caused by de novo DNA mutations that are not present in their parents and therefore not inherited. Although environmental issues were not considered in this study, these unexpected results are consistent with the possibility that toxic DNA-damaging exposures could have occurred during germline, prenatal, or postnatal development that altered fetal but not parental DNA. If proved to be true, idiopathic autism would not be heritable genetic disorder in these cases.

Oxidative Stress During Neurodevelopment

Pro-oxidant exposures that increase oxygen- or nitrogen-free radical generation and/or deplete glutathione are well known to induce DNA damage and to disrupt normal developmental processes especially during critical developmental windows. Although the blood–brain barrier (BBB) protects adults from toxic exposures, it is not completely formed until about 6 months of age creating a less effective barrier in the infant [124]. At the BBB, the high activity of the Phase II detoxification enzyme, glutathione-S-transferase (GST), in the epithelial cells of the choroid plexus provides a major mechanism for detoxification and neuroprotection [125]. This enzyme conjugates metals and hydrophobic toxic compounds with glutathione rendering them inactive and water-soluble for excretion in the bile and urine. GST activity depends on adequate glutathione availability, and low levels of glutathione and/or pro-oxidant conditions would further compromise BBB neuroprotection in the infant or fetus. Maternal exposure to pro-oxidant heavy metals constitutes another risk factor during development because these chemicals cross the placenta and are often found at higher levels in the umbilical cord than in maternal blood [126]. Prenatal viral infection is another major risk factor for developmental pathology and neuropsychiatric disorders. Viral infection induces intracellular glutathione export and oxidative stress in order to facilitate its replication [127]. For example, viral infection in pregnant rats triggers a decrease in GSH/GSSG ratio and oxidative stress in the hippocampus of fetal male rats [128]. Of possible relevance, prenatal infection with rubella or cytomegalovirus is one of few the established risk factors for autism.

Microglial maturation and axon myelination are incomplete at birth and continue to mature throughout early childhood and into adolescence. Relative to other CNS cells, glial cells have lower intracellular GSH and increased pro-oxidant iron levels [129]. This combination interacts with a high rate of oxidative metabolism to render these cells more sensitive to pro-oxidant exposures. Recent evidence indicates that a modest shift in the intracellular redox microenvironment can have a major impact on whether oligodendrocyte precursors undergo differentiation or proliferative self-renewal [122]. A more reducing environment is associated with precursor proliferation whereas a more oxidizing environment promotes differentiation in these cells. Because GSH/GSSG is the major determinant of intracellular redox status, a small shift in the redox ratio could have a profound effect on neurodevelopmental fate if precursor cells failed to proliferate or differentiate at the appropriate time during the developmental program.

Genes, Environment, and the Autism Metabolic Phenotype

An overview of the complex interplay between genes and environment and autistic behavior is presented in Fig. 11.4. Although genes determine the vulnerability or resistance to environmental toxicants, environmental toxicants alter the expression of genes creating a bidirectional interaction. As shown in Fig. 11.4, although both genes and environment are thought to contribute to autistic behavior, both involve multiple, additive, and variable factors that

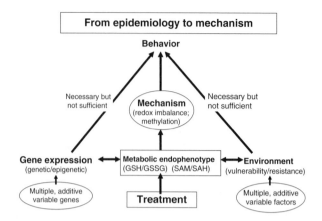

Fig. 11.4 Although both genes and environmental factors appear to be necessary, in the majority of cases neither is independently sufficient for the autistic phenotype. A metabolic endophenotype provides an intermediate biomarker that is influenced by both genes and environment and can offer insights into relevant candidate genes and pathways. The metabolic redox ratio (GSH/GSSG) and the methylation ratio (SAM/SAH) are relevant examples of metabolic endophenotypes that could provide insights into potential mechanisms (oxidative stress and methylation, respectively) that contribute to neurologic and behavioral abnormalities associated with autism

greatly complicate and confound the identification of responsible genes and the relevant environmental factors. Although both genes and environmental factors appear to be necessary, in the majority of cases neither is independently sufficient for the autistic phenotype. A metabolic endophenotype provides an intermediate biomarker that is influenced by both genes and environment and can offer insights into relevant candidate genes and pathways. The metabolic redox ratio (GSH/GSSG) and the methylation ratio (SAM/SAH) are relevant examples of metabolic endophenotypes that could provide insights into potential mechanisms (oxidative stress and methylation, respectively) that contribute to neurologic and behavioral abnormalities associated with autism. More importantly, because oxidative inhibition of metabolic pathways is inherently reversible, candidate pathways can provide targets for intervention and treatment strategies.

Systemic Implications of Oxidative Stress and Autism Metabolic Pathology

Although cell-specific differences in metabolism account for tissue-specific functions, there are also interwoven metabolic pathways present in all cells that are fundamental for cell survival. Metabolically based vital cell functions include pathways for the stress response, nutrient uptake and metabolism, redox homeostasis, energy production, and regulation of gene expression, proliferation and cell death. Perturbation of these common pathways by genetic, nutritional, or environmental factors would not target a single system in isolation but would be expected to have a broader multi-system impact in the body. Chronic metabolic imbalance will leave a metabolic footprint that can be followed analytically to gain new insights into the pathophysiology and pathogenesis of complex disease and thereby open new windows for therapeutic intervention.

Autism Metabolic Phenotype

The metabolic footprint we have observed in many autistic children [7] will negatively affect proliferative potential (immune function, DNA synthesis, and repair), essential methylation reactions (DNA, RNA, protein, lipids, neurotransmitters, creatine), and glutathione-dependent redox homeostasis (detoxification, signal transduction, gene expression, cell cycle control, and cell death). The link between these basic functions and folate-dependent methylation and glutathione metabolic pathways is shown diagrammatically in Fig. 11.1. Because the autism diagnosis is based solely on behavioral symptoms, it is logical although perhaps too simplistic to assume that brain abnormalities are the sole cause of autism. Clearly brain involvement is a major consideration; however, many autistic children also exhibit evidence of GI and immunologic pathology that is consistent with systemic involvement.

Based on escalating direct and indirect evidence, it is plausible to propose that some forms of autism could be a manifestation of a genetically based inability to control or resolve oxidative stress. Metabolic evidence from an increasing number of studies supports the possibility that a significant proportion of autistic children may be under chronic oxidative stress. Because the response to oxidative stress is programmed into every cell, it is likely that the plasma biomarkers observed in autistic children reflect a systemic redox imbalance rather than tissue-specific response. A systems' approach to autism biology suggests the provocative possibility that some autistic behaviors could, in part or in parallel, reflect a neurologic manifestation of a genetically based systemic metabolic derangement. Such a paradigm shift from a primary neurodevelopmental disorder to a broader systemic disorder with neurologic sequelae would widen the net in the quest to understand the biologic basis of autism [130]. A more systemic approach would encompass not only the neurologic manifestations but also the evidence for GI and immunologic pathology associated with autism [131, 132, 133, 134]. In addition to impaired CNS function, abnormalities in folate-dependent methionine and glutathione metabolism also occur with GI and immunologic dysfunction [36, 114, 135, 136]. Moreover, pro-oxidant environmental exposures are known to affect all three systems [137].

Gut–Brain–Immune Connection Applied to Autism

The bidirectional arrows in Fig. 11.5 illustrate the functional interdependence and continuous crosstalk that occurs between the brain, the gut, and the immune system to maintain functional homeostasis in all three systems. The messengers of this communication are hormones, neuropeptides, neurotransmitters, and cytokines. Of interest, all three systems are developmentally immature at birth and require appropriate environmental signals to mature normally. The brain requires sensory input from the environment, the immune system requires antigenic stimuli, and the gut requires microbial colonization and dietary substrate. The mutual interdependence between these systems implies that an environmental insult to one system could indirectly affect the other two, depending on severity and timing. The developmental trajectory of all three systems has been shown to be highly sensitive to oxidative stress especially during critical developmental periods. Thus, genetic predisposition, nutritional deficiencies, or environmental exposures that negatively affect maternal or perinatal glutathione redox capacity could be expected to have broad developmental consequences. The established risk factors for autism including maternal viral infection, valproate and thalidomide exposure, Fragile X and Rett syndrome, all involve a component of oxidative stress, suggesting the possibility that disparate origins may converge onto a final common pathway that contributes to the risk of autism. Moreover, a failure to control pro-oxidant insults during prenatal or postnatal development would promote a self-amplifying vicious cycle leading to chronic inflammation.

The autism triad: brain-gut-immune axis

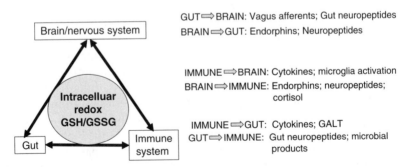

Fig. 11.5 The bidirectional arrows illustrate the functional interaction and continuous crosstalk that occurs between the brain, the gut, and the immune system to maintain functional homeostasis in all three systems. The messengers of this communication are hormones, neuropeptides, neurotransmitters, and cytokines. The interdependence between these systems implies that an environmental insult to one system could indirectly affect the other two, depending on severity and timing

Future Prospects

Although considerations outlined in this chapter are logical projections based on available scientific evidence, the hypothesis presented is nonetheless biologically plausible and intended to stimulate further thought and experimental investigation. The scientific question posed by this discussion is whether the genetic component of autism could involve multiple susceptibility alleles that interact to create a fragile, environmentally sensitive metabolic imbalance that results in a vulnerable phenotype with diminished ability to control and resolve pro-oxidant encounters.

If oxidative stress and impaired methylation are genetically determined and/or precipitated by prenatal/perinatal environmental exposures, the abnormal metabolic profile would be present in infants before the diagnosis of autism. Furthermore, if the abnormal metabolic profile is present in the children who undergo autistic "regression" at approximately 18–24 months, it would suggest that the abnormal metabolism precedes the regressive onset and maybe underlie the predisposition. Thus, for example, the presence of low cysteine and GSH/GSSG or low methionine and SAM/SAH at 12–18 months of age would be an important new finding that would strengthen the possibility that the abnormal metabolism is associated with the development and progression of the disorder. Furthermore, early detection of the abnormal profile that is predictive of a subsequent diagnosis of autism would provide a valuable biochemical test to expedite the early identification of children at high risk for developing autism in order to schedule earlier intervention strategies. Finally, and possibly most importantly, the early identification of a specific metabolite imbalance would provide insights into targeted intervention strategies to improve outcome and possibly prevent the progression into autism. Although clearly still in the realm of theory, if proven correct in future studies, the impact would be immense.

References

1. Miller AL. The methionine-homocysteine cycle and its effects on cognitive diseases. *Altern Med Rev* 2003; 8(1):7–19.
2. Serra JA, Dominguez RO, de Lustig ES, Guareschi EM, Famulari AL, Bartolome EL, et al. Parkinson's disease is associated with oxidative stress: Comparison of peripheral antioxidant profiles in living Parkinson's, Alzheimer's and vascular dementia patients. *J Neural Transm* 2001; 108(10):1135–1148.
3. Schulz JB, Lindenau J, Seyfried J, Dichgans J. Glutathione, oxidative stress and neurodegeneration. *Eur J Biochem* 2000; 267(16):4904–4911.
4. Muntjewerff JW, Van der Put N, Eskes T, Ellenbroek B, Steegers E, Blom H, et al. Homocysteine metabolism and B-vitamins in schizophrenic patients: Low plasma folate as a possible independent risk factor for schizophrenia. *Psychiatry Res* 2003; 121(1):1–9.
5. Pogribna M, Melnyk S, Pogribny I, Chango A, Yi P, James SJ. Homocysteine metabolism in children with Down syndrome: In vitro modulation. *Am J Hum Genet* 2001; 69(1):88–95.

6. James SJ, Cutler P, Melnyk S, Jernigan S, Janak L, Gaylor DW, et al. Metabolic biomarkers of increased oxidative stress and impaired methylation capacity in children with autism. *Am J Clin Nutr* 2004; 80(6):1611–1617.

7. James SJ, Melnyk S, Jernigan S, Cleves MA, Halsted CH, Wong DH, et al. Metabolic endophenotype and related genotypes are associated with oxidative stress in children with autism. *Am J Med Genet B Neuropsychiatr Genet* 2006; 141(8):947–956.

8. Banerjee RV, Matthews RG. Cobalamin-dependent methionine synthase. *FASEB J* 1990; 4(5):1450–1459.

9. Vitvitsky V, Mosharov E, Tritt M, Ataullakhanov F, Banerjee R. Redox regulation of homocysteine-dependent glutathione synthesis. *Redox Rep* 2003; 8(1):57–63.

10. McQuillen PS, Ferriero DM. Selective vulnerability in the developing central nervous system. *Pediatr Neurol* 2004; 30(4):227–235.

11. Rougemont M, Do KQ, Castagne V. New model of glutathione deficit during development: Effect on lipid peroxidation in the rat brain. *J Neurosci Res* 2002; 70(6):774–783.

12. Dringen R, Gutterer JM, Hirrlinger J. Glutathione metabolism in brain metabolic interaction between astrocytes and neurons in the defense against reactive oxygen species. *Eur J Biochem* 2000; 267(16):4912–4916.

13. Dringen R, Gutterer JM, Hirrlinger J. Glutathione metabolism in brain metabolic interaction between astrocytes and neurons in the defense against reactive oxygen species. *Eur J Biochem* 2000; 267(16):4912–4916.

14. Bains JS, Shaw CA. Neurodegenerative disorders n humans: The role of glutathione in oxidative atress-mediated neuronal death. *Brain Res Rev* 1997; 25:335–358.

15. Chauhan A, Chauhan V. Oxidative stress in autism. *Pathophysiology* 2006; 13(3):171–181.

16. Kern JK, Jones AM. Evidence of toxicity, oxidative stress, and neuronal insult in autism. *J Toxicol Environ Health B Crit Rev* 2006; 9(6):485–499.

17. Zoroglu SS, Armutcu F, Ozen S, Gurel A, Sivasli E, Yetkin O, et al. Increased oxidative stress and altered activities of erythrocyte free radical scavenging enzymes in autism. *Eur Arch Psychiatry Clin Neurosci* 2004; 254(3):143–147.

18. Zoroglu SS, Yurekli M, Meram I, Sogut S, Tutkun H, Yetkin O, et al. Pathophysiological role of nitric oxide and adrenomedullin in autism. *Cell Biochem Funct* 2003; 21(1):55–60.

19. Yorbik O, Sayal A, Akay C, Akbiyik DI, Sohmen T. Investigation of antioxidant enzymes in children with autistic disorder. *Prostaglandins Leukot Essent Fatty Acids* 2002; 67(5):341–343.

20. Vargas DL, Nascimbene C, Krishnan C, Zimmerman AW, Pardo CA. Neuroglial activation and neuroinflammation in the brain of patients with autism. *Ann Neurol* 2005; 57(1):67–81.

21. Pardo CA, Vargas DL, Zimmerman AW. Immunity, neuroglia and neuroinflammation in autism. *Int Rev Psychiatry* 2005; 17(6):485–495.

22. Ming X, Stein TP, Brimacombe M, Johnson WG, Lambert GH, Wagner GC. Increased excretion of a lipid peroxidation biomarker in autism. *Prostaglandins Leukot Essent Fatty Acids* 2005; 73(5):379–384.

23. Chauhan A, Chauhan V, Brown WT, Cohen I. Oxidative stress in autism: Increased lipid peroxidation and reduced serum levels of ceruloplasmin and transferrin – the antioxidant proteins. *Life Sci* 2004; 75(21):2539–2549.

24. Schafer FQ, Buettner GR. Redox environment of the cell as viewed through the redox state of the glutathione disulfide/glutathione couple. *Free Radic Biol Med* 2001; 30(11):1191–1212.

25. Lu SC. Regulation of hepatic glutathione synthesis. *Semin Liver Dis* 1998; 18(4):331–343.

26. Klatt P, Lamas S. Regulation of protein function by S-glutathiolation in response to oxidative and nitrosative stress. *Eur J Biochem* 2000; 267(16):4928–4944.

27. Dickinson DA, Forman HJ. Cellular glutathione and thiols metabolism. *Biochem Pharmacol* 2002; 64(5–6):1019–1026.

28. Paolicchi A, Dominici S, Pieri L, Maellaro E, Pompella A. Glutathione catabolism as a signaling mechanism. *Biochem Pharmacol* 2002; 64(5–6):1027–1035.

29. Rahman I, Marwick J, Kirkham P. Redox modulation of chromatin remodeling: Impact on histone acetylation and deacetylation, NF-kappaB and pro-inflammatory gene expression. *Biochem Pharmacol* 2004; 68(6):1255–1267.

30. Pastore A, Federici G, Bertini E, Piemonte F. Analysis of glutathione: Implication in redox and detoxification. *Clin Chim Acta* 2003; 333(1):19–39.

31. Jonas CR, Ziegler TR, Gu LH, Jones DP. Extracellular thiol/disulfide redox state affects proliferation rate in a human colon carcinoma (Caco2) cell line. *Free Radic Biol Med* 2002; 33(11):1499–1506.

32. Bains JS, Shaw CA. Neurodegenerative disorders in humans: The role of glutathione in oxidative stress-mediated neuronal death. *Brain Res Brain Res Rev* 1997; 25(3):335–358.

33. Horvath K, Perman JA. Autism and gastrointestinal symptoms. *Curr Gastroenterol Rep* 2002; 4(3):251–258.

34. McFadden SA. Phenotypic variation in xenobiotic metabolism and adverse environmental response: Focus on sulfur-dependent detoxification pathways. *Toxicology* 1996; 111(1–3):43–65.

35. Dringen R, Hirrlinger J. Glutathione pathways in the brain. *Biol Chem* 2003; 384(4):505–516.

36. Droge W, Breitkreutz R. Glutathione and immune function. *Proc Nutr Soc* 2000; 59(4):595–600.

37. Sen CK. Cellular thiols and redox-regulated signal transduction. *Curr Top Cell Regul* 2000; 36:1–30.

38. Filomeni G, Rotilio G, Ciriolo MR. Cell signalling and the glutathione redox system. *Biochem Pharmacol* 2002; 64(5–6):1057–1064.

39. Dickinson DA, Forman HJ. Cellular glutathione and thiols metabolism. *Biochem Pharmacol* 2002; 64(5–6):1019–1026.

40. Janaky R, Ogita K, Pasqualotto BA, Bains JS, Oja SS, Yoneda Y, et al. Glutathione and signal transduction in the mammalian CNS. *J Neurochem* 1999; 73(3):889–902.

41. Meister A. Glutathione metabolism. *Methods Enzymol* 1995; 251:3–7.

42. Thomas JA, Mallis RJ. Aging and oxidation of reactive protein sulfhydryls. *Exp Gerontol* 2001; 36(9):1519–1526.

43. Dargel R. Lipid peroxidation – a common pathogenetic mechanism? *Exp Toxicol Pathol* 1992; 44(4):169–181.

44. Shih AY, Erb H, Sun X, Toda S, Kalivas PW, Murphy TH. Cystine/glutamate exchange modulates glutathione supply for neuroprotection from oxidative stress and cell proliferation. *J Neurosci* 2006; 26(41):10514–10523.

45. Himi T, Ikeda M, Yasuhara T, Nishida M, Morita I. Role of neuronal glutamate transporter in the cysteine uptake and intracellular glutathione levels in cultured cortical neurons. *J Neural Transm* 2003; 110(12):1337–1348.

46. Machlin LJ, Bendich A. Free radical tissue damage: Protective role of antioxidant nutrients. *FASEB J* 1987; 1:441–445.

47. Yao JK, Leonard S, Reddy R. Altered glutathione redox state in schizophrenia. *Dis Markers* 2006; 22(1–2):83–93.

48. Frey BN, Andreazza AC, Kunz M, Gomes FA, Quevedo J, Salvador M, et al. Increased oxidative stress and DNA damage in bipolar disorder: A twin-case report. *Prog Neuropsychopharmacol Biol Psychiatry* 2007; 31(1):283–285.

49. Tilleux S, Hermans E. Neuroinflammation and regulation of glial glutamate uptake in neurological disorders. *J Neurosci Res* 2007; 85(10):2059–2070.

50. Yao JK, Reddy RD, van Kammen DP. Oxidative damage and schizophrenia: An overview of the evidence and its therapeutic implications. *CNS Drugs* 2001; 15(4):287–310.

51. Yamamoto BK, Bankson MG. Amphetamine neurotoxicity: Cause and consequence of oxidative stress. *Crit Rev Neurobiol* 2005; 17(2):87–117.

52. Loeken MR. Free radicals and birth defects. *J Matern Fetal Neonatal Med* 2004; 15(1):6–14.
53. Dringen R. Glutathione metabolism and oxidative stress in neurodegeneration. *Eur J Biochem* 2000; 267(16):4903.
54. Kovacic P, Jacintho JD. Systemic lupus erythematosus and other autoimmune diseases from endogenous and exogenous agents: Unifying theme of oxidative stress. *Mini Rev Med Chem* 2003; 3(6):568–575.
55. Muhle R, Trentacoste SV, Rapin I. The genetics of autism. *Pediatrics* 2004; 113(5):e472–e486.
56. Risch N, Spiker D, Lotspeich L, Nouri N, Hinds D, Hallmayer J, et al. A genomic screen of autism: Evidence for a multilocus etiology. *Am J Hum Genet* 1999; 65(2):493–507.
57. Wiznitzer M. Autism and tuberous sclerosis. *J Child Neurol* 2004; 19(9):675–679.
58. Humphrey A, Higgins JN, Yates JR, Bolton PF. Monozygotic twins with tuberous sclerosis discordant for the severity of developmental deficits. *Neurology* 2004; 62(5):795–798.
59. Gomez MR, Kuntz NL, Westmoreland BF. Tuberous sclerosis, early onset of seizures, and mental subnormality: Study of discordant homozygous twins. *Neurology* 1982; 32(6):604–611.
60. Mueller SG, Trabesinger AH, Boesiger P, Wieser HG. Brain glutathione levels in patients with epilepsy measured by in vivo (1)H-MRS. *Neurology* 2001; 57(8):1422–1427.
61. Bellissimo MI, Amado D, Abdalla DS, Ferreira EC, Cavalheiro EA, Naffah-Mazzacoratti MG. Superoxide dismutase, glutathione peroxidase activities and the hydroperoxide concentration are modified in the hippocampus of epileptic rats. *Epilepsy Res* 2001; 46(2):121–128.
62. Sudha K, Rao AV, Rao A. Oxidative stress and antioxidants in epilepsy. *Clin Chim Acta* 2001; 303(1–2):19–24.
63. Bolton PF, Park RJ, Higgins JN, Griffiths PD, Pickles A. Neuro-epileptic determinants of autism spectrum disorders in tuberous sclerosis complex. *Brain* 2002; 125(Pt 6):1247–1255.
64. Hrdlicka M, Komarek V, Propper L, Kulisek R, Zumrova A, Faladova L, et al. Not EEG abnormalities but epilepsy is associated with autistic regression and mental functioning in childhood autism. *Eur Child Adolesc Psychiatry* 2004; 13(4):209–213.
65. Sierra C, Vilaseca MA, Brandi N, Artuch R, Mira A, Nieto M, et al. Oxidative stress in Rett syndrome. *Brain Dev* 2001; 23(Suppl 1):S236–S239.
66. Piven J, Gayle J, Landa R, Wzorek M, Folstein S. The prevalence of fragile X in a sample of autistic individuals diagnosed using a standardized interview. *J Am Acad Child Adolesc Psychiatry* 1991; 30(5):825–830.
67. Valinluck V, Tsai HH, Rogstad DK, Burdzy A, Bird A, Sowers LC. Oxidative damage to methyl-CpG sequences inhibits the binding of the methyl-CpG binding domain (MBD) of methyl-CpG binding protein 2 (MeCP2). *Nucleic Acids Res* 2004; 32(14):4100–4108.
68. Berry-Kravis E. Epilepsy in fragile X syndrome. *Dev Med Child Neurol* 2002; 44(11):724–728.
69. Moretti P, Zoghbi HY. MeCP2 dysfunction in Rett syndrome and related disorders. *Curr Opin Genet Dev* 2006; 16(3):276–281.
70. Sandberg G, Schalling M. Effect of *in vitro* promoter methylation and CGG repeat expansion on *FMR-1* expression. *Nucleic Acids Res* 1997; 25(14):2883–2887.
71. Amir RE, Van dV I, Wan M, Tran CQ, Francke U, Zoghbi HY. Rett syndrome is caused by mutations in X-linked MECP2, encoding methyl-CpG-binding protein 2. *Nat Genet* 1999; 23(2):185–188.
72. Kent L, Evans J, Paul M, Sharp M. Comorbidity of autistic spectrum disorders in children with Down syndrome. *Dev Med Child Neurol* 1999; 41(3):153–158.
73. Zitnanova I, Korytar P, Sobotova H, Horakova L, Sustrova M, Pueschel S, et al. Markers of oxidative stress in children with Down syndrome. *Clin Chem Lab Med* 2006; 44(3):306–310.

74. Zana M, Janka Z, Kalman J. Oxidative stress: A bridge between Down's syndrome and Alzheimer's disease. *Neurobiol Aging* 2007; 28(5):648–676.
75. Pastore A, Tozzi G, Gaeta LM, Giannotti A, Bertini E, Federici G, et al. Glutathione metabolism and antioxidant enzymes in children with Down syndrome. *J Pediatr* 2003; 142(5):583–585.
76. Chango A, Abdennebi-Najar L, Tessier F, Ferre S, Do S, Gueant JL, et al. Quantitative methylation-sensitive arbitrarily primed PCR method to determine differential genomic DNA methylation in Down Syndrome. *Biochem Biophys Res Commun* 2006; 349(2):492–496.
77. Baieli S, Pavone L, Meli C, Fiumara A, Coleman M. Autism and phenylketonuria. *J Autism Dev Disord* 2003; 33(2):201–204.
78. Kienzle Hagen ME, Pederzolli CD, Sgaravatti AM, Bridi R, Wajner M, Wannmacher CM, et al. Experimental hyperphenylalaninemia provokes oxidative stress in rat brain. *Biochim Biophys Acta* 2002; 1586(3):344–352.
79. Sierra C, Vilaseca MA, Moyano D, Brandi N, Campistol J, Lambruschini N, et al. Antioxidant status in hyperphenylalaninemia. *Clin Chim Acta* 1998; 276(1):1–9.
80. Martinez-Cruz F, Pozo D, Osuna C, Espinar A, Marchante C, Guerrero JM. Oxidative stress induced by phenylketonuria in the rat: Prevention by melatonin, vitamin E, and vitamin C. *J Neurosci Res* 2002; 69(4):550–558.
81. Tchantchou F, Graves M, Ortiz D, Chan A, Rogers E, Shea TB. S-adenosyl methionine: A connection between nutritional and genetic risk factors for neurodegeneration in Alzheimer's disease. *J Nutr Health Aging* 2006; 10(6):541–544.
82. Purohit V, Abdelmalek MF, Barve S, Benevenga NJ, Halsted CH, Kaplowitz N, et al. Role of S-adenosylmethionine, folate, and betaine in the treatment of alcoholic liver disease: Summary of a symposium. *Am J Clin Nutr* 2007; 86(1):14–24.
83. Williams G, King J, Cunningham M, Stephan M, Kerr B, Hersh JH. Fetal valproate syndrome and autism: Additional evidence of an association. *Dev Med Child Neurol* 2001; 43(3):202–206.
84. Stromland K, Philipson E, Andersson GM. Offspring of male and female parents with thalidomide embryopathy: Birth defects and functional anomalies. *Teratology* 2002; 66(3):115–121.
85. Holmes AS, Blaxill MF, Haley BE. Reduced levels of mercury in first baby haircuts of autistic children. *Int J Toxicol* 2003; 22(4):277–285.
86. Blaxill MF, Redwood L, Bernard S. Thimerosal and autism? A plausible hypothesis that should not be dismissed. *Med Hypotheses* 2004; 62(5):788–794.
87. Alonso-Aperte E, Ubeda N, Achon M, Perez-Miguelsanz J, Varela-Moreiras G. Impaired methionine synthesis and hypomethylation in rats exposed to valproate during gestation. *Neurology* 1999; 52(4):750–756.
88. Hishida R, Nau H. VPA-induced neural tube defects in mice. I. Altered metabolism of sulfur amino acids and glutathione. *Teratog Carcinog Mutagen* 1998; 18(2): 49–61.
89. Tong V, Teng XW, Chang TK, Abbott FS. Valproic acid I: Time course of lipid peroxidation biomarkers, liver toxicity, and valproic acid metabolite levels in rats. *Toxicol Sci* 2005; 86(2):427–435.
90. Sinn DI, Kim SJ, Chu K, Jung KH, Lee ST, Song EC, et al. Valproic acid-mediated neuroprotection in intracerebral hemorrhage via histone deacetylase inhibition and transcriptional activation. *Neurobiol Dis* 2007; 26(2):464–472.
91. Schulpis KH, Lazaropoulou C, Regoutas S, Karikas GA, Margeli A, Tsakiris S, et al. Valproic acid monotherapy induces DNA oxidative damage. *Toxicology* 2006; 217(2–3):228–232.
92. Marchion DC, Bicaku E, Daud AI, Sullivan DM, Munster PN. Valproic acid alters chromatin structure by regulation of chromatin modulation proteins. *Cancer Res* 2005; 65(9):3815–3822.

93. Hansen JM, Harris KK, Philbert MA, Harris C. Thalidomide modulates nuclear redox status and preferentially depletes glutathione in rabbit limb versus rat limb. *J Pharmacol Exp Ther* 2002; 300(3):768–776.

94. Parman T, Wiley MJ, Wells PG. Free radical-mediated oxidative DNA damage in the mechanism of thalidomide teratogenicity. *Nat Med* 1999; 5(5):582–585.

95. James SJ, Slikker W III, Melnyk S, New E, Pogribna M, Jernigan S. Thimerosal neurotoxicity is associated with glutathione depletion: protection with glutathione precursors. *Neurotoxicology* 2005; 26(1):1–8.

96. Baskin DS, Ngo H, Didenko VV. Thimerosal induces DNA breaks, caspase-3 activation, membrane damage, and cell death in cultured human neurons and fibroblasts. *Toxicol Sci* 2003; 74(2):361–368.

97. Makani S, Gollapudi S, Yel L, Chiplunkar S, Gupta S. Biochemical and molecular basis of thimerosal-induced apoptosis in T cells: A major role of mitochondrial pathway. *Genes Immun* 2002; 3(5):270–278.

98. Kenchappa RS, Diwakar L, Annepu J, Ravindranath V. Estrogen and neuroprotection: Higher constitutive expression of glutaredoxin in female mice offers protection against MPTP-mediated neurodegeneration. *FASEB J* 2004; 18(10):1102–1104.

99. Green PS, Simpkins JW. Neuroprotective effects of estrogens: Potential mechanisms of action. *Int J Dev Neurosci* 2000; 18(4–5):347–358.

100. Behl C, Skutella T, Lezoualc'h F, Post A, Widmann M, Newton CJ, et al. Neuroprotection against oxidative stress by estrogens: structure-activity relationship. *Mol Pharmacol* 1997; 51(4):535–541.

101. Roof RL, Hall ED. Gender differences in acute CNS trauma and stroke: Neuroprotective effects of estrogen and progesterone. *J Neurotrauma* 2000; 17(5):367–388.

102. Behl C. Estrogen can protect neurons: Modes of action. *J Steroid Biochem Mol Biol* 2002; 83(1–5):195–197.

103. Vina J, Sastre J, Pallardo F, Borras C. Mitochondrial theory of aging: importance to explain why females live longer than males. *Antioxid Redox Signal* 2003; 5(5):549–556.

104. Wu GY, Fang YZ, Yang S, Lupton JR, Turner ND. Glutathione metabolism and its implications for health. *J Nutr* 2004; 134(3):489–492.

105. Ibim S, Randall R, Han P. Modulation of hepatic glucose-6-phosphate dehydrogenase activity in male and female rats by estrogen. *Life Sci* 1994; 45:1559–1565.

106. Lavoie JC, Rouleau T, Truttmann AC, Chessex P. Postnatal gender-dependent maturation of cellular cysteine uptake. *Free Radic Res* 2002; 36(8):811–817.

107. Lavoie JC, Chessex P. Gender and maturation affect glutathione status in human neonatal tissues. *Free Radic Biol Med* 1997; 23(4):648–657.

108. Du L, Bayir H, Lai Y, Zhang X, Kochanek PM, Watkins SC, et al. Innate gender-based proclivity in response to cytotoxicity and programmed cell death pathway. *J Biol Chem* 2004; 279(37):38563–38570.

109. Dimitrova KR, DeGroot K, Myers AK, Kim YD. Estrogen and homocysteine. *Cardiovasc Res* 2002; 53:577–588.

110. Dimitrova KR, DeGroot KW, Suyderhoud JP, Pirovic EA, Munro TJ, Wieneke J, et al. 17-beta estradiol preserves endothelial cell viability in an in vitro model of homocysteine-induced oxidative stress. *J Cardiovasc Pharmacol* 2002; 39(3):347–353.

111. White JF. Intestinal Pathology in Autism. *Exp Biol Med* 2003; 228:639–649.

112. White JF. Intestinal pathophysiology in autism. *Exp Biol Med (Maywood)* 2003; 228(6):639–649.

113. Jefferies H, Bot J, Coster J, Khalil A, Hall JC, McCauley RD. The role of glutathione in intestinal dysfunction. *J Invest Surg* 2003; 16(6):315–323.

114. Martensson J, Jain A, Meister A. Glutathione is required for intestinal function. *Proc Natl Acad Sci USA* 1990; 87(5):1715–1719.

115. Deykin EY, MacMahon B. Viral exposure and autism. *Am J Epidemiol* 1979; 109(6):628–638.

116. Nicolson GL, Gan R, Nicolson NL, Haier J. Evidence for Mycoplasma ssp., Chlamydia pneunomiae, and human herpes virus-6 coinfections in the blood of patients with autistic spectrum disorders. *J Neurosci Res* 2007; 85(5):1143–1148.
117. Cai J, Chen Y, Seth S, Furukawa S, Compans RW, Jones DP. Inhibition of influenza infection by glutathione. *Free Radic Biol Med* 2003; 34(7):928–936.
118. Kalebic T, Kinter A, Poli G, Anderson ME, Meister A, Fauci AS. Suppression of human immunodeficiency virus expression in chronically infected monocytic cells by glutathione, glutathione ester, and N-acetylcysteine. *Proc Natl Acad Sci USA* 1991; 88(3):986–990.
119. Bailey A, Le Couteur A, Gottesman I, Bolton P, Simonoff E, Yuzda E, et al. Autism as a strongly genetic disorder: Evidence from a British twin study. *Psychol Med* 1995; 25(1):63–77.
120. Kates WR, Burnette CP, Eliez S, Strunge LA, Kaplan D, Landa R, et al. Neuroanatomic variation in monozygotic twin pairs discordant for the narrow phenotype for autism. *Am J Psychiatry* 2004; 161(3):539–546.
121. Grandjean P, Landrigan PJ. Developmental neurotoxicity of industrial chemicals. *Lancet* 2006; 368(9553):2167–2178.
122. Zaibo L, Dong T, Proschel C, Noble M. Chemically diverse toxicants converge on Fyn and c-Clb to disrupt precursor cell function. *PLoS Biol* 2007; 5:212–231.
123. Zhao X, Leotta A, Kustanovich V, Lajonchere C, Geschwind DH, Law K, et al. A unified genetic theory for sporadic and inherited autism. *Proc Natl Acad Sci USA* 2007; 104(31):12831–12836.
124. Adinolfi M. The development of the human blood-CSF-brain barrier. *Dev Med Child Neurol* 1985; 27(4):532–537.
125. Ghersi-Egea JF, Strazielle N, Murat A, Jouvet A, Buenerd A, Belin MF. Brain protection at the blood-cerebrospinal fluid interface involves a glutathione-dependent metabolic barrier mechanism. *J Cereb Blood Flow Metab* 2006; 26(9):1165–1175.
126. Sakamoto M, Kubota M, Liu XJ, Murata K, Nakai K, Satoh H. Maternal and fetal mercury and n-3 polyunsaturated fatty acids as a risk and benefit of fish consumption to fetus. *Environ Sci Technol* 2004; 38(14):3860–3863.
127. Cai J, Chen Y, Seth S, Furukawa S, Compans RW, Jones DP. Inhibition of influenza infection by glutathione. *Free Radic Biol Med* 2003; 34(7):928–936.
128. Lante F, Meunier J, Guiramand J, Maurice T, Cavalier M, Jesus Ferreira MC, et al. Neurodevelopmental damage after prenatal infection: role of oxidative stress in the fetal brain. *Free Radic Biol Med* 2007; 42(8):1231–1245.
129. Thorburne SK, Juurlink BH. Low glutathione and high iron govern the susceptibility of oligodendroglial precursors to oxidative stress. *J Neurochem* 1996; 67(3):1014–1022.
130. Herbert MR. Autism: A brain disorder or a disorder that affects the brain. *Clin Neuropsych* 2005; 2:2354–2379.
131. Horvath K, Perman JA. Autistic disorder and gastrointestinal disease. *Curr Opin Pediatr* 2002; 14(5):583–587.
132. Jyonouchi H, Sun SN, Itokazu N. Innate immunity associated with inflammatory responses and cytokine production against common dietary proteins in patients with autism spectrum disorder. *Neuropsychobiology* 2002; 46(2):76–84.
133. Ashwood P, Murch SH, Anthony A, Pellicer AA, Torrente F, Thomson MA, et al. Intestinal lymphocyte populations in children with regressive autism: Evidence for extensive mucosal immunopathology. *J Clin Immunol* 2003; 23(6):504–517.
134. Wakefield AJ, Ashwood P, Limb K, Anthony A. The significance of ileo-colonic lymphoid nodular hyperplasia in children with autistic spectrum disorder. *Eur J Gastroenterol Hepatol* 2005; 17(8):827–836.
135. Shaw CA, Bains JS. Synergistic *versus* antagonistic actions of glutamate and glutathione: The role of excitotoxicity and oxidative stress in neuronal disease. *Cell Mol Biol* 2002; 48(2):127–136.

136. Bains JS, Shaw CA. Neurodegenerative disorders in humans: The role of glutathione in oxidative stress-mediated neuronal death. *Brain Res Brain Res Rev* 1997; 25(3):335–358.
137. Risher JF, Murray HE, Prince GR. Organic mercury compounds: Human exposure and its relevance to public health. *Toxicol Ind Health* 2002; 18(3):109–160.
138. Finkelstein JD. Methionine metabolism in mammals. *J Nutr Biochem* 1990; 1:228–237.
139. Mato JM, Corrales FJ, Lu SC, Avila MA. S-adenosylmethionine: A control switch that regulates liver function. *FASEB J* 2002; 16(1):15–26.
140. Finkelstein JD. The metabolism of homocysteine: Pathways and regulation. *Eur J Pediatr* 1998; 157:S40–S44.

Color Plates

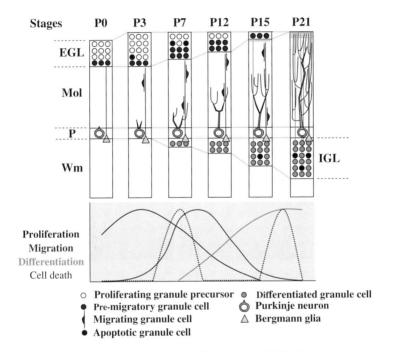

Color Plate 1. Granule neuron precursors proliferate superficially in the external germinal layer (EGL), then migrate inwardly along Bergmann glia. (Chapter 1, Fig. 2; *see* complete caption and discussion on p. 9)

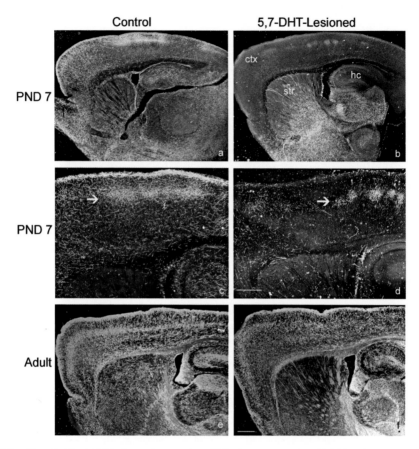

Color Plate 2. Dark field photomicrographs of 5-hydroxytryptophan (5-HT) staining in parasagittal sections from control [age normal (**a**, **e**), saline injected (**c**)] and 5,7-dihydroxydopamine (5,7-DHT)-lesioned (**b**, **d**, **f**) mice at postnatal day (PND) 7 (**a–d**) and in the adult (**e**, **f**). (Chapter 5, Fig. 1; *see* complete caption and discussion on p. 117)

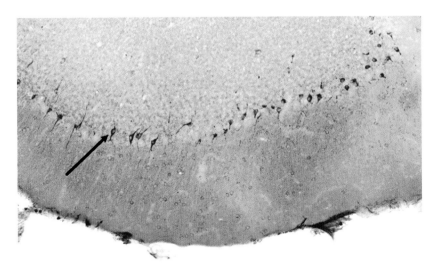

Color Plate 3. Immunohistochemistry performed on monkey cerebellum with plasma from a subject with autism demonstrating intense cytoplasmic staining of Golgi cells (arrow). (Chapter 12, Fig. 1; *see* complete caption and discussion on p. 274)

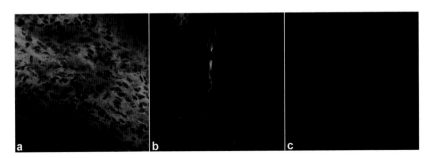

Color Plate 4. Immunofluorescence using FITC anti-human IgG against sectioned mouse brain from newborn pups of dams injected with: **(a)** IgG from mothers of children with autistic disorder (MCAD); **(b)** IgG from mothers with unaffected children (MUC); and **(c)** vehicle (Chapter 14, Fig. 1; *see* discussion on p. 318)

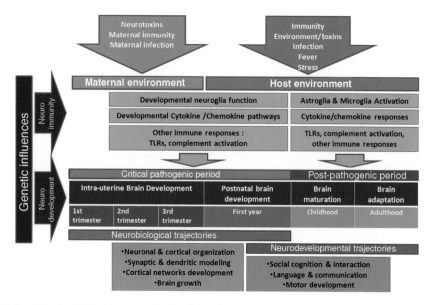

Color Plate 5. Mechanisms involved in the pathogenesis of autism spectrum disorders (ASD) likely affect brain development during the critical pathogenic period of intrauterine life and the first postnatal year. (Chapter 15, Fig. 3; *see* complete caption on p. 338 and discussion on p. 337)

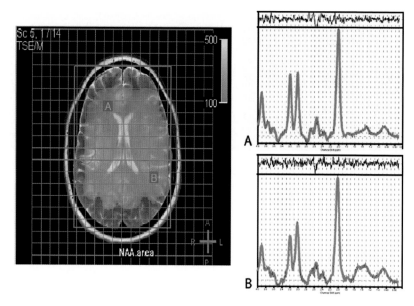

Color Plate 6. Example of MRS imaging techniques (MRSI) acquisition overlaid with NAA chemical map; representative gray matter **(A)** and white matter **(B)** spectra on right (Chapter 17, Fig. 1; *see* discussion on p. 366)

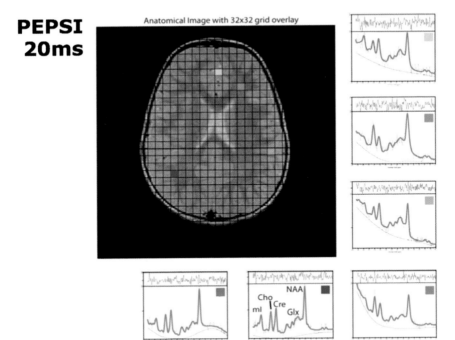

Color Plate 7. A magnetic resonance spectroscopy (MRS) imaging techniques (MRSI) slab location (20 mm thick) overlaid by the 32×32 MRSI axial matrix (Chapter 17, Fig. 2; *see* discussion on p. 366)

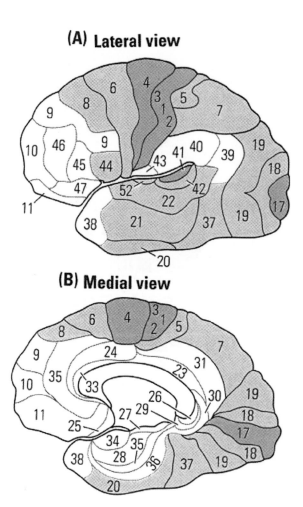

Color Plate 8. A map of Brodmann's areas that divides the cortex into primary sensory and motor cortices, and association cortex into unimodal and heteromodal cortices and assigns numbers to specialized regions. (Chapter 18, Fig. 1; *see* discussion on p. 387)

Part IV
Immunology, Maternal-Fetal Interaction, and Neuroinflammation

Chapter 12
The Immune System in Autism

Is There a Connection?

Luke Heuer, Paul Ashwood and Judy Van de Water

Abstract Autistic spectrum disorders (ASD) are a broad spectrum of hetero-geneous neurodevelopmental disorders, most with unknown etiology. Although the development of autism is suspected to be the result of a complex combination of genetic, environmental, neurobiological, and immunological factors, the role of each of these factors is still under investigation. In this chapter, we review current concepts regarding the immunological aspects of autism as they pertain to areas of our own research. Three specific areas are covered including auto-antibodies in children with ASD, maternal antibodies directed against fetal brain, and immune dysfunction in children with ASD.

Keywords Autism · immunology · maternal · autoantibodies · fetus

Introduction

Autistic spectrum disorders (ASD) are a broad spectrum of heterogeneous neurodevelopmental disorders known as pervasive developmental disorders (PDD), which include autism, Asperger's syndrome, Rett's disorder, and child-hood disintegrative disorder. By definition, ASD are characterized by distur-bances and impairments in social interaction, verbal and non-verbal communication, and imagination [1] with onset usually occurring in the first 36 months of childhood. Repetitive and stereotyped behaviors as well as attention and sensory abnormalities are common findings in patients with ASD. Recently, the prevalence of ASD has increased dramatically [2]. Reports estimate ASD to affect approximately one per 150 persons, with a biased male to female ratio of three or four to one (3–4:1) [3]. The exact etiologies of autism and ASD remain largely unknown, although they are likely to result from a complex combination of environmental, neurobiological, immunological and genetic factors.

J. Van de Water
Division of Rheumatology/Allergy and Clinical Immunology, University of California, 451 E. Health Sciences Drive, Suite 6510 GBSF, Davis, CA 95616, USA
e-mail: javandewater@ucdavis.edu

A.W. Zimmerman (ed.), *Autism*, DOI: 10.1007/978-1-60327-489-0_12,
© Humana Press, Totowa, NJ 2008

Strong genetic links have been shown for cases with tuberous sclerosis, Fragile X syndrome, neurofibromatosis, and chromosomal abnormalities [4, 5, 6]. Population-based twin studies have demonstrated a higher concordance rate for ASD among monozygotic twins compared with dizygotic twins [7]. The concordance rate in monozygotic twins is estimated to be approximately 90% when considering a broader phenotype of ASD, whereas in dizygotic twins this rate is between 0 and 24% [8, 9, 10, 11, 12]. Furthermore, the familial risk is five to ten times higher than the general population, with the ASD rate in siblings of ASD children estimated between 2 and 6% [5]. One of the most confounding aspects of ASD is the phenotypic heterogeneity that encompasses these disorders, suggesting that ASD may actually comprise several disorders with separate and specific etiologies that all share a common behavioral phenotype. As many as 15 weak gene interactions are thought to play a role in their etiologies [7, 13]. These include genes involved in the patterning of the central nervous system (CNS) (including reelin, bcl-2, engrailed-2,Wnt), those that govern biochemical pathways (such as serotonin transporter gene variants, PTEN, MET, and Ca(V)1.2), those responsible for the development of dendrites and synapses (BDNF, MeCP2, neuroligin), and genes associated with the immune system and autoimmune disorders (chromosome 6, HLA-DRB1*04, complement component C4B) [14, 15, 16, 17, 18, 19, 20, 21, 22]. However, the recent dramatic increase in prevalence as well as the lack of 100% concordance in monozygotic twins indicates that other factors such as environmental influences could potentially play a role in the etiology of autism.

Numerous physical and chemical environmental agents have been shown to produce developmental toxicity in humans. These include organic mercury compounds and pharmaceuticals such as thalidomide and the anti-seizure medication valproic acid [23, 24, 25, 26]. In addition, chemical insults from the environment, prenatal rubella infection, perinatal hypoxia, post-natal infections such as encephalitis, and metabolic disorders such as phenylketonuria have been associated with ASD [27]. Although there have been several genetic linkage studies published, no one gene or set of genes has been associated with 100% of autism cases [28, 29, 30, 31, 32, 33, 34, 35, 36, 37, 38, 39, 40, 41, 42, 43]. In this sense, the reported weak gene interactions that are thought to play a role in the etiology of autism, while probably contributory, are likely not the sole risk factors.

Interestingly, several of the chromosomal regions reported to be associated with autism include genes critical to immune function [28, 40, 43]. In addition, environmental agents associated with autism have the ability to affect both the nervous system and the immune system. There is emerging evidence and growing concern that a dysregulated or abnormal immune response may be involved in ASD. In general, the links between the immune and the nervous systems are becoming increasingly well known. Cytokines and other products of immune activation have widespread effects on neuronal pathways and can alter behaviors such as mood and sleep [44, 45]. Aberrant immune activity during critical periods of brain and neuronal development could potentially play a role in neural dysfunction that is typical of autism.

Immune findings in various studies regarding patients with ASD are often inconsistent because of many factors, most notably the use of inappropriate control groups and heterogenous, ill-defined subject groups. The interpretation of results may be further complicated because of comorbidities, such as mental retardation, epilepsy, sleep disorders, and gastrointestinal symptoms often found in ASD. Various immune system components and mediators including cytokine and immunoglobulin levels, cellular numbers and responsiveness, monocytes/macrophages, and natural killer cells have all been investigated in ASD [46, 47, 48, 49, 50]. Despite the lack of consensus, it is widely agreed that a subset of patients with ASD demonstrate abnormal or dysregulated immunity. In this chapter, we will explore the findings of our laboratory in the context of the current field of immunology research in autism and discuss their implications with respect to the pathogenesis of ASD. We will cover three distinct areas within our current research, which include autoantibodies in children with ASD, maternal antibodies directed against fetal brain, and immune dysfunction in children with ASD.

Autoantibodies in Autism Spectrum Disorder

One of the most commonly reported immune system-related abnormalities in subjects with ASD is the presence of autoantibodies to various nervous system "self" components. These include antibodies to myelin basic protein, a serotonin receptor, neurofilament proteins, brain endothelial cell proteins, heat shock protein as well as simultaneous elevation of cerebellar and dietary opioid peptide autoantibodies [51, 52, 53, 54, 55, 56, 57, 58, 59, 60]. However, it is important to note that many of these autoantibodies are not unique to individuals with ASD, nor are they found in all subjects. For example, MBP self-reactive antibodies are also found in patients with multiple sclerosis [61] and normal controls [51]. More specifically, a recent study by Singer et al. demonstrated serum anti-brain autoantibodies using both ELISA and western blots. Interestingly, the subjects with autism as well as their non-autistic siblings had bands with greater density at 73 kDa in the cerebellum and cingulate gyrus when compared with controls [60].

In our approach to this body of evidence, we decided to look not for antibodies directed against specific brain proteins, but rather at targeted brain regions and whether children with ASD had any antibodies directed against proteins specific to that region. By western blot, 29% of subjects with ASD were observed to have intense specific reactivity against proteins from the human thalamus at ~52 kDa compared with 8% of typically developing and sibling controls. No reactivity was noted in the developmental delay group [62]. Previous studies utilizing western blots have pointed researchers in the direction of what brain regions may be affected in ASD, but conformational changes that occur during this process may mistakenly omit proteins reactive to autoantibodies in the plasma of patients. Low-abundance proteins, such as Golgi cell proteins, may also fail to be recognized by western blot. The use of immunohistochemistry (IHC) circumvents both of these issues. In this study, immunohistochemical staining showed that intense Golgi cell staining was observed in 21% (7) of subjects with autism, compared to 0% of typically developing controls and 0% of subjects with developmental delay. Intense staining was found to be present in 7% (1) sibling of a subject with ASD (Fig. 12.1 and see Color Plate 3, following p. XX). Such reactivity was not restricted by diagnosis; staining was distributed equally among patients diagnosed with early onset autism and those diagnosed with autism following regression [63]. The presence of autoantibodies in subjects with autism as well as their typically developing siblings may denote possible environmental or genetic factors in autoantibody development.

One must ask what the role of such autoantibodies may be in autism. The specific area of the brain targeted by the autoantibodies described above suggests that the Golgi cells may be at the very least indirectly affected. The Golgi cell of the cerebellum is a large interneuron located in the granular layer that acts as a brake, modulating the activity of the mossy fiber to granule cell relay, whose input will leave the cerebellum through the Purkinje cell [64, 65]. Damage to this regulatory system may disrupt normal developmental signaling, or result in an excess of excitatory signals relayed to Purkinje cells, resulting in

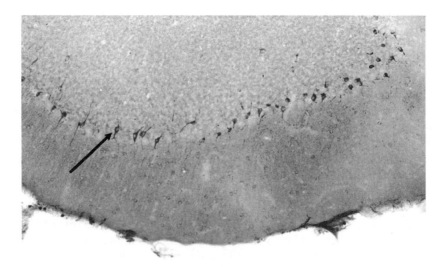

Fig. 12.1 Immunohistochemistry performed on monkey cerebellum with plasma from a subject with autism demonstrating intense cytoplasmic staining of Golgi cells (arrow). This staining pattern was rarely seen in control subjects, and never at this intensity (*see* Color Plate 3)

neuronal death or damage. Interestingly, abnormal sensitivity to light and/or sound has been reported in some individuals with ASD [66]. Areas in the brain involved in the auditory and visual circuits, such as the thalamus, have been reported as abnormal in some high-functioning subjects with ASD [67]. This could be achieved by autoantibody acting as a ligand, causing hyperactivity or possibly excitotoxic death through excessive signaling. These alterations could result in atypical development, resulting in neuroanatomical changes, stereotypical behavior, and epilepsy, pathologies that are present in a subset of patients with ASD. Alternatively, if the autoantibody acts as an antagonist, the absence of necessary function or signaling during neurodevelopment is also capable of leading to abnormalities in the developing nervous system. Either of these scenarios may result in permanent alteration in receptor density as well as in neurotransmitter/cytokine release in both systems (Fig. 12.2).

Elevated levels of circulating autoantibodies to nervous system components in comparison with normal controls have been reported in a number of psychiatric disorders including schizophrenia (brain structures, lymphocytes), obsessive-compulsive disorder (brain), neuropsychiatric symptoms associated with systemic lupus erythematosus (SLE) (cross-reactivity of anti-DNA antibodies with NMDA receptors) and Tourette syndrome (TS) (neural proteins), Sydenham chorea and pediatric autoimmune neuropsychiatric disorders associated with streptococcal infection (PANDAS) [68, 69, 70, 71, 72, 73]. In the case of SLE, it was demonstrated in a mouse model that with blood–brain

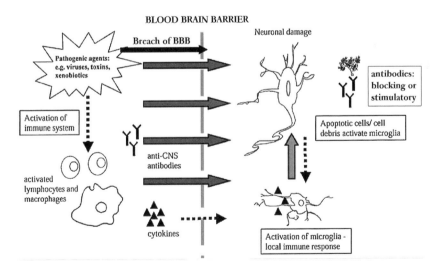

Fig. 12.2 In this model, two insults are required to occur: (1) The peripheral immune system must be activated to produce autoantibodies directed against brain proteins. (2) The blood-brain barrier (BBB) must be abrogated to allow these immune mediators access to brain proteins. Once granted access, the autoantibodies can act as antagonists to block normal neurodevelopment, or they may act as agonists to over stimulate neurons and cause excitation-induced damage to the area

barrier (BBB) abrogation, in this case induced by LPS, systemic autoantibodies associated with SLE (anti-dsDNA and anti-NR2 antibodies, a subunit of the NMDA receptor) were capable of causing excitotoxic neuronal death in the hippocampus and cognitive deficits (memory loss) *in the absence of an inflammatory response* (i.e., there was no complement deposition, immune cell infiltration, or microglial/astrocytic activation) [71]. Changes in BBB permeability can be caused by a number of factors, including stress, subclinical infection (either directly or through cytokines), nicotine or epinephrine exposure [74, 75, 76, 77]. Some of the above findings are controversial for various reasons, but the observations each point toward the prospect of autoantibodies causing neural dysfunction, leading to cognitive changes.

More recent findings from Huerta et al. have shown that antibody-mediated damage is also capable of causing emotional changes, in particular, the conditioned fear response in a mouse model [78]. Together, these studies demonstrate that antibodies in the circulation are capable of reaching the brain when there is a BBB disturbance and that the area affected may be dictated by the factor responsible for BBB abrogation. This prospect is of interest in ASD because of the heterogeneity of the symptoms and neuroanatomical abnormalities seen in subjects with these disorders. Additionally, autoantibodies to intracellular antigens penetrating the cell membrane (in vitro and in vivo) and interacting with their antigens have been well documented in the literature in autoimmune diseases such as Sjøgren's syndrome and SLE [79, 80, 81]. Mechanistic studies have proposed that these IgG antibodies can enter cells via cross reactivity with various cell surface proteins resulting in apoptosis of the cell [82, 83, 84].

The autoantibodies reported in patients with ASD thus far have not been directly associated with pathology, and may represent epiphenomena. However, the identification and characterization of such brain-specific autoantibodies may be of importance in the discovery of which neurodevelopmental processes and/or brain structures are disturbed in individuals with ASD. Additionally, it is important to keep in mind that the presence of neural autoantibodies is not abnormal and has been reported in normal, healthy controls. It is possible that the absence of certain genetic and environmental factors reduces the potential of these antibodies to interact with their respective antigens. Elevated levels present in the plasma of subjects with ASD compared with controls may be indicative of altered development or injury of the CNS, changes in the immune system, or both.

The brain is commonly described as an *immune-privileged* site, but better terminology would describe it as an *immunologically customized* site. Immune reactivity is tightly regulated in the CNS, including the presence of autoreactive T cells, which have been shown to have the ability to maintain, protect, as well as repair the CNS, including following injury. This has been demonstrated to occur through the production of neurotrophic molecules in a spinal cord crush injury in a mouse model [85, 86]. As in other areas of the body, immune surveillance also occurs routinely in the CNS, even in the presence of an intact

BBB [87]. A change in this carefully balanced system may be capable of inducing aberrant immune responses, leading to autoimmunity.

The increased recognition of immune reactivity to what morphologically appear to be Golgi cells of the cerebellum in plasma from subjects with ASD compared with controls alerts us to potential insults taking place in at least one region of the brain. Should this occur during a vulnerable time period in neurodevelopment, such an insult may lead to neurochemical, neuroanatomical, and/or immune-related irregularities that could then result in one of the many phenotypes that are categorized as ASD. Autoantibodies may be either the cause, or a result, of abnormal CNS development that takes place in children with ASD. The correlation between the presence of the ~52-kDa band and the presence of Golgi cell staining in subjects with ASD points us toward the identification of a potentially relevant neural autoantigen.

As mentioned previously, a single insult during a critical period of development is capable of causing a variety of changes that may permanently modify further neurodevelopment. Such insults include a disturbed cytokine milieu, or altered neuropeptide and/or neurochemical levels. Strict classification of subject and control groups, as well as further analysis of the connections between disturbances in both the nervous and the immune systems, will help us to better interpret study results and understand this enigmatic spectrum of disorders.

Maternal Antibodies to Fetal Brain

The role of the maternal immune system in fetal neurodevelopment is an area of active research. It has long been known that in humans, maternal IgG isotype antibodies readily cross the placenta and equip the immunologically naïve fetus with a subset of the maternal adaptive humoral immune system proteins [88]; these maternal IgG antibodies are known to persist for up to 6 months postnatally [89]. However, together with IgG antibodies that are immunoprotective, autoantibodies that react to fetal "self" proteins can also cross the placenta. A recent report demonstrated maternal IgG antibody reactivity to rodent Purkinje cells in serum from a mother of one child with autism and another with dyslexia; when injected into gestating mice, there were behavioral deficits in the pups [90]. In another small study, mothers of children with autism and their affected children were found to have consistent patterns of antibody reactivity against rat prenatal (day 18) brain proteins. In contrast, unaffected children and control mothers had alternative patterns of reactivity [91].

The preponderance of evidence suggests a prenatal or early postnatal etiology for autism, potentially involving errant developmental cues. Advances in understanding the role of immune system components during fetal neurodevelopment combined with the cross-talk between the maternal and the fetal immune systems led us to investigate the profiles of autoantibody reactivity in mothers of children with autism and to compare them with profiles from mothers of typically developing children and from mothers of children with other developmental disorders excluding autism.

Our studies have identified the presence of autoantibodies to proteins at 37 and 73 kDa in human fetal brain, which occur significantly more often in mothers of children with autism when compared with two distinct control populations [92]. Bands that were shown to be different between autism and controls were seen in more than 25% of mothers who had children with autism. The fact that these bands are not found in all mothers of children with autism further emphasizes the heterogeneity that is widely reported in autism, and the variety of potential etiologic mechanisms that may exist [93]. Previous studies have also suggested a role for maternal antibodies in the etiology of some cases of autism [90]. Moreover as noted above, recently Zimmerman et al. reported differing patterns of serum immunoreactivity to prenatal rat brain between mothers of children with autism and mothers of control children. Furthermore, the authors demonstrated that immunoreactivity persisted in maternal circulation for up to 18 years post-delivery [91]. Interestingly, the group differences in brain reactivity patterns were observed only with pre-natal rat brain protein and not post-natal (day 8) rat brain protein. The patterns described in this study in rat brain differ from our observations, possibly because of disparities between rat and human brain proteins or differences in sample processing. However, the presence of maternal antibody reactivity against brain proteins and the association with an outcome of autism in the offspring is consistent between the two studies.

The transplacental passage of maternal IgG isotype antibodies has long been known as a mechanism for fetal immune instruction [88] and protection [94, 95]. A recently described organelle in the placental epithelium, which expresses the low-affinity IgG receptor, FcγRIIb, as well as the IgG receptor and transport protein FcRn, appears to provide a dedicated transport mechanism for maternal IgG to enter the fetal circulation [96]. Detectable levels of maternal IgG are present in fetal circulation as early as 18 weeks gestation, and by 38 weeks gestation, fetal levels are comparable with maternal levels. Interestingly, neonatal IgG, which is overwhelmingly maternal in origin, is seen at levels exceeding the maternal concentration at delivery and persists at detectable levels up to 6 months post-delivery [88]. Thus, the window of exposure to maternal IgG coincides substantially with critical periods of early neurodevelopment and continues well past gestation.

Despite the beneficial nature of the majority of maternal IgG received by the fetus, a number of neonatal autoimmune diseases have been demonstrated to result from pathogenic maternal IgG. Notably, the presence of maternal anti-Ro/SS-A and anti-La/SS-B antibodies cause neonatal lupus syndrome, often leading to congenital heart block [97]. In addition, cases of neonatal anti-phospholipid syndrome (APS), mediated through maternal autoantibodies, have been observed in the newborn infants of mothers with primary APS [98]. Finally, abnormal thyroid function is often noted in infants born to mother with Hashimoto's thyroiditis or Graves' disease, caused by placental transfer of maternal anti-thyroid antibodies [99]. Typically, symptoms of neonatal thyroiditis resolve as maternal antibodies are cleared from the circulation of the infant during the first 9 months following birth.

In the original study by Dalton et al., the effects of such maternal antibodies on behavior were addressed in a mouse model [90]. When gestating female mice were exposed to the serum from the mother of multiple children with autism, changes in behavior were noted in the offspring. These data suggest the potential for exposure to autoantibodies during critical phases of neurodevelopment to affect behavioral outcome.

Increasing attention has been given to the notion that autism, as a spectrum of disorders, likely encompasses numerous, etiologically distinct behavioral phenotypes. We also observe that maternal reactivity to fetal brain proteins occurs more frequently in mothers of autistic children who exhibit behavioral regression than in those with early onset autism. This may help to elucidate the biologic mechanisms contributing to phenotypic variance in ASD. Assuming that the observed maternal autoantibody reactivity was also present during the prenatal and/or early postnatal period, the association of autoantibodies to neural antigens with delayed onset autism appears paradoxical. Perhaps this may be explained by a pathogenic mechanism involving the interference of maternal autoantibodies with neurodevelopmental pathways for which compensatory mechanisms exist, but are ultimately overwhelmed, leading to disease symptoms. A similar developmental progression is noted in Rett syndrome, in which mutations in the gene *Mecp2* manifest in behavioral regression around 18 months of age [100]. Finally, it is important to note that the presence of maternal autoantibodies to both the 37- and the 73-kDa proteins does not provide an etiologic mechanism for all cases of regressive autism, and their presence is strongly associated with the regressive phenotype only in a subpopulation of individuals.

In summary, there exists substantial evidence for an association between the presence of maternal immune system biomarkers and a diagnosis of autism in a subset of children. The presence of specific anti-fetal brain antibodies in the circulation of mothers during pregnancy may be a potential trigger that, when paired with genetic susceptibility, is sufficient to induce a downstream effect on neurodevelopment leading to autism.

Immune Dysfunction in Children with ASD

Although the presence of autoantibodies appears to be noteworthy in a subset of individuals with autism, there have been several studies describing additional alterations in the immune system of children with ASD. These studies include reports of decreased peripheral lymphocyte numbers [101], decreased response to T-cell mitogens [102, 103], incomplete or partial T-cell activation evinced by increased numbers of DR + T cells without the expression of the IL-2 receptor

(IL-2R) [53, 103, 104], dysregulated apoptosis mechanisms [46], and the imbalance of serum Ig levels [101, 105]. In particular, our laboratory has observed significantly decreased levels of total IgG and IgM but not IgA or IgE, in children with ASD compared with typically developing age-matched controls [106]. Preliminary studies in our laboratory as well as published work by other researchers support the notion of an altered cytokine profile in ASD patients [39]. In addition, several publications have associated autism with immune-based genes including class II HLA-DRB1 alleles, class III complement C4 alleles, and HLA extended haplotypes [107, 108]. Moreover, evidence for an immune role in autism comes from recent animal models, which indicate that the maternal immune response to infection can influence fetal brain development through increased levels of circulating cytokines [109, 110]. For example, infection of neonatal rats with Borna disease virus (BDV) leads to neuronal death in the hippocampus, cerebellum and neocortex, and a behavioral syndrome that has similarities to autism [111]. These abnormalities correlate with major alterations of cytokine expression in various brain regions, indicating a likely role for cytokines as mediators of CNS injury in this model [112, 113]. Mouse models of maternal influenza virus infection at mid-gestation have similar neuropathological and behavioral abnormalities in the offspring, which are consistent with those seen in autism and were again suggestive of a strong immune component [110, 114]. Vargas et al. recently reported findings of an ongoing immune cytokine activation in the postmortem brains of patients with ASD [115].

The immune system and the nervous system are highly interconnected, and there is evidence that alterations of the immune system, especially during gestation, have the potential to alter behavioral phenotypes. Beginning early in development, the relationship between the immune and the nervous systems is exceedingly complex, continuing into adulthood and mediated mainly through the hypothalamo–pituitary–adrenal (HPA) axis [116, 117, 118]. Immune system factors, such as major histocompatibility complex I, cytokines and chemokines are important in many stages of neurodevelopment and CNS plasticity, functioning and maintenance. Several proteins associated with the nervous system, such as neuropeptides, have a broad range of effects on the development of the immune system and its function (suppression as well as activation), including the innervation of immune system associated organs, such as the lymph nodes and spleen [119, 120, 121, 122, 123, 124]. A carefully established equilibrium and timing of the previously mentioned parameters is vital for normal immune and CNS functioning. Changes incurred during development could cause alterations that are life long, such as alterations in receptor distribution and/or number in both systems and modifications in neuropeptides, cytokines, hormones, or neurotransmitter release both during and following neurodevelopment.

Not only do the nervous system and the immune system share extensive interconnectivity, but perhaps because of it, they also share several signaling pathways. Several molecules involved in these pathways have recently been associated with autism. These include phosphatase and tensin homologue

(PTEN), a regulatory molecule involved in the PI3K/AKT signaling pathway, that has recently been associated with a subset of autistic individuals with macrocephaly [16, 17, 18]. Selective deletion of PTEN in neurons of mice results in altered social interaction and neuropathology consistent with that reported in human ASD patients [21]. In addition, the PI3K/AKT pathway and PTEN have also been reported to be crucial to immune function as reviewed by Koyasu [20]. A genetic variant in subjects with autism has been reported to disrupt MET transcription. The MET gene codes for a receptor tyrosine kinase, a receptor that is responsible for signaling involved in neocortical and cerebellar growth and maturation, immune function, as well as gastrointestinal repair [19]. Lastly, a missense mutation in the L-type calcium channel Ca(V)1.2 results in Timothy syndrome [22]. This multi-organ disorder is characterized by a plethora of symptoms including autism and immune deficiency, as calcium channels are critical signaling components in both the immune and the nervous systems. Thus, in addition to the ability of independent changes in immune function to alter neurodevelopment and vice versa, the fact that these two systems share several signaling pathways leaves open the possibility that a common defect at this level may affect both systems, but perhaps in different ways.

Making the Final Connections

ASD is an extremely heterogeneous group of disorders with multiple phenotypes and subgroups that share behavioral commonalities. This inherent complexity has made deciphering the etiology of the broad spectrum of ASD extremely difficult. On its most basic level, we know that for ASD there are abnormalities in both the nervous system and the immune system, and that both systems are prone to genetic susceptibilities and potential environmental insults. However, because of the complex interaction between the two systems during development and throughout adulthood, there are multiple avenues for disease pathogenesis as presented in Fig. 12.3. As is represented in this model, there are three potential explanations for the observed neuroimmune phenotypes of ASD.

In the first scenario, genetic susceptibility and environmental insults result in alterations of neurodevelopment. The dysfunctional nervous system then acts on the immune system through neurotransmitters and neuroendocrine hormones to modulate its activity, resulting in the observed immune phenotype. In this setting, immune dysregulation would be considered an epiphenomenon, in that immune dysregulation is not a causative factor, but rather a side effect of an aberrant nervous system. This possibility would seem to negate the merit of studying immune dysregulation in ASD. However, even if it is an epiphenomenon, immune dysregulation is still intimately connected to alterations in the nervous system and elucidating the causative mechanisms responsible for immune dysregulation would provide invaluable insight to the true etiologies of ASD.

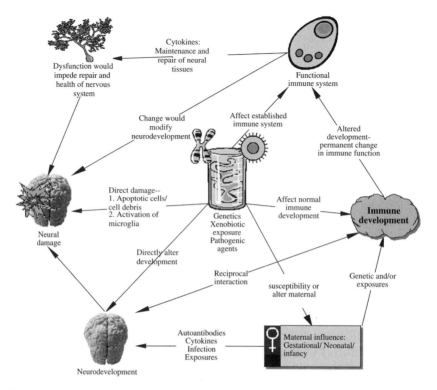

Fig. 12.3 Schematic of potential interactions between the immune and nervous systems. Possible roles of genes and environmental factors on neurologic and immunologic development are illustrated

In the second case, genetic susceptibility and environmental insults result in alterations of the immune system. An alteration in both the maternal and the neonatal immune system has the capability of affecting neurodevelopment through both cytokine production and pathogenic autoantibodies. Although cytokines are part of the normal developmental process and alterations in both timing and quantity could affect critical steps, autoantibodies would present a different etiology. Antibodies are highly specific molecules that have the ability to act as an *agonist*, to activate receptors on cell surfaces, or as an *antagonist* to block the cell from receiving a signal. Additionally, antibodies are capable of inducing cellular damage alone or in conjunction with immune cells such as microglia within the brain. In this manner, autoantibodies directed against cellular proteins and present during critical phases of neurodevelopment would alter brain physiology and subsequently behavior. Although we have shown some evidence for a pathogenic role of autoantibodies in our monkey model, there remains the argument that the presence of these autoantibodies is just an epiphenomenon. Even if this is true, the increased presence of auto-antibodies indicates a dysregulated immune response to injury in the brain. By studying both the time course of autoantibody generation and the cellular protein that is being targeted, issues that are currently being addressed, we may better understand the pathogenesis of disease progression.

In the third scenario, genetic susceptibility and environmental insults would affect a biochemical pathway common to both the nervous system and the immune system. Cells of the immune and nervous system share several common signaling pathways that include molecules such as PTEN, MET, and Ca(V)1.2, which have already been described to be dysfunctional in some cases of ASD. In this scenario, immune dysfunction may again be considered an epiphenomenon in that a shared dysfunctional signaling pathway causes altered neurodevelopment but at the same time causes an unrelated alteration in immune activity. The observed pathogenesis of ASD may, however, require dysfunction at some level in both systems. Regardless, immune cells are significantly easier to both gain access to and assay ex vivo, making the immune system a valuable tool to study dysfunction at the molecular level even if it is not a contributing factor.

Given the broad spectrum of behavioral phenotypes, as well the heterogeneity of physiological abnormalities associated with ASD, the likelihood of a single basic explanation is highly improbable. It is our contention that the pathogenesis of ASD is more likely a result of dysfunction, at some level, in both the nervous system and the immune system.

References

1. Association AP. American Psychiatric Association. *Diagnostic and Statistical Manual of Mental Disorders*. Washington, DC; 1994.
2. Fombonne E. The prevalence of autism. *JAMA* 2003; 289(1): 87–9.
3. CDC (Centers for Disease Control). Prevalence of autism spectrum disorders – autism and developmental disabilities monitoring network, 14 sites, United States, 2002. *MMWR Surveill Summ* 2007; 56(1): 12–28.
4. Wiznitzer M. Autism and tuberous sclerosis. *J Child Neurol* 2004; 19(9): 675–9.
5. Rutter M. Genetic studies of autism: From the 1970s into the millennium. *J Abnorm Child Psychol* 2000; 28(1): 3–14.
6. Cohen D, Pichard N, Tordjman S, et al. Specific genetic disorders and autism: Clinical contribution towards their identification. *J Autism Dev Disord* 2005; 35(1): 103–16.
7. Cook EH, Jr. Genetics of autism. *Child Adolesc Psychiatr Clin N Am* 2001; 10(2): 333–50.
8. Coleman M, Gillberg C. *The Biology of the Autistic Syndromes*. New York: Praeger Publishers; 1985.
9. Bailey A, Le Couteur A, Gottesman I, et al. Autism as a strongly genetic disorder: Evidence from a British twin study. *Psychol Med* 1995; 25(1): 63–77.
10. Steffenburg S, Gillberg C, Hellgren L, et al. A twin study of autism in Denmark, Finland, Iceland, Norway and Sweden. *J Child Psychol Psychiatry* 1989; 30(3): 405–16.
11. Ritvo ER, Freeman BJ, Mason-Brothers A, Mo A, Ritvo AM. Concordance for the syndrome of autism in 40 pairs of afflicted twins. *Am J Psychiatry* 1985; 142(1): 74–7.
12. Hallmayer J, Glasson EJ, Bower C, et al. On the twin risk in autism. *Am J Hum Genet* 2002; 71(4): 941–6.
13. Risch N, Spiker D, Lotspeich L, et al. A genomic screen of autism: Evidence for a multilocus etiology. *Am J Hum Genet* 1999; 65(2): 493–507.
14. Polleux F, Lauder JM. Toward a developmental neurobiology of autism. *Ment Retard Dev Disabil Res Rev* 2004;10(4): 303–17.

15. Burger RA, Warren R.P. Possible Immunogenetic Basis For Autism. *Ment Retard Dev Disabil Res Rev* 1998; 4: 137–41.

16. Boccone L, Dessi V, Zappu A, et al. Bannayan-Riley-Ruvalcaba syndrome with reactive nodular lymphoid hyperplasia and autism and a PTEN mutation. *Am J Med Genet A* 2006; 140(18): 1965–9.

17. Butler MG, Dasouki MJ, Zhou XP, et al. Subset of individuals with autism spectrum disorders and extreme macrocephaly associated with germline PTEN tumour suppressor gene mutations. *J Med Genet* 2005; 42(4): 318–21.

18. Goffin A, Hoefsloot LH, Bosgoed E, Swillen A, Fryns JP. PTEN mutation in a family with Cowden syndrome and autism. *Am J Med Genet* 2001; 105(6): 521–4.

19. Kobayashi R. Perception metamorphosis phenomenon in autism. *Psychiatry Clin Neurosci* 1998; 52(6): 611–20.

20. Koyasu S. The role of PI3K in immune cells. *Nat Immunol* 2003; 4(4): 313–9.

21. Kwon CH, Luikart BW, Powell CM, et al. Pten regulates neuronal arborization and social interaction in mice. *Neuron* 2006; 50(3): 377–88.

22. Splawski I, Timothy KW, Sharpe LM, et al. Ca(V)1.2 calcium channel dysfunction causes a multisystem disorder including arrhythmia and autism. *Cell* 2004; 119(1): 19–31.

23. Hess EV. Environmental chemicals and autoimmune disease: Cause and effect. *Toxicology* 2002; 181–182: 65–70.

24. Sever LE. Looking for causes of neural tube defects: where does the environment fit in? *Environ Health Perspect* 1995;103 Suppl 6(2):165-71.

25. Rodier PM, Ingram JL, Tisdale B, Nelson S, Romano J. Embryological origin for autism: Developmental anomalies of the cranial nerve motor nuclei. *J Comp Neurol* 1996; 370(2): 247–61.

26. Ingram JL, Peckham SM, Tisdale B, Rodier PM. Prenatal exposure of rats to valproic acid reproduces the cerebellar anomalies associated with autism. *Neurotoxicol Teratol* 2000; 22(3): 319–24.

27. Baird G, Cass H, Slonims V. Diagnosis of autism. *BMJ* 2003; 327(7413): 488–93.

28. Ma DQ, Cuccaro ML, Jaworski JM, et al. Dissecting the locus heterogeneity of autism: Significant linkage to chromosome 12q14. *Mol Psychiatry* 2007; 12(4): 376–84.

29. Curran S, Powell J, Neale BM, et al. An association analysis of candidate genes on chromosome 15 q11–13 and autism spectrum disorder. *Mol Psychiatry* 2006; 11(8): 709–13.

30. Gauthier J, Joober R, Dube MP, et al. Autism spectrum disorders associated with X chromosome markers in French-Canadian males. *Mol Psychiatry* 2006; 11(2): 206–13.

31. Philippi A, Roschmann E, Tores F, et al. Haplotypes in the gene encoding protein kinase c-beta (PRKCB1) on chromosome 16 are associated with autism. *Mol Psychiatry* 2005; 10(10): 950–60.

32. Buxbaum JD, Silverman J, Keddache M, et al. Linkage analysis for autism in a subset families with obsessive-compulsive behaviors: Evidence for an autism susceptibility gene on chromosome 1 and further support for susceptibility genes on chromosome 6 and 19. *Mol Psychiatry* 2004; 9(2): 144–50.

33. Bacchelli E, Blasi F, Biondolillo M, et al. Screening of nine candidate genes for autism on chromosome 2q reveals rare nonsynonymous variants in the cAMP-GEFII gene. *Mol Psychiatry* 2003; 8(11): 916–24.

34. Auranen M, Varilo T, Alen R, et al. Evidence for allelic association on chromosome 3q25–27 in families with autism spectrum disorders originating from a subisolate of Finland. *Mol Psychiatry* 2003; 8(10): 879–84.

35. Jamain S, Quach H, Quintana-Murci L, et al. Y chromosome haplogroups in autistic subjects. *Mol Psychiatry* 2002; 7(2): 217–9.

36. Badner JA, Gershon ES. Regional meta-analysis of published data supports linkage of autism with markers on chromosome 7. *Mol Psychiatry* 2002; 7(1): 56–66.

37. De Braekeleer M, Tremblay M, Thivierge J. Genetic analysis of genealogies in mentally retarded autistic probands from Saguenay Lac-Saint-Jean (Quebec, Canada). *Ann Genet* 1996; 39(1): 47–50.

38. Fisher SE, Francks C, McCracken JT, et al. A genomewide scan for loci involved in attention-deficit/hyperactivity disorder. *Am J Hum Genet* 2002; 70(5): 1183–96.
39. Korvatska E, Van de Water J, Anders TF, Gershwin ME. Genetic and immunologic considerations in autism. *Neurobiol Dis* 2002; 9(2): 107–25.
40. Warren RP, Singh VK, Cole P, et al. Possible association of the extended MHC haplotype B44-SC30-DR4 with autism. *Immunogenetics* 1992; 36(4): 203–7.
41. Chudley AE. Genetic landmarks through philately – autism spectrum disorders: A genetic update. *Clin Genet* 2004; 65(5): 352–7.
42. Weyand CM, McCarthy TG, Goronzy JJ. Correlation between disease phenotype and genetic heterogeneity in rheumatoid arthritis. *J Clin Invest* 1995; 95(5): 2120–6.
43. Warren RP, Odell JD, Warren WL, et al. Strong association of the third hypervariable region of HLA-DR beta 1 with autism. *J Neuroimmunol* 1996; 67(2): 97–102.
44. Hogan D, Morrow JD, Smith EM, Opp MR. Interleukin-6 alters sleep of rats. *J Neuroimmunol* 2003; 137(1–2): 59–66.
45. Larson SJ. Behavioral and motivational effects of immune-system activation. *J Gen Psychol* 2002; 129(4): 401–14.
46. Engstrom AH, Ohlson S, Stubbs EG, et al. Decreased expression of CD95 (FAS/APO-1) on CD4+ T-lymphocytes from participants with autism. *J Develop Phys Disabil* 2003; 15(2): 155–63.
47. Sweeten TL, Bowyer SL, Posey DJ, Halberstadt GM, McDougle CJ. Increased prevalence of familial autoimmunity in probands with pervasive developmental disorders. *Pediatrics* 2003; 112(5): e420.
48. Jyonouchi H, Sun S, Le H. Proinflammatory and regulatory cytokine production associated with innate and adaptive immune responses in children with autism spectrum disorders and developmental regression. *J Neuroimmunol* 2001; 120(1–2): 170–9.
49. Warren RP, Foster A, Margaretten NC. Reduced natural killer cell activity in autism. *J Am Acad Child Adolesc Psychiatry* 1987; 26(3): 333–5.
50. Warren RP, Yonk LJ, Burger RA, et al. Deficiency of suppressor-inducer (CD4+CD45RA+) T cells in autism. *Immunol Invest* 1990; 19(3): 245–51.
51. Singh VK, Warren RP, Odell JD, Warren WL, Cole P. Antibodies to myelin basic protein in children with autistic behavior. *Brain Behav Immun* 1993; 7(1): 97–103.
52. Singh VK, Warren R, Averett R, Ghaziuddin M. Circulating autoantibodies to neuronal and glial filament proteins in autism. *Pediatr Neurol* 1997; 17(1): 88–90.
53. Plioplys AV, Greaves A, Kazemi K, Silverman E. Lymphocyte function in autism and Rett syndrome. *Neuropsychobiology* 1994; 29(1): 12–6.
54. Singh VK, Lin SX, Yang VC. Serological association of measles virus and human herpesvirus-6 with brain autoantibodies in autism. *Clin Immunol Immunopathol* 1998; 89(1): 105–8.
55. Singh VK, Singh EA, Warren RP. Hyperserotoninemia and serotonin receptor antibodies in children with autism but not mental retardation. *Biol Psychiatry* 1997; 41(6): 753–5.
56. Todd RD, Hickok JM, Anderson GM, Cohen DJ. Antibrain antibodies in infantile autism. *Biol Psychiatry* 1988; 23(6): 644–7.
57. Connolly AM, Chez MG, Pestronk A, et al. Serum autoantibodies to brain in Landau-Kleffner variant, autism, and other neurologic disorders. *J Pediatr* 1999; 134(5): 607–13.
58. Silva SC, Correia C, Fesel C, et al. Autoantibody repertoires to brain tissue in autism nuclear families. *J Neuroimmunol* 2004; 152(1–2): 176–82.
59. Vojdani A, Campbell AW, Anyanwu E, et al. Antibodies to neuron-specific antigens in children with autism: Possible cross-reaction with encephalitogenic proteins from milk, Chlamydia pneumoniae and Streptococcus group A. *J Neuroimmunol* 2002; 129(1–2): 168–77.
60. Singer HS, Morris CM, Williams PN, et al. Antibrain antibodies in children with autism and their unaffected siblings. *J Neuroimmunol* 2006; 178: 149–155.
61. Egg R, Reindl M, Deisenhammer F, Linington C, Berger T. Anti-MOG and anti-MBP antibody subclasses in multiple sclerosis. *Mult Scler* 2001; 7(5): 285–9.

62. Cabanlit M, Wills S, Goines P, Ashwood P, Van de Water J. Brain-specific autoantibodies in the plasma of subjects with autistic spectrum disorder. *Ann NY Acad Sci* 2007; 1107: 92–103.

63. Wills S, Cabanlit M, Bennett J, et al. Detection of autoantibodies to neural cells of the cerebellum in the plasma of patients with autism spectrum disorders (ASD). In preparation 2008. In Press Brain, Behavior and Immunity.

64. Hirano T, Watanabe D, Kawaguchi SY, Pastan I, Nakanishi S. Roles of inhibitory interneurons in the cerebellar cortex. *Ann NY Acad Sci* 2002; 978: 405–12.

65. De Schutter E, Vos B, Maex R. The function of cerebellar Golgi cells revisited. *Prog Brain Res* 2000; 124: 81–93.

66. Rogers SJ, Ozonoff S. Annotation: What do we know about sensory dysfunction in autism? A critical review of the empirical evidence. *J Child Psychol Psychiatry* 2005; 46(12): 1255–68.

67. Tsatsanis KD, Rourke BP, Klin A, et al. Reduced thalamic volume in high-functioning individuals with autism. *Biol Psychiatry* 2003; 53(2): 121–9.

68. Hoekstra PJ, Kallenberg CG, Korf J, Minderaa RB. Is Tourette's syndrome an autoimmune disease? *Mol Psychiatry* 2002; 7(5): 437–45.

69. Lang B, Dale RC, Vincent A. New autoantibody mediated disorders of the central nervous system. *Curr Opin Neurol* 2003; 16(3): 351–7.

70. Rothermundt M, Arolt V, Bayer TA. Review of immunological and immunopathological findings in schizophrenia. *Brain Behav Immun* 2001; 15(4): 319–39.

71. Kowal C, DeGiorgio LA, Nakaoka T, et al. Cognition and immunity; antibody impairs memory. *Immunity* 2004; 21(2): 179–88.

72. Jones AL, Mowry BJ, Pender MP, Greer JM. Immune dysregulation and self-reactivity in schizophrenia: Do some cases of schizophrenia have an autoimmune basis? *Immunol Cell Biol* 2005; 83(1): 9–17.

73. Kirvan CA, Swedo SE, Heuser JS, Cunningham MW. Mimicry and autoantibody-mediated neuronal cell signaling in Sydenham chorea. *Nat Med* 2003; 9(7): 914–20.

74. Abdel-Rahman A, Shetty AK, Abou-Donia MB. Disruption of the blood-brain barrier and neuronal cell death in cingulate cortex, dentate gyrus, thalamus, and hypothalamus in a rat model of Gulf-War syndrome. *Neurobiol Dis* 2002; 10(3): 306–26.

75. Esposito P, Chandler N, Kandere K, et al. Corticotropin-releasing hormone and brain mast cells regulate blood-brain-barrier permeability induced by acute stress. *J Pharmacol Exp Ther* 2002; 303(3): 1061–6.

76. Friedman A, Kaufer D, Shemer J, et al. Pyridostigmine brain penetration under stress enhances neuronal excitability and induces early immediate transcriptional response. *Nat Med* 1996; 2(12): 1382–5.

77. Xaio H, Banks WA, Niehoff ML, Morley JE. Effect of LPS on the permeability of the blood-brain barrier to insulin. *Brain Res* 2001; 896(1–2): 36–42.

78. Huerta PT, Kowal C, DeGiorgio LA, Volpe BT, Diamond B. Immunity and behavior: Antibodies alter emotion. *Proc Natl Acad Sci USA* 2006; 103(3): 678–83.

79. Schmidt-Acevedo S, Perez-Romano B, Ruiz-Arguelles A. 'LE cells' result from phagocytosis of apoptotic bodies induced by antinuclear antibodies. *J Autoimmun* 2000; 15(1): 15–20.

80. Portales-Perez D, Alarcon-Segovia D, Llorente L, et al. Penetrating anti-DNA monoclonal antibodies induce activation of human peripheral blood mononuclear cells. *J Autoimmun* 1998; 11(5): 563–71.

81. Sisto M, Lisi S, Castellana D, et al. Autoantibodies from Sjogren's syndrome induce activation of both the intrinsic and extrinsic apoptotic pathways in human salivary gland cell line A-253. *J Autoimmun* 2006; 27: 38–49.

82. Alarcon-Segovia D, Ruiz-Arguelles A, Fishbein E. Antibody to nuclear ribonucleoprotein penetrates live human mononuclear cells through Fc receptors. *Nature* 1978; 271(5640): 67–9.

83. Madaio MP, Yanase K. Cellular penetration and nuclear localization of anti-DNA antibodies: Mechanisms, consequences, implications and applications. *J Autoimmun* 1998; 11(5): 535–8.

84. Raz E, Ben-Bassat H, Davidi T, Shlomai Z, Eilat D. Cross-reactions of anti-DNA autoantibodies with cell surface proteins. *Eur J Immunol* 1993; 23(2): 383–90.

85. Cohen IR, Schwartz M. Autoimmune maintenance and neuroprotection of the central nervous system. *J Neuroimmunol* 1999; 100(1–2): 111–4.

86. Schwartz M, Cohen IR. Autoimmunity can benefit self-maintenance. *Immunol Today* 2000; 21(6): 265–8.

87. Hickey WF. Basic principles of immunological surveillance of the normal central nervous system. *Glia* 2001; 36(2): 118–24.

88. Garty BZ, Ludomirsky A, Danon YL, Peter JB, Douglas SD. Placental transfer of immunoglobulin G subclasses. *Clin Diagn Lab Immunol* 1994; 1(6): 667–9.

89. Heininger U, Desgrandchamps D, Schaad UB. Seroprevalence of Varicella-Zoster virus IgG antibodies in Swiss children during the first 16 months of age. *Vaccine* 2006; 24(16): 3258–60.

90. Dalton P, Deacon R, Blamire A, et al. Maternal neuronal antibodies associated with autism and a language disorder. *Ann Neurol* 2003; 53(4): 533–7.

91. Zimmerman AW, Connors SL, Matteson KJ, et al. Maternal antibrain antibodies in autism. *Brain Behav Immun* 2007; 21: 351–7.

92. Braunschweig D, Ashwood P, Hertz-Picciotto I, et al. Maternal serum antibodies to fetal brain in autism. Manuscript in preparation 2007. *Neurotoxicology* 2008 Mar; 29(2): 226–31.

93. Hertz-Picciotto I, Croen LA, Hansen R, et al. The CHARGE study: An epidemiologic investigation of genetic and environmental factors contributing to autism. *Environ Health Perspect* 2006; 114(7): 1119–25.

94. Simister NE. Placental transport of immunoglobulin G. *Vaccine* 2003; 21(24): 3365–9.

95. Harris NL, Spoerri I, Schopfer JF, et al. Mechanisms of neonatal mucosal antibody protection. *J Immunol* 2006; 177(9): 6256–62.

96. Mishima T, Kurasawa G, Ishikawa G, et al. Endothelial expression of Fc gamma receptor IIb in the full-term human placenta. *Placenta* 2007; 28(2–3): 170–4.

97. Tincani A, Nuzzo M, Motta M, et al. Autoimmunity and pregnancy: Autoantibodies and pregnancy in rheumatic diseases. *Ann NY Acad Sci* 2006; 1069: 346–52.

98. Soares A, Schoffen JP, De Gouveia EM, Natali MR. Effects of the neonatal treatment with monosodium glutamate on myenteric neurons and the intestine wall in the ileum of rats. *J Gastroenterol* 2006; 41(7): 674–80.

99. Fu J, Jiang Y, Liang L, Zhu H. Risk factors of primary thyroid dysfunction in early infants born to mothers with autoimmune thyroid disease. *Acta Paediatr* 2005; 94(8): 1043–8.

100. Williamson SL, Christodoulou J. Rett syndrome: New clinical and molecular insights. *Eur J Hum Genet* 2006; 14(8): 896–903.

101. Ashwood P, Anthony A, Pellicier AA, et al. Intestinal lymphoctye populations in children with regressive autism: Evidence for extensive mucosal immunopathology. *J Clin Immunol* 2003; 23: 504–17.

102. Stubbs EG, Crawford ML. Depressed lymphocyte responsiveness in autistic children. *J Autism Child Schizophr* 1977; 7(1): 49–55.

103. Warren RP, Margaretten NC, Pace NC, Foster A. Immune abnormalities in patients with autism. *J Autism Dev Disord* 1986; 16(2): 189–97.

104. Denney DR, Frei BW, Gaffney GR. Lymphocyte subsets and interleukin-2 receptors in autistic children. *J Autism Dev Disord* 1996; 26(1): 87–97.

105. Krause I, He XS, Gershwin ME, Shoenfeld Y. Brief report: Immune factors in autism: A critical review. *J Autism Dev Disord* 2002; 32(4): 337–45.

106. Heuer L, Schauer J, Goines P, et al. Reduced levels of immunoglobulin in children with autism correlates with behavioral symptoms. 2008. Submitted Autism Research.

107. Odell D, Maciulis A, Cutler A, et al. Confirmation of the association of the C4B null allelle in autism. *Hum Immunol* 2005; 66(2): 140–5.

108. Torres AR, Maciulis A, Odell D. The association of MHC genes with autism. *Front Biosci* 2001; 6: D936–43.

109. Yamashita Y, Fujimoto C, Nakajima E, Isagai T, Matsuishi T. Possible association between congenital cytomegalovirus infection and autistic disorder. *J Autism Dev Disord* 2003; 33(4): 455–9.

110. Patterson PH. Maternal infection: Window on neuroimmune interactions in fetal brain development and mental illness. *Curr Opin Neurobiol* 2002; 12(1): 115–8.

111. Hornig M, Solbrig M, Horscroft N, Weissenbock H, Lipkin WI. Borna disease virus infection of adult and neonatal rats: models for neuropsychiatric disease. *Curr Top Microbiol Immunol* 2001; 253: 157–77.

112. Plata-Salaman CR, Ilyin SE, Gayle D, Romanovitch A, Carbone KM. Persistent Borna disease virus infection of neonatal rats causes brain regional changes of mRNAs for cytokines, cytokine receptor components and neuropeptides. *Brain Res Bull* 1999; 49(6): 441–51.

113. Sauder C, de la Torre JC. Cytokine expression in the rat central nervous system following perinatal Borna disease virus infection. *J Neuroimmunol* 1999; 96(1): 29–45.

114. Shi L, Fatemi SH, Sidwell RW, Patterson PH. Maternal influenza infection causes marked behavioral and pharmacological changes in the offspring. *J Neurosci* 2003; 23(1): 297–302.

115. Vargas DL, Nascimbene C, Krishnan C, Zimmerman AW, Pardo CA. Neuroglial activation and neuroinflammation in the brain of patients with autism. *Ann Neurol* 2005; 57(1): 67–81.

116. Wrona D. Neural-immune interactions: An integrative view of the bidirectional relationship between the brain and immune systems. *J Neuroimmunol* 2006; 172(1–2): 38–58.

117. Haddad JJ, Saade NE, Safieh-Garabedian B. Cytokines and neuro-immune-endocrine interactions: A role for the hypothalamic-pituitary-adrenal revolving axis. *J Neuroimmunol* 2002; 133(1–2): 1–19.

118. Steinman L. Elaborate interactions between the immune and nervous systems. *Nat Immunol* 2004; 5(6): 575–81.

119. Marques-Deak A, Cizza G, Sternberg E. Brain-immune interactions and disease susceptibility. *Mol Psychiatry* 2005; 10(3): 239–50.

120. Rothwell NJ, Luheshi G, Toulmond S. Cytokines and their receptors in the central nervous system: Physiology, pharmacology, and pathology. *Pharmacol Ther* 1996; 69(2): 85–95.

121. Mignini F, Streccioni V, Amenta F. Autonomic innervation of immune organs and neuroimmune modulation. *Auton Autacoid Pharmacol* 2003; 23(1): 1–25.

122. Huh GS, Boulanger LM, Du H, et al. Functional requirement for class I MHC in CNS development and plasticity. *Science* 2000; 290(5499): 2155–9.

123. Biber K, Zuurman MW, Dijkstra IM, Boddeke HWGM. Chemokines in the brain: Neuroimmunology and beyond. *Curr Opin Pharmacol* 2002; 2(1): 63.

124. Mehler MF, Kessler JA. Cytokines in brain development and function. *Adv Protein Chem* 1998; 52: 223–51.

Chapter 13
Maternal Immune Activation, Cytokines and Autism

Paul H. Patterson, Wensi Xu, Stephen E.P. Smith and Benjamin E. Devarman

Abstract Normal pregnancy involves an elevated inflammatory state, both systemically in the mother and in the placenta. However, further increases in inflammation, as with maternal infection, can enhance the risk of autism and schizophrenia in the offspring. Animal studies show that maternal immune activation (MIA) increases inflammatory cytokines in the fetal environment, as well as in the fetal brain. Since the adult autistic brain and cerebrospinal fluid (CSF) exhibit high levels of inflammatory cytokines, we hypothesize that MIA sets in motion a self-perpetuating cycle of subacute inflammation in the brain that not only affects neural development, but also acutely influences ongoing postnatal behavior. Experiments aimed at testing the effects of preventing or interrupting this inflammatory cycle are possible with available animal models.

Keywords IL-6 · TNF · IL-10 · influenza · schizophrenia · maternal infection · preterm birth · LPS · poly(I:C)

Introduction: Genes Versus Environment in Autism

Although it is well-known that susceptibility genotype is a key feature of autism risk, this is sometimes taken to the extreme, when autism is referred to as a "genetic disorder," as if it is an autosomal-dominant disorder. Thus, it is worth emphasizing that only a small fraction of autism cases can be attributed to known chromosomal abnormalities or single gene mutations, and that the prevalence in siblings of autistic children is only 5–10%. Moreover, although much emphasis has been placed on early studies that estimated the concordance for autism in monozygous twins as high as 90%, recent work puts this figure at only 50–70% [1]. This is in the same range as the concordance for schizophrenia (50%), where further analysis of monozygotic twins revealed an important finding: the

P.H. Patterson
Biology Division, California Institute of Technology, 216-76, Caltech, Pasadena, CA 91125, USA
e-mail: php@caltech.edu

A.W. Zimmerman (ed.), *Autism*, DOI: 10.1007/978-1-60327-489-0_13,
© Humana Press, Totowa, NJ 2008

concordance for twins sharing a placenta is 60%, while that for twins with separate placentas is 11% [2]. Moreover, the concordance for dizygotic twins is twice as high as for siblings in schizophrenia, and these groups share the same level of genetic similarity. Thus, the fetal environment is a key component in the outcome of twin concordance results. Although a similar study of chorion type in twins has not been reported for autism, it seems possible that the fetal environment plays an important role in this disorder as well.

The importance of placenta as an environmental factor is further illustrated by studies of animals with multiple fetuses; there are gradients of nutrient exchange according to the position along the uterus, and masculine and feminine traits can be influenced by intrauterine hormonal signals among fetuses [3]. In addition, as discussed below, mouse studies have provided evidence of placental heterogeneity within a given litter, leading to differential responses to maternal infection. Thus, animal studies support the point that two-thirds of monozygotic human twins share the same placenta, which means that they are sharing key environmental factors that are not shared by most dizygotic twins.

Environmental Factors in Autism

It is clear that specific environmental factors can increase the incidence of autism. For instance, maternal exposure to thalidomide or valproic acid strongly increases the risk of autism in the offspring [4, 5]. In addition, maternal infection with rubella virus increases the risk of autism in the offspring >200-fold, and other maternal infections can also increase the risk [6, 7, 8, 9]. Although these specific environmental risk factors no longer contribute significantly to the incidence of autism, this evidence provides the proof-of-principle for a role of the fetal environment. It also suggests that other types of infection and environmental exposure could influence the current rates of autism incidence. Another example serving as proof-of-principle for the role of the fetal environment in a mental disorder is the observation that maternal respiratory infection increases the risk of schizophrenia in the offspring three- to sevenfold [10, 11]. We review here various animal models to explore how maternal immune activation (MIA) may contribute to autism.

Maternal Immune Activation: Poly(Inosine:Cytosine)

Animal studies have broadened our perspective on the ways that maternal infection can alter fetal brain development with relevance to autism. Study of the adult offspring of pregnant mice given an influenza respiratory infection revealed a series of behavioral abnormalities that are related to behaviors found in autism and schizophrenia [12]. This led to the question, is this effect specific to influenza, or would a more general activation of the maternal immune system cause similar changes in the fetus? To distinguish between these two possibilities, we stimulated the maternal immune system, in the absence of virus, by injecting pregnant mice on E9.5 or E12.5 i.p. with a synthetic double-stranded RNA [poly(inosine:cytosine), poly(I:C)], which acts through the toll-like receptor 3 (TLR3) and evokes an inflammatory response resembling that seen with viral infection. Remarkably, poly(I:C) treatment causes a deficit in prepulse inhibition (PPI) of the acoustic startle in the offspring similar to that seen in the offspring of infected mothers [12]. A PPI deficit is also observed in autism [13]. Subsequently, we and others have found that MIA by poly(I:C) causes many other behavioral abnormalities in the offspring. These are analogous to abnormalities in autism and schizophrenia, and include deficits in social interaction, latent inhibition (LI), working memory and novel object exploration, as well as increased amphetamine-induced locomotion (Table 13.1). It is noteworthy that a number of these behavioral findings (deficits in PPI and LI, increased anxiety in the open field, and enhanced reversal learning) have been replicated in several laboratories, despite using different species of animals (rats and mice), different stages of administration, and different doses and routes of administration of poly(I:C). It is clear, however, that the time of administration can influence the behavioral phenotype of the offspring [14].

In addition to behavioral changes, poly(I:C)-induced MIA causes neuropathology in the brains of the offspring. Most relevant to autism is a localized deficit of Purkinje cells. Such a deficit has been repeatedly observed in autopsy studies of autism [15], and there is a strong inverse correlation between the magnitude of cerebellar lobules VI and VII hypoplasia and the degree of novel object exploration and stereotypic behavior in autistic children [16, 17]. Moreover, functional magnetic resonance imaging reveals abnormal cerebellar activation during motor and cognitive tasks in autistic subjects [18, 19], and autistic subjects also exhibit abnormalities in eyeblink conditioning and visual saccades, which are behaviors particularly relevant for the known functions of lobules VI and VII [20, 21]. There are, however, reports of negative findings regarding cerebellar pathology and autism [15, 22]. There is also evidence for cerebellar pathology in schizophrenia, including Purkinje cell loss and deficits in cerebellar volume, as well as behavioral

Table 13.1 Poly(I:C)-induced maternal immune activation causes numerous, reproducible changes in the offspring related to autism and schizophrenia

Behavioral abnormality	Brain pathology	Reference
PPI, open field, novel object, LI, social interaction	Large adult brain, Purkinje cell deficit	[12, 28, 54]
LI, reversal learning	Hippocampal necrosis, increased DA release	[30, 83]
PPI, open field, LI, amphetamine-induced locomotion, working memory	GABA$_A$ receptor increase	[14, 31, 84]
PPI, open field, amphetamine-induced locomotion	DA turnover increase	[29]
Open field, novel object, reversal learning	n.d.	[85]

The different rows represent various research groups. DA, dopamine; GABA, γ-aminobutyric acid; LI, latent inhibition; n.d., not determined; PPI, prepulse inhibition evidence (saccades and eyeblink conditioning) that points to pathology in lobules VI and VII [23, 24, 25]. Thus, it is clinically relevant that we find the adult offspring of pregnant mice given a midgestation respiratory infection, or a poly(I:C) injection, display a localized deficit in Purkinje cells, specifically in lobule VII [26]. This deficit is also seen at postnatal day 11 (P11), indicating that it occurs as a result of abnormal cerebellar development rather than neurodegeneration in the adult.

Brain size has also been found to be abnormal in autism, with macrocephaly occurring during postnatal development [27]. We reported a similar finding for the offspring of mice that received a respiratory infection [28].

Poly(I:C)-induced MIA results in increased dopamine (DA) release and turnover in adult offspring [29, 30], which is of interest in the context of the DA theory of schizophrenia. The importance of this observation for autism is not clear. These offspring also display a substantial increase in γ-aminobutyric acid ($GABA_A$) receptor expression [31], which is also relevant for schizophrenia and possibly for autism [32]. Severe necrosis in the hippocampus of these offspring was also reported [30], but this has not been reproduced by several other groups [31] (L. Shi and S. Smith, unpublished observations).

Maternal Immune Activation: Lipopolysaccharide

A different method of MIA is used to mimic bacterial infection. In this protocol, pregnant mice or rats are injected (intrauterine or intraperitoneal, i.p.) with lipopolysaccharide (LPS), which acts through the TLR4. Intrauterine bacterial infection or fetal or early postnatal exposure to hypoxia can result in white matter injury and risk of cerebral palsy [33]. Although obstetrical complications have been associated with autism, a specific link between maternal bacterial infection and autism in the offspring has not yet been made. Nonetheless, some of the same behavioral abnormalities seen in the offspring of poly(I:C)-treated mothers have been observed in the offspring of LPS-treated mothers (Table 13.2). For instance, a severe schedule of LPS administration i.p. causes a PPI deficit in the offspring [34]. Moreover, single or double maternal injections of LPS can cause increased anxiety, deficits in social interaction and learning, and increased amphetamine-induced locomotion in the adult offspring (see Table 13.2).

Also in common with the poly(I:C) findings are observations of brain inflammation, as defined by astrogliosis [enhanced glial fibrillary acid protein (GFAP) staining] and altered microglial immunostaining. A GFAP increase was also reported for neonatal offspring of influenza-infected mothers [35]. This evidence of inflammatory changes in the brain is consistent with the striking findings of immune dysregulation in autism reported by Vargas et al. [36]. They observed marked astrogliosis, microglial activation, and dramatically upregulated cytokines in the brain and cerebrospinal fluid (CSF) in samples from 11 autistic subjects, aged 5–44 years. It is also relevant that a similarly impressive dysregulation in immune-related genes has recently been reported for adult schizophrenia brains [37].

Table 13.2 Lipopolysaccharide-induced maternal immune activation causes numerous changes in the offspring related to autism and schizophrenia

Behavioral abnormality	Brain pathology	Reference
n.d.	Less MBP staining; GFAP increase; altered microglial immunostaining	[86]
n.d.	Decreased TH + neurons in mesencephalon; reduced striatal DA; increased microglial staining	[87]
PPI	TH increase in BNST and shell nucl accumb; MHCII staining of microglia; GFAP increase	[34]
n.d.	*Less PLP and CNPase staining	[88]
Elevated plus maze; beam walking	n.d.	[89]
Amphetamine-induced locomotion	n.d.	[90]
n.d.	Less MBP, PLP, and myelin staining; more microglial staining	[76]
Water maze; object recognition; passive avoidance; social interaction; elevated plus maze	Hippocampal pyramidal neurons fewer, smaller and more densely packed	[91, 92, 93]
Motor function normal	*Less CNPase and PLP staining	[94]
n.d.	**Less MBP in internal capsule; cell death in striatum, white matter and ventricular zone; GFAP increase, hypomelination, enlarged ventricles	[95, 96]
n.d.	**Increased GFAP; less PLP	[48]
n.d.	***Microglial activation; axonal injury; reduction in oligodendrocytes	[97]

LPS administered i.p., except where noted (* cervix; ** uterine horn; *** amnion). BNST, basal nucleus of striatal terminalis; DA, dopamine; GFAP, glial fibrillary acidic protein; MBP, myelin basic protein; n.d., not determined; nucl accumb, shell of nucleus accumbens; PLP, proteolipid protein; PPI, prepulse inhibition; LPS, lipopolysaccharide; TH, tyrosine hydroxylase

Interestingly, the brain inflammatory changes reported thus far for the LPS and poly(I:C) models have not been as dramatic as those found by Vargas et al. and Arion et al. [37] for the human disorders. The closest to the human cases are the results of Borrell et al. [34], who used the most severe maternal LPS protocol (injection every other day throughout pregnancy).

It would be informative to directly compare the most aggressive LPS and poly(I:C) MIA protocols, both in terms of behavioral outcome and in the neuropathology and markers of brain immune dysregulation.

Maternal Immune Activation: Periodontal Bacteria

Intrauterine infections are prevalent among women who give birth prematurely, and very low birth weight is correlated with perinatal mortality and neonatal morbidity, including serious neurological disorders. Among the microorganisms isolated from the preterm placenta, amniotic fluid and chorioamnion are gram-negative bacteria that are known to be involved in periodontal disease, *Fusobacterium nucleatum* and *Porphyromonas gingivalis*. Moreover, epidemiological evidence has linked periodontal disease with premature delivery [38]. In a pregnant mouse model, i.v. injection of *F. nucleatum* results in premature delivery, stillbirths and nonsustained live births [39]. Interestingly, the bacterial infection is confined to the uterus. An alternative mouse model using *P. gingivalis* has produced several further results of interest. In this case, systemic induction of MIA leads to fetal growth restriction in every litter, but not in every fetus. Importantly, *P. gingivalis* DNA is found only in the placentas of affected fetuses, and these placentas show elevation of the proinflammatory cytokines interferon (IFN)-γ, interleukin (IL)-2 and IL-12, and reduction of mRNAs for the anti-inflammatory cytokines IL-4 and IL-10 [40]. These results not only link key cytokine changes with fetal morbidity, but also highlight the importance of heterogeneity among placentas in the same uterus, which was alluded to in the introduction. It will be of interest to examine the behavior and neuropathology of the affected offspring. Testing a possible association of maternal periodontal disease and autism could also be worthwhile.

Other Protocols of Immune Activation

A much broader discussion of various different types of animal models of various features of autism was recently published [41]. Not covered in the present review is the work on direct injection of LPS into the fetus, which was reviewed by Wang et al. [42]. Alternative protocols involving injecting LPS, viruses, or cytokines during the early postnatal period have been well reviewed by others [42, 43, 44].

Maternal Immune Activation Elevates Cytokines in the Fetal Environment

In addition to the alterations in placental cytokines by periodontal bacterial infection, there is a significant literature on the cytokines induced in the fetal environment by LPS-induced MIA. It is clear that despite various doses and injection schedules employed, several cytokines are elevated in the placenta [IL-1β, IL-6, tumor necrosis factor (TNF-α)] and amniotic fluid (IL-6, TNF-α) (Table 13.3). The more difficult and interesting issue of whether cytokines are altered in fetal brain is not completely clear. Results include a small increase [45], a significant increase [46], or undetectable [47] TNF-α protein or mRNA. A striking recent report describes significant increases of mRNAs for IL-1β, IL-6, TNF-α, and IFN-γ (as well as other cytokines) in fetal brain [48]. In this

Table 13.3 Lipopolysaccharide-induced maternal immune activation increases cytokine levels in fetal environment

Cytokines increased	Location	Reference
IL-1α, IL-6	Amniotic fluid	[98]
IL-1β, TNF-α	Fetal brain	[86]
IL-1α, IL-6, TNF-α	Placenta	[45]
IL-6, TNF-α	Amniotic fluid	
****[δ]IL-1β, IL-6, IL-8, TNF-α	Choriamnion	[97, 99]
****[δ]IL-6, IL-8	Amniotic fluid	
****IL-1α, IL-1β	Placenta	[100]
****IL-1α, IL-1β, IL-6, TNF-α	Fetal membranes	
IL-1β, IL-6, TNF-α	Amniotic fluid	[101]
IL-1α, IL-6, TNF-α	Placenta	
**IL-6, TNF-α, IL-10	Placenta	[46]
**TNF-α, IFN-γ	Fetal brain	
IL-1β, TNF-α	Fetal brain	[76]
IL-1β, IL-6, TNF-α	Placenta	[47]
ND: IL-1β, IL-6, TNF-α	Fetal brain	
IL-1β	Fetal plasma	
***IL-1β, IL-6, TNF-α, IFN-γ**	Placenta	[48]
***IL-1β, IL-6, TNF-α, IFN-γ**	Fetal brain	
TNF-α	Amniotic fluid	[51]
*TNF-α	Placenta	
IL-6	Placenta	[102]
IL-6	Fetal plasma	
*IL-6, IL-1β	Fetal brain	[104]

Assays were for cytokine proteins, except where noted (*mRNA assayed). LPS was administered i.p., except where noted (**cervix administration; ***uterine administration; ****amniotic sac administration). Rodents were used except where noted ([δ]ewes). ND, not detected

case, LPS was delivered into the uterine horn, which is presumably a more aggressive delivery method than i.p. injection.

Similar studies are underway for poly(I:C)-induced MIA, and several of the same cytokines appear to be increased in the fetal environment (Table 13.4;

Table 13.4 Poly(I:C)-induced maternal immune activation alters cytokine levels in the fetal environment

Cytokines increased	Location	Reference
TNF-α	Amniotic fluid	[103]
TNF-α	Placenta	
TNF-α*	Fetal brain	
IL-1β**, IL-6**, IL-10**, TNF-α**	Fetal brain	[49]
IL-6, KC, TNF-α	Placenta + decidua	Fig. 13.1

Levels are increased except where noted (*protein decreased; **levels either increased or decreased depending on the stage of poly(I:C) administration and interval between MIA and assay; see text). Cytokine proteins were assayed, except where noted (**both protein and mRNA assayed). Poly(I:C) was administered i.p. KC, keratinocyte-derived chemokine

Fig. 13.1). In addition, there is one extensive study of cytokine changes in fetal brain [49]. In that study, at 6-h postinjection, IL-1β, IL-6, IL-10, and TNF-α are all significantly increased, although the extent of the effects depends on the embryonic stage at which poly(I:C) was injected. This group also investigated whether these increases are due to local synthesis by assaying cytokine mRNAs in the fetal brain. The most striking results are significant increases in IL-6, IL-10, and TNF-α at 6-h postinjection on E17. It is notable that both pro- and anti-inflammatory cytokines are increased, although this depends again on the embryonic stage of injection [49]. It will be important to determine what drives these fetal brain gene expression changes – are the cytokines in the fetal environment responsible? Another observation of interest is our finding that IL-6 mRNA may be elevated in the placenta (see Fig. 13.1).

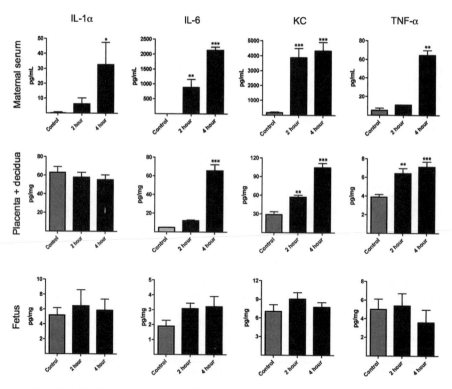

Fig. 13.1 Poly(I:C)-induced maternal immune activation (MIA) elevates cytokine levels in the maternal fetal unit. The ELISA method was used to quantify cytokines in the maternal serum, fetal environment (placenta plus decidua), and whole fetus. Poly(I:C) was injected i.p. on E9.5 and tissues were taken at two subsequent time points. Although all cytokines tested increase dramatically in maternal serum, only IL-6, KC, and TNF increase in the placenta, and only IL-6 shows a tendency to increase in the fetus. ***$p < 0.001$; **$p < 0.01$; *$p < 0.05$ (one-way ANOVA with Tukey's posthoc test). KC, keratinocyte-derived chemokine

Pro/Anti-Inflammatory Balance Mediates Maternal Immune Activation Effects on the Fetus

Cytokines are induced by MIA, but what effects do they have on the fetus? Two approaches have been taken toward answering this question: injecting or upregulating cytokines during pregnancy in the absence of MIA and blocking endogenous cytokines or preventing their induction during MIA. Investigations of the role of TNF-α in LPS-induced fetal loss and growth restriction have shown that injection of anti-TNF-α antibodies or an inhibitor of TNF-α synthesis (pentoxifylline) can reduce these effects of LPS. Conversely, injection of TNF-α alone can induce fetal loss [50, 51]. Fetal loss induced by intrauterine LPS is exacerbated significantly in IL-18 knockout mice, but not in IL-1α/β knockout mice [52].

Investigation of cytokine mediation of MIA effects on neuropathology and behavior in the offspring has focused on IL-6. Samuelsson et al. [53] injected IL-6 i.p. in pregnant rats multiple times, on E8, 10, and 12 for their "early" protocol and on E16, 18, and 20 for their "late" protocol. Both protocols had profound effects on the offspring. One remarkable finding was that IL-6 levels remain elevated in the hippocampus of the offspring at 4 and 24 weeks of age. This is reminiscent of the ongoing, permanent state of immune dysregulation seen in adult autistic brains [36]. Further evidence of this parallel is the astrogliosis and elevated GFAP levels in the adult hippocampus of the IL-6-exposed offspring and in autistic brains. It seems likely that fetal exposure to IL-6 sets in motion a self-propagating, subclinical inflammatory cascade. The consequences of this IL-6 challenge include dysmorphic neurons in the adult hippocampus, with elevated caspase-3 mRNA and procaspase-3 protein. Interestingly, as with poly(I:C)-induced MIA, GABA$_A$ receptor mRNA is increased in the adult hippocampus of the IL-6-exposed offspring [53]. One hippocampal-dependent behavior was monitored in this study, spatial memory in the water maze, and the IL-6-exposed offspring displayed increased escape latency and time spent near the pool wall. Thus, prolonged exposure to elevated IL-6 during fetal development causes a deficit in working memory, as is also seen in LPS- induced MIA.

In exploring which of the cytokines to pursue for effects on the fetus, our group tested several cytokines that are elevated in maternal serum following respiratory infection or poly(I:C)-induced MIA. These cytokines were injected in pregnant mice, and PPI assayed in the adult offspring. The only cytokine that yielded consistent perturbation results was IL-6, and we went on to show that a single maternal injection of IL-6 at E12.5 yields offspring with deficits in both PPI and LI [54]. A more compelling experimental approach is, however, blocking the endogenous IL-6 that is induced by MIA. We did this in two ways: by maternal injection of a function-blocking, anti-IL-6 monoclonal antibody and by inducing MIA in IL-6 knockout mice. In the former approach, we found that coinjecting pregnant mice with poly(I:C) and anti-IL-6 almost completely eliminates the behavioral deficits caused by poly(I:C)-induced MIA. These include PPI, LI, anxiety in the open field, and social interaction. Control experiments included coinjection of poly(I:C) with anti-INF-γ, which is not able to block the effects of MIA. In an independent approach to eliminating the effects of IL-6, we injected poly(I:C) in pregnant IL-6 knockout mice. Here also, MIA does not cause PPI and open-field deficits in the adult offspring [54].

Induction of MIA, either by respiratory infection [55] or by poly(I:C) injection [54], also alters gene expression in the brains of the offspring. We tested whether coinjection of anti-IL-6 plus poly(I:C) could block the effect of MIA on gene expression in the adult offspring. Microarray analyses were performed on mRNA extracted from the rostral 2 mm of the frontal cortex. This area corresponds to the human prefrontal cortex, which shows molecular, functional, and microanatomical alterations in human cognitive disorders. Combining data from two experiments, we found that of 94 significant changes in gene expression ($p < 0.01$), 87 were normalized by the IL-6 antibody. That is, 92% of the changes caused by MIA were prevented by coinjection of anti-IL-6 [54]. Moreover, in a two-way (samples and genes), unsupervised, hierarchical clustering of the gene expression intensities, the three experimental groups (injection with saline, poly(I:C), or poly(I:C) + anti-IL-6) separate into distinct clusters. Remarkably, four out of the five mice in the poly(I:C) + anti-IL-6 group cluster with the saline group (rather than with the poly(I:C) group). Therefore, maternal anti-IL-6 treatment not only prevents the development of abnormal behaviors in the offspring, but also prevents the gene expression changes caused by MIA. These results do not rule out a role for other cytokines, or even other types of factors, but the data suggest that such factors may act through the IL-6 pathway by increasing IL-6 production, or by inducing other effectors upstream of IL-6. That is, since blocking IL-6 function or eliminating its synthesis blocks the effects of MIA, it seems unlikely that other factors act independently of IL-6.

Where is the site, or sites, of IL-6 action that is responsible for altering fetal brain development? A first obvious possibility is the fetal brain itself, since IL-6 is known to be able to act directly on neurons [56]. In rats, radiolabeled IL-6 enters the fetus during mid, but not late, gestation [57], which fits with human data showing influenza infection increasing risk of schizophrenia in the second trimester [10]. Furthermore IL-6 acts through the STAT pathway, which regulates the balance between neurogenesis and gliogenesis [56]. Thus, altered STAT signaling could cause some of the types of changes observed in exposed offspring. Second, the placenta is also a target of interest, because IL-6, or a downstream effector, could alter the transfer of nutrients or other materials. Interleukin-6 alters vascular permeability in the adult brain after bacterial challenge [58], and a similar change in the permeability of the placenta could have significant effects on transfer of potentially harmful proteins (i.e., antibodies) into the fetal environment, or could allow maternal immune cells to infiltrate and bind alloantigens in the fetus. A third hypothesis for IL-6 action involves the adjustments made by the immune system during pregnancy to

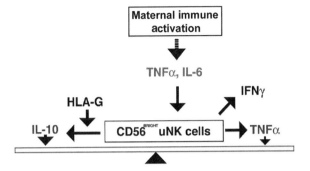

Fig. 13.2 The balance of pro- and anti-inflammatory cytokines influences pregnancy outcome. Pro- (TNF-α) and anti-inflammatory (IL-10) cytokines produced by a subclass of natural killer cells in the uterus and placenta (CD56^{BRIGHT}uNK) participate in the regulation of maternal tolerance of the fetus and in fighting infection. Imbalance toward either pro- or anti-inflammatory direction can endanger pregnancy and/or influence fetal development. Cytokines that are elevated by maternal immune activation (MIA), including IL-6 and TNF-α, can tilt this balance in the proinflammatory direction. The nonclassical histocompatibility leukocyte antigen (HLA)-G can stimulate mononuclear cells to secrete IL-10 and inhibit the secretion of TNF-α. Moreover, some HLA-G alleles are associated with low levels of HLA-G and increased miscarriage [60]

prevent rejection of the fetus. Normal pregnancy can be viewed as a state of controlled inflammation [59]. Uterine natural killer (uNK) cells, important for maintaining pregnancy, may be the primary regulators of inflammation at the fetomaternal interface [60] (Fig. 13.2). Injection of high doses of poly(I:C) causes loss of pregnancy in rats, and pretreatment with an anti-uNK antibody prevents this loss, implicating uNK cells in the miscarriage [61]. Thus, IL-6 could act on maternal immune cells, making them less tolerant of the fetus. Although this may increase the possibility of fetal loss or low birth weight, it is not known if this will also alter fetal brain development and lead to the types of abnormal behaviors described above. Interleukin-6 could also enhance the production of maternal antibodies, which could cross-react with the fetal brain, as has been proposed for autism [62] (see chapters by Singer and Van de Water). Future research can therefore productively focus on the effects of IL-6 on both the fetus and the placenta.

Cytokines and Autism

Given the strong evidence for involvement of cytokines in MIA on fetal brain development in animals, and the evidence that maternal infection can increase the risk of autism, it is important to consider what is known about cytokines in autism. Although a number of studies have assayed cytokines and other inflammatory markers in serum and blood, the most impressive data were obtained from brain homogenates. In a postmortem study of seven autistic and seven control subjects, numerous pro- and anti-inflammatory cytokines and chemokines are highly upregulated in autism [36]. For instance, IL-6 is increased more than 30-fold in anterior cingulate gyrus and more than sixfold in midfrontal gyrus. Such results are similar to those from the multiple IL-6 injection model of MIA, where IL-6 is elevated in the hippocampus of adult offspring [53]. Even more striking changes were seen in the CSF of six autistic subjects where, for instance, IFN-γ is increased more than 230-fold [36]. A recent finding of elevated TNF-α in CSF from 10 autistic children with a history of regression supports the view of the autistic brain as being in a chronic, subclinical inflammatory state [63]. In addition, a recent microarray study comparing gene expression in the superior temporal gyrus of six autistic subjects with matched controls found increased transcript levels in many immune system-related genes. Among other cytokines and chemokines, these included IL-6 and genes of the IL-6 pathway (105).

Mechanisms of Cytokine Action and Potential Therapies

The evidence of elevated cytokine proteins and mRNAs in the fetal brain following MIA raises the obvious possibility that cytokines act directly on developing neurons and glia. It is known, for instance, that transgenic, early overexpression of IL-6 in astrocytes causes major neuropathology and decreases seizure threshold [64], and seizures are a common symptom in autism. Interleukin-6 and related cytokines strongly influence neural stem cell proliferation and differentiation choices, as well as other features of brain development [56].

In addition to altering development, it is likely that the permanent elevation of cytokines in the brain seen in autism directly affects ongoing postnatal behavior. It is clear, for instance, that exogenous as well as endogenous IL-6 and IL-1 regulate neuronal excitability, long-term potentiation, and learning [56, 65]. Interleukin-6 and related cytokines also regulate the stress response, feeding, sleep, and depressive behaviors in the adult brain, and injections of certain cytokines can induce psychiatric symptoms in adult humans [56, 66, 67, 68]. Such acute effects of elevated cytokines could perhaps explain a series of puzzling case studies that report the sudden onset of autistic symptoms in children and adults following encephalitis or infection with herpes simplex, varicella, or cytomegalovirus [69]. Central nervous system infections of this type are known to rapidly induce proinflammatory cytokine expression. On the other hand, a prospective study of 30 children with autism spectrum disorders reported that the frequency of aberrant behavioral symptoms decline during periods of sickness and fever (106). Clearly, this association reguires further study.

A key point about the hypothesis of cytokines directly inducing or influencing autistic behavior is that it raises the possibility of developing treatments based on anticytokine or anti-inflammatory agents. Such treatments using antibodies against cytokines of the IL-6 family are in clinical trials for arthritis and inflammatory bowel disease, and anti-inflammatory drugs are being tested in Alzheimer's disease. In fact, a preliminary study in 25 autistic children of the thiazolidinedione, pioglitazone, which has anti-inflammatory properties, revealed a significant decrease in irritability, lethargy, stereotypy, and hyperactivity, with greater effects on the younger patients [70].

Potential treatments during pregnancy to block the developmental effects of maternal infection on the fetus present a far more difficult challenge. For instance, blocking IL-6 function during maternal influenza infection results in more severe symptoms and miscarriage [54]. Thus, there may be negative effects in lowering the inflammatory response too much during pregnancy. In the other her, administration of the anti-inflammatory cytokine IL-10 to pregnant rats given a uterine bacterial infection prevents fetal loss and white matter damage [71]. Administration of IL-10 also has a protective effect in pregnant mice given LPS [72] (see Fig. 13.2). Moreover, genetically enforced expression of IL-10 in macrophages attenuates the effects of MIA by poly(I:C) as measured by assays of PPI, LI, and open-field anxiety in adult offspring [73]. An attractive feature of this potential therapeutic is that endogenous IL-10 is essential for resistance to LPS-induced preterm labor and fetal loss. Thus, administration of this cytokine enhances the natural protective mechanism by attenuating the production of proinflammatory cytokines such as IL-6, IL-1α and TNF-α [72]. Subcutaneous administration of IL-10 was found to be nontoxic in long-term, preclinical studies in mice and monkeys, although testing was not done during pregnancy [74]. A cautionary observation is, however, that enhanced levels of IL-10 in the absence of MIA in pregnant mice can lead to behavioral abnormalities in the adult offspring [73]. In a different approach, oral administration of the antioxidant N-acetylcysteine in the LPS-induced MIA model suppresses inflammatory cytokines and prevents the characteristic white matter pathology [75, 76]. It is important to keep in mind, however, that normal pregnancy also involves increased systemic inflammation [77]. Low-dose LPS administration tips the balance toward exaggerated inflammation, including hypertension and proteinuria, in pregnant but not nonpregnant rats [78]. Infusion of TNF-α also causes hypertension in pregnant but not nonpregnant rats [79].

Perspectives

It is clear that virally induced MIA can dramatically increase the risk of autism. More epidemiological studies are needed to determine the current importance of this risk factor and to identify which infectious agents are most relevant. It is estimated, for instance, that 14–21% of schizophrenia can be attributed to maternal respiratory infection [11]. In addition, since animal studies highlight the importance of cytokines in mediating the effects of MIA on fetal brain development, it may also be likely that other factors, not related to infection, could elevate key cytokines and thereby imitate features of MIA. Stress is an example of such a factor that can influence cytokine levels, corticosteroids and the immune system [80] (see chapter by Connors et al.). Moreover, stress during pregnancy is known to alter neuronal development and behavior of the off-spring [78, 81]. Another example comes from studies of preterm birth where increased levels of proinflammatory cytokines in amniotic fluid and cord blood are associated with adverse neonatal neurological outcome. However, infectious agents are usually not found and antibiotics do not decrease the rate of preterm birth in patients with preterm labor and intact membranes. Thus, it appears that inflammation, rather than infection, plays the critical role in outcome [82] (see Fig. 13.2).

The evidence of a permanent, abnormal inflammatory state in the autistic brain, coupled with the MIA results from animal studies, suggests that a novel type of self-propagating immune dysregulation begins during fetal development. This hypothesis can be tested in animal models of MIA: can this cycle of immune dysregulation be interrupted in the postnatal or mature brain? If successful in altering abnormal behavior, such an intervention would have clear clinical implications. Another corollary of this hypothesis is that ongoing immune dysregulation in the brain and CSF may be detectable in the blood. Identification of peripheral biomarkers of the relevant features of brain inflammation in autism would have important implications for diagnosis and treatment. Given the face and construct validity of the current rodent (and potential future nonhuman primate) models of MIA, testing these hypotheses appears to be feasible and essential. Moreover, the MIA protocols can be applied to emerging mouse models of genetic risk factors. This will provide a test of the dominant paradigm of autism: environmental factors act in the context of genetic susceptibility.

Acknowledgments We thank Janet Baer, Kathleen Hamilton, Bill Lease, and Doreen McDowell for assistance with the experiments described from this laboratory. Funding for our recent work comes from the NIMH, and the Autism Speaks, Cure Autism Now and McKnight Foundations.

References

1. Lemery-Chalfant K, Goldsmith HH, Schmidt NL, Arneson CL, Van Hulle CA. Wisconsin Twin Panel: current directions and findings. *Twin Res Hum Genet* (2006) 9:1030–7.
2. Phelps JA, Davis JO, Schartz KM. Nature, Nurture, and Twin Research Strategies. *Curr Dir Psychol Sci* (1997) 6:117–31.
3. Ryan BC, Vandenbergh JG. Intrauterine position effects. *Neurosci Biobehav Rev* (2002) 26:665–78.
4. Christianson AL, Chesler N, Kromberg JG. Fetal valproate syndrome: clinical and neuro-developmental features in two sibling pairs. *Dev Med Child Neurol* (1994) 36:361–9.
5. Stromland K, Nordin V, Miller M, Akerstrom B, Gillberg C. Autism in thalidomide embryopathy: a population study. *Dev Med Child Neurol* (1994) 36:351–6.
6. Chess S, Fernandez P, Korn S. Behavioral consequences of congenital rubella. *J Pediatr* (1978) 93:699–703.

7. Patterson PH. Maternal infection: window on neuroimmune interactions in fetal brain development and mental illness. *Curr Opin Neurobiol* (2002) 12:115–8.
8. Sweeten TL, Posey DJ, McDougle CJ. Brief report: autistic disorder in three children with cytomegalovirus infection. *J Autism Dev Disord* (2004) 34:583–6.
9. Yamashita Y, Fujimoto C, Nakajima E, Isagai T, Matsuishi T. Possible association between congenital cytomegalovirus infection and autistic disorder. *J Autism Dev Disord* (2003) 33:455–9.
10. Brown AS. Prenatal infection as a risk factor for schizophrenia. *Schizophr Bull* (2006) 32:200–2.
11. Brown AS, Begg MD, Gravenstein S, et al. Serologic evidence of prenatal influenza in the etiology of schizophrenia. *Arch Gen Psychiatry* (2004) 61:774–80.
12. Shi L, Fatemi SH, Sidwell RW, Patterson PH. Maternal influenza infection causes marked behavioral and pharmacological changes in the offspring. *J Neurosci* (2003) 23:297–302.
13. Perry W, Minassian A, Lopez B, Maron L, Lincoln A. Sensorimotor gating deficits in adults with autism. *Biol Psychiatry* (2007) 61:482–6.
14. Meyer U, Feldon J, Schedlowski M, Yee BK. Towards an immuno-precipitated neuro-developmental animal model of schizophrenia. *Neurosci Biobehav Rev* (2005) 29:913–47.
15. Palmen SJ, van Engeland H, Hof PR, Schmitz C. Neuropathological findings in autism. *Brain* (2004) 127:2572–83.
16. Akshoomoff N, Lord C, Lincoln AJ, et al. Outcome classification of preschool children with autism spectrum disorders using MRI brain measures. *J Am Acad Child Adolesc Psychiatry* (2004) 43:349–57.
17. Pierce K, Courchesne E. Evidence for a cerebellar role in reduced exploration and stereotyped behavior in autism. *Biol Psychiatry* (2001) 49:655–64.
18. Allen G, Courchesne E. Differential effects of developmental cerebellar abnormality on cognitive and motor functions in the cerebellum: an fMRI study of autism. *Am J Psychiatry* (2003) 160:262–73.
19. Kates WR, Burnette CP, Eliez S, et al. Neuroanatomic variation in monozygotic twin pairs discordant for the narrow phenotype for autism. *Am J Psychiatry* (2004) 161:539–46.
20. Nowinski CV, Minshew NJ, Luna B, Takarae Y, Sweeney JA. Oculomotor studies of cerebellar function in autism. *Psychiatry Res* (2005) 137:11–9.
21. Takarae Y, Minshew NJ, Luna B, Sweeney JA. Oculomotor abnormalities parallel cerebellar histopathology in autism. *J Neurol Neurosurg Psychiatry* (2004) 75:1359–61.
22. Kaufmann WE, Cooper KL, Mostofsky SH, et al. Specificity of cerebellar vermian abnormalities in autism: a quantitative magnetic resonance imaging study. *J Child Neurol* (2003) 18:463–70.
23. Bottmer C, Bachmann S, Pantel J, et al. Reduced cerebellar volume and neurological soft signs in first-episode schizophrenia. *Psychiatry Res* (2005) 140:239–50.
24. Brown SM, Kieffaber PD, Carroll CA, et al. Eyeblink conditioning deficits indicate timing and cerebellar abnormalities in schizophrenia. *Brain Cogn* (2005) 58:94–108.
25. Ho BC, Mola C, Andreasen NC. Cerebellar dysfunction in neuroleptic naive schizophrenia patients: clinical, cognitive, and neuroanatomic correlates of cerebellar neurologic signs. *Biol Psychiatry* (2004) 55:1146–53.
26. Patterson PH. Pregnancy, immunity, sehizophrenia and autism. *Engineering Sci* (2006) 69:10–21.
27. Courchesne E, Redcay E, Kennedy DP. The autistic brain: birth through adulthood. *Curr Opin Neurol* (2004) 17:489–96.
28. Fatemi SH, Earle J, Kanodia R, et al. Prenatal viral infection leads to pyramidal cell atrophy and macrocephaly in adulthood: implications for genesis of autism and schizophrenia. *Cell Mol Neurobiol* (2002) 22:25–33.
29. Ozawa K, Hashimoto K, Kishimoto T, Shimizu E, Ishikura H, Iyo M. Immune activation during pregnancy in mice leads to dopaminergic hyperfunction and cognitive impairment in the offspring: a neurodevelopmental animal model of schizophrenia. *Biol Psychiatry* (2006) 59:546–54.

30. Zuckerman L, Rehavi M, Nachman R, Weiner I. Immune activation during pregnancy in rats leads to a postpubertal emergence of disrupted latent inhibition, dopaminergic hyperfunction, and altered limbic morphology in the offspring: a novel neurodevelopmental model of schizophrenia. *Neuropsychopharmacology* (2003) 28:1778–89.

31. Nyffeler M, Meyer U, Yee BK, Feldon J, Knuesel I. Maternal immune activation during pregnancy increases limbic GABAA receptor immunoreactivity in the adult offspring: implications for schizophrenia. *Neuroscience* (2006) 143:51–62.

32. Schmitz C, van Kooten IA, Hof PR, van Engeland H, Patterson PH, Steinbusch HW. Autism: neuropathology, alterations of the GABAergic system, and animal models. *Int Rev Neurobiol* (2005) 71:1–26.

33. Hagberg H, Mallard C. Effect of inflammation on central nervous system development and vulnerability. *Curr Opin Neurol* (2005) 18:117–23.

34. Borrell J, Vela JM, Arevalo-Martin A, Molina-Holgado E, Guaza C. Prenatal immune challenge disrupts sensorimotor gating in adult rats. Implications for the etiopathogenesis of schizophrenia. *Neuropsychopharmacology* (2002) 26:204–15.

35. Fatemi SH, Araghi-Niknam M, Laurence JA, Stary JM, Sidwell RW, Lee S. Glial fibrillary acidic protein and glutamic acid decarboxylase 65 and 67 kDa proteins are increased in brains of neonatal BALB/c mice following viral infection in utero. *Schizophr Res* (2004) 69:121–3.

36. Vargas DL, Nascimbene C, Krishnan C, Zimmerman AW, Pardo CA. Neuroglial activation and neuroinflammation in the brain of patients with autism. *Ann Neurol* (2005) 57:67–81.

37. Arion D, Unger T, Lewis DA, Levitt P, Mirnics K. Molecular evidence for increased expression of genes related to immune and chaperone function in the prefrontal cortex in schizophrenia. *Biol Psychiatry* (2007) 62:711–21.

38. Bobetsis YA, Barros SP, Offenbacher S. Exploring the relationship between periodontal disease and pregnancy complications. *J Am Dent Assoc* (2006) 137:7S–13S.

39. Han YW, Redline RW, Li M, Yin L, Hill GB, McCormick TS. *Fusobacterium nucleatum* induces premature and term stillbirths in pregnant mice: implication of oral bacteria in preterm birth. *Infect Immun* (2004) 72:2272–9.

40. Lin D, Smith MA, Elter J, et al. *Porphyromonas gingivalis* infection in pregnant mice is associated with placental dissemination, an increase in the placental Th1/Th2 cytokine ratio, and fetal growth restriction. *Infect Immun* (2003) 71:5163–8.

41. Patterson P. *Modeling Features of Autism n Animals.* Boca Raton: Taylor & Francis; 2005.

42. Wang X, Rousset CI, Hagberg H, Mallard C. Lipopolysaccharide-induced inflammation and perinatal brain injury. *Semin Fetal Neonatal Med* (2006) 11:343–53.

43. Nawa H, Takei N. Recent progress in animal modeling of immune inflammatory processes in schizophrenia: implication of specific cytokines. *Neurosci Res* (2006) 56:2–13.

44. Pearce B. Modeling the role of infections in the etiology of mental illness. *Clin Neurosci Res* (2003) 3:271–82.

45. Urakubo A, Jarskog LF, Lieberman JA, Gilmore JH. Prenatal exposure to maternal infection alters cytokine expression in the placenta, amniotic fluid, and fetal brain. *Schizophr Res* (2001) 47:27–36.

46. Bell MJ, Hallenbeck JM, Gallo V. Determining the fetal inflammatory response in an experimental model of intrauterine inflammation in rats. *Pediatr Res* (2004) 56:541–6.

47. Ashdown H, Dumont Y, Ng M, Poole S, Boksa P, Luheshi GN. The role of cytokines in mediating effects of prenatal infection on the fetus: implications for schizophrenia. *Mol Psychiatry* (2006) 11:47–55.

48. Elovitz MA, Mrinalini C, Sammel MD. Elucidating the early signal transduction pathways leading to fetal brain injury in preterm birth. *Pediatr Res* (2006) 59:50–5.

49. Meyer U, Nyffeler M, Engler A, et al. The time of prenatal immune challenge determines the specificity of inflammation-mediated brain and behavioral pathology. *J Neurosci* (2006) 26:4752–62.

50. Silver RM, Lohner WS, Daynes RA, Mitchell MD, Branch DW. Lipopolysaccharide-induced fetal death: the role of tumor-necrosis factor alpha. *Biol Reprod* (1994) 50:1108–12.

51. Xu DX, Chen YH, Wang H, Zhao L, Wang JP, Wei W. Tumor necrosis factor alpha partially contributes to lipopolysaccharide-induced intra-uterine fetal growth restriction and skeletal development retardation in mice. *Toxicol Lett* (2006) 163:20–9.

52. Wang X, Hagberg H, Mallard C, et al. Disruption of interleukin-18, but not interleukin-1, increases vulnerability to preterm delivery and fetal mortality after intrauterine inflammation. *Am J Pathol* (2006) 169:967–76.

53. Samuelsson AM, Jennische E, Hansson HA, Holmang A. Prenatal exposure to interleukin-6 results in inflammatory neurodegeneration in hippocampus with NMDA/GABA(A) dysregulation and impaired spatial learning. *Am J Physiol Regul Integr Comp Physiol* (2006) 290:R1345–56.

54. Smith SE, Li J, Garbett K, Mirnics K, Patterson PH. Maternal immune activation alters fetal brain development through interleukin-6. *J Neurosci* (2007) 27:10695–702.

55. Fatemi SH, Pearce DA, Brooks AI, Sidwell RW. Prenatal viral infection in mouse causes differential expression of genes in brains of mouse progeny: a potential animal model for schizophrenia and autism. *Synapse* (2005) 57:91–9.

56. Bauer S, Kerr BJ, Patterson PH. The neuropoietic cytokine family in development, plasticity, disease and injury. *Nat Rev Neurosci* (2007) 8:221–32.

57. Dahlgren J, Samuelsson AM, Jansson T, Holmang A. Interleukin-6 in the maternal circulation reaches the rat fetus in mid-gestation. *Pediatr Res* (2006) 60:147–51.

58. Paul R, Koedel U, Winkler F, et al. Lack of IL-6 augments inflammatory response but decreases vascular permeability in bacterial meningitis. *Brain* (2003) 126:1873–82.

59. Sargent IL, Borzychowski AM, Redman CW. NK cells and human pregnancy – an inflammatory view. *Trends Immunol* (2006) 27:399–404.

60. Christiansen OB, Nielsen HS, Kolte AM. Inflammation and miscarriage. *Semin Fetal Neonatal Med*(2006) 11:302–8.

61. Arad M, Atzil S, Shakhar K, Adoni A, Ben-Eliyahu S. Poly I-C induces early embryo loss in f344 rats: a potential role for NK cells. *Am J Reprod Immunol* (2005) 54:49–53.

62. Dalton P, Deacon R, Blamire A, et al. Maternal neuronal antibodies associated with autism and a language disorder. *Ann Neurol* (2003) 53:533–7.

63. Chez MG, Dowling T, Patel PB, Khanna P, Kominsky M. Elevation of tumor necrosis factor-alpha in cerebrospinal fluid of autistic children. *Pediatr Neurol* (2007) 36:361–5.

64. Samland H, Huitron-Resendiz S, Masliah E, Criado J, Henriksen SJ, Campbell IL. Profound increase in sensitivity to glutamatergic – but not cholinergic agonist-induced seizures in transgenic mice with astrocyte production of IL-6. *J Neurosci Res* (2003) 73:176–87.

65. Jankowsky JL, Patterson PH. Cytokine and growth factor involvement in long-term potentiation. *Mol Cell Neurosci* (1999) 14:273–86.

66. Capuron L, Dantzer R. Cytokines and depression: the need for a new paradigm. *Brain Behav Immun* (2003) 17 Suppl 1:S119–24.

67. Schiepers OJ, Wichers MC, Maes M. Cytokines and major depression. *Prog Neuropsychopharmacol Biol Psychiatry* (2005) 29:201–17.

68. Theoharides TC, Weinkauf C, Conti P. Brain cytokines and neuropsychiatric disorders. *J Clin Psychopharmacol* (2004) 24:577–81.

69. Libbey JE, Sweeten TL, McMahon WM, Fujinami RS. Autistic disorder and viral infections. *J Neurovirol* (2005) 11:1–10.

70. Boris M, Kaiser CC, Goldblatt A, et al. Effect of pioglitazone treatment on behavioral symptoms in autistic children. *J Neuroinflammation* (2007) 4:3.

71. Pang Y, Rodts-Palenik S, Cai Z, Bennett WA, Rhodes PG. Suppression of glial activation is involved in the protection of IL-10 on maternal *E. coli* induced neonatal white matter injury. *Brain Res Dev Brain Res* (2005) 157:141–9.

72. Robertson SA, Skinner RJ, Care AS. Essential role for IL-10 in resistance to lipopolysaccharide-induced preterm labor in mice. *J Immunol* (2006) 177:4888–96.

73. Meyer U, Murray PJ, Urwyler A, Yee BK, Schedlowski M, Feldon J. Adult behavioral and pharmacological dysfunctions following disruption of the fetal brain balance

between pro-inflammatory and IL-10-mediated anti-inflammatory signaling. *Mol Psychiatry* (2008) 13:208–21.

74. Rosenblum IY, Johnson RC, Schmahai TJ. Preclinical safety evaluation of recombinant human interleukin-10. *Regul Toxicol Pharmacol* (2002) 35:56–71.

75. Beloosesky R, Gayle DA, Ross MG. Maternal N-acetylcysteine suppresses fetal inflammatory cytokine responses to maternal lipopolysaccharide. *Am J Obstet Gynecol* (2006) 195:1053–7.

76. Paintlia MK, Paintlia AS, Barbosa E, Singh I, Singh AK. N-acetylcysteine prevents endotoxin-induced degeneration of oligodendrocyte progenitors and hypomyelination in developing rat brain. *J Neurosci Res* (2004) 78:347–61.

77. Borzychowski AM, Sargent IL, Redman CW. Inflammation and pre-eclampsia. *Semin Fetal Neonatal Med* (2006) 11:309–16.

78. Faas MM, Schuiling GA, Baller JF, Visscher CA, Bakker WW. A new animal model for human preeclampsia: ultra-low-dose endotoxin infusion in pregnant rats. *Am J Obstet Gynecol* (1994) 171:158–64.

79. Alexander BT, Cockrell KL, Massey MB, Bennett WA, Granger JP. Tumor necrosis factor-alpha-induced hypertension in pregnant rats results in decreased renal neuronal nitric oxide synthase expression. *Am J Hypertens* (2002) 15:170–5.

80. Pace TW, Hu F, Miller AH. Cytokine-effects on glucocorticoid receptor function: relevance to glucocorticoid resistance and the pathophysiology and treatment of major depression. *Brain Behav Immun* (2007) 21:9–19.

81. Kofman O. The role of prenatal stress in the etiology of developmental behavioural disorders. *Neurosci Biobehav Rev* (2002) 26:457–70.

82. Elovitz MA. Anti-inflammatory interventions in pregnancy: now and the future. *Semin Fetal Neonatal Med* (2006) 11:327–32.

83. Zuckerman L, Weiner I. Maternal immune activation leads to behavioral and pharmacological changes in the adult offspring. *J Psychiatr Res* (2005) 39:311–23.

84. Meyer U, Feldon J, Schedlowski M, Yee BK. Immunological stress at the maternal–foetal interface: a link between neurodevelopment and adult psychopathology. *Brain Behav Immun* (2006) 20:378–88.

85. Wolff AR CK, Bilkey DK. Hippocampal dysfunction in an animal model of schizophrenia. Int Brain Res Org (2007) SYM-25-04.

86. Cai Z, Pan ZL, Pang Y, Evans OB, Rhodes PG. Cytokine induction in fetal rat brains and brain injury in neonatal rats after maternal lipopolysaccharide administration. *Pediatr Res* (2000) 47:64–72.

87. Ling Z, Chang QA, Tong CW, Leurgans SE, Lipton JW, Carvey PM. Rotenone potentiates dopamine neuron loss in animals exposed to lipopolysaccharide prenatally. *Exp Neurol* (2004) 190:373–83.

88. Bell MJ, Hallenbeck JM. Effects of intrauterine inflammation on developing rat brain. *J Neurosci Res* (2002) 70:570–9.

89. Bakos J, Duncko R, Makatsori A, Pirnik Z, Kiss A, Jezova D. Prenatal immune challenge affects growth, behavior, and brain dopamine in offspring. *Ann N Y Acad Sci* (2004) 1018:281–7.

90. Fortier ME, Joober R, Luheshi GN, Boksa P. Maternal exposure to bacterial endotoxin during pregnancy enhances amphetamine-induced locomotion and startle responses in adult rat offspring. *J Psychiatr Res* (2004) 38:335–45.

91. Golan H, Lev V, Mazar Y. Alterations in behavior in adult offspring mice following maternal inflammation during pregnancy. *Dev Psychobiol* (2006) 48:162–8.

92. Golan H, Levav T, Mendelsohn A, Huleihel M. Involvement of tumor necrosis factor alpha in hippocampal development and function. *Cereb Cortex* (2004) 14:97–105.

93. Golan HM, Lev V, Hallak M, Sorokin Y, Huleihel M. Specific neurodevelopmental damage in mice offspring following maternal inflammation during pregnancy. *Neuropharmacology* (2005) 48:903–17.

94. Poggi SH, Park J, Toso L, et al. No phenotype associated with established lipopolysaccharide model for cerebral palsy. *Am J Obstet Gynecol* (2005) 192:727–33.
95. Rousset CI, Chalon S, Cantagrel S, et al. Maternal exposure to LPS induces hypomyelination in the internal capsule and programmed cell death in the deep gray matter in newborn rats. *Pediatr Res* (2006) 59:428–33.
96. Wang X, Hagberg H, Zhu C, Jacobsson B, Mallard C. Effects of intrauterine inflammation on the developing mouse brain. *Brain Res* (2007) 1144:180–5.
97. Nitsos I, Rees SM, Duncan J, et al. Chronic exposure to intra-amniotic lipopolysaccharide affects the ovine fetal brain. *J Soc Gynecol Investig*(2006) 13:239–47.
98. Fidel PL Jr., Romero R, Wolf N, et al. Systemic and local cytokine profiles in endotoxin-induced preterm parturition in mice. *Am J Obstet Gynecol* (1994) 170:1467–75.
99. Kramer BW, Moss TJ, Willet KE, et al. Dose and time response after intraamniotic endotoxin in preterm lambs. *Am J Respir Crit Care Med* (2001) 164:982–8.
100. Rounioja S, Rasanen J, Glumoff V, Ojaniemi M, Makikallio K, Hallman M. Intra-amniotic lipopolysaccharide leads to fetal cardiac dysfunction. A mouse model for fetal inflammatory response. *Cardiovasc Res* (2003) 60:156–64.
101. Gayle DA, Beloosesky R, Desai M, Amidi F, Nunez SE, Ross MG. Maternal LPS induces cytokines in the amniotic fluid and corticotropin releasing hormone in the fetal rat brain. *Am J Physiol Regul Integr Comp Physiol* (2004) 286:R1024–9.
102. Beloosesky R, Gayle DA, Amidi F, et al. *N*-Acetyl-cysteine suppresses amniotic fluid and placenta inflammatory cytokine responses to lipopolysaccharide in rats. *Am J Obstet Gynecol* (2006) 194:268–73.
103. Gilmore JH, Jarskog LF, Vadlamudi S. Maternal poly I:C exposure during pregnancy regulates TNF alpha, BDNF, and NGF expression in neonatal brain and the maternal-fetal unit of the rat. *J Neuroimmunol* (2005) 159:106–12.
104. Liverman CS, Kaftan HA, Cui L Hersperger SG, Taboada E. Klein RM, Berman NEJ. Altered expression of pro-inflammatory and developmental genes in the fetal brain in a mouse model of maternal infection. Neurosci Lett (2006) 399:220–5.
105. Garbett K. Ebert PJ, Mitchell A, Lintas C, Manzi B, Mlrnics K. Persico AM Immune transcriptome alterations in the temporal cortex of subjects with autism Neurobiol Dis (2008) 30:303–11.
106. Curran LK, Newschaffer CJ, Lee L-C, Crawford SO, Johnston MV, Zimmerman AW. Behaviors associated with fever in children with autism spectrum disorders. Pediatrics (2007) 120:e1386–92.

Chapter 14
Maternal Antibodies and the Placental–Fetal IgG Transfer Theory

Christina M. Morris, Mikhail Pletnikov, Andrew W. Zimmerman and Harvey S. Singer

Abstract We hypothesize that maternal autoimmunity is a contributing factor to etiology in up to 40% of pregnancies that lead to autism. More specifically, we propose that the transplacental transfer of maternal antibodies to the fetus alters fetal brain development, and, on a background of genetic susceptibility, ultimately leads to the postnatal emergence of autism. Circumstantial evidence for this hypothesis includes human studies showing that maternal autoimmune disorders can adversely affect fetal brain development, as well as studies of human leukocyte antigens (HLA). To date, two groups of investigators have identified differential patterns of specific antibodies directed to human fetal brain in sera from mothers of children with autistic disorder (MCAD) as compared with mothers of unaffected children. In both studies, specific maternal antibodies correlated with the presence of developmental regression in offspring. Lastly, the pregnant dam mouse model has shown that MCAD IgG can cross the placenta, enter embryonic brain, induce an immune response, and cause behavioral changes. In this chapter, we review circumstantial and direct evidence for, and future requirements necessary to confirm, the placental–fetal IgG transfer theory.

Keywords Maternal antibodies · autoimmune hypothesis in autism · autistic disorder · anti-fetal brain antibodies · placental–fetal transfer of antibodies

Introduction

Autism is the most commonly diagnosed developmental disability among children, with an apparent prevalence of autistic spectrum disorders (ASDs) being one in 150 [1]. Despite extensive research, the causes for autism in most affected children remain elusive. A variety of genetic, biochemical, and

H.S. Singer
Division of Pediatric Neurology and Department of Psychiatry and Behavioral
Sciences, Johns Hopkins University School of Medicine; and Kennedy Krieger
Institute, Baltimore, MD, USA
e-mail: hsinger@jhmi.edu

A.W. Zimmerman (ed.), *Autism*, DOI: 10.1007/978-1-60327-489-0_14,
© Humana Press, Totowa, NJ 2008

environmental factors have been proposed to have roles in causing autism [2, 3, 4, 5, 6, 7, 8, 9, 10, 11, 12, 13, 14], but definitive evidence for each is lacking. Because of the recent trend of increasing incidence – over 900% between 1992 and 2001 [15, 16] – it is imperative to uncover the component etiologies of autism, which likely involve complex interrelated factors.

Autistic phenotypes include a broad range of symptoms and severity. For example, a child with high functioning autism or Asperger syndrome falls under the same umbrella as a child with severe mental retardation and almost no language skills. One affected child may exhibit frequent motor stereotypies, whereas another may merely have very focused interests. This clinical diversity among the neurodevelopmental disorders categorized as "autism" implies that their causes also reflect similar complexity and variability, and are likely due to factors that modify gene expression and development in utero that extend postnatally. Several theories gaining increased attention suggest that autism may be associated with abnormalities of the immune system. Supporting evidence, circumstantial and direct, for immune dysfunction in ASDs is growing, but definitive evidence and the underlying pathophysiologic mechanism remain to be elucidated [17, 18, 19]; see Chapter 12 by Van de Water and Ashwood).

Immune findings have been described in both the peripheral blood and the central nervous system (CNS) in patients with autism; however, their pathogenicity has been difficult to discern [20]. In most cases, they do not accompany recognized immunodeficiency syndromes or autoimmune disorders. An important question is whether these findings are pivotal to the underlying biology of autism, or epiphenomena of more fundamental causes. Examples include immunohistochemical examination of postmortem autistic brains that have identified an active, and apparently chronic, process of immune activation in the cerebral cortex, white matter, and cerebellum [21]. Furthermore, elevated concentrations of pro-inflammatory cytokines were found in autistic brains and CSF, whereas high levels of anti-inflammatory cytokines are thought to counterbalance the inflammation [21, 22, 23]. There are other reports of increased concentrations of cytokines in peripheral blood that mediate both innate and adaptive responses [24, 25, 26, 27]. In addition, a variety of serum antibodies have been detected in individuals with the diagnosis of ASD that are directed to components of adult as well as fetal brain epitopes [11, 28, 29], most of which have not been identified. Some systemic immune findings in autism correspond with both T-helper (T_H-1 and T_H-2-like) adaptive immune responses and may reflect characteristics of both autoimmune and atopic diseases [30], whereas others suggest abnormalities in the innate immune system [27], consistent with findings in the CNS [21]. Taken together, the findings in children with autism suggest widespread dysregulation among various components of the immune system, none of which, as yet, has been identified as a biomarker or part of a final common pathway of causation, such as infection, allergy, autoimmune disease, or dysregulated development. Although it is possible that an immune pathogenesis for autism could develop during infancy or early childhood, we believe another and more likely possibility is that it develops in utero.

Hypothesis

We hypothesize that immune factors have a role in autism. However, in contrast to other hypotheses, we emphasize that the process begins in utero and is associated with the placental transfer of maternal antibodies that, in turn, interfere with fetal brain development. On the basis of requirements proposed by Archelos and Hartung [31], confirmation of this maternal–fetal IgG transfer hypothesis requires the following: 1) detection of specific antibodies against fetal brain in mothers of children with autistic disorder (MCAD) that differ from those in mothers with unaffected children (MUC); 2) confirming that maternal immunoglobulins pass through the placenta and fetal blood–brain barrier and bind to epitopes in fetal brain; 3) demonstration that maternal IgG induces an immune response in the developing fetal brain; 4) documentation that exposed offspring have signs of disordered development that mimic that seen in autistic disorder; and 5) showing that the process can be altered with immunomodulatory therapy [31]. In the following sections, we present supporting evidence, both circumstantial and direct, that begins to fulfill the first 4 of these requirements and support the theory of maternal–fetal IgG transfer as a cause for autism.

Supporting Evidence

Circumstantial Evidence

HLA Studies

The human leukocyte antigen (HLA) system (part of the major histocompatibility complex) is composed of a group of genes (on chromosome 6) that encode a series of cell-surface antigen-presenting proteins that are important factors in immune function. Class I molecules (A, B, C) typically present antigens that invade cells whereas Class II molecules (DR, DP, DQ) present antigens to T lymphocytes that influence B-cell production of antibodies. The Class II molecule HLA-DR4, known to be associated with autoimmune disorders such as rheumatoid arthritis (RA), pemphigus vulgaris, and Type I insulin-dependent diabetes mellitus, has been connected with autism [32, 33, 34]. In a study of 16 families in East Tennessee, researchers found an increased frequency of HLA-DR4 in mothers and their sons with autism [34]. High-resolution typing of HLA-DR4 in families with autism plus increased rates of autoimmune disorders has shown large amounts of both 0401 and 0404 alleles, previously associated with RA [2, 34]. One interpretation is that having HLA-DR4 or related Class II alleles may increase one's environmental susceptibility within a geographical region, by predisposing mothers and their children with DR4 to develop unusual immune reactivity. Whether this altered immune function includes the production of antibodies that cross-react with fetal brain antigens is undetermined. Class I antigens, HLA-A2 [35], HLA-A1, and

HLA-B8, have also been associated with ASDs. For example, HLA-B8 has been found in autistic males whereas HLA-A68 was found in their fathers [10].

Maternal Influences

Studies have suggested that the risk of developing an autoimmune disease tends to increase with age. Several investigators have suggested that older women are more likely to give birth to an autistic child than younger mothers. Research done to support this claim is abundant but not entirely conclusive. Croen et al. found a significant increase in relative risk, at 1.31 and 1.28 for occurrence of ASD in offspring with each 10-year increase in maternal and paternal age, respectively [36]. Furthermore, these authors identified a correlation between increased parental age and the incidence/prevalence of three main autism subgroups, autistic disorder, pervasive developmental disorder-not otherwise specified (PDD-NOS), and Asperger syndrome [36]. Further studies are required to confirm this association as well as to determine whether maternal antibodies directed to the fetal brain may be more likely to form with increasing maternal age.

A further potential link between maternal autoimmunity and autism includes the increased occurrence of asthma, allergy, autoimmune psoriasis, and Type I diabetes in the mothers of children with ASD [37]. Other investigators have found that first-degree relatives of children with autism and Asperger syndrome were more likely to have an autoimmune disease, primarily hypothyroidism, Hashimoto's thyroiditis, or rheumatic fever, as compared with controls [2, 38]. Of ASD children with a positive history of autoimmunity in the family, the mother was the most frequently identified first-degree relative. The aforementioned studies remain controversial, however, because other investigators have not identified increased rates of autoimmune disorders in families with autism [39].

Hygiene Hypothesis

In the last 50 years, there has been a dramatic increase in asthma, allergies, and autoimmune diseases, currently affecting almost half the population of industrialized nations [40, 41]. In contrast, allergies and autoimmune diseases remain rare in developing countries. Developed in 1989, the hygiene hypothesis explains this discrepancy as being due to the increased use of vaccines, antibiotics, general hygiene, and availability of healthcare in the wealthier industrialized areas of the world [42]. More specifically, improved care is claimed to reduce the body's exposure to infections, decrease stimulation of the immune system, and possibly allow the immune system to develop improperly (i.e., thymic development), all leading to inappropriate reactivity of the immune system to non-pathogenic agents [43, 44, 45, 46]. The hygiene hypothesis has been cited to explain an increased incidence of autoimmune disorders [42], including autism [46]. Whether this specific hypothesis has direct relevance is unknown, but one could speculate that it influences the process of maternal autoimmunity and microchimerism (see Section *Stimuli for Maternal Antibody Formation*).

Clinical Examples of Maternal Antibodies Altering Fetal Human Development

Rh Disease

Because the fetus develops within the uterus and receives vital nutrients from the mother through the placenta, it is at continued risk from factors that affect the mother. Hence, if the mother's immune system is activated and produces immunoglobulins that can cross into the fetal circulation, the fetus can potentially develop adverse effects. Hemolytic disease of the newborn, or Rh disease, serves as a model disorder in which the maternal immune system produces antibodies that can subsequently have significant effects on fetal development. In brief, Rh, or Rhesus factor, represents a group of protein markers, or antigens, found on human erythrocytes, which are thought to play a role in ammonium transport [47]. In this disorder, the immune system in an RhD-negative mother is exposed to RhD, through transplacental exposure from a positive fetus, and mounts a potent immune response. The mother's antibodies to RhD attack the erythrocytes of the developing fetus and depending on severity can result in stillbirth. Although Rh disease does not serve as a model for maternal antibodies causing autism [48], it exemplifies that maternal antibody development can have serious consequences for the fetus.

Terbutaline

This drug for asthma, also used for tocolysis (to stop preterm labor), is a β2-adrenoceptor agonist with a proposed causal association with autism in dizygotic twins [49]. In a rodent study, terbutaline administered to postnatal rats (equivalent to late 2nd to early 3rd trimester human gestation) demonstrated increased microglial activation in the cerebral cortex, cerebellar white matter, and cerebrocortical white matter. Behavioral evaluations showed hyper-reactive behaviors analogous to those seen in autism [50]. This model emphasizes that neuroglial activation during a critical period of brain development can be a marker of subsequent behavioral abnormalities.

Poly(I:C)

Poly(I:C)-induced changes represent an additional example of maternal immune activation influencing fetal neurodevelopment. Pregnant rodents injected with poly(I:C), a synthetic double-stranded RNA molecule, develop an inflammatory response through the activation of the toll-like receptor 3. Toll-like receptors, part of the innate immune system, are located on dendritic cells and activate effector cells, including natural killer cells and natural killer T cells. The resulting offspring showed altered behavioral development, including abnormal tests of social interactions, and a localized deficit of Purkinje cells [51].

Systemic Lupus Erythematosus

Although results differ [52], the predominance of evidence suggests that SLE autoantibodies against ribosomal P proteins are linked to neuropsychiatric manifestations [53, 54, 55, 56, 57]. For example, the prevalence of anti-P antibodies has been related to disease activity, whereas disease remission was associated with disappearance of these antibodies [55, 58, 59]. Anti-P antibodies have been shown to bind to neurons, T cells, monocytes, and hepatocytes, enhance the production of proinflammatory cytokines, and cause CNS and liver damage [55, 60, 61]. The effects of maternal antibodies in SLE on neonates have also been well recognized, especially neonatal heart block (see below). An increased risk of learning disabilities was found in one study of children born to mothers with active SLE [62].

In a study designed to assess the brain-binding effect of anti-P antibodies, affinity-purified human anti-ribosomal P antibodies were injected into the cerebral ventricles of 3-month-old female mice [63]. Behavior was evaluated by the forced swimming test, motor deficits by rotarod, grip strength, and staircase tests, and cognitive deficits by T-maze alternation and passive avoidance tests. Antibodies induced depression-like behavior (increased immobility time in the forced swimming test) in mice that was partially blocked by antiidiotypic antibody and pharmacotherapy. There were no motor or cognitive effects. Anti-ribosomal P antibodies specifically stained neurons in the hippocampus, cingulate gyrus, and olfactory piriform cortex. Because the anti-ribosomal P antibody did not stain all nucleated cells and binding involved membrane proteins, it is believed that the antibody actually binds to a surface antigen [59, 60, 61, 64].

Direct Evidence

Presence of Anti-Fetal Brain Antibodies in MCAD

Three separate research groups have evaluated the sera of mothers who have had children with autism in search of anti-neuronal antibodies that differ from controls. Although methodologies and brain tissues differ, all have reported the presence of unique maternal antibody patterns. In a study performed by the authors [65], serum antibodies in 100 MCAD were compared with 100 age-matched MUC using as antigenic substrates human fetal and adult brain [caudate, frontal lobe (BA9), cerebellum, and cingulate gyrus], glial fibrillary acidic protein (GFAP), and myelin basic protein (MBP). Reactive bands, determined by Western immunoblotting, were considered atypical if present in more MCAD compared with MUC, or if the optical density of the band (peak height), corrected for maternal IgG content, was greater in MCAD. Using these criteria, results showed that sera from MCAD contain antibodies that differ from controls against prenatally expressed brain antigens. Focusing on results with fetal epitopes, MCAD had significantly more individuals with immunoblot bands against human fetal tissue at 36 kDa; 10% of MCAD compared with only 2% MUC ($p = 0.017$). Similar findings at 36 kDa were also identified in rodent embryonic brain tissue (MCAD, $n = 48$; MUC, $n = 31$; $p = 0.010$). Whether these two bands represent the same antigenic protein in two different species, however, remains undetermined. MCAD also showed band specificity against rodent embryonic tissue at 73 kDa ($n = 47$; MUC, $n = 31$; $p = 0.015$). In studies seeking differences between MCAD and MUC based on the density of immunoblot bands against human fetal brain, corrected for IgG content, MCAD had a significantly denser band at 61 kDa ($p = 0.037$) and at 39 kDa where their corrected peak height was higher (trend, $p = 0.085$). Furthermore, as will be discussed in subsequent sections, MCAD with offspring having developmental regression possessed greater reactivity against human fetal brain at 36 and 39 kDa. Maternal IgG from MCAD also showed significant reactivity against human adult brain epitopes in caudate (100 kDa), cerebellum (73 kDa), and cingulate gyrus (73 kDa). Nevertheless, the only possible identified overlap was between bands at 61 kDa in human fetal brain and 63 kDa in adult human frontal cortex. Altogether, more MCAD had reactivity, and more intense IgG bands, to fetal brain antigens, compared with MUC, as summarized in Table 14.1. Our estimation that maternal autoimmunity may be a contributing factor in up to 40% of autism cases comes from our finding that 40/100 MCAD had anti-fetal brain antibody reactivity at one or more of the identified sites of importance; 61, 39, or 36 kDa.

Table 14.1 Band specificity (number of subjects with band) and band density (corrected peak height)*: Significant differences and trends between mothers of children with autistic disorder (MCAD) and mothers with unaffected children (MUC) with human tissue as the epitope

	MCAD (m ± SEM)	MUC (m ± SEM)	P-values
Fetal human			
61 kDa			
Number with	30	31	ns
Peak hgt	14.1 ± 2.0	9.2 ± 1.2	0.037
39 kDa	14	15	ns
Peak hgt	10.9 ± 1.3	7.5 ± 1.3	0.085
36 kDa			
Number with	10	2	0.017
Fetal rodent			
73 kDa			
Number with	47	31	0.015
36 kDa			
Number with	48	31	0.010
31 kDa			
Number with	25	35	ns
Peak hgt	10.7 ± 5.5	7.6 ± 0.9	0.028

*All peak height values were corrected for sample specific IgG content. Peak height values are expressed as mean ± SEM. Abbreviation, hgt = height; ns = not significant.

Braunschweig and colleagues have also conducted a maternal antibody study assaying reactivity of MCAD IgG ($n = 61$) against human fetal brain tissue [66]. As presented in abstract form, maternal reactivity was more prevalent at approximately 32, 37, 73, and 100 kDa. In these studies the band at 37 kDa conferred the greatest risk for autism. This finding suggests a possible overlap with the 36-kDa band seen in the Singer study, because western blot standards and SDS–PAGE methods localize proteins to only an approximate molecular weight.

In a small pilot study, Zimmerman and colleagues identified serum antibodies against embryonic rodent brain in mothers with autistic offspring that were absent or reduced in mothers of unaffected children [10]. More specifically, serum from mothers of children with ASD ($n = 11$) contained extra bands identified in the 75- to 100-kDa range. These data have some overlap with the band specificity reported in embryonic rodent brain at 73 kDa by Singer and colleagues.

Stimuli for Maternal Antibody Formation

The maternal–fetal transplacental IgG hypothesis proposes that the mothers of some children with autistic disorder possess antibodies that cross-react against fetal brain. The inciting factor for the production of these anti-fetal brain antibodies remains unclear. One possibility is the process of microchimerism, i.e., exposure of the mother to fetal (non-self) proteins. During pregnancy, small fragments of DNA, and even some fetal cells, pass through umbilical cord blood into the mother's body, and have been detected up to 38 years after pregnancy [67, 68]. Evidence for microchimerism includes detection of male cells in a female who has been pregnant with no history of blood transfusion or organ transplantation. We propose that these cells may trigger the maternal immune system to recognize (fetal) cells or components as foreign, i.e., brain proteins or neuronal growth factors. Thus, "fetomaternal microchimerism" may serve a role in the generation of anti-fetal brain antibodies in MCAD. Another possible stimulus for antibody production may be maternal infection, which could lead to antibodies that cross-react with fetal antigens by a process of molecular mimicry. Prenatal infections may also affect the developing fetal brain directly (e.g., rubella) or by placental transfer of maternal immune factors (e.g., cytokines; see Chapter 13 by Patterson et al.). The antigens to which the maternal antibodies are directed may be affiliated with growth hormone signaling, myelin development, or some other fundamental pathway, which either immediately or indirectly affects social development, language skills, and other autistic traits.

Association Between Presence of Maternal Antibodies Against Fetal Brain Tissue and Developmental Regression in Offspring with Autism

Two groups of investigators have shown a correlation between specific maternal antibody reactivity and the appearance of developmental regression in their autistic offspring. In the study by Singer et al. [65], analyses were performed to determine whether there was an association between the presence of specific antibodies against fetal epitopes at 36, 39, and 61 kDa in the mothers and the clinical regression in their offspring. Of 100 MCAD, 47 had autistic offspring with a clinical history of regression in language and social skills. Results showed that there was a significant association between a mother possessing a serum antibody against human fetal brain at 36 and 39 kDa, and having a child with autistic disorder plus developmental regression. Similarly, Croen et al. [3] have discovered an association between the presence of regression in a child with ASD, defined as positive responses to questions 11 and 25 on the ADI-R, and his/her mother having anti-fetal human brain antibodies in plasma at 37 kDa. Lastly, Molloy and colleagues have identified a correlation between autism with regression, as defined by standardized methods, i.e., word loss, and maternal first- and second-degree relatives with autoimmune thyroid disease [69].

Pregnant Dam (Mouse) Model

The pregnant dam mouse model has been used in several studies to assess the effect of maternal factors on fetal brain and subsequent behavioral effects. The model consists of the intraperitoneal injection of an agent (e.g., maternal IgG in our studies) during the final trimester, the sacrifice of fetal pups to assess the effect on immune or other markers, and the observation of behavior during early postnatal periods and adulthood.

Effects of Placenta and Fetal Blood–Brain Barrier

The maternal transplacental hypothesis requires that maternally produced antibodies traverse the placenta, enter the fetal circulation and subsequently, fetal brain. In rodents, the maternal layers of the placenta are absent and thus present no barrier. In humans, maternal immunoglobulins can pass directly from maternal blood through two layers of the placenta into the fetus [70]. Antibodies pass through the first layer, the synctiotrophoblast, by MHC-like neonatal Fc receptors, and through the second layer, the vascular endothelium, by an unclear process, possibly involving a novel Fc receptor (FcR gamma IIb)-expressing organelle [71]. During the 3rd through 5th months of human pregnancy, the fetus can also swallow maternal antibodies in amniotic fluid and absorb them in the small intestine by a similar Fc-mediated process [72, 73]. In humans, maternal anti-streptococcal surface antigen IgG, Dengue HAI, and anti-pneumococcal antibodies all "naturally" cross the placenta from mother to child [74, 75, 76].

The capacity for human maternal antibodies to pass from fetal circulation to fetal brain has been confirmed in the pregnant dam mouse model. For example, in a study conducted by Dalton and colleagues [8], pup brains produced from mice injected with serum from a mother of children with ASDs showed marked cerebellar human-IgG binding, specifically to Purkinje cells. The Singer laboratory has also confirmed the ability of human maternal IgG, following intraperitoneal injections into pregnant mice, to pass into the fetal circulation and subsequently bind to fetal mouse brain (see Fig. 14.1 and see Color Plate 4, following p. XX). More specifically, pregnant mice were injected with purified human IgG throughout gestational days 14–18, mouse pups were killed after delivery, and brains were immediately dissected (Singer et al., unpublished). Fresh frozen sections of brains stained with FITC-labeled anti-human IgG showed that pups whose mothers received IgG from MCAD fluoresced more brightly than those receiving injections of IgG from control mothers. This preliminary finding confirms that maternal IgG binds to fetal mouse brain and MCAD mothers have higher concentrations of anti-brain antibody reactivity, compared with MUC.

Detection of an Immune Response in Fetal Mouse Brain

Although antibodies may cross the placenta and enter the developing fetal brain, to date, it remains undetermined whether these antibodies can elicit a neuroimmune response in the offspring. In preliminary analyses, the peritoneal injection of IgG from mothers of children with autism into pregnant mouse dams does appear to induce microglial activation in pup brains as detected at birth by increased anti-Ibal staining (Singer et al, unpublished). Although further studies are needed in autism, in the neonatal lupus syndrome maternal anti-Ro/SSA and anti-La/SSB anti-apoptotic cell autoantibodies cause fetal cardiac inflammation and tissue damage [77].

IgG Causes Behavioral Alterations in Offspring

Confirmation of the maternal–placental transfer antibody hypothesis requires proof that the antibody, or other influencing factor, is capable of causing behavioral alterations in offspring. Preliminary data derived from several sources indeed

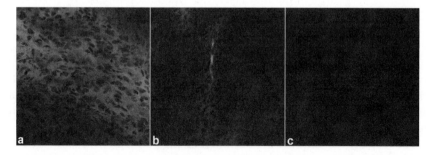

Fig. 14.1 Immunofluorescence using FITC anti-human IgG against sectioned mouse brain from newborn pups of dams injected with: **(a)** IgG from mothers of children with autistic disorder (MCAD); **(b)** IgG from mothers with unaffected children (MUC); and **(c)** vehicle (*see* Color Plate 4)

appear to support this possibility: i) prior studies showing that maternal sera are capable of inducing behavioral changes and ii) preliminary studies investigating the effect of MCAD sera on behavioral paradigms in postnatal and adult mice.

Prior Studies in Rodent Models

Rodent models have been previously used to confirm that maternal IgG injected into pregnant dams is capable of altering development in offspring. Vincent et al. [9] investigated the possibility that maternal antibodies, which cross the placenta and bind to fetal antigens, could be responsible for familial dyslexia. Serum samples from mothers with two or more children with dyslexia were injected into pregnant mice and offspring were evaluated for behavioral abnormalities. Results demonstrated that mouse offspring had deficits in motor tests that correlated with changes of cerebellar metabolites as measured by magnetic resonance spectroscopy. Jacobson and colleagues [78] injected pregnant mice with plasma from four fetal anti-acetylcholine receptor antibody (fAChR)-positive women whose fetuses had severe arthrogryposis multiplex congenita. Transfer of human IgG to mouse fetuses occurred after E15 and reached appreciable levels in the fetus by E18. Fetuses exposed in utero to fAChR antibodies had reduced movements and deformities. Control IgG-treated pups were born spontaneously and were normal. As cited above, Dalton et al. injected serum-containing anti-Purkinje antibodies from a mother of an autistic child into gestating mice and showed the offspring had a variety of behavioral changes including reduced stationary rod performance [8]. Lastly, anti-cardiolipin antibodies injected into pregnant mice caused a reduced number of pregnancies and fetal size [79, 80].

Preliminary Behavioral Results from Injections of MCAD IgG into Pregnant Mouse Dams

A small number of offspring from pregnant dams injected with either pooled IgG from MCAD or control sera (MUC) have undergone behavioral testing at various stages of development, days 1–21 and in adulthood (Singer, unpublished). Studies are ongoing, and some of these Preliminary results are presented.

Early Development

Using a standard battery of neurodevelopmental tests, no gross physical abnormalities nor differences in righting reflex, fore limb and hind limb placing responses, rooting reflex, vibrissa placing response, and negative geotaxis were identified.

Hyperactivity

During the first two postnatal weeks, neuromotor patterns (pivoting, walking forward and backward, running and jumping) were delayed in offspring from

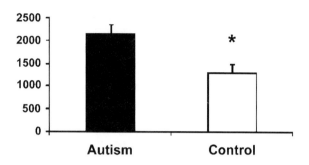

Fig. 14.2 Greater magnitude of hyper-reactivity to aversive stimuli in the offspring of dams prenatally treated with IgG from autism mothers (MCAD) than IgG of maternal controls (MUC) *P < 0.01

mothers prenatally treated with IgG compared with those from MCAD. Increased activity was found in older animals using paradigms assessing reactivity to aversive stimuli (acoustic startle response, see Fig. 14.2), horizontal and vertical (rearing) locomotor activities (novel open field). Prenatal treatment with "autistic" IgG significantly increased ambulatory activity in the open field in adult mice (Fig. 14.3, below); a two-way repeated measures ANOVA of the data showed a significant effect of treatment, $F(1,299) = 3.5$, $p < 0.05$, and a significant effect of intervals, $F(5,299) = 36.8$, $p < 0.001$.

Rearing activity, a measure of exploratory activity, in mice prenatally treated with "control" (MUC) serum significantly declined over time. In contrast, rearing activity in animals prenatally treated with "autistic" serum (MCAD) did not exhibit habituation in the open field (Fig. 14.4, below). Two-way repeated measures ANOVA of the data for rearing showed a significant effect of intervals, $F(5,299) = 3.6$, $p = 0.004$, and the interval by treatment interaction, $F(5,299) = 3.1$, $p = 0.01$. Taken together, initial startle data and the results from open field tests suggest that prenatal treatment with maternal "autistic"

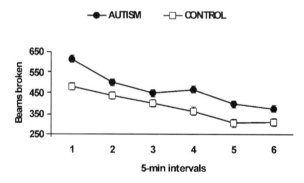

Fig. 14.3 The open field ambulatory activities in adult mice prenatally treated with serum from mothers of children with autistic disorder (MCAD) and serum from mothers with unaffected children (MUC). The Y axis – infrared beams broken; the X axis – 5-min intervals at which the activity was recorded. Note greater locomotor activity in the autism group compared to the control group

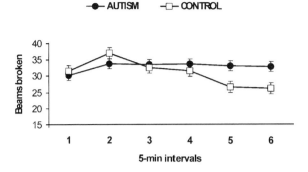

Fig. 14.4 The rearing activities in adult mice prenatally treated with serum from mothers of children with autistic disorder (MCAD) or serum from mothers with unaffected children (MUC). The Y axis – infrared beams broken; the X axis – 5-min intervals at which the activity was recorded. Note the lack of habituation of rearing in the autism group compared to the control group

serum produces hyper-reactivity to novel and aversive stimuli similar to responses observed in autistic patients to novelty.

Sociability and Social Recognition

A pilot study of the effects of prenatal treatments with "autistic" or control human maternal sera on sociability and social recognition in adult male mice offspring was performed using a modified Crawley's laboratory methodology [81]. Briefly, experimental or control mice were initially habituated for 5 min to the open field box. Afterwards, two identical mesh enclosures were placed in opposite corners of the box, with one enclosure containing a live unfamiliar mouse and the other one being empty. Experimental or control mice were allowed to explore the two enclosures for 10 min. Time spent in the area with the enclosure containing the live mouse was compared with the time spent in the area containing the empty enclosure. Immediately after this sociability test, an additional new mouse was placed in the previously empty enclosure and experimental or control mice were allowed to explore the two enclosures for another 10 min. Time spent exploring the new or familiar mouse was recorded. We found that mice prenatally treated with "autistic" maternal serum spent an equivalent amount of time exploring a live mouse kept in the enclosure and an inanimate object (i.e., an empty enclosure). In contrast, control mice spent significantly more time exploring a live mouse compared with time spent exploring an inanimate object (Fig. 14.5, below, left panel, Student's t test, $p < 0.05$). Both groups of mice, however, spent significantly more time exploring an unfamiliar (new) mouse compared with time spent exploring a familiar mouse. These results suggest that prenatal treatment with "autistic" maternal serum might affect (i.e., reduce) a natural preference in mice to explore and

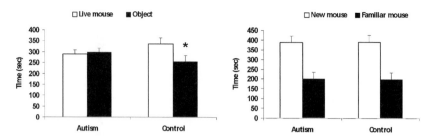

Fig. 14.5 Sociability (the left panel) and social recognition (the right panel) tests in adult mice prenatally treated with serum from mothers of children with autistic disorder (MCAD) and serum from mothers with unaffected children (MUC). The Y axis – time in sec; the X axis – groups of mice. Note a greater amount of time spent exploring an inanimate object in the autism group (*–$p < 0.05$ vs. object) (left). In contrast, both groups showed the similar increases in time spent exploring a new mouse (right)

interact with live mice compared with inanimate objects. In contrast, no alterations in the ability to recognize new live partners were found in mice treated with "autistic" maternal serum. Thus, prenatal treatment with "autistic" maternal serum led to decreased sociability in adult mice without affecting their ability to recognize new animals of the same species.

In summary, pilot experiments suggest that prenatal treatment of mice with serum drawn from mothers with autistic children produce behavioral alterations in the offspring reminiscent of some abnormalities observed in patients with autism spectrum disorders.

Discussion

In autism, there is often a period of apparently normal neurodevelopment before the identification of disrupted social and language skills or appearance of stereotypic behaviors. This preceding period of normalcy suggests either a) that an adverse insult occurs during infancy or early childhood and alters the normal developmental remodeling of existing neural networks [82] or b) that an intrauterine insult affects early neurodevelopment, but remains clinically silent until the expected postnatal maturation and appearance of cognitive and social processes. In the majority of prior studies, investigators seeking evidence for an autoimmune mechanism in autism have focused on the presence of abnormal serum antibodies in subjects with this disorder and suggested a postnatal insult [5, 11, 29]. In contrast, we emphasize findings, both direct and circumstantial, that support a prenatal contributing mechanism and specifically the placental–fetal IgG transfer theory.

Direct support for an intrauterine exposure hypothesis includes documentation of anti-brain antibodies in the sera of mothers with autistic offspring that are absent or reduced in mothers of unaffected children and, most importantly, preliminary evidence showing that the injection of serum from mothers with anti-brain antibodies into pregnant mice can cause behavioral alterations in offspring. Whether maternal sensitization alone is capable of accounting for autism in children is unknown. At this juncture however, recognizing that serum antibodies reactive against fetal brain do not in and of themselves predict autism, we believe there exists a complex association between maternal anti-fetal brain antibodies, co-existing host factors/environment, and clinical outcome. Nevertheless, although some investigators have claimed that the decline in language skills and social behaviors during developmental regression represents epileptiform abnormalities, with or without epilepsy [83, 84], our findings in MCAD suggest a possible autoimmune mechanism. More specifically, mothers having autistic children with developmental regression were more likely to have serum antibodies with reactivity against human fetal brain at 36 and 39 kDa than were mothers of autistic children without regression.

Prior clinical and postmortem investigations of brain development in autistic subjects have failed to clarify the timing of the insult. At birth, autistic brains tend to be smaller than those of healthy infants, but from 6 to 14 months of age undergo a period of accelerated growth [85]. We speculate that the fetal brain tissue epitopes detected in sera from MCAD will be identified as required developmental growth factors. Although some studies have suggested that sera of children with autism contain antibodies against GFAP and MBP [7], we have confirmed that antibodies against these specific antigens are not increased in sera from MCAD [10]. Similarly, an observed association of higher neonatal blood concentrations of brain derived neurotrophic factor (BDNF) and autism [86, 87] was not reflected in the sera of MCAD. Whether a common immune pattern is shared by mothers and their affected, but not their unaffected, offspring remains unknown.

Further analyses are required before the placental–fetal IgG transfer theory is confirmed. For example, prior studies have included only the analysis of antibody patterns obtained years after the delivery of the affected offspring, rather than prospective longitudinal evaluations in mothers and offspring, including cord blood samples. Antibody alterations in sera from MCAD have not been correlated with antibody findings in their offspring. Data analyses need to be performed on a larger number of subjects containing reactivity at specific molecular weights and discrepancies of reported molecular weight reactivity need to be clarified. Clearly, identification of the fetal protein epitope(s) involved in immune reactivity would be a major breakthrough. Additionally, although identified autoantibody repertoires in fetal tissue are considered to be inciting pathological factor(s), they could also be secondary immune responses to components of previously damaged tissue [88]. Lastly, preliminary findings of hyperactivity and altered sociability and social recognition in offspring of pregnant dams receiving intraperitoneal MCAD IgG are intriguing, but additional studies are required.

Confirmation in our animal model that an immune mechanism plays a role in the development of autism, triggered by maternal antibodies crossing the placenta to fetal brain and leading to behavioral dysfunction, could have several major implications including the establishment of an immune biological mechanism in autism and the potential development of diagnostic screening tests and new therapeutic interventions.

Summary

The possibility that a maternal immune component may contribute to the etiology of autism is an intriguing possibility. We have presented emerging evidence in support for a theory that maternal antibodies, present in mothers of children with autism, can affect fetal brain development and behavior in animal models. If the materno–feto–placental transfer theory is confirmed, and the specific maternal anti-fetal brain antigen-antibody mechanisms, along with its genetic, immunological, and other associated risk factors are elucidated, such knowledge may eventually lead to the development of screening techniques before or during pregnancy, and institution of preventive methods. We speculate that such methods will be analogous to screening for the Rh factor, although there may be several antigens involved that increase the risk for autism. Once the antigens are identified, immunomodulatory therapies – like Rho(D) immune globulin to prevent hemolytic disease of the newborn – may provide a practical approach to the prevention of this disabling and increasingly prevalent disease.

Acknowledgments The authors acknowledge the research support of the National Alliance for Autism Research, The Hussman Foundation, Dr. Barry and Mrs. Renee Gordon, and Michael and Lou Ann Grier.

References

1. Autism and Developmental Disabilities Monitoring Network Surveillance Year 2002 Principal Investigators; Centers for disease Control and Prevention. Prevalence of autism spectrum disorders –autism and developmental disabilities monitoring network, 14 sites, United States, 2002. *MMWR Surveill Summ* 2007; 56(1):12–28.
2. Comi AM, Zimmerman AW, Frye VH, Law PA, Peeden JN. Familial clustering of autoimmune disorders and evaluation of medical risk factors in autism. *J Child Neurol* 1999; 14(6):388–94.
3. Croen LA, Grether JK, Yoshida CK, Odouli R, Van de Water J. Maternal autoimmune diseases, asthma, and allergies, and childhood autism spectrum disorders. *Arch Pediatr Adolesc Med* 2005; 159(2):151–7.
4. Gupta S. Immunological treatments for autism. *J Autism Dev Disord* 2000; 30(5):475–9.
5. Singh VK, Rivas WH. Prevalence of serum antibodies to caudate nucleus in autistic children. *Neurosci Lett* 2004; 355(1–2):53–6.
6. Singh VK, Warren RP, Odell JD, Warren WL, Cole P. Antibodies to myelin basic protein in children with autistic behavior. *Brain Behav Immun* 1993; 7(1):97–103.
7. Singh VK, Warren R, Averett R, Ghaziuddin M. Circulating autoantibodies to neuronal and glial filament proteins in autism. *Pediatr Neurol* 1997; 17(1):88–90.
8. Dalton P, Deacon R, Blamire A, et al. Maternal neuronal antibodies associated with autism and a language disorder. *Ann Neurol* 2003; 53(4):533–7.

9. Vincent A, Deacon R, Dalton P, et al. Maternal antibody-mediated dyslexia ? Evidence for a pathogenic serum factor in a mother of two dyslexic children shown by transfer to mice using behavioural studies and magnetic resonance spectroscopy. *J Neuroimmunol* 2002; 130(1–2):243–7.

10. Zimmerman AW, Connors SL, Matteson KJ, et al. Maternal antibrain antibodies in autism. *Brain Behav Immun* 2007; 21(3):351–7.

11. Singer HS, Morris CM, Williams PN, Yoon DY, Hong JJ, Zimmerman AW. Antibrain antibodies in children with autism and their unaffected siblings. *J Neuroimmunol* 2006; 178(1–2):149–55.

12. Singer HS, Morris CM, Gause CD, Gillin P, Lee L-C, Zimmerman AW. Serum antibrain antibody differences in mothers of children with autism disorder: A study with fetal human and rodent tissue. International Meeting for Autism Research 2007; 6.

13. Morris CM, Gause CD, Gillin PK, Crawford S, Zimmerman AW, Singer HS. Fetal antibrain antibodies in serum of mothers of children with autistic disorder: correlation with developmental regression. *Ann Neurol* 2007; 62(S11):S116–7.

14. Lawler CP, Croen LA, Grether JK, Van de Water J. Identifying environmental contributions to autism: provocative clues and false leads. *Ment Retard Dev Disabil Res Rev* 2004; 10(4):292–302.

15. U.S. Department of Education, Office of Special Education Programs. 25th Annual (2003) Report to Congress on the Implementation of the Individuals with Disabilities Education Act, vol. 1. Washington, D.C.; 2005.

16. Newschaffer CJ, Croen LA, Daniels J, et al. The epidemiology of autism spectrum disorders. *Annu Rev Public Health* 2007; 28:235–58.

17. Korvatska E, Van de Water J, Anders TF, Gershwin ME. Genetic and immunologic considerations in autism. *Neurobiol Dis* 2002; 9(2):107–25.

18. Cohly HH, Panja A. Immunological findings in autism. *Int Rev Neurobiol* 2005; 71:317–41.

19. Ashwood P, Van de Water J. A review of autism and the immune response. *Clin Dev Immunol* 2004; 11(2):165–74.

20. Zimmerman AW. The Immune System. In: Bauman ML, Kemper TL, eds. The *Neurobiology of Autism, 2nd ed.* 2 ed. Baltimore: The Johns Hopkins University Press; 2005:371–86.

21. Vargas DL, Nascimbene C, Krishnan C, Zimmerman AW, Pardo CA. Neuroglial activation and neuroinflammation in the brain of patients with autism. *Ann Neurol* 2005; 57(1):67–81.

22. Zimmerman AW, Jyonouchi H, Comi AM, et al. Cerebrospinal fluid and serum markers of inflammation in autism. *Pediatr Neurol* 2005; 33(3):195–201.

23. Chez MG, Dowling T, Patel PB, Khanna P, Kominsky M. Elevation of tumor necrosis factor-alpha in cerebrospinal fluid of autistic children. *Pediatr Neurol* 2007; 36(6):361–5.

24. Croonenberghs J, Bosmans E, Deboutte D, Kenis G, Maes M. Activation of the inflammatory response system in autism. *Neuropsychobiology* 2002; 45(1):1–6.

25. Singh VK. Plasma increase of interleukin-12 and interferon-gamma. Pathological significance in autism. *J Neuroimmunol* 1996; 66(1–2):143–5.

26. Molloy CA, Morrow AL, Meinzen-Derr J, et al. Elevated cytokine levels in children with autism spectrum disorder. *J Neuroimmunol* 2006; 172(1–2):198–205.

27. Jyonouchi H, Sun S, Le H. Proinflammatory and regulatory cytokine production associated with innate and adaptive immune responses in children with autism spectrum disorders and developmental regression. *J Neuroimmunol* 2001; 120(1–2):170–9.

28. Connolly AM, Chez MG, Pestronk A, Arnold ST, Mehta S, Deuel RK. Serum autoantibodies to brain in Landau-Kleffner variant, autism, and other neurologic disorders. *J Pediatr* 1999; 134(5):607–13.

29. Silva SC, Correia C, Fesel C, et al. Autoantibody repertoires to brain tissue in autism nuclear families. *J Neuroimmunol* 2004; 152(1–2):176–82.

30. Gupta S, Aggarwal S, Rashanranvan B, Lee T. Th1- and Th2-like cytokines in CD4 + and CD8 + T cells in autism. *J Neuroimmunol* 1998; 85(1):106–9.

31. Archelos JJ, Hartung HP. Pathogenetic role of autoantibodies in neurological diseases. *Trends Neurosci* 2000; 23(7):317–27.
32. Warren RP, Singh VK, Cole P, et al. Possible association of the extended MHC haplotype B44-SC30-DR4 with autism. *Immunogenetics* 1992; 36(4):203–7.
33. Torres AR, Maciulis A, Stubbs EG, Cutler A, Odell D. The transmission disequilibrium test suggests that HLA-DR4 and DR13 are linked to autism spectrum disorder. *Hum Immunol* 2002; 63(4):311–6.
34. Lee LC, Zachary AA, Leffell MS, et al. HLA-DR4 in families with autism. *Pediatr Neurol* 2006; 35(5):303–7.
35. Torres AR, Sweeten TL, Cutler A, et al. The association and linkage of the HLA-A2 class I allele with autism. *Hum Immunol* 2006; 67(4–5):346–51.
36. Croen LA, Najjar DV, Fireman B, Grether JK. Maternal and paternal age and risk of autism spectrum disorders. *Arch Pediatr Adolesc Med* 2007; 161(4):334–40.
37. Croen LA, Yoshida CK, Odouli R, Grether JK. Maternal autoimmune and allergic diseases and childhood autism. In: International Meeting for Autism Research; 2004 May 7–8; Sacramento, CA; 2004.
38. Sweeten TL, Bowyer SL, Posey DJ, Halberstadt GM, McDougle CJ. Increased prevalence of familial autoimmunity in probands with pervasive developmental disorders. *Pediatrics* 2003; 112(5):e420.
39. Micali N, Chakrabarti S, Fombonne E. The broad autism phenotype: findings from an epidemiological survey. *Autism* 2004; 8(1):21–37.
40. Holgate ST. The epidemic of allergy and asthma. *Nature* 1999; 402(6760 Suppl):B2–4.
41. Jacobson DL, Gange SJ, Rose NR, Graham NM. Epidemiology and estimated population burden of selected autoimmune diseases in the United States. *Clin Immunol Immunopathol* 1997; 84(3):223–43.
42. Fleming J, Fabry Z. The hygiene hypothesis and multiple sclerosis. *Ann Neurol* 2007; 61(2):85–9.
43. Strachan DP. Hay fever, hygiene, and household size. *Br Med J* 1989; 299(6710):1259–60.
44. Bach JF. Infections and autoimmune diseases. *J Autoimmun* 2005; 25 Suppl:74–80.
45. Vercelli D. Mechanisms of the hygiene hypothesis – molecular and otherwise. *Curr Opin Immunol* 2006; 18(6):733–7.
46. Becker KG. Autism, asthma, inflammation, and the hygiene hypothesis. *Med Hypotheses* 2007; 69(4):731–40.
47. Westhoff CM. The Rh blood group system in review: a new face for the next decade. *Transfusion* 2004; 44(11):1663–73.
48. Miles JH, Takahashi TN. Lack of association between Rh status, Rh immune globulin in pregnancy and autism. *Am J Med Genet A* 2007; 143(13):1397–407.
49. Connors SL, Crowell DE, Eberhart CG, et al. Beta2-adrenergic receptor activation and genetic polymorphisms in autism: data from dizygotic twins. *J Child Neurol* 2005; 20(11):876–84.
50. Zerrate MC, Pletnikov M, Connors SL, et al. Neuroinflammation and behavioral abnormalities after neonatal terbutaline treatment in rats: implications for autism. *J Pharmacol Exp Ther* 2007; 322(1):16–22.
51. Shi L, Fatemi SH, Sidwell RW, Patterson PH. Maternal influenza infection causes marked behavioral and pharmacological changes in the offspring. *J Neurosci* 2003; 23(1):297–302.
52. Gerli R, Caponi L, Tincani A, et al. Clinical and serological associations of ribosomal P autoantibodies in systemic lupus erythematosus: prospective evaluation in a large cohort of Italian patients. *Rheumatology (Oxford)* 2002; 41(12):1357–66.
53. Lu XY, Ye S, Wang Y, et al. Clinical significance of neuro-reactive autoantibodies in neuro-psychiatric systemic lupus erythematosus. *Zhonghua Yi Xue Za Zhi* 2006; 86(35):2462–6.
54. Eber T, Chapman J, Shoenfeld Y. Anti-ribosomal P-protein and its role in psychiatric manifestations of systemic lupus erythematosus: myth or reality? *Lupus* 2005; 14(8):571–5.
55. Toubi E, Shoenfeld Y. Clinical and biological aspects of anti-P-ribosomal protein autoantibodies. *Autoimmun Rev* 2007; 6(3):119–25.

56. Muscal E, Myones BL. The role of autoantibodies in pediatric neuropsychiatric systemic lupus erythematosus. *Autoimmun Rev* 2007; 6(4):215–7.
57. Tzioufas AG, Tzortzakis NG, Panou-Pomonis E, et al. The clinical relevance of antibodies to ribosomal-P common epitope in two targeted systemic lupus erythematosus populations: a large cohort of consecutive patients and patients with active central nervous system disease. *Ann Rheum Dis* 2000; 59(2):99–104.
58. Shovman O, Zandman-Goddard G, Gilburd B, et al. Restricted specificity of anti-ribosomal P antibodies to SLE patients in Israel. *Clin Exp Rheumatol* 2006; 24(6):694–7.
59. Reichlin M. Autoantibodies to the ribosomal P proteins in systemic lupus erythematosus. *Clin Exp Med* 2006; 6(2):49–52.
60. Koren E, Reichlin MW, Koscec M, Fugate RD, Reichlin M. Autoantibodies to the ribosomal P proteins react with a plasma membrane-related target on human cells. *J Clin Invest* 1992; 89(4):1236–41.
61. Koscec M, Koren E, Wolfson-Reichlin M, et al. Autoantibodies to ribosomal P proteins penetrate into live hepatocytes and cause cellular dysfunction in culture. *J Immunol* 1997; 159(4):2033–41.
62. Neri F, Chimini L, Bonomi F, et al. Neuropsychological development of children born to patients with systemic lupus erythematosus. *Lupus* 2004; 13(10):805–11.
63. Katzav A, Solodeev I, Brodsky O, et al. Induction of autoimmune depression in mice by anti-ribosomal P antibodies via the limbic system. *Arthritis Rheum* 2007; 56(3):938–48.
64. Zhang W, Reichlin M. Production and characterization of a human monoclonal anti-idiotype to anti-ribosomal P antibodies. *Clin Immunol* 2005; 114(2):130–6.
65. Singer HS, Morris CM, Gause CD, Gillin PK, Crawford S, Zimmerman AW. Antibodies against fetal brain in sera of mothers with autistic children. *J Neuroimmunol* 2008; 194(1–2):165–72.
66. Braunschweig DA, Krakowiak P, Ashwood P, et al. Maternal plasma antibodies to human fetal brain in autism. In: International Meeting for Autism Research; 2007 May 2007; Seattle, Washington; 2007.
67. Bianchi DW. Robert E. Gross Lecture. Fetomaternal cell trafficking: a story that begins with prenatal diagnosis and may end with stem cell therapy. *J Pediatr Surg* 2007; 42(1):12–8.
68. Adams KM, Nelson JL. Microchimerism: an investigative frontier in autoimmunity and transplantation. *JAMA* 2004; 291(9):1127-31.
69. Molloy CA, Morrow AL, Meinzen-Derr J, et al. Familial autoimmune thyroid disease as a risk factor for regression in children with Autism Spectrum Disorder: a CPEA Study. *J Autism Dev Disord* 2006; 36(3):317–24.
70. Baintner K. Transmission of antibodies from mother to young: Evolutionary strategies in a proteolytic environment. *Vet Immunol Immunopathol* 2007; 117(3–4):153–61.
71. Takizawa T, Anderson CL, Robinson JM. A novel Fc gamma R-defined, IgG-containing organelle in placental endothelium. *J Immunol* 2005; 175(4):2331–9.
72. Israel EJ, Simister N, Freiberg E, Caplan A, Walker WA. Immunoglobulin G binding sites on the human foetal intestine: a possible mechanism for the passive transfer of immunity from mother to young. *Immunology* 1993; 79(1):77–81.
73. Shah U, Dickinson BL, Blumberg RS, Simister NE, Lencer WI, Walker WA. Distribution of the IgG Fc receptor, FcRn, in the human fetal intestine. *Pediatr Res* 2003; 53(2):295–301.
74. Baril L, Briles DE, Crozier P, et al. Natural materno-fetal transfer of antibodies to PspA and to PsaA. *Clin Exp Immunol* 2004; 135(3):474–7.
75. Perret C, Chanthavanich P, Pengsaa K, et al. Dengue infection during pregnancy and transplacental antibody transfer in Thai mothers. *J Infect* 2005; 51(4):287–93.
76. Quiambao BP, Nohynek HM, Kayhty H, et al. Immunogenicity and reactogenicity of 23-valent pneumococcal polysaccharide vaccine among pregnant Filipino women and placental transfer of antibodies. *Vaccine* 2007; 25(22):4470–7.
77. Reed JH, Neufing PJ, Jackson MW, et al. Different temporal expression of immunodominant Ro60/60 kDa-SSA and La/SSB apotopes. *Clin Exp Immunol* 2007; 148(1):153–60.

78. Jacobson L, Polizzi A, Morriss-Kay G, Vincent A. Plasma from human mothers of fetuses with severe arthrogryposis multiplex congenita causes deformities in mice. *J Clin Invest* 1999; 103(7):1031–8.
79. Blank M, Cohen J, Toder V, Shoenfeld Y. Induction of anti-phospholipid syndrome in naive mice with mouse lupus monoclonal and human polyclonal anti-cardiolipin antibodies. *Proc Natl Acad Sci USA* 1991; 88(8):3069–73.
80. Shoenfeld Y, Sherer Y, Blank M. Antiphospholipid antibodies in pregnancy. *Scand J Rheumatol* 1998; 107:33–6.
81. Crawley JN. Designing mouse behavioral tasks relevant to autistic-like behaviors. *Ment Retard Dev Disabil Res Rev* 2004; 10(4):248–58.
82. Grossman AW, Churchill JD, Bates KE, Kleim JA, Greenough WT. A brain adaptation view of plasticity: is synaptic plasticity an overly limited concept? *Prog Brain Res* 2002; 138:91–108.
83. McVicar KA, Ballaban-Gil K, Rapin I, Moshe SL, Shinnar S. Epileptiform EEG abnormalities in children with language regression. *Neurology* 2005; 65(1):129–31.
84. Tuchman R. Autism and epilepsy: what has regression got to do with it? *Epilepsy Curr* 2006; 6(4):107–11.
85. Courchesne E, Carper R, Akshoomoff N. Evidence of brain overgrowth in the first year of life in autism. *JAMA* 2003; 290(3):337–44.
86. Nelson KB, Grether JK, Croen LA, et al. Neuropeptides and neurotrophins in neonatal blood of children with autism or mental retardation. *Ann Neurol* 2001; 49(5):597–606.
87. Miyazaki K, Narita N, Sakuta R, et al. Serum neurotrophin concentratins in autism and mental retardation: a pilot study. *Brain Dev* 2004; 26:292–5.
88. Bornstein NM, Aronovich B, Korczyn AD, Shavit S, Michaelson DM, Chapman J. Antibodies to brain antigens following stroke. *Neurology* 2001; 56(4):529–30.

Chapter 15
Can Neuroinflammation Influence the Development of Autism Spectrum Disorders?

Carlos A. Pardo-Villamizar

Abstract Autism is a complex neurodevelopmental disorder of early onset that is highly variable in its clinical presentation. Although the causes of autism in most patients remain unknown, several lines of research support the view that both polygenic and environmental factors influence the development of abnormal cortical circuitry that underlies autistic cognitive processes and behaviors. The role of the immune system in the development of autism is controversial, but it is clear now that immune factors are involved in the modeling of the central nervous system (CNS) during prenatal and postnatal stages, during which neuroimmune responses may disrupt normal neurodevelopment and lead to the neuropathological abnormalities characteristic of autism. Several studies showing peripheral immune abnormalities support immune hypotheses; however, until recently there has been no demonstration of immune abnormalities within the CNS. Recently, our laboratory demonstrated the presence of neuroglial and innate neuroimmune system activation in brain tissue and cerebrospinal fluid of patients with autism, findings that support the view that neuroimmune abnormalities occur in the brain of patients with autism and may contribute to the diversity of the autistic phenotypes. The role of neuroglial activation and neuroinflammation is still uncertain but could be critical in maintaining, if not also initiating, some of the CNS abnormalities present in autism. A better understanding of the role of neuroinflammation in the pathogenesis of autism may have important clinical and therapeutic implications. Future studies should focus on the actions of neuroimmune factors during brain development as important components involved in the pathogenesis of autism.

Keywords Neuroimmunity · neuroinflammation · astrocytes · microglia · neuroglia · cytokines · neurodevelopment

C.A. Pardo-Villamizar
Department of Neurology, Division of Neuroimmunology and Neuroinfectious Disorders, Johns Hopkins University School of Medicine, Baltimore, MD 21287, USA
e-mail: cpardov1@jhmi.edu

A.W. Zimmerman (ed.), *Autism*, DOI: 10.1007/978-1-60327-489-0_15,
© Humana Press, Totowa, NJ 2008

Introduction

Immune system responses have been proposed in the past several years as potential pathogenic factors in the development of autism spectrum disorders (ASDs) [1, 2]. Evidence for this has been derived from the effects of maternal viral infectious disorders during pregnancy [3], an excess of autoimmune disorders in mothers of subjects with ASD or their families [4], and the effects of environmental factors on the immune system, all of which support the view that disturbances of immune function play a role in the pathogenesis of ASD and perpetuation of their associated behavioral and neurological abnormalities. The increased frequency of immunological abnormalities in patients with ASD [5] also indicate that, in addition to the spectrum of neurological and behavioral problems exhibited by patients with ASD, other systems, including the immune system and gastrointestinal tract, are also affected. The causes and effects of the immune system abnormalities in autism are unknown but could be critical for maintaining, if not initiating, some of the abnormalities in central nervous system (CNS) function. These abnormalities likely have polygenic and environmental bases that will have important clinical and therapeutic implications. Current evidence suggests that neurobiological abnormalities in ASD are associated with changes in cytoarchitectural and neuronal organization that may be determined by genetic, environmental, immunological, and toxic factors [6, 7]. Since neuroimmune pathways and CNS cell populations such as the neuroglia (astroglia and microglia) play central roles during brain development, in cortical organization, neuronal function, and modulation of immune responses, it is quite possible that these factors, acting in concert with host immunogenetic factors, may contribute to the pathogenesis of ASD. This review examines the current knowledge and evidence that support the view that neuroinflammation has a role in the pathogenesis of autism.

Immune Responses and the Central Nervous System

Although the CNS had been considered an immune-privileged organ, several studies have recently demonstrated that such a concept is no longer valid [8]. It is now very clear that the immune system and the CNS have constant interaction between them that preserves their homeostasis and function (Fig. 15.1). Both systems are highly hierarchical in structure and maintain their functions based on complex interactions among specialized cells and chemical mediators. Although the central role of the immune response is to react to tissue injury and maintain an active process of surveillance for detection of infection and noxious factors, there is growing evidence that the immune system in the brain also contributes to fundamental processes of brain modeling and plasticity. Because of the diversity in the microenvironments and cellular processes involved in immune and nervous system activity, the elements and mechanisms involved in neuroimmune responses in the CNS are different from those that operate in

Fig. 15.1 The immune
system maintains a constant
interplay with the nervous
system to preserve normal
homeostasis by modulating
the processes of inflammation
and neuronal–neuroglial
interaction. These two
processes are fundamental
to maintaining equilibrium
between normal function
and dysfunction

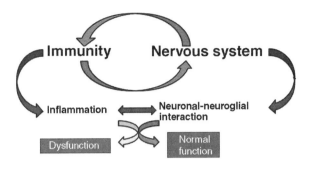

non-CNS tissues. Some of these mechanisms are modulated by the interplay of chemical mediators such as cytokines and chemokines as well as immune cells (e.g., leukocytes and monocytes), together with elements of the CNS such as neurons, neurotransmitters, neuroglia, and blood vessels. In the CNS environment, neuroglial cell populations and the blood–brain barrier (BBB) function as a neurovascular–immune unit that facilitates the continuous functional association of the CNS with the immune system and constitutes the main modulator of neuroimmune interactions [9, 10] (Fig. 15.2).

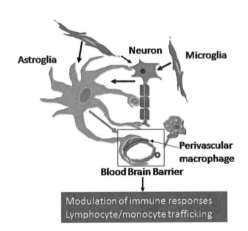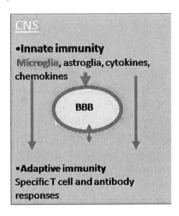

Fig. 15.2 The interaction between the immune system and the nervous systems is facilitated by a neurovascular–immune unit that allows continuous functional communication between the two systems and constitutes the main modulator of neuroimmune interactions. The blood–brain barrier (BBB) and its elements, such as endothelia, pericytes, perivascular macrophages, and astroglia, modulate both innate and adaptive immune responses in the CNS. A complex interplay of cellular and molecular processes involved in neuronal activity and neuronal–neuroglia interaction is facilitated by mediators such as cytokines and chemokines, produced by astroglia and microglia, which influence the function of the BBB and modulate the trafficking of immune cells such as monocytes and lymphocytes. Disruption of the BBB function is frequently associated with adaptive neuroimmune responses that increase trafficking of leukocytes and monocytes into the CNS. Innate neuroimmune responses are mainly mediated through activation of neuroglial elements of the CNS, such as microglia and astrocytes, and are enhanced by the active production of immune mediators (e.g., cytokines, chemokines, complement, and metalloproteinases)

What is Neuroinflammation

Inflammation in the CNS (neuroinflammation) is a host mechanism that involves elements of the immune and nervous systems in response to injury, dysfunction, or infection of the CNS or its components. Neuroinflammation involves the interplay of elements of the immune system such as cellular responses by monocytes, B, or T cells as well chemical mediators such as cytokines, chemokines and their receptors, complement and other mediators, along with the elements of the CNS such as neuroglia, the BBB, and chemical mediators (e.g., neurotransmitters and metalloproteinases) [11, 12]. Similar to immune activation in other organs or tissues, neuroinflammation may involve both innate and adaptive immune responses. Most of the adaptive neuroimmune responses are triggered by infectious processes (e.g., viral encephalitis) or external injury to the CNS (e.g., head trauma) and involve disruption of the BBB, infiltration of T cells into the brain and spinal cord structures, or production of specific antibodies that target CNS antigens or other proteins involved in disease processes. Innate neuroimmune responses mostly involve intrinsic responses by neuroglia, such as microglial and astroglial activation along with increased production of cytokines, chemokines, and other neuroimmune mediators [e.g., metalloproteinases, toll-like receptors (TLRs)] by CNS cells (neurons, neuroglia, and elements of the BBB). Innate neuroimmune responses occur in neurodegenerative diseases or abnormalities of CNS homeostasis such as those that occur in metabolic or seizure disorders. Frequently, both innate and adaptive neuroimmune responses occur concurrently; however, they may have distinctive pathogenic roles at different stages of neurological dysfunction. There is a growing controversy about whether the roles of neuroinflammatory responses are deleterious or protective in the setting of neurological dysfunction or injury, and it is now evident from different experimental approaches that neuroinflammation may play dual roles, both in providing neuroprotection and producing injury in the CNS [11, 12, 13].

The Influence of Neuroimmune Responses in Brain Development

Brain development is a complex process in which different molecular pathways overlap to produce a well–organized, compartmentalized structure. The development of the cerebral cortex, for example, involves highly complex and coordinated phases of cellular organization that includes four overlapping stages: neurogenesis, fate determination, migration, and differentiation, which finally result in the highly organized circuitry and architecture that characterize the mammalian neocortex. The role of neuroimmune factors is essential for all processes of neuronal migration, axonal growth, neuronal positioning, cortical lamination as well as dendritic and synaptic formation. All elements of the neuroimmune network, including astrocytes and microglia, immune mediators such as cytokines, chemokines, soluble factors produced by neuroglia (e.g., growth factors and neurotrophins) as well as other classical immune pathways such as those associated with the major histocompatibility complex (e.g., MHC class I) and complement, participate in mechanisms of neurodevelopment during intrauterine and postnatal stages. Therefore, the dual roles, developmental and immunological, are now well-recognized features of all elements of the neuroimmune responses [14, 15].

Roles of Neuroglia During Brain Development

Both microglial and astroglial functions appear during the first and second trimesters of intrauterine brain development [16, 17, 18, 19]. At these stages, neuroglia mainly facilitate processes of brain modeling rather than immune function. Migration of microglia into the cerebral wall originates from the ventricular lumen and leptomeninges and progresses in a radial and tangential pattern toward the immature white matter, subplate layer, and cortical plate. An intraparenchymal vascular route is also evident by week 12 of gestation [19]. Major waves of colonization of the CNS by microglia coincide with the early stages of vascularization, formation of radial glial, and neuronal migration during the fourth and fifth months of gestation and beyond. The patterns of migration and spatial distribution of microglia along the radial glia appear to be coordinated and regulated by the expression of subsets of chemokines such as RANTES and MCP-1, and possibly other soluble factors such as integrins and adhesion molecules (e.g., ICAM-2 and PECAM) [20]. As compared with rodents, in which microglial colonization occurs at late embryonic or early postnatal stages, the expansion of microglia within the brain is well established at weeks 20 to 22 of intrauterine brain development [17, 21]. Because of the monocytic lineage of microglia, their developmental function has been assumed to be mostly associated with debris removal. This concept has been modified by observations that microglia exert regulatory function and control of neuronal death, as well as axonal pathfinding roles, synaptic building, and production of immune regulators and neurotrophins [17, 18, 22]. Astroglial function during brain development appears in the form of a highly specialized form of astroglia, the radial glia, a diverse group of cells involved in processes of neuronal migration in different regions of the brain and spinal cord [23, 24, 25] that participate in fundamental mechanisms of area specification and cortical expansion [26]. Radial glia, along with microglia, act in very well-coordinated processes that facilitate migration and positioning of neurons into the cerebral cortex, mechanisms that are also synchronized with the developmental function of subsets of cytokines and chemokines and their receptors in the modeling of the cerebral cortex [20, 27]. Radial glia transform into astroglia after the completion of the phases of brain development, but it is now believed that subsets of radial glia persist even during postnatal stages to give origin to neurons and modulate the production of stem cells in neurogenic areas of the brain during adulthood [28].

Neurodevelopmental Roles of Cytokines and Chemokines

Along with the neuroglia, another important element of the neuroimmune response that contributes to brain development is a complex network of cytokines, chemokines, and their receptors. Both cytokines and chemokines are well-known mediators of immunological activity, maturation and trafficking of immune cells, and responses to injury [29, 30]. Although few human studies are available, neuropathological studies have demonstrated that subsets of chemokines are present in the CNS and exert their modulatory and developmental functions on neurons, neuroglia, and the BBB by interacting with membrane receptors, even at early stages of brain development [31]. Some chemokines are critical for neuroimmune function and are involved early in neurodevelopment. For example, CCL2 (also known as macrophage chemoattractant protein-1, MCP-1) and CCL5 (also known as RANTES) are important cues for processes of microglial migration and colonization of the CNS [20]. Experimental models in rodents have demonstrated the crucial function of CXCL12 (also known as stromal-derived factor-1, SDF-1) and its receptor, CXCR4, in mechanisms of cerebellar development, neuronal migration, and axonal path finding [32, 33, 34, 35]. CXCL12 appear to have an important function in mechanisms that control migration of Cajal–Retzius cells, important sources of the glycoprotein reelin, and other factors that support radial glia in the critical process of cortical lamination [36]. Disturbances in reelin function and Cajal–Retzius cells result in abnormalities of neuronal positioning, cortical lamination and columnar organization of the cortical neurons, and putative factors that may contribute to the neuropathological abnormalities in autism [37]. Similarly, cytokines have been recognized as important factors during brain development [38, 39]. Cytokines traditionally recognized as "proinflammatory," such interleukin (IL)-6, tumor necrosis factor (TNF)-α, and IL-1β, as well as "anti-inflammatory" cytokines, such as transforming growth factor (TGF)-β, are involved in pathways that contribute to CNS development [40, 41, 42, 43]. The TGF-β cytokine family, for example, is involved in essential nervous tissue remodeling, including cell-cycle control, regulation of early development and differentiation, neuron survival, and astrocyte differentiation [44]. Cytokines such IL-6 appear to have central roles as mediators of behavioral and transcriptional changes during maternal immune activation [45].

Major Histocompatibility Complex and Complement Cascades in Brain Development and Synaptogenesis

Although not directly involved in mechanisms of migration and patterning of the CNS, other classical immune pathways such as the class I MHC and the complement cascade have been shown to be involved in important processes of synaptic modeling. Recent findings in models of retina and brain development have shown that complement proteins such as C1q are involved in processes of synaptic elimination and early postnatal pruning, a critical stage of cortical modeling [46]. Interestingly, another classical immune pathway involved in mechanisms of T-cell activation and antigen presentation, the class I MHC, is associated with processes of synaptic plasticity and refinement of connectivity [14, 47]. These processes are important for the optimal formation of activity-dependent synaptic connections and the elimination of silent nonfunctional synapses [14].

Neuroglia and Neuroimmune Responses During Adulthood

In the CNS, neuroglial cells such as astrocytes and microglia, along with perivascular macrophages and endothelial cells, are important for maintaining normal neuronal function and homeostasis [22, 48, 49, 50, 51, 52, 53, 54]. As such, these CNS elements are also involved in immune function and the inter-action of the immune and nervous systems. Neuroglial cells also contribute in a number of ways to the regulation of immune responses and neuronal activity in the CNS. Both microglia and astrocytes are involved in crucial neurobiological functions and contribute extensively to processes of cortical organization, neuroaxonal guidance, and synaptic plasticity [55, 56]. Astrocytes, for example, play an important role in the detoxification of excess excitatory amino acids [57], maintenance of the integrity of the BBB [51], production of neurotrophic factors, [53] and the metabolism of glutamate [11, 57]. Under normal homeo-static conditions, astrocytes facilitate neuronal survival by producing growth factors and mediating the uptake and removal of excitotoxic neurotransmitters, such as glutamate, from the synaptic microenvironment [57, 58]. Both astroglia and microglia are involved in pathogenic inflammatory mechanisms that are common to diverse disorders of the CNS and are important factors in the neuroimmune response that occurs in response to disruption of homeostasis in the CNS. During neuroglial activation secondary to injury or neuronal dysfunction, astrocytes and microglia produce several factors that modulate inflammatory responses. For example, they secrete proinflammatory cytokines, chemokines, and metalloproteinases that can magnify and accelerate immune reactions within the CNS [53, 59, 60]. Microglial cells, for example, are involved in synaptic stripping, cortical plasticity as well as immune surveillance [48, 54, 61, 62]. Perivascular astrocytes and microglia modulate BBB activity and secrete factors that regulate trafficking of immune cells into the CNS.

In several degenerative and immune-mediated disorders of the CNS, such as Alzheimer's disease (AD), HIV dementia, epilepsy, and multiple sclerosis, astroglia and microglia associated with innate neuroimmune responses are cen-tral to the pathogenic mechanisms of neurodegeneration and appear to mediate important processes that lead to neuronal dysfunction [48, 63]. For example, in HIV dementia, microglial activation and infiltration by macrophages contribute to the neuronal damage responsible for dementia [64]. In other disorders such as epilepsy, and particularly in Rasmussen's syndrome, a rare pediatric epileptic disorder, both astroglial and microglial reactions, occur in parallel to adaptive neuroimmune responses mediated by T-cell infiltration of the cerebral cortex [65]. In neurodegenerative disorders such as Parkinson's disease (PD) and AD, the role of the neuroimmune response has been exposed as a critical part of the cascade of molecular and cellular events leading to either dysfunction of specific neuronal populations, such as nigral cells in PD, or processing and degradation of amyloid in AD [66, 67, 68]. Because of the central importance of neuroin-flammatory pathways and neuroglia in response to neuronal dysfunction and their pathogenic roles in diverse neurological disorders, my colleagues and I have hypothesized that neuroimmune responses and neuroinflammation are associated with the pathogenic mechanisms involved in cortical and neuronal dysfunction observed in ASD.

Neuroglia and Synaptic Plasticity

Neuroglia are fundamental components of communication in neuronal networks by virtue of their contribution to the modulation of synaptic and dendritic function and by the generation of responses to neuronal activity. These functions are facilitated by complex neuroglia–neuronal interactions in which both microglia and astroglia maintain a dynamic structural and functional association with synapses and dendrites [56, 69]. Neuronal–neuroglial interactions are regionally specific and occur in all regions of the CNS. Astroglia exhibit extensive and elaborate interplay with both presynaptic and postsynaptic structures in a highly organized way, as individual astrocytes are associated with specific microdomains and local circuits. In the hippocampus, for example, individual astrocytes maintain a functional association with a specific neuronal territory that modulates the physiology of a well-defined subset of synapses. It has been calculated that in the hippocampus, an individual astrocyte domain may modulate the function of 140,000 synapses [70, 71]. Similarly, Bergmann glia in the cerebellum have specific local interactions with individual synapses in discrete microdomains [72]. Recent morphological and neurophysiological studies have also demonstrated that individual astroglia in the cerebral cortex interact with specific islands of functional synapses and that a single astrocyte may enwrap from four to eight neurons and establish contacts with 300–600 dendrites [73]. Astroglial responses to neuronal activity are facilitated by calcium signaling, triggered by factors that include glutamate and ATP, and also involve purinergic receptors and gap-junctional signaling [74]. The demonstration of the functional role of calcium signaling in the gliotransmitter environment has exposed the critical role of astrocytes in modulating synaptic function [69, 75]. The secretion of astroglial factors with neuromodulatory function is central to the regulation of synaptic efficacy and long-term synaptic plasticity

[76, 77, 78]. Among these factors, cytokines such as TNF-α appear to be involved in mechanisms of glia-mediated homeostasis during activity-dependent remodeling of the developing and established neuronal circuits that follow brain injury [79, 80]. These observations support the important role of the gliotransmitter environment in the modulation of neuronal function.

How May Neuroimmune Responses Influence the Pathogenesis of Autism Spectrum Disorders

Neuroimmune mechanisms may participate in the pathogenesis of ASD, since many of the molecular and cellular processes involved in intrauterine and postnatal brain development engage elements of the neuroimmune system, including neuroglia, immune mediators such as cytokines and chemokines, and other elements of the immune response, such as complement activation and TLRs (Fig. 15.3 and see Color Plate 5, following p. XX). Because disorganization of cortical neurons, abnormalities in minicolumnar organization and subcortical white matter, as well as brain growth abnormalities, are the dominant features of the neuropathology of ASD [81, 82], it is likely that the critical period of pathogenesis occurs during fetal brain development and the first year of life [6]. During this critical pathogenic period, neurodevelopmental processes such as neuronal migration, cortical lamination, synaptic and dendritic modeling, and the establishment of neuronal and cortical networks are influenced not only by genetic factors, but also by important neuroimmune mechanisms that involve astrocytes and microglia, interactions of cytokines, chemokines, and their receptors, developmental expression of TLRs as well as complement activation (see Fig. 15.3) [43, 46, 83]. These neurobiological processes, along with the influence of the neuroimmune system, are crucial for the development of neurological and behavior trajectories, such as social cognition, language and communication, and motor development [6]. Disturbances of specific neurobiological trajectories triggered by abnormalities in the maternal environment or by genetic influences may eventually translate into abnormalities in the neurodevelopmental trajectories and behaviors that characterize ASD. In addition to the critical period of pathogenesis, neuroimmune responses may also be activated during the postpathogenic period, when neurological regression, abnormal neuronal activity (e.g., seizures), and aberrant neural networks may trigger these responses as part of deleterious reactions or even as part of neuroprotective pathways. It is unclear at present whether neuroimmune responses or neuroinflammation are common occurrences in the brain of patients with ASD during postnatal stages or adulthood; however, our neuroimmunopathological studies suggest that a chronic stage of immune activation or neuroinflammation occurs in subsets of patients with ASD, and the presence of these changes is not determined by age or duration of the disorder, and may be an ongoing, long-term neuropathological process [84].

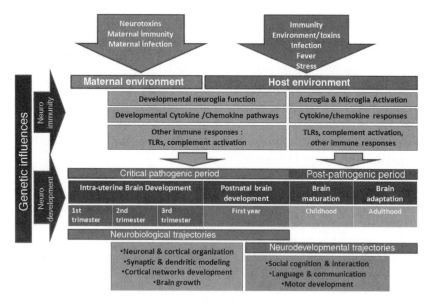

Fig. 15.3 Mechanisms involved in the pathogenesis of autism spectrum disorders (ASD) likely affect brain development during the critical pathogenic period of intrauterine life and the first postnatal year. Genetic, as well as maternal and environmental, factors likely influence neuroimmune pathways and eventually the normal progression of neurobiological trajectories that determine neuronal and cortical organization, synaptic modeling, cortical networking, and brain growth. The aberrant development of neurobiological trajectories modifies neurodevelopmental trajectories of social interaction, language, and motor development, which trigger the onset of stereotypic behaviors that characterize the clinical profile of ASD. Neuroimmune activation may persist during the postpathogenic period and continue to affect brain maturation and adaptation in children and adults with ASD (*see* Color Plate 5)

Maternal Factors in the Modulation of Neuroimmune Responses

There is growing evidence that the maternal immune environment affects the developing fetal CNS and determines specific patterns of inflammation-mediated brain and behavioral pathology [85, 86, 87]. Experimental models of maternal infections and immune challenges have demonstrated the effects of immune responses on the development of neurobiological and behavioral abnormalities that are consistent with known features of neuropsychiatric disorders such as autism and schizophrenia [88, 89]. Although the gestational periods of vulnerability may determine specific outcomes, it appears that infection-associated immunological events in early stages of fetal life may have a greater impact on the occurrence and progression of neurodevelopmental abnormalities than later stages [86]. In animal models, the disruption of normal cytokine/chemokine developmental pathways, triggered by either infections or immune challenges, appears to modify the expression patterns of these immune mediators in the fetal brain, which determine the development of neuropathological changes and the behavioral dysfunction that emerge later in life [45, 85, 90]. Interleukin-6, a well-known proinflammatory cytokine during adulthood, appears to be a key factor in the development of pathogenic effects of maternal immune challenges [45]. Other immune regulators such as bone morphogenetic proteins and TGF-β proteins also appear to play important roles during stages of maternal immune reactions that challenge the developing fetal brain [91]. These experimental observations point out that future studies on the pathogenesis of autism ASD should focus on a more detailed analysis of maternal infections as potential risk factors associated with the development of the disorders.

Other interesting observations that may expose the critical role of the maternal intrauterine environment in the pathogenesis of ASD are the demonstration of autoantibodies in the serum of some mothers of patients with autism that cross-react with neural epitopes in the fetal brain [92, 93, 94]. These findings suggest that, at least in a subset of patients with autism, the potential passive transfer of maternal antibodies during fetal life may have played a pathogenic role in the presence of the disorder. It is unclear, however, what the specific targets of these autoantibodies are or what triggers their production. These observations support the relevance of maternal immune states and may also be concordant with previous descriptions of an increased frequency of familial or autoimmune disorders in mothers of subjects with autism [4].

Chronic Neuroimmune Reactions in Autism Spectrum Disorder

Although most of the focus of the role of neuroinflammation is directed toward the critical periods of prenatal and postnatal development, there is growing evidence that neuroimmune reactions are also involved in neuropathological and behavioral disturbances during postnatal stages and the adulthood of patients with autism. Our neuropathological studies of postmortem brain tissues from patients with autism demonstrate an active and ongoing neuroinflammatory process in the cerebral cortex and white matter, characterized by astroglial and neuroglial activation. These findings support a role for neuroimmune responses in the pathogenesis and persistence of abnormalities in ASD [84]. Since both astroglia and microglia are involved in pathogenic inflammatory mechanisms common to many different disorders of the CNS, it is possible that different factors (e.g., genetic susceptibility, maternal factors, and prenatal environmental exposures) may trigger the development of these neuroglial reactions. Furthermore, more detailed studies of immune factors performed by protein array techniques demonstrated that cytokines/chemokines such as MCP-1, IL-6, and TGF-β1, which are mainly derived from activated neuroglia, are the most prevalent cytokines in brain tissues [84]. Similar observations are also seen in the CSF from patients with autism. These findings strongly support the theory that neuroimmune reactions are part of the neuropathological processes in ASD and that immune responses are among the factors that may contribute to CNS dysfunction in ASD. However, the significance of the neuroinflammatory response to specific neuropathologies and behavioral disruptions in ASD and its relevance to the etiology of ASD requires further exploration.

Genetic Susceptibility and Immune Responses in Neurological Disease

A critical issue in the natural history of a disease is the influence that the immune system and immunogenetic host factors may have on its pathogenesis. The pathogenic mechanisms of many neurological diseases, including neurodegenerative and neuroimmunological disorders, are influenced by the spectrum and functions of the proteins associated with immunological pathways. The expression of proteins such as MHC antigens, cytokines, chemokines, and integrins is closely associated with the function of these immunological pathways and may determine the patterns of susceptibility and severity of disease. Allelic variations in regulatory regions of cytokine genes, point mutations, or single nucleotide substitutions have been shown to affect gene transcription and levels of cytokine expression and to produce interindividual variation in cytokine production [95]. Single nucleotide polymorphisms (SNPs) and haplotypes of cytokine and chemokine genes produce genetic differences in cytokine expression that predispose to disease or confer resistance in immune-mediated disorders by influencing the strength and duration of the immune response [96, 97, 98]. The clinical outcomes of autoimmune, inflammatory, and neurodegenerative disorders appear to be influenced by the balance between proinflammatory and anti-inflammatory pathways, modulated by cytokines and chemokines. This relationship has been supported by numerous reports of the association of some cytokine alleles and expression phenotypes with immune-mediated or autoimmune disorders [96, 97, 98]. The association of SNPs of genes associated with inflammation has been investigated in inflammatory and neurodegenerative disorders such as AD and PD, as well as neuroinflammatory disorders such as multiple sclerosis and HIV-associated dementia [99, 100, 101, 102, 103, 104]. These polymorphisms may modify the natural history of the disease by producing increased or decreased susceptibility to immune-mediated responses or other pathogenic factors. For example, a polymorphism of the MCP-1 gene (A-2518G), is associated with an increased risk of developing HIV dementia [104], early onset of PD [105], increased resistance to antipsychotic therapy in schizophrenic patients [106], and increased levels of MCP-1 in serum in AD patients [107, 108]. In other disorders, SNPs of cytokine genes have been shown to be protective against the onset or increased severity of disease. For example, a polymorphism in position -1082 of the IL-10 gene promoter has been shown to be protective against severe forms of multiple sclerosis, with the effect increasing over the years [109]. More recently, the potential association of an SNP of IL-6 has been associated with the development of PD [102, 103]. At present the clearest evidence of the role of immunogenetic factors in autism is derived from studies of subsets of families in which HLA-DR4 frequencies of mothers, fathers, and children in were compared with a reference series of normal, unrelated controls. Studies of HLA typing indicated that mothers and their sons in a geographically defined group had a significantly higher frequency of DR4 than normal control subjects [110]. However, future studies of potential immunogenetic factors should also focus on defining the presence of specific SNPs and haplotypes in cytokines, chemokines, their receptors, and other immune factors, such as complement, integrins, and matrix metalloproteinases (MMPs).

Conclusion

It is clear that neuroimmune factors may affect brain development as well as CNS function during adulthood. Because defined immune challenges occur during neurodevelopment, the neuroinflammation and immune responses that follow such challenges, whether environmental, maternal, or neurogenetic, may constitute critical pathogenic factors in the development of ASD. The role of neuroglial activation and neuroinflammation is still uncertain but could be critical in maintaining, if not initiating, some of the abnormalities present in the CNS in autism. A better understanding of the role of neuroimmune reactions in the pathogenesis of ASD may have important clinical and therapeutic implications. Future studies should focus on the interaction of neuroimmune factors with brain development in the pathogenesis of autism.

Acknowledgments Dr. Pardo is supported by the Peter Emch Fund for Autism Research, The Bart A. McLean Fund for Neuroimmunology Research, Cure Autism Now and NIH-NIDA (K08-DA16160). The author expresses appreciation to Drs Andrew Zimmerman and Diana Vargas for their contributions to the discussion of the ideas discussed in this review.

References

1. Zimmerman AW: The Immune System; in: Bauman M, Kemper TL, (eds): *The Neurobiology of Autism*. Baltimore, The Johns Hopkins University Press, 2005, pp. 371–386.
2. Pardo CA, Vargas DL, Zimmerman AW: Immunity, neuroglia and neuroinflammation in autism. *Int Rev Psychiatry* 2005; 17(6):485–495.
3. Chess S, Fernandez P, Korn S: Behavioral consequences of congenital rubella. *J Pediatr* 1978; 93(4):699–703.
4. Comi AM, Zimmerman AW, Frye VH, Law PA, Peeden JN: Familial clustering of autoimmune disorders and evaluation of medical risk factors in autism. *J Child Neurol* 1999; 14(6):388–394.

5. Ashwood P, Wills S, Van de Water J: The immune response in autism: A new frontier for autism research. *J Leukoc Biol* 2006; 80(1):1–15.

6. Pardo CA, Eberhart CG: The neurobiology of autism. *Brain Pathol* 2007; 17:434–447.

7. Cicco-Bloom E, Lord C, Zwaigenbaum L, Courchesne E, Dager SR, Schmitz C, Schultz RT, Crawley J, Young LJ: The developmental neurobiology of autism spectrum disorder. *J Neurosci* 2006; 26(26):6897–6906.

8. Zipp F, Aktas O: The brain as a target of inflammation: common pathways link inflammatory and neurodegenerative diseases. *Trends Neurosci* 2006; 29(9):518–527.

9. Kim JH, Kim JH, Park JA, Lee SW, Kim WJ, Yu YS, Kim KW: Blood-neural barrier: Intercellular communication at glio–vascular interface. *J Biochem Mol Biol* 2006; 39(4):339–345.

10. Abbott NJ, Ronnback L, Hansson E: Astrocyte-endothelial interactions at the blood–brain barrier. *Nat Rev* 2006; 7(1):41–53.

11. Tilleux S, Hermans E: Neuroinflammation and regulation of glial glutamate uptake in neurological disorders. *J Neurosci Res* 2007; 85(10):2059–2070.

12. Griffiths M, Neal JW, Gasque P: Innate immunity and protective neuroinflammation: New emphasis on the role of neuroimmune regulatory proteins. *Int Rev Neurobiol* 2007; 82:29–55.

13. Skaper SD: The brain as a target for inflammatory processes and neuroprotective strategies. *Ann N Y Acad Sci* 2007; 1122:23–34.

14. Boulanger LM, Shatz CJ: Immune signalling in neural development, synaptic plasticity and disease. *Nat Rev* 2004; 5(7):521–531.

15. Tonelli LH, Postolache TT, Sternberg EM: Inflammatory genes and neural activity: Involvement of immune genes in synaptic function and behavior. *Front Biosci* 2005; 10:675–680.

16. Wierzba-Bobrowicz T, Gwiazda E, Poszwinska Z: Morphological study of microglia in human mesencephalon during the development and aging. *Folia Neuropathol* 1995; 33(2):77–83.

17. Rezaie P, Male D: Colonisation of the developing human brain and spinal cord by microglia: A review. *Microsc Res Tech* 1999; 45(6):359–382.

18. Streit WJ: Microglia and macrophages in the developing CNS. *Neurotoxicology* 2001; 22(5):619–624.

19. Monier A, dle-Biassette H, Delezoide AL, Evrard P, Gressens P, Verney C: Entry and distribution of microglial cells in human embryonic and fetal cerebral cortex. *J Neuropathol Exp Neurol* 2007; 66(5):372–382.

20. Rezaie P, Cairns NJ, Male DK: Expression of adhesion molecules on human fetal cerebral vessels: Relationship to microglial colonisation during development. *Brain Res Dev Brain Res* 1997; 104(1–2):175–189.

21. Perry VH, Hume DA, Gordon S: Immunohistochemical localization of macrophages and microglia in the adult and developing mouse brain. *Neuroscience* 1985; 15(2):313–326.

22. Bessis A, Bechade C, Bernard D, Roumier A: Microglial control of neuronal death and synaptic properties. *Glia* 2007; 55(3):233–238.

23. Levitt P, Rakic P: Immunoperoxidase localization of glial fibrillary acidic protein in radial glial cells and astrocytes of the developing rhesus monkey brain. *J Comp Neurol* 1980; 193(3):815–840.

24. Rakic P: Elusive radial glial cells: Historical and evolutionary perspective. *Glia* 2003; 43(1):19–32.

25. Bentivoglio M, Mazzarello P: The history of radial glia. *Brain Res bull* 1999; 49(5):305–315.

26. Rakic P: Evolving concepts of cortical radial and areal specification. *Prog Brain Res* 2002; 136:265–280.

27. Rakic P: Developmental and evolutionary adaptations of cortical radial glia. *Cereb Cortex* 2003; 13(6):541–549.

28. Fricker-Gates RA: Radial glia: A changing role in the central nervous system. *Neuroreport* 2006; 17(11):1081–1084.
29. Ransohoff RM, Liu L, Cardona AE: Chemokines and chemokine receptors: Multipurpose players in neuroinflammation. *Int Rev Neurobiol* 2007; 82:187–204.
30. Engelhardt B, Ransohoff RM: The ins and outs of T-lymphocyte trafficking to the CNS: Anatomical sites and molecular mechanisms. *Trends Immunol* 2005; 26(9):485–495.
31. Rostene W, Kitabgi P, Parsadaniantz SM: Chemokines: A new class of neuromodulator? *Nat Rev* 2007; 8(11):895–903.
32. Zou YR, Kottmann AH, Kuroda M, Taniuchi I, Littman DR: Function of the chemokine receptor CXCR4 in haematopoiesis and in cerebellar development. *Nature* 1998; 393(6685):595–599.
33. Ma Q, Jones D, Borghesani PR, Segal RA, Nagasawa T, Kishimoto T, Bronson RT, Springer TA: Impaired B-lymphopoiesis, myelopoiesis, and derailed cerebellar neuron migration in C. *Proc Natl Acad Sci USA* 1998; 95(16):9448–9453.
34. Lazarini F, Tham TN, Casanova P, renzana-Seisdedos F, Dubois-Dalcq M: Role of the alpha-chemokine stromal cell-derived factor (SDF-1) in the developing and mature central nervous system. *Glia* 2003; 42(2):139–148.
35. Stumm R, Hollt V: CXC chemokine receptor 4 regulates neuronal migration and axonal pathfinding in the developing nervous system: Implications for neuronal regeneration in the adult brain. *J Mol Endocrinol* 2007; 38(3):377–382.
36. Paredes MF, Li G, Berger O, Baraban SC, Pleasure SJ: Stromal-derived factor-1 (CXCL12) regulates laminar position of Cajal-Retzius cells in normal and dysplastic brains. *J Neurosci* 2006; 26(37):9404–9412.
37. Fatemi SH, Snow AV, Stary JM, aghi-Niknam M, Reutiman TJ, Lee S, Brooks AI, Pearce DA: Reelin signaling is impaired in autism. *Biol psychiatry* 2005; 57(7):777–787.
38. Pousset F, Fournier J, Keane PE: Expression of cytokine genes during ontogenesis of the central nervous system. *Ann N Y Acad Sci* 1997; 814:97–107.
39. Pousset F: Developmental expression of cytokine genes in the cortex and hippocampus of the rat central nervous system. *Brain Res* 1994; 81(1):143–146.
40. Munoz-Fernandez MA, Fresno M: The role of tumour necrosis factor, interleukin 6, interferon-gamma and inducible nitric oxide synthase in the development and pathology of the nervous system. *Prog Neurobiol* 1998; 56(3):307–340.
41. Nakashima K, Taga T: Mechanisms underlying cytokine-mediated cell-fate regulation in the nervous system. *Mol Neurobiol* 2002; 25(3):233–244.
42. Taga T, Fukuda S: Role of IL-6 in the neural stem cell differentiation. *Clin Rev Allergy Immunol* 2005; 28(3):249–256.
43. Bauer S, Kerr BJ, Patterson PH: The neuropoietic cytokine family in development, plasticity, disease and injury. *Nat Rev* 2007; 8(3):221–232.
44. Gomes FC, Sousa VO, Romao L: Emerging roles for TGF-beta1 in nervous system development. *Int J Dev Neurosci* 2005; 23(5):413-424.
45. Smith SE, Li J, Garbett K, Mirnics K, Patterson PH: Maternal immune activation alters fetal brain development through interleukin-6. *J Neurosci* 2007; 27(40):10695–10702.
46. Stevens B, Allen NJ, Vazquez LE, Howell GR, Christopherson KS, Nouri N, Micheva KD, Mehalow AK, Huberman AD, Stafford B, Sher A, Litke AM, Lambris JD, Smith SJ, John SW, Barres BA: The classical complement cascade mediates CNS synapse elimination. *Cell* 2007; 131(6):1164–1178.
47. Boulanger LM, Huh GS, Shatz CJ: Neuronal plasticity and cellular immunity: Shared molecular mechanisms. *Curr Opin Neurobiol* 2001; 11(5):568–578.
48. Aloisi F: Immune function of microglia. *Glia* 2001; 36(2):165–179.
49. Dong Y, Benveniste EN: Immune function of astrocytes. *Glia* 2001; 36(2):180–190.
50. Williams KC, Hickey WF: Central nervous system damage, monocytes and macrophages, and neurological disorders in AIDS. *Annu Rev Neurosci* 2002; 25:537–562.

51. Prat A, Biernacki K, Wosik K, Antel JP: Glial cell influence on the human blood–brain barrier. *Glia* 2001; 36(2):145–155.
52. Neumann H: Control of glial immune function by neurons. *Glia* 2001; 36(2):191–199.
53. Bauer J, Rauschka H, Lassmann H: Inflammation in the nervous system: The human perspective. *Glia* 2001; 36(2):235–243.
54. Trapp BD, Wujek JR, Criste GA, Jalabi W, Yin X, Kidd GJ, Stohlman S, Ransoh-off R: Evidence for synaptic stripping by cortical microglia. *Glia* 2007; 55(4):360–368.
55. Fields RD, Stevens-Graham B: New insights into neuron-glia communication. *Science* 2002; 298(5593):556–562.
56. Murai KK, Van Meyel DJ: Neuron glial communication at synapses: Insights from vertebrates and invertebrates. *Neuroscientist* 2007; 13(6):657–666.
57. Nedergaard M, Takano T, Hansen AJ: Beyond the role of glutamate as a neurotransmitter. *Nat Rev Neurosci* 2002; 3(9):748–755.
58. Ransom B, Behar T, Nedergaard M: New roles for astrocytes (stars at last). *Trends Neurosci* 2003; 26(10):520–522.
59. Rosenberg GA: Matrix metalloproteinases in neuroinflammation. *Glia* 2002; 39(3):279–291.
60. Benn T, Halfpenny C, Scolding N: Glial cells as targets for cytotoxic immune mediators. *Glia* 2001; 36(2):200–211.
61. Graeber MB, Bise K, Mehraein P: Synaptic stripping in the human facial nucleus. *Acta Neuropathol (Berl)* 1993; 86(2):179–181.
62. Ambrosini E, Aloisi F: Chemokines and glial cells: A complex network in the central nervous system. *Neurochem Res* 2004; 29(5):1017–1038.
63. Golde TE: Inflammation takes on Alzheimer disease. *Nat Med* 2002; 8(9):936–938.
64. Kaul M, Garden GA, Lipton SA: Pathways to neuronal injury and apoptosis in HIV-associated dementia. *Nature* 2001; 410(6831):988–994.
65. Pardo CA, Vining EP, Guo L, Skolasky RL, Carson BS, Freeman JM: The pathology of Rasmussen syndrome: Stages of cortical involvement and neuropathological studies in 45 hemispherectomies. *Epilepsia* 2004; 45(5):516–526.
66. McGeer PL, McGeer EG: Local neuroinflammation and the progression of Alzheimer's disease. *J Neurovirol* 2002; 8(6):529–538.
67. Blasko I, Grubeck-Loebenstein B: Role of the immune system in the pathogenesis, prevention and treatment of Alzheimer's disease. *Drugs Aging* 2003; 20(2):101–113.
68. Nagatsu T, Sawada M: Inflammatory process in Parkinson's disease: Role for cytokines. *Curr Pharm Des* 2005; 11(8):999–1016.
69. Volterra A, Meldolesi J: Astrocytes, from brain glue to communication elements: The revolution continues. *Nat Rev* 2005; 6(8):626–640.
70. Bushong EA, Martone ME, Jones YZ, Ellisman MH: Protoplasmic astrocytes in CA1 stratum radiatum occupy separate anatomical domains. *J Neurosci* 2002; 22(1):183–192.
71. Kirov SA, Harris KM: Dendrites are more spiny on mature hippocampal neurons when synapses are inactivated. *Nat Neurosci* 1999; 2(10):878–883.
72. Grosche J, Matyash V, Moller T, Verkhratsky A, Reichenbach A, Kettenmann H: Microdomains for neuron–glia interaction: Parallel fiber signaling to Bergmann glial cells. *Nat Neurosci* 1999; 2(2):139–143.
73. Halassa MM, Fellin T, Takano H, Dong JH, Haydon PG: Synaptic islands defined by the territory of a single astrocyte. *J Neurosci* 2007; 27(24):6473–6477.
74. Scemes E, Giaume C: Astrocyte calcium waves: What they are and what they do. *Glia* 2006; 54(7):716–725.
75. Nedergaard M, Ransom B, Goldman SA: New roles for astrocytes: Redefining the functional architecture of the brain. *Trends Neurosci* 2003; 26(10):523–530.
76. Montana V, Malarkey EB, Verderio C, Matteoli M, Parpura V: Vesicular transmitter release from astrocytes. *Glia* 2006; 54(7):700–715.

77. Pascual O, Casper KB, Kubera C, Zhang J, Revilla-Sanchez R, Sul JY, Takano H, Moss SJ, McCarthy K, Haydon PG: Astrocytic purinergic signaling coordinates synaptic networks. *Science* 2005; 310(5745):113–116.
78. Panatier A, Theodosis DT, Mothet JP, Touquet B, Pollegioni L, Poulain DA, Oliet SH: Glia-derived D-serine controls NMDA receptor activity and synaptic memory. *Cell* 2006; 125(4):775–784.
79. Stellwagen D, Malenka RC: Synaptic scaling mediated by glial TNF-alpha. *Nature* 2006; 440(7087):1054–1059.
80. Beattie EC, Stellwagen D, Morishita W, Bresnahan JC, Ha BK, Von ZM, Beattie MS, Malenka RC: Control of synaptic strength by glial TNFalpha. *Science* 2002; 295(5563):2282–2285.
81. Casanova MF: The neuropathology of autism. *Brain Pathol* 2007; 17(4):422–433.
82. Bauman ML, Kemper TL: Structural Brain Anatomy in Autism: What is the evidence? in: Bauman ML, Kemper TL, (eds): *The Neurobiology of Autism*. Baltimore, The Johns Hopkins University Press, 2005, pp. 136–149.
83. Lagercrantz H, Ringstedt T: Organization of the neuronal circuits in the central nervous system during development. *Acta Paediatr* 2001; 90(7):707–715.
84. Vargas DL, Nascimbene C, Krishnan C, Zimmerman AW, Pardo CA: Neuroglial activation and neuroinflammation in the brain of patients with autism. *Ann Neurol* 2005; 57(1):67–81.
85. Meyer U, Nyffeler M, Engler A, Urwyler A, Schedlowski M, Knuesel I, Yee BK, Feldon J: The time of prenatal immune challenge determines the specificity of inflammation-mediated brain and behavioral pathology. *J Neurosci* 2006; 26(18):4752–4762.
86. Meyer U, Yee BK, Feldon J: The neurodevelopmental impact of prenatal infections at different times of pregnancy: The earlier the worse? *Neuroscientist* 2007; 13(3):241–256.
87. Hagberg H, Mallard C: Effect of inflammation on central nervous system development and vulnerability. *Curr Opin Neurol* 2005; 18(2):117–123.
88. Patterson PH: Maternal infection: Window on neuroimmune interactions in fetal brain development and mental illness. *Curr Opin Neurobiol* 2002; 12(1):115–118.
89. Patterson PH: Neuroscience. Maternal effects on schizophrenia risk. *Science* 2007; 318(5850):576–577.
90. Shi L, Fatemi SH, Sidwell RW, Patterson PH: Maternal influenza infection causes marked behavioral and pharmacological changes in the offspring. *J Neurosci* 2003; 23(1):297–302.
91. Jonakait GM: The effects of maternal inflammation on neuronal development: Possible mechanisms. *Int J Dev Neurosci* 2007; 25(7):415–425.
92. Singer HS, Morris CM, Williams PN, Yoon DY, Hong JJ, Zimmerman AW: Antibrain antibodies in children with autism and their unaffected siblings. *J Neuroimmunol* 2006; 178(1–2):149–155.
93. Braunschweig D, Ashwood P, Krakowiak P, Hertz-Picciotto I, Hansen R, Croen LA, Pessah IN, Van de WJ: Autism: Maternally derived antibodies specific for fetal brain proteins. *Neurotoxicolog* 2008; 29(2):226–231.
94. Zimmerman AW, Connors SL, Matteson KJ, Lee LC, Singer HS, Castaneda JA, Pearce DA: Maternal antibrain antibodies in autism. *Brain Behav Immun* 2007; 21(3):351–357.
95. Meenagh A, Williams F, Ross OA, Patterson C, Gorodezky C, Hammond M, Leheny WA, Middleton D: Frequency of cytokine polymorphisms in populations from western Europe, Africa, Asia, the Middle East and South America. *Hum Immunol* 2002; 63(11):1055–1061.
96. Bidwell J, Keen L, Gallagher G, Kimberly R, Huizinga T, McDermott MF, Oksenberg J, McNicholl J, Pociot F, Hardt C, D'Alfonso S: Cytokine gene polymorphism in human disease: On-line databases: Supplement 1. *Genes Immun* 2001; 2(2):61–70.

97. Bidwell JL, Wood NA, Morse HR, Olomolaiye OO, Keen LJ, Laundy GJ: Human cytokine gene nucleotide sequence alignments: Supplement 1. *Eur J Immunogenet* 1999; 26(2–3):135–223.
98. Haukim N, Bidwell JL, Smith AJ, Keen LJ, Gallagher G, Kimberly R, Huizinga T, McDermott MF, Oksenberg J, McNicholl J, Pociot F, Hardt C, D'Alfonso S: Cytokine gene polymorphism in human disease: On-line databases: Supplement 2. *Genes Immun* 2002; 3(6):313–330.
99. Crusius JB, Pena AS, van Oosten BW, Bioque G, Garcia A, Dijkstra CD, Polman CH: Interleukin-1 receptor antagonist gene polymorphism and multiple sclerosis. *Lancet* 1995; 346(8980):979.
100. Epplen C, Jackel S, Santos EJ, D'Souza M, Poehlau D, Dotzauer B, Sindern E, Haupts M, Rude KP, Weber F, Stover J, Poser S, Gehler W, Malin JP, Przuntek H, Epplen JT: Genetic predisposition to multiple sclerosis as revealed by immunoprinting. *Ann Neurol* 1997; 41(3):341–352.
101. Mycko M, Kowalski W, Kwinkowski M, Buenafe AC, Szymanska B, Tronczynska E, Plucienniczak A, Selmaj K: Multiple sclerosis: The frequency of allelic forms of tumor necrosis factor and lymphotoxin-alpha. *J Neuroimmunol* 1998; 84(2):198–206.
102. Hakansson A, Westberg L, Nilsson S, Buervenich S, Carmine A, Holmberg B, Sydow O, Olson L, Johnels B, Eriksson E, Nissbrandt H: Investigation of genes coding for inflammatory components in Parkinson's disease. *Mov Disord* 2005; 20(5):569–573.
103. Hakansson A, Westberg L, Nilsson S, Buervenich S, Carmine A, Holmberg B, Sydow O, Olson L, Johnels B, Eriksson E, Nissbrandt H: Interaction of polymorphisms in the genes encoding interleukin-6 and estrogen receptor beta on the susceptibility to Parkinson's disease. *Am J Med Genet B Neuropsychiatr Genet* 2005; 133(1):88–92.
104. Gonzalez E, Rovin BH, Sen L, Cooke G, Dhanda R, Mummidi S, Kulkarni H, Bamshad MJ, Telles V, Anderson SA, Walter EA, Stephan KT, Deucher M, Mangano A, Bologna R, Ahuja SS, Dolan MJ, Ahuja SK: HIV-1 infection and AIDS dementia are influenced by a mutant MCP-1 allele linked to increased monocyte infiltration of tissues and MCP-1 levels. *Proc Natl Acad Sci USA* 2002; 99(21):13795–13800.
105. Nishimura M, Kuno S, Mizuta I, Ohta M, Maruyama H, Kaji R, Kawakami H: Influence of monocyte chemoattractant protein 1 gene polymorphism on age at onset of sporadic Parkinson's disease. *Mov Disord* 2003; 18(8):953–955.
106. Mundo E, Altamura AC, Vismara S, Zanardini R, Bignotti S, Randazzo R, Montresor C, Gennarelli M: MCP-1 gene (SCYA2) and schizophrenia: A case-control association study. *Am J Med Genet B Neuropsychiatr Genet.* 2005; 132B(1):1–4.
107. Fenoglio C, Galimberti D, Lovati C, Guidi I, Gatti A, Fogliarino S, Tiriticco M, Mariani C, Forloni G, Pettenati C, Baron P, Conti G, Bresolin N, Scarpini E: MCP-1 in Alzheimer's disease patients: A-2518G polymorphism and serum levels. *Neurobiol Aging* 2004; 25(9):1169–1173.
108. Pola R, Flex A, Gaetani E, Proia AS, Papaleo P, Giorgio AD, Straface G, Pecorini G, Serricchio M, Pola P: Monocyte chemoattractant protein-1 (MCP-1) gene polymorphism and risk of Alzheimer's disease in Italians. *Exp Gerontol* 2004; 39(8):1249–1252.
109. Luomala M, Lehtimaki T, Huhtala H, Ukkonen M, Koivula T, Hurme M, Elovaara I: Promoter polymorphism of IL-10 and severity of multiple sclerosis. *Acta Neurol Scand* 2003; 108(6):396–400.
110. Lee LC, Zachary AA, Leffell MS, Newschaffer CJ, Matteson KJ, Tyler JD, Zimmerman AW: HLA-DR4 in families with autism. *Pediatr Neurol* 2006; 35(5):303–307.

Part V
Neuroanatomy and Neural networks

Chapter 16
The Significance of Minicolumnar Size Variability in Autism

A Perspective from Comparative Anatomy

Manuel F. Casanova

Abstract Recent postmortem studies indicate the presence of diminished minicolumnar size in the cortex of patients with autism as compared to controls. A diminution in minicolumnar width in autism restricts the absolute span of this module's variability in both size and associated circuitry. Anatomically, minicolumns can be divided into cell core and peripheral neuropil space compartments. Development of the pyramidal cell core is constrained by radial cell migration and their attendant radially oriented axons and dendrite bundles. A major portion of a minicolumn's variability resides in its peripheral neuropil space where its constituent cells and process are more heterogeneous regarding their sources. This heterogeneity may have provided brains, in both evolution and development, with a way of adapting the function of minicolumns within specific networks. We surmise that minicolumnar variability is the result of genetic and epigenetic influences that provide for combinatorial diversity within overlapping networks resulting in behavioral flexibility.

Keywords Autism · brain · minicolumns · variability · cortex

Introduction

Autism is defined clinically by abnormalities in social interaction and communication, and restricted and stereotyped patterns of behaviors and activities. Some of the earliest problems relate to joint attention and reciprocity behaviors, for example, resistance to being held and inability to follow another person's gaze. Infants establish eye contact for instrumental purposes only, that is, not to share an experience but to obtain something they need. Furthermore, autistic patients have difficulties in interpreting facial and body language and fail to

M.F. Casanova
Gottfried and Gisela Endowed Chair in Psychiatry, University of Louisville,
Department of Psychiatry, 500 South Preston Street, Bldg. 55A, Ste. 217, Louisville,
KY 40292, USA
e-mail: m0casa02@louisville.edu

A.W. Zimmerman (ed.), *Autism*, DOI: 10.1007/978-1-60327-489-0_16,
© Humana Press, Totowa, NJ 2008

develop peer relationships. These symptoms are noticed before the child is 3 years old and range in a spectrum from mild to severe. Not all of the symptoms of autism are necessarily expressed in each patient. Those patients at the severe end of the spectrum are perceived as having a more homogeneous clinical presentation. Higher functioning autistic patients and those falling within the broader spectrum of symptomatology have a heterogeneous presentation and tend to blend with the general population. For most patients, autism represents a significant and life-long impairment in social and occupational functioning.

Research suggests that autism is a multifactorial or complex trait. Genetic studies report that monozygotic twins have a concordance rate of 75% as opposed to 3% for dizygotic (fraternal) twins. Families having an autistic patient have a 10–40% increased incidence of other developmental disorders. Researchers suggest the influence of 3–10 different genes in the expression of autism. Each gene (from both parents) acts in an additive fashion along with the environment to produce the final phenotype. It is implied that the underlying pathology(ies) to a multifactorial trait should exhibit a continuous distribution of changes.

The most reproducible macroscopic feature of autism is increased brain size. This finding is observed early in life and later disappears with aging as brain growth plateaus relative to controls [1]. Enlarged brain volume in autistic children is the result of increased cerebral gray/white matter and cerebellum; however, increases in white matter volume seems disproportionate to brain weight. A recent report on magnetic resonance imaging (MRI) relates the white matter increase primarily to the outer myelinated cortical compartment [2]. This compartment is made of late myelinating corticocortical fibers. Long and short association (corticocortical) fibers are some of the efferent projections within the myelinated bundles of cortical minicolumns. Further elucidation of these white matter findings requires techniques with a higher level of resolution (see below).

Microscopic findings in autism have been non-specific and, in most cases, uncorroborated. Early reports of smaller neurons within the limbic system suggested a neurodevelopmental origin, but the alternate diagnosis of simple atrophy was never considered or eliminated [3, 4]. Smaller neurons may represent a "developmental phenotype" but also a stage within a metabolic and morphological spectrum leading to cell death, *aposklesis* (cell withering associated with neurodegeneration), or a type of non-apoptotic dark degenerating cell. Loss of Purkinje and granule cells suggested a putative role for the cerebellum in autism [3, 4]. A hypoxic etiology may account for these findings, that is, Purkinje cell loss may be the result of unrecognized seizures, anticonvulsant medications (e.g., Dilantin), preagonal conditions, or prolonged postmortem intervals. These early reports on cell size and numbers were based on qualitative observations or relied on biased (non-stereological) assumptions.

More recently, Bailey et al. [5] drew attention to widespread cortical abnormalities in autism, for example, disruption of laminar organization and heterotopias. The findings of Bailey et al. [5] along with the occasional presence in autism of polymicrogyria, macrogyria, and schizencephaly suggest disordered migration during corticogenesis [6]. Recent findings of normal cortical thickness and preservation of number of cells per minicolumn (E40–100) challenge the argument of a migrational abnormality in autism [7, 8] (see below under Delaunay triangulation).

A possible time frame of susceptibility for autism is derived from the concurrence of cranial nerve palsies (Moebius syndrome or congenital oculofacial paralysis). Moebius syndrome is defined during brainstem formation in the first trimester of pregnancy. The window of susceptibility in autism is most consistent with a gestational lesion during the period of asymmetrical divisions of perivetricular germinal cells that define the total number of minicolumns in the cortex (E < 40 days).

Neurologists traditionally consider autism as a disease of the gray matter (as opposed to the white matter). The tentative localization is supported by evidence of seizures in a significant percentage of cases and the absence of spasticity or vision loss. Despite suggested findings in the gastrointestinal tract [9], there is no firm evidence that other organs aside from the brain are involved. Similarly, there is no indication of peripheral nervous system involvement. Dysfunction of higher cognitive functions further pinpoints a deficit to the isocortex, more specifically to the modular organization of the cortex that provides for emergent cognitive properties. The smallest cortical module is known as the minicolumn.

Minicolumnar Pathology in Autism

Autism stands as a neurodevelopmental condition defined early on during gestation and capable of affecting brain growth. One possible explanation for this is the recently described presence of minicolumnar abnormalities in autism [10]. In this study, nine autistic patients and an equal number of controls were studied for morphometric features of pyramidal cell arrays (Brodmann areas 9, 21, and 22). The feature extraction properties of the algorithm were corrected for minicolumnar fragmentation (z-axis artifact), curvature of tissue sections, and 3D proportions (stereological modeling) [11]. Significant findings were reported in the spacing of columns and in their internal structure, that is, the relative dispersion of cells. Minicolumns in autistic patients were smaller and more numerous per linear length of distance measured. The same population was used to replicate the findings by using a different neuronomorphometric measure, that is, the Grey Level Index (GLI) [12]. The GLI is defined as the ratio of the area covered by Nissl-stained elements to unstained area in postmortem samples. The results revealed significant diagnosis-dependent effects in feature distance (minicolumnar width) between autistic patients and controls. More recent studies have corroborated the presence of a minicolumnopathy in autism within an independent sample [7]. In this latter study, the mean neuronal and nucleolar cross sections were smaller in autistic patients as compared to controls, while the neuronal density in autism exceeded the comparison group by 23%. Analysis of a inter- and intracluster distances of a Delaunay triangulation suggested that that the increased cell density was the result of more minicolumns in the case of autistic patients as opposed to more cells per minicolumns. The authors concluded that a "reduction in both somatic and nucleolar cross sections could reflect a bias towards shorter connecting fibers, which favors local computation at the expense of interareal and callosal connectivity" [7].

The findings of a minicolumnopathy in autism appear to be region specific. In a topographical study that involved serial sections of cellodin embedded brain hemispheres, digital photomicrographs were taken to examine nine representative sections of four cortical types: (1) paralimbic, (2) high-order (heteromodal) association, (3) modality-specific (unimodal) association, and (4) idiotypic areas (primary sensory/motor cortices) [13]. A computerized image analysis algorithm was implemented to calculate the mean peripheral neuropil space of minicolumns in each of the aforementioned brain regions. The areas mostly affected in this study exhibited different degrees of cytoarchitectural differentiation but were strongly interconnected: Brodmann areas 10 (frontopolar, high-order heteromodal) and 24 (anterior cingulate, paralimbic). No significant findings were reported for idiotypic cortex (primary sensory/motor cortices).

Additional studies have shown the specificity of minicolumnar findings in autism by eliminating the possible confound of mental retardation. In a series of Down syndrome brains, minicolumnar size reached adult size earlier than normal, a phenomenon attributed to accelerated aging [14, 15]. The normal-sized minicolumns in Down syndrome was all the more striking when considering the small brain size observed in this condition. Similarly, a different minicolumnar morphometry has been observed in other conditions that provide for an autism-like phenotype, for example, rubella babies and tuberous sclerosis [11].

Minicolumns

A primary feature of encephalization among primates in comparison with other mammalian species is the increased volume of cortex relative to subcortical structures. As cortical thickness varies little relative to surface area among mammals and less so among catarrhine primates (Cercopithecoidea and Hominoidea) [16], this increase largely reflects the expansion of the cortical sheet. Mammalian isocortex comprises large numbers of re-iterative radially oriented neuronal cell columns, or minicolumns, which are hypothesized to be an elemental modular microcircuit [15, 17]. Ontogenetic cell columns are the developmental precursors of mature cortical minicolumns [7]. These structures are derived from clones of neuronal precursors migrating from germinal zones to the forming cortical plate along common radially oriented glial fibers. The phylogenetic expansion of the cortical sheet is hypothesized to be a consequence of an increase in the numbers and tangential extent of germinal zone clonal progenitors via selective changes in the regulation of their mitotic activity [18].

Minicolumnar development is regulated by genetic and epigenetic mechanisms controlling progenitor proliferation, migration, specification, and formation of connections. Selection acting on these mechanisms will affect expansion and differentiation of minicolumn populations which may provide a basis for functional subdivision within a given area. Aggregates of minicolumns with common sensory response properties may be subdivided into subnetworks which process specific submodalities in parallel within a receptive field. Functional subdivisions of isocortex differ in degree of minicolumnar morphometric variability. It is therefore noteworthy that primates possess a great number of cytoarchitectural divisions in comparison with other species such as rodents [16].

With allometric expansion of prefrontal, temporal, and other areas, striate (visual) cortex represents a relatively smaller proportion of cortex in humans than in non-human primates. Its size in humans, relative to overall body proportions and thalamocortical inputs, is comparable to other primates [19]. Nevertheless, expansion and differentiation of connectivity between visual and other cortical areas argues for increased functional differentiation within human striate cortex. The following sections of this chapter will examine how minicolumnar variability may affect functional and connectional differentiation among cortical networks.

Methodological Considerations

It is unlikely that increased minicolumnar variability is the result of z-axis artifacts, that is, columnar fragmentation caused by section thickness [11]. A z-axis artifact can be produced by variations in the central positioning of a three-dimensional cell column and its subsequent sectioning and study in two dimensions. Such stereological bias should not account for a significant portion of the findings, as measures of center-to-center spacing for vertical bundles done in both the horizontal and the tangential plane reveal a similar degree of variability [20]. More likely, morphological variation of minicolumns is a function of their intrinsic anatomy, for example, the width of afferent/efferent fiber bundles and the size/number of cell somas within a given minicolumn.

Recent studies employing computer-assisted morphometric imaging reveal differences between individual brains that reflect variability of columnar organization within each area. Investigations of macaque visual cortex have revealed variability in spacing of discrete components of the minicolumn, including apical dendritic bundles [21] and double-bouquet cell axons [22, 23], as well as differences in distribution of those elements within primary and secondary visual areas [24]. Within macaque striate cortex, center-to-center spacing of myelinated axon bundles exhibited a coefficient of variation of 0.31 [25]. Variation in minicolumnar width may be principally determined by number and composition of cellular components within the peripheral neuropil space surrounding the core pyramidal cell array [26]. Hendry and colleagues [27] showed significant intra-areal variation in number and laminar distribution of GABAergic inhibitory interneurons within 50-μm wide intervals in different areas of monkey isocortex. This variation demonstrated area-specific differences in organization, for example, within area 18 patterns of variation in minicolumnar width occurred in narrow and wide bands from 150- to 700-μm wide. The significance of these bands in terms of functional columnar architecture remains obscure.

Among primates, data from recent studies suggest a trend toward increased variability in spacing of minicolumnar components with encephalization, both within and between cortical areas. Vellate astroglia have superficially situated cell bodies and radially oriented interlaminar processes that serve to provide a matrix for the structural and physiological support of supragranular pyramidal cell arrays [28]. In the cortex of Old World monkeys, the astroglial palisade exhibits regular spacing of interlaminar processes [29]. In contrast, among great apes, spacing of astrocytic processes is less uniform among chimpanzees in comparison with gorillas and orangutans [30]. This suggests a phylogenetic trend of increasing within-area minicolumnar variability. This trend is further supported by evidence that human brains exhibit inter- and intra-area variability with respect to double-bouquet axon type, thickness, density, and in the number and pattern of minicolumns not paired to them, in contrast to the more uniform and less extensive distribution found in non-primates [31]. GABAergic interneuron double-bouquet cells project bundles of axons radially within peripheral neuropil space of minicolumns [22] and in general exhibit a one-to-one correspondence with myelinated axon bundles in minicolumns [32].

Increased within-area variability is likely extended to variability in cerebral dominance. Lateralization in minicolumnar morphometry of isocortical areas, most notably in auditory cortex, represents an important contributor to increased variability in humans relative to non-human primates. In humans, minicolumnar width, spacing, pyramidal cell size and axon bundle thickness are all greater in left than in right hemisphere posterior auditory cortical areas [33, 34, 35]. Increased laterality of minicolumnar width and width variability in humans relative to chimpanzees was confirmed with computer assisted morphometric methods [33, 34, 35, 36, 37].

Trends for Between-Area Minicolumnar Variability

The number and density of neurons per minicolumn varies across cortical areas within species. Early concepts of columnar uniformity [38, 39] have been challenged by subsequent studies employing unbiased stereological methods. These revealed that the number and density of neurons per minicolumnar volume vary between different areas of the cortex. Beaulieu [40] estimated the total number of neurons under a given area of pial surface in the monocular visual, barrel field, and primary motor regions of the rat and showed that neuron density derived from the unbiased estimates of number differed between cortical regions. Comparable differences were observed in the GABAergic subpopulation, which comprised 15% of all neurons regardless of region. Similarly, Skoglund and colleagues [41], using the optical dissector counting method to measure neuronal density in primary motor, primary somatosensory, and secondary visual cortex in the rat, obtained comparable densities and cell count differences.

Minicolumnar Variability Differences Between Species

An influential early perspective in comparative neuroanatomy argued for a basic uniformity of cortical microstructure among mammalian and especially primate taxa [42]. This idea contrasted with the work of classical neuroanatomists who identified radially oriented columns of neurons in Nissl-stained sections as specific to primate temporal cortex [43]. More contemporary studies have overturned the assumption that measures of neuronal structure and density are uniform among mammals.

Haug [44] and Stolzenburg and colleagues [45] each reported that counts of neurons beneath a defined area of pial surface yielded different values for each species studied. Likewise, White and Peters (1993) [46]showed that for minicolumns, defined as cellular constituents organized around a common apical dendritic bundle, both minicolumnar size and total number of neurons differed substantially between monkey, cat, rat, and mouse. For example, minicolumns in rat V1 were 60 μm in width and contained 355 cells; in comparison monkey V1 minicolumns were 31 μm in diameter and had on average 142 cells. In mouse SM1, minicolumns averaged 25 μm in diameter and comprised 53 neurons.

In a study of striate cortex in cats, Kaschube and colleagues [47] showed that measures of orientation column shape and size variability clustered according to degree of genetic relatedness among animals and also between hemispheres within each brain. Ohki and colleagues (2005) [48] have provided direct physiological evidence for minicolumnar-scaled architecture underlying vision processing. Employing two-photon calcium-activated fluorescent imaging in area 18 of the cat, they were able to obtain single cell-width resolution of direction and orientation tuning. Parameter values varied continuously from neuron to neuron across the area map, while fractures were evident as discontinuous shifts in stimulus response between adjoining groups of cells. Comparable observations in rats revealed minicolumnar-scaled tuning, but without the ordered response progression seen in the cat. Rather, no trend was observed for orientation or direction stimulus response values among neighboring neurons. This suggests that in cat visual cortex adjoining minicolumns within a common receptive field segregate into interposed subnetworks engaged in parallel processing of stimulus features. Previous studies clearly indicate significant minicolumnar variability within and across areas in individual brains and within and across mammalian species [15, 17]. Two-photon imaging provides the means for relating physiological responses of individual minicolumns to their morphometry, anatomy, and molecular features. Future work employing this technique in other mammalian taxa holds promise for further establishing minicolumnar variability as an evolutionary character.

Functional Implications of Increased Minicolumnar Variability

To what extent is the morphometric variability in minicolumns related to systematic differences in the cytoarchitectonics of the pyramidal cell core and its peripheral neuropil space? A recent study of minicolumnar morphometry in autistic patients demonstrated increased numbers of narrower minicolumns associated with decreased peripheral neuropil space [10]. While development of the pyramidal cell core is constrained by radial cell migration and formation of radially oriented axon and dendrite bundles, the constituents of the peripheral neuropil space are more heterogeneous and their sources, modes of migration, morphogenesis, and synaptogenesis more varied. This heterogeneity may provide the basis both in evolution and development for adapting minicolumns to function within specific networks. Early work by Seldon [35] showed increased minicolumnar size and intercolumnar spacing in posterior language processing areas of left in comparison to right hemisphere temporal cortex. Basal dendritic arbors connecting neighboring minicolumns were not proportionately as large in language association areas, suggesting greater segregation of groups of minicolumns into local subnetworks. Increased spacing between larger functional macrocolumns is also left-lateralized [49] consistent with an increase in the number of minicolumnar subnetworks within each macrocolumn.

Variability among the multiple components of the minicolumn (e.g., number of neurons and amount of synapses) may contribute to the fault tolerance of larger networks such as macrocolumns. Redundant systems can be reliable even when the underlying components are error-prone; this is the basis for majority voting circuits [50]. McCulloch (1959) [56] characterized unreliable networks of threshold elements as "logically stable" when elements' thresholds could vary in tandem—not changing the function computed by the network as a whole—and "logically unstable" when thresholds could vary independently. He showed that, provided that failures of individual components occur with probability less than $p = 0.5$, redundant systems of unstable nets could be designed for greater reliability than redundant systems of stable nets of the same size.

However paradoxical it appears at first for the functional plasticity of mini-columns, their ability to compute different functions within the same module can be used to their advantage with respect to sensitivity to error.

Plasticity is associated not only with the tuning of synaptic activity states but also optimal selection among alternate subnetworks of microcircuits develop-ing within a given context. Such parallel subnetworks may process complemen-tary submodalities within a defined receptive or associative field; alternatively they may provide overlapping response characteristics to a common input. Competition among networks allows for circuit optimization, in particular by means of learning, within an individual's lifespan. We hypothesize that mini-columnar diversity provides the substrate for this competition and the basis for adapting learned behavior to context. During development, neurogenetic pro-grams interact with epigenetic factors to regulate formation of cortical micro-circuit templates, which are then shaped and pruned by differential patterns of sensory activity. Incipient behavioral patterns in turn constrain selection for mechanisms of plasticity, establishing a dynamic, mutually-informing loop, a process referred to as the *Baldwin effect* [51, 52]. Thus, increased minicolumnar diversity may give rise to greater potential for combinatorial activity of micro-circuits within overlapping networks, resulting in enhanced learning and behavioral flexibility. This process may provide the basis for the adaptation of non-human primate homologues of Brodmann areas 44, 45, and 47 to the specific requirements of language processing. While gross morphological asym-metries of areas related to language processing in humans have been identified in great apes [53, 54], no direct relationship exists between gross morphological boundaries and the characteristic cytoarchitecture of these areas [55]. Com-parative studies of minicolumnar variability in these areas between human and non-human primates may contribute to understanding of phylogenetic trends underlying the emergence of language.

Conclusions

Postmortem studies demonstrate diminished minicolumnar size in the cortex of autistic patients as compared to controls. Diminished size restricts the absolute span of minicolumnar variability and its associated circuitry. Studies on com-parative anatomy suggest that minicolumnar variability reflects a robust het-erogeneity of cortical minicolumnar architecture in humans: Functionally, this is consistent with an increased capacity for connectional plasticity at the circuit level. A review of the literature establishes evolutionary trends of minicolumnar variability within and between cortical areas and differences in degrees of variability between mammalian taxa, while among primates, humans display the greatest degree of variability. Taken as a whole, these findings indicate that minicolumnar variability is likely a phenotypic character under selection in the truest Darwinian sense. We surmise that the mechanism of selection is related to neurological processes concerning cognitive and behavioral flexibility specific to adaptive learning and behaviors. Subsequent subdivision and specialization of cortical hierarchies are the result of the variable cytoarchitecture and functionality of the individual minicolumns constituting them and allows for increases in complexity and flexibility of the behavioral repertoire. These observations provide important clinicopathological correlates to autistic symptomatology.

Acknowledgments This chapter is based upon work supported by the National Alliance for Autism Research (NAAR) and National Institute of Mental Health grants mh61606, mh62654, and mh69991. The authors express their gratitude for his helpful advice and use of facilities to Mr. Archibald Fobbs (YAKOVLEV-HALEEM BRAIN COLLECTION, ARMED FORCES INSTITUTE of PATHOLOGY).

References

1. Dawson G, Munson J, Webb SJ, Nalty T, Abbott R, Toth K. Rate of head growth decelerates and symptoms worsen in the second year of life in autism. *Biological Psychiatry* 2007;61(4):458.
2. Herbert MR, Ziegler DA, Makris N, et al. Localization of white matter volume increase in autism and developmental language disorder. *Annals Neurology* 2004;55(4):530–540.
3. Bauman M, Kemper T. *The Neurobiology of Autism,* 2nd ed. Baltimore, MD: The Johns Hopkins University Press; 1996.
4. Bauman M, Kemper TL. Histoanatomic observations of the brain in early infantile autism. *Neurology* 1985;35(6):866–874.
5. Bailey A, Luthert P, Dean A, et al. A clinicopathological study of autism. *Brain* 1998;121(5):889–905.
6. Piven J, Arndt S, Bailey J, Havercamp S, Andreasen N, Palmer P. An MRI study of brain size in autism. *American Journal of Psychiatry* 1995;152(8):1145–1149.
7. Casanova MF, van Kooten IA, Switala AE, et al. Minicolumnar abnormalities in autism. *ACTA Neuropathologica (Berlin)* 2006;112(3):287–303.
8. Hutsler JJ, Love T, Zhang H. Histological and magnetic resonance imaging assessment of cortical layering and thickness in autism spectrum disorders. *Biological Psychiatry* 2007;61(4):449.
9. Casanova MF. The minicolumnopathy of autism: a link between migraine and gastrointestinal symptoms. *Medical Hypothesis* 2008;70(1):73–80.
10. Casanova M, Buxhoeveden D, Switala A, Roy E. Minicolumnar pathology in autism. *Neurology* 2002;58(3):428–432.
11. Casanova MF, Switala AE. Minicolumnar morphometry: computerized image analysis. In: Casanova MF, (ed.) *Neocortical Modularity and the Cell Minicolumn.* New York, NY: Nova Science Publishers, Inc.; 2005:161–180.
12. Casanova MF, Buxhoeveden DP, Switala AE, Roy E. Neuronal density and architecture (Gray Level Index) in the brains of autistic patients. *Journal of Child Neurology* 2002;17(7):515–521.
13. Casanova MF, Van Kooten IA, Switala AE, et al. Abnormalities of cortical minicolumnar organization in the prefrontal lobes of autistic patients. *Clinical Neuroscience Research* 2006;6(3–4):127–133.
14. Buxhoeveden D, Casanova MF. Accelerated maturation in brains of patients with Down syndrome. *Journal of Intellectual Disability Research* 2004;48(Pt 7):704–705.

15. Buxhoeveden DP, Casanova MF. The minicolumn hypothesis in neuroscience. *Brain* 2002;125(5):935–951.
16. Northcutt RG, Kaas JH. The emergence and evolution of mammalian neocortex. *Trends in Neurosciences* 1995;18(9):373.
17. Mountcastle VB. The columnar organization of the neocortex. *Brain* 1997;120(4):701–722.
18. Rakic P. A small step for the cell, a giant leap for mankind: a hypothesis of neocortical expansion during evolution. *Trends in Neurosciences* 1995;18(9):383.
19. Streidter GF. *Principles of Brain Evolution*. Sunderland, MA: Sinauer Associates, Inc.; 2005.
20. Feldman ML, Peters A. A study of barrels and pyramidal dendritic clusters in the cerebral cortex. *Brain Research* 1974;77(1):55.
21. Peters A, Sethares C. Layer IVA of rhesus monkey primary visual cortex. *Cerebral Cortex* 1991;1(6):445–462.
22. DeFelipe J, Hendry SHC, Hashikawa T, Molinari M, Jones EG. A microcolumnar structure of monkey cerebral cortex revealed by immunocytochemical studies of double bouquet cell axons. *Neuroscience* 1990;37(3):655.
23. Peters A. The organization of the primary visual cortex in the macaque. In: Peters A, Rockland KS, (eds.) *Primary Visual Cortex in Primates*. New York: Plenum Press; 1994:1–35.
24. Peters A, Cifuentes JM, Sethares C. The organization of pyramidal cells in area 18 of the rhesus monkey. *Cerebral Cortex* 1997;7(5):405–421.
25. Peters A, Sethares C. Myelinated axons and the pyramidal cell modules in monkey primary visual cortex. *Journal of Comparative Neurology* 1996;365(2):232–255.
26. Buxhoeveden D, Casanova MF. Encephalization, minicolumns, and hominid evolution. In: Casanova M, (ed.) *Neocortical Modularity and the Cell Minicolumn*. New York, NY: Nova Science Publishers, Inc.; 2005:117–136.
27. Hendry SH, Schwark HD, Jones EG, Yan J. Numbers and proportions of GABA-immunoreactive neurons in different areas of monkey cerebral cortex. *Journal of Neuroscience* 1987;7(5):1503–1519.
28. Reisin HD, Colombo JA. Considerations on the astroglial architecture and the columnar organization of the cerebral cortex. *Cellular and Molecular Neurobiology* 2002;22(5–6):633–644.
29. Colombo JA, Hartig W, Lipina S, Bons N. Astroglial interlaminar processes in the cerebral cortex of prosimians and Old World monkeys. *Anatomy and Embryology (Berlin)* 1998;197(5):369–376.
30. Colombo JA, Sherwood CC, Hof PR. Interlaminar astroglial processes in the cerebral cortex of great apes. *Anatomy and Embryology (Berlin)* 2004;208(3):215–218.
31. Yáñez IB, Muñoz A, Contreras J, Gonzalez J, Rodriguez-Veiga E, DeFelipe J. Double bouquet cell in the human cerebral cortex and a comparison with other mammals. *Journal of Comparative Neurology* 2005;486(4):344–360.
32. del Rio MR, DeFelipe J. Double bouquet cell axons in the human temporal neocortex: relationship to bundles of myelinated axons and colocalization of calretinin and calbindin D-28 k immunoreactivities. *Journal of Chemical Neuroanatomy* 1997;13(4):243.
33. Anderson B, Southern BD, Powers RE. Anatomic asymmetries of the posterior superior temporal lobes: a postmortem study. *Neuropsychiatry, Neuropsychology, and Behavioural Neurology* 1999;12(4):247–254.
34. Hutsler JJ, Gazzaniga MS. Acetylcholinesterase staining in human auditory and language cortices: regional variation of structural features. *Cerebral Cortex* 1996;6(2):260–270.
35. Seldon HL. Structure of human auditory cortex. II. Axon distributions and morphological correlates of speech perception. *Brain Research* 1981;229(2):295.
36. Buxhoeveden D, Lefkowitz W, Loats P, Armstrong E. The linear organization of cell columns in human and nonhuman anthropoid Tpt cortex. *Anatomy and Embryology* 1996;194(1):23.
37. Buxhoeveden DP, Switala AE, Litaker M, Roy E, Casanova MF. Lateralization of minicolumns in human planum temporale is absent in nonhuman primate cortex. *Brain Behavior and Evolution* 2001;57(6):349–358.

38. Creutzfeldt OD. Generality of the functional structure of the neocortex. *Naturwissenschaften* 1977;64(10):507–517.
39. Rockel AJ, Hiorns RW, Powell TPS. The basic uniformity in structure of the neocortex. *Brain* 1980;103(2):221–244.
40. Beaulieu C. Numerical data on neocortical neurons in adult rat, with special reference to the GABA population. *Brain Research* 1993;609(1–2):284.
41. Skoglund TS, Pascher R, Berthold CH. Aspects of the quantitative analysis of neurons in the cerebral cortex. *Journal of Neuroscience Methods* 1996;70(2):201.
42. Jerison HJ. *Evolution of the Brain and Intelligence*. New York: Academic Press; 1973.
43. Von Economo CF, Koskinas GN. *Die Cytoarchitektonic der Hirnrinde des erwachsenen Menschen*. Wien: Springer; 1925.
44. Haug H. Brain sizes, surfaces, and neuronal sizes of the cortex cerebri: a stereological investigation of man and his variability and a comparison with some mammals (primates, whales, marsupials, insectivores, and one elephant). *The American Journal of Anatomy* 1987;180(2):126–142.
45. Stolzenburg JU, Reichenbach A, Neumann M. Size and density of glial and neuronal cells within the cerebral neocortex of various insectivorian species. *GLIA* 1989;2(2):78–84.
46. White EL, Peters A. Cortical modules in the posteromedial barrel subfield (Sml) of the mouse. *Journal of Comparative Neurology* 1993;334(1):86–96.
47. Kaschube M, Wolf F, Geisel T, Lowel S. Genetic influence on quantitative features of neocortical architecture. *Journal of Neuroscience* 2002;22(16):7206–7217.
48. Ohki K, Chung S, Ch'ng YH, Kara P, Reid RC. Functional imaging with cellular resolution reveals precise micro-architecture in visual cortex. *Nature* 2005;433(7026):597–603.
49. Galuske RAW, Schlote W, Bratzke H, Singer W. Interhemispheric asymmetries of the modular structure in human temporal cortex. *Science* 2000;289(5486):1946–1949.
50. Stroud CE. Reliability of majority voting based VLSI fault-tolerant circuits. *IEEE Transactions VLSI Systems* 1994;2:516–521.
51. Baldwin JM. A new factor in evolution. *The American Naturalist* 1896;30:441–451.
52. Krubitzer L, Kaas J. The evolution of the neocortex in mammals: how is phenotypic diversity generated? *Current Opinion in Neurobiology* 2005;15(4):444.
53. Cantalupo C, Hopkins WD. Asymmetric Broca's area in great apes. *Nature* 2001;414(6863):505.
54. Gannon PJ, Holloway RL, Broadfield DC, Braun AR. Asymmetry of chimpanzee planum temporale: humanlike pattern of wernicke's brain language area homolog. *Science* 1998;279(5348):220–222.
55. Sherwood CC, Broadfield DC, Holloway RL, Gannon PJ, Hof PR. Variability of Broca's area homologue in African great apes: implications for language evolution. The *Anatomical Record. Part A, Discoveries in Molecular, Cellular, and Evolutionary Biology* 2003;271(2):276–285.
56. McCulloch WS. Agatha Tyche of nervous nets – the lucky reckoners. In: Warren S. McCulloch. Embodiments of mind. Cambridge: MIT Press, 1965;203–215. Reprint of: National Physical Laboratory. Mechanisation of thought processes. London: H.M. Stationery Office, 1959;611–625.

Chapter 17
Imaging Evidence for Pathological Brain Development in Autism Spectrum Disorders

Stephen R. Dager, Seth D. Friedman, Helen Petropoulos and Dennis W.W. Shaw

Abstract Though much has been learned about autism spectrum disorders (ASD) during the past decade, the mechanisms underlying ASD remain an enigma. One of the more consistent brain anatomical findings associated with ASD has been larger brains, on average 10–15% enlargement by magnetic resonance imaging (MRI), in preschool-aged children evaluated soon after clinical diagnosis. It is the premise of this chapter that research investigating cellular composition in ASD in vivo can elucidate aspects of the underlying pathophysiology, such as the phenomenon of early brain enlargement in ASD, and help guide ongoing theoretical model development. This chapter will review applications of magnetic resonance spectroscopy (MRS) and magnetic resonance transverse relaxation techniques (T2r) used in an attempt to elucidate developmental mechanisms underlying brain structural alterations in ASD. For example, one hypothesis put forth to explain findings of brain enlargement in ASD has implicated alterations in the complicated biochemistry governing apoptosis and/or synaptic pruning, with resultant neuronal "overgrowth." However, studies from our laboratory of brain chemical alterations in preschool-aged ASD children instead have found evidence for decreased neuronal compactness or density, contradictory to such theories. Another theory to explain observations of early but not later brain enlargement in ASD suggests accelerated "normal" brain growth, which then decelerates or plateaus before the time course of brain growth for typically developing children. Contrary to this theory, our recent quantitative T2r study, designed to characterize the temporal progression of brain maturation, implicates mechanism(s) other than more rapid growth to account for larger brains in ASD. Although these research findings cannot be considered diagnostic, they do provide new insights for pursuing mechanisms underlying ASD.

Keywords Autism · magnetic resonance · spectroscopy · brain · growth · chemistry

S.R. Dager
Department of Radiology, Interim Director, University of Washington Autism Center, University of Washington, 1100 NE 45th Str, Suite 555, Seattle, WA, USA 98105
e-mail: srd@u.washington.edu

A.W. Zimmerman (ed.), *Autism*, DOI: 10.1007/978-1-60327-489-0_17,
© Humana Press, Totowa, NJ 2008

Clinical Presentation

Diagnosis of ASD, typically differentiated clinically into autistic disorder (AD) and pervasive developmental disorder-not otherwise specified (PDD-NOS), is based on core clinical features that include abnormal interpersonal and emotional interactions, disordered language and communication, and repetitive and stereotypic behaviors [1] Through systematic and extensive clinical evaluation, the earliest age for reliable diagnosis of ASD has been progressively pushed back to where it can now be established by 2 years of age [2]. Data from behavioral studies of infant siblings of older children with ASD suggest that some of the defining features of this condition are not present at 6 months of age (e.g., social deficits) but have their first appearance around 12 months or later postpartum [3, 4]. Specific physical stigmata are not associated with ASD, and there are no established biomarkers to confirm the diagnosis [5, 6, 7]. ASD is frequently associated with mental retardation, evident in 30–70% of individuals, the variability of which may be related to the recognized extent of ASD symptom expression [8, 9, 10]. There also is a substantial risk of seizures in ASD that have a bimodal pattern of onset either in the first 2 years of life or, more typically, as the child enters adolescence, with prevalence rates variously estimated at between 15 and 38% [11, 12].

At one time blamed largely on poor mothering skills, the characteristic clinical expression of ASD is now generally agreed to reflect a complex, but poorly understood, interaction between genetic and environmental factors [5, 6, 13]. The early presentation and chronic clinical course that characterize ASD provide a strong rationale for investigating abnormal brain developmental mechanisms presumed to underlie the disorder. Although much progress has been made in understanding brain developmental processes and vulnerable periods during development in general, the biological mechanisms underlying ASD, and related risks of mental retardation and seizures, remain elusive. Autism is thought to most likely be due to early-onset brain differences in cellular development, with subsequent differences in cell density, connectivity, and ultimately, function. Therefore, non-invasive in vivo measures of neuronal development and integrity, applied at targeted developmental time-points, may be instructive to better understand the nature of abnormal brain development and function in ASD.

Structural Imaging Relationships

Non-invasive magnetic resonance imaging techniques provide a window to the brain that can characterize not only structural abnormalities but, potentially, neurochemical and functional mechanisms underlying abnormal developmental processes in ASD. Magnetic resonance imaging (MRI) studies of brain structure in ASD have been extensively reviewed elsewhere [14]. One of the more intriguing and consistent MRI findings, suggested by earlier head circumference (HC) studies, has been strong evidence for cerebral enlargement in ASD, at least during early childhood. The three published MRI investigations to date that examined preschool-aged children (ages 2–4 years) studied soon after clinical diagnosis of ASD all found enlarged cerebral volumes, on average 10–15%, excluding cerebrospinal fluid, at this age [15, 16, 17]. This finding also appears to be more specific to a younger developmental age range based on cross-sectional age-nested MRI studies [15, 18].

Indirect evidence from studies evaluating the trajectory of HC growth, which in early childhood is closely correlated to brain size (excluding those subjects with increased cerebrospinal spaces), suggests that the onset of cerebral enlargement among individuals with ASD occurs after birth and before the age of 12 months [19, 20, 21, 22, 23]. One recent study has also reported a somewhat later onset between 12 and 18 months for HC divergence [17]. MRI findings from several cross-sectional studies, stratified by age across a broad age range of affected individuals, support the notion of early accelerated brain growth, or other factors responsible for brain enlargement, from 2 to 4 years of age, which then levels off after approximately 6 years of age [15, 18]. In a meta-analysis using HC (converted to brain volume), brain volumes measured from MRI, and brain weight from autopsy studies, it was found that brain size changes from 13% smaller than controls at birth to 10% greater than controls at 1 year, and only 2% greater by adolescence [24]. Within individual subjects, the growth curve can be quite variable. Furthermore, although the distribution of brain size within autistic samples across the age range remains to be further characterized, imaging and HC studies, as well as the original observations by Kanner [25], clearly support the idea that some adults with ASD also have enlarged brains or macrocephaly [19, 26, 27, 28, 29, 30]. What remains unclear is what mechanisms might account for this apparent brain overgrowth [31].

Numerous factors have been postulated to underlie the brain volume increases in ASD, including theories of accelerated early brain growth [15, 32], or a failure of apoptosis and/or synaptic pruning [33]. Although the heterogeneity of brain volumes in clinical samples precludes an assessment of underlying factors at the level of the individual, as has been approached in histological inquiry [26, 34], evaluating micro-structural features within groups may provide some insight into what mechanisms could account for such brain enlargement.

Our laboratory's interest has been to systematically apply magnetic resonance (MR) methods to characterize MRI observations of brain anatomical variations, which may result from underlying cellular alterations, in children with ASD. Two main MR approaches, MR spectroscopy (MRS) quantification of brain chemical concentrations and quantitative measures of chemical and water transverse (T2) relaxation, will be further described. These non-structural imaging approaches can provide a non-invasive, quantitative appraisal of the cytoarchitecture, developmental stage, and brain tissue chemical composition that underlie brain alterations in ASD.

Magnetic Resonance Spectroscopy

MRS provides a non-invasive method for characterizing tissue-based chemistry and cellular features in vivo [35]. Although MRI is sensitive to changes in tissue water characteristics used to define anatomy at a macroscopic level, it is less specific to what may be occurring at a cellular level. In this regard, MRS has been used to detect abnormalities in brain regions that are normal appearing by MRI, as well as to elucidate pathology underlying MRI-visible abnormalities [36, 37, 38, 39, 40]. Numerous studies have utilized MRI to investigate autism, but there have been relatively few using MRS.

In brain, the MRS-visible concentrations and mobility of low-molecular weight chemicals can be measured as spectral peaks, allowing noninvasive characterization of tissue-based chemical or metabolic abnormalities in specific neuropsychiatric disorders [41, 42, 43]. It is beyond the scope of this chapter to describe in detail, but certain atomic nuclei possess magnetic properties because of unpaired protons or neutrons that have potential neuropsychiatric research applications, including hydrogen (^{1}H), phosphorus (^{31}P), and fluorine (^{19}F) [42]. MRS investigations using these nuclei have all been undertaken in the field of ASD research, but the majority of work, and the focus of this chapter, has been applications of ^{1}H MRS. Quantifiable chemicals by ^{1}H MRS include N-acetyl aspartate (NAA, most often measured as the total of NAA + N-acetyl aspartyl glutamate), creatine (CRE; composed of creatine and phosphocreatine), choline (CHO; includes multiple resonances primarily consisting of four membrane/myelin-related chemicals: phosphorylethanolamine, phosphorylcholine, glycerophosphorylethanolamine, and glycerophosphorylcholine), myo-Inositol (mI) and lactate (LAC; difficult to detect at rest in the normal brain at 1.5 T). Glutamate, γ-aminobutyric acid (GABA), and glutamine have complicated peak shapes and resonate at overlapping spectral locations, resulting in the use of "GLX" as a common description of the combined peaks [41].

Found only in the nervous system, and primarily in neurons, NAA appears to be a sensitive marker for both neuronal integrity and neuronal–glial homeostasis, as there is a complex NAA shuttle between neurons and oligodendrocytes [44, 45, 46, 47, 48]. Presumed roles for NAA include regulating synaptogenesis, synaptic maintenance, myelination, the regulation of cellular osmolarity, and neuronal metabolism. During normal development, NAA increases dramatically over the first year of life then at a slower rate during the next 2–3 years, gradually plateauing into early adulthood [41]. Increases in NAA, paralleling myelination, are thought to reflect neuronal maturation because of increasing synaptic complexity as well as increased dendritic and axonal projections [41, 49, 50, 51]. The spectral CHO peak reflects only the mobile fraction of choline-containing compounds present in tissue, with the larger fraction of choline being incorporated into molecules comprising the cell membranes that are MR-invisible [41]. In contrast to NAA, CHO levels decrease rapidly over the first year of life, also paralleling myelination and reflecting decreases in the PE and PC components of the CHO peak [41, 49, 50, 51]. Bulk CRE in the proton spectrum provides some index of energy metabolism, although ^{31}P MRS provides an unambiguous measure of phosphocreatine, a more straightforward measure of high-energy phosphate level [36, 41, 49]. Lastly, mI is an important regulator of brain osmotic balance and, as a precursor for phosphoinositides, involved in the cell membrane second messenger system [52].

MRS measurements of NAA reduction have been reported to be a marker of brain abnormalities in children at risk for DD. For example, in one study of neurologically at-risk children, those who subsequently manifested DD had reduced NAA/CHO ratios; MRS measures proved to be a better predictor of outcome than structural MRI [51]. Further evidence of the potential utility of NAA comes from a relationship demonstrated between concentrations and IQ in healthy adult samples [53]. In contrast to decreased NAA, ^{1}H MRS visible CHO increases, at least acutely, in most pathological states with cellular disruption as a result of membrane and myelin breakdown [36, 37, 54]. LAC has an important role in brain bioenergetics and is often elevated in the setting of inborn errors of metabolism or hypoxia but also under conditions of moderate reduction in cerebral blood flow and/or increased brain metabolism as occurs, for example, as a consequence of hyperventilation or caffeine ingestion [36, 55, 56, 57]. Small LAC elevations, in conjunction with other chemical alterations, such as GLX elevation, may also reflect subtle impairment of brain energetics, such as a shift in redox state, providing a potential marker for mitochondrial compromise in specific psychiatric disorders, such as bipolar disorder [40].

Comparison of ^{1}H MRS findings across published studies of ASD is challenging because of the relatively small sample sizes, differences in methodology of MRS acquisition and brain regions studied, and differences in populations sampled (e.g., age range, autistic subtype, IQ, and health) [33, 58, 59, 60, 61, 62, 63, 64, 65, 66, 67, 68, 69, 70]. However, patterns of abnormal ^{1}H MRS found in many, although not all, studies of ASD suggest differences in tissue chemical composition or neuronal integrity. There is also some suggestion of age-related MRS differences. For example, reduced NAA levels tend to be more consistently found in younger populations with ASD [33, 59, 60, 61, 62, 63, 64, 66, 68, 69], although some studies of older populations also report reduced levels of NAA [70].

^{1}H MRS studies of autism have generally employed single-voxel techniques that acquire information from a single brain area, typically of large volume (on the order of $2 \times 2 \times 2 = 8 \text{ cm}^3$) of interest. Most ^{1}H MRS studies of ASD have been performed using long-echo time acquisition parameters because of technical feasibility. Very-short echo times can be acquired that allow quantitative assessment of chemicals with short T2 not present at long echo times (e.g., mI and GLX), and alleviating some relaxation evolution that can present challenges for chemical quantification [33, 71]. Furthermore, acquiring both short and long echo chemical measurements can provide an estimate of chemical transverse relaxation (T2 relaxation), a measure of a molecule's mobility within its cellular environment [33].

The implementation of multi-voxel MRS imaging techniques (MRSI), also referred to as chemical shift imaging, has allowed greater spatial coverage of implicated brain regions in ASD. MRSI techniques can be used to acquire 2-D or 3-D voxel arrays for quantitative chemical assessment and to systematically map regional brain chemistry, similar to mapping the distribution of water protons in a proton density MRI. Rapid ^{1}H MRSI techniques, such as proton echo-planar spectroscopic imaging (PEPSI) [72], allow multiple regions to be simultaneously acquired in clinically reasonable time frames (e.g., 1024 voxels in 5–8 min), making such an approach feasible in studies of sleeping children. Fig. 17.1 (see Color Plate 6, following p. XX) shows a 16×16 MRSI acquisition from a Philips 3T scanner wherein the superimposed NAA chemical map and characteristic spectra from gray and white matter voxels are demonstrated. In Fig. 17.2 (see Color Plate 7, following p. XX), a representative PEPSI slab location (20 mm thick) is overlaid by the 32×32 PEPSI axial voxel matrix. Coupled with volumetric MRI using regression analysis, one also can take advantage of the large number of voxel samples obtained to calculate estimates of "pure" gray or white matter neurochemistry [40, 66], as depicted for a single subject in Fig. 17.3.

The first published ^{1}H MRS study of autism [73] was a single-voxel study, using long echo-acquisition parameters, of a small cohort of children and adolescents with autism that sampled the right parieto-occipital white matter region with no alterations detected in metabolite ratios (e.g., NAA/CRE, CHO/ CRE, NAA/CHO). Using short echo-acquisition methods, a subsequent single-voxel ^{1}H MRS study detected reduced NAA (referenced to brain water) in the right medial temporal lobe region and left cerebellar hemisphere for a sample of individuals with autism who ranged in age from childhood to early adulthood [59]. Findings reported by this group for an expanded sample of children and adults with autism, ranging in age from 2 to 21 years, and studied using single-voxel ^{1}H MRS techniques to sample multiple brain regions (frontal, parietal, temporal, brain stem, and cingulate) also revealed reduced NAA in the temporal lobes bilaterally, more consistently observed for younger subjects [60, 61]. Although one study to date [58], using single-voxel techniques, has reported elevated brain LAC in an individual subject with autism, possibly reflecting the involvement of mitochondrial abnormalities, this finding has not been observed in other studies.

^{1}H MRSI techniques have been used to date in three studies of ASD [33, 66, 63, 68]. Using a combination of both short-echo (20-ms) and long-echo (272-ms) acquisition parameters, in conjunction with water referencing, we utilized PEPSI to study a sample of forty-five 3- to 4-year-old children with ASD compared with age-matched control groups of delayed development (DD) and typically developing (TD) children [33, 66]. The ASD sample exhibited a widespread regional pattern of reduced CHO, CRE, NAA, and mI concentrations with particular involvement of frontal, cingular, thalamic, insular, superior and medial temporal, callosal, and parietal regions [33]. Prolonged chemical T2 expressed as a percentage relative to the TD group (T2r) were also observed for CHO, CRE, and NAA

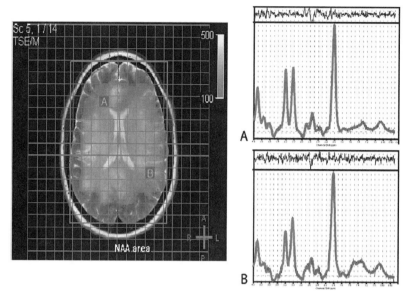

Fig. 17.1 Example of MRS imaging techniques (MRSI) acquisition overlaid with NAA chemical map; representative gray matter **(A)** and white matter **(B)** spectra on right (*see* Color Plate 6)

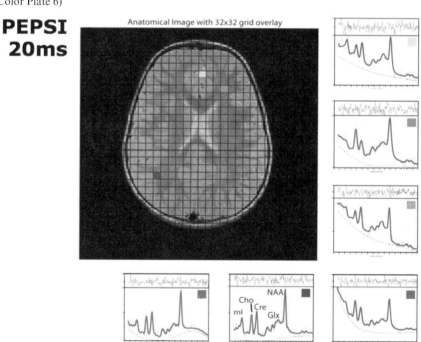

Fig. 17.2 A magnetic resonance spectroscopy (MRS) imaging techniques (MRSI) slab location (20 mm thick) overlaid by the 32×32 MRSI axial matrix (*see* Color Plate 7)

Neurochemical x Tissue Type

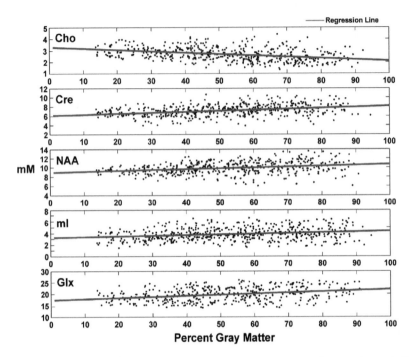

Fig. 17.3 Regression techniques using the percentage of gray matter from each voxel allows estimates of "pure" gray and white matter chemical concentrations to be calculated, as depicted for a single autism spectrum disorder (ASD) subject

across the same widely distributed brain regions [33]. Overall, the direction and widespread distribution of chemical alterations observed was not consistent with theoretical models of diffusely increased neuronal packing density in the children with ASD. In that study population, and as described further on, tissue segmentation and application of regression techniques further identified predominantly gray matter involvement for the pattern of reduced chemical concentrations and prolonged chemical T2 relaxation [66]. A subsequent ^{1}H MRSI study, from another group but also using similar regression techniques, replicated findings of gray matter, but not white matter, reductions in NAA [68]. An additional ^{1}H MRSI study conducted at a long TE (272 ms) and using a multisectional approach with quantification based on acquisition parameters, examined 22 children and adolescents with autism compared with age-matched healthy control subjects and found decreased CHO levels within the left anterior cingulate, left caudate, and occipital regions, and increased levels of CHO and CRE within the right caudate nucleus [63].

Whereas two studies in older children [63] or adults [62] with high functioning autism or Asperger syndrome have found localized brain regions of increased neurochemical concentrations, including NAA in the later study, the majority of studies have found decreased neurochemical concentrations [33, 59, 60, 61, 66, 68, 69, 70] or no differences [58, 64,65]. In our work, we did not observe increased levels of any neurochemicals in preschool-aged children with ASD [33, 66]. Although differences in ASD samples (e.g., high versus low IQ and age-range) and methodological factors (e.g., short versus long echo acquisitions) represent possible methodological considerations, it is notable that the two regions identified with increased neurometabolites in those studies (L caudate and R frontal lobe) were not found to be reduced in our initial report of a younger sample [33], and leaves open the possibility that these regions show increasing levels with age. In this context, a recent single-voxel ^{1}H MRS report studying an adult sample with ASD has reported elevated GLX in the amygdala-hippocampal region [67]. Although this finding in older individuals with ASD remains to be replicated, in the context of a recent study of pre-adolescents with ASD that reported lower GLX levels [68], increasing GLX levels with increasing age has implications for seizure onset in adolescents with ASD and for theories of glutamate imbalance in this population [74, 75] (see Chapter 6 by Evers & Hollander).

Testing models of apoptotic/synaptic pruning deficits in ASD

Combining MRSI with MRI can allow hypothesis-driven approaches toward a better appreciation of mechanisms underlying specific anatomical alterations. To account for the approximately 10% cerebral enlargement observed in children with ASD at 3–4 years of age [16], our PEPSI study from the same cohort of children tested an exploratory hypothesis of regionally increased NAA concentrations posited to reflect increased or more densely packed neurons or connections arising from disturbances of normal neuronal apoptosis or synaptic pruning during early development [33]. In this model, molecular T2 relaxation times were also predicted to be shortened as a result of these densely packed molecules having increased interactions; thus, a faster dispersion from the coherent magnetized state, following excitation by the RF pulse. Instead, findings described above of regional reductions in NAA and other tissue-based neurochemicals, as well as increased neurochemical T2, indicative of increased molecular mobility, were in direct contrast to these a priori predictions [33]. Combining neurochemical measurements with structural MRI data, we localized these neurochemical alterations primarily to the gray matter, distinct from age-matched non-autistic DD children [66]. As summarized in Table 17.1, our findings support an altered gray matter neurochemical environment for ASD children at 3–4 years of age. Specifically, ASD children demonstrated overall reductions in gray matter NAA, CHO, CRE, and mI compared with TD controls, and decreased NAA, CHO, and mI relative to the DD children. Decreased white matter NAA and mI were also observed for both ASD and

Table 17.1 ASD brain chemical relationships at age 3–4 years

3–4 Years of Age		Statistics				Post-Hoc Tests		
Tissue	Metabolite	F	Covariates		P	ASD-TD	DD-TD	ASD-DD
			Gender	CBV				
Gray	Cho	19.97	–	0.07	<0.001	−17.42	−11.46	−6.73
	Cre	2.93	–	–	0.061	−6.51		
	NAA	3.69	–	–	0.031	−4.84		−3.54
	ml	6.12	–	–	0.004	−10.30		−10.63
	Glx	0.84	–	–	–			
	Lac	0.05	–	–	–			
White	Cho	1.18	–	–	–			
	Cre	1.40	–	–	–			
	NAA	3.07	–	–	0.054	−5.08	−6.41	
	ml	2.44	0.093	–	0.096	−10.68	−12.32	
	Glx	1.98	–	–	–			
	Lac	1.25	–	–	–			
T2r-Gray	Cho	3.67	–	–	0.031	12.37		
	Cre	0.63	–	0.008	–			
	NAA	0.68	–	0.001	–			
T2r-White	Cho	2.01	0.042	–	–			
	Cre	1.86	–	0.007	–			
	NAA	1.51	–	0.033	–			

CBV = Cerebral brain volume

DD compared with TD children. These findings suggest that cellular densities are not increased at this age point and, thus, do not support models of failed or reduced apoptosis and/or delayed synaptic pruning, nor provide evidence for neuronal "overgrowth," during the preschool years of development soon after emergence of clinical symptoms diagnostic for ASD.

Testing Models of Accelerated Brain Growth in ASD

If cellular densities are not increased in ASD, what features might underlie these larger brains? One emerging idea has been that postpartum brain growth in ASD may be on a more rapid trajectory, which then peaks earlier than in TD children [15, 19, 20, 21]. In this model, an increase in brain volume of children with ASD is posited to occur because of accelerated "normal" brain growth between birth and the first 4 years of life, after which a plateau in growth or development takes place, allowing for a relative "catch up" growth of TD children. Although this model does not speak directly to the cellular features underlying enlarged cerebral volumes, understanding whether children with ASD demonstrate physiological changes consistent with accelerated growth may inform possible mechanisms.

In our 3- to 4-year-old sample, we speculated that a specific MRI tissue property, quantitative water T2 relaxation, could help to shed light on whether the brains of young children with ASD showed accelerated development. Two types of relaxation can be measured with MRI: T1 (longitudinal or spin–lattice relaxation, which is determined by the magnetic field strength and the composition of the surrounding lattice) and T2 (transverse or spin–spin relaxation that reflects the loss of phase coherence as protons precess, primarily as a result of interactions with adjacent protons from nearby tissues) [35]. Both can provide a biomarker of neuronal development and myelination. In a newborn child, gray matter is hypointense relative to white matter on a T2-weighted image, whereas the reverse is true on a T1-weighted image. Brain development is characterized by shortening of both T1 and T2 relaxation times, with corresponding intensity changes on MR images. Other evolving physical phenomena during this time period, such as increasing magnetization transfer that reflects the exchange between pools of highly mobile-free water protons and less mobile macromolecule-bound protons, also affect the observed signal intensities. The increasing concentrations in myelin of cholesterol and glycolipids, which contain hydroxyl and ketone moieties that interact with free water, have been attributed to early T1 shortening on the MR image [76]. In the first few months of development, the evolution of relative gray and white matter T1 image intensities occurs more rapidly than for T2. There is generally a reversal of the relative gray and white matter T1 intensities to the adult pattern by approximately 6 months of age, although contrast continues to increase between gray and white matter beyond this age. T2 image intensity changes, which also reverse as a result of decreasing T2 relaxation times, occur over a longer time frame.

The mechanisms underlying changes in T2 relaxation are complex, but largely attributed to chemical maturation of the myelin sheath, saturation of fatty acids within myelin membranes, displacement of free water in the extracellular space because of increases in axonal diameter, the elaboration of neuronal anatomy (such as synaptogenesis and dendritic arborization), and the development of glial cells [76]. T2 relaxation changes during early life brain development likely occur as an exponential or bi-exponential process, with rapid shortening in the first 6 months of life and, thereafter, a slower progression until approximately 18–24 months of age [77]. By that age, relative gray and white matter intensities have reversed in comparison with the MRI of a newborn, and white matter, because of greater T2 shortening, has become hypointense relative to gray matter. By about 18 months of age, a child's MRI is qualitatively similar to what is observed in an adult because most myelination, and much of the age-related T2 intensity changes, have already occurred, although more subtle changes continue into adulthood. These relative T2 intensity changes in gray and white matter have been used to track the course of brain development in healthy children [78, 79] and have provided evidence of delayed myelination in children with developmental delay [80].

To test the hypothesis of accelerated early brain growth in ASD, we quantitatively assessed the T2 relaxation of brain water, using the well-documented increasing restriction of mobility during development [77, 78, 79, 81] as a temporal marker for the stage of brain maturation. In this model, reduced water T2 relaxation (shortened T2) would be expected if there were accelerated early brain maturation in association with ASD. Given the evidence for gray matter chemical abnormalities in young children with ASD [66], T2 relaxation was measured separately in both gray and white matter. This necessitated segmentation to delineate gray matter from white matter, and exclude CSF, to provide T2 measures of compartmentalized tissue water content, and assist in measuring regions of interest [82].

In adopting this approach of using quantitative water T2 relaxation as a relative measure of brain developmental stage, we quantitatively assessed global white matter and cortical gray matter T2 relaxation in children with ASD compared with age-matched populations of TD children and children with developmental delay. These children comprised an expanded sample of the 3- to 4-year-old children whose T1 volumetric and spectroscopic imaging data were previously described [16, 33, 66]. We hypothesized that if the enlarged cerebral volume previously reported in ASD was due to accelerated normal brain growth, gray and white matter T2 in the ASD children would be decreased relative to the TD children, reflecting more advanced brain maturation. The DD children would be expected to demonstrate prolonged T2 relaxation times in conjunction with delayed brain development.

In contrast to what would have been expected if their larger brains represented acceleration of normal brain growth, gray matter T2 was *prolonged* in the ASD children when comparing group means at 2–4 years of age with age-matched control groups (Fig. 17.4; see ref [82]). Stratifying the samples by age, it becomes apparent that the increased T2 observed in the ASD group is consistent across this age range, suggesting that the underlying maturational process parallels but lags the normal developing brain (Fig. 17.5).

These observations and other reports similarly implicating gray matter alterations in ASD [83, 84], and as well our findings of disproportionate decreases in gray matter neurochemical concentrations and increased T2 relaxation [66], support the possibility that gray matter is critically affected in ASD children at

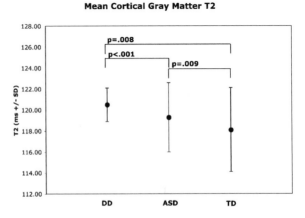

Fig. 17.4 Cortical gray matter age-adjusted T2 relaxation (mean ± SD) for 2- to 4-year-old children with autism spectrum disorder (ASD), idiopathic developmental delay (DD), or typical development (TD) [ref 82]

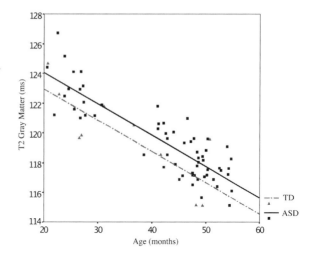

Fig. 17.5 Age relationships for cortical gray matter T2 relaxation from individual subjects with autism spectrum disorder (ASD) or typical development (TD)

this early age. Although these combined findings support a primary gray matter process in ASD, there remains the possibility of greater white matter involvement later in development. In support of this consideration, a recent report, while observing regional gray matter T2 increases, found an overall preponderance of altered white matter T2 relaxation in localized regions of the brain for older children with ASD [85]. Alternatively, if pathology in ASD is specific to the neuron itself and not the myelin, T2 findings in white matter may be masked by the volume of the myelin, at least in measurements of whole brain T2.

There is mounting evidence for other abnormal neurochemical and histologic processes early in the course of autism that could produce an aberrant neuronal structural configuration and account for the differential gray matter T2 findings in the ASD children. Deficits in the neuroregulatory proteins Reelin and Bcl-2 implicate abnormal pathways for neuronal migration or synaptic plasticity in autism [86, 87]. Additionally, increased levels of whole blood and platelet serotonin, reported in an estimated 30% of affected individuals with ASD, may lead to depleted development of sertonergic fibers, decreased numbers of serotonin terminals, and decreased synaptogenesis [88]. In the postnatal developing brain, decreased serotonin synthesis activity has also been linked to aberrations in cortical columnar development and connectivity [89]. Consistent with this observation are postmortem studies of autism that report higher numbers of cortical minicolumns, which are smaller in size, but with greater intercellular dispersion [90]. These accumulated observations in autism converge toward a model that may be characterized most simply as increased gray matter with a reduced unit density. As a possible mediator or alternative consideration, our evidence suggests early inflammatory processes in autism that could also contribute to these measured findings [82]. For example, microglial activity and elevated cytokines in various brain regions have been reported across a broad age range of individuals with autism [91]. Additionally, increases in glial fibrillary acidic protein found in postmortem samples of frontal and parietal cerebral cortex and cerebellum from older (>18 years old) individuals with autism also suggest the presence of immune activity [92]. There is more evidence for abnormal innate immune system activity in ASD, both peripherally and in the CNS, as opposed to adaptive immunity, with implications for how the immune system might be involved with developmental processes [93, 94, 95]. Future work merging immune profiles and MRI biomarkers will be helpful to examine these relationships in vivo.

Summary

Although the cellular mechanisms underlying brain changes in ASD remain unknown, the available in vivo imaging data have helped to inform some potential hypotheses. At the time of our initial MRS study, which began in 1996, we hypothesized that brain enlargement in ASD might be related to abnormal apoptosis and/or synaptic pruning, which would manifest as increased neurochemical concentrations and shortened metabolite relaxation. Instead, our MRS findings supported an opposite cellular environment, leading to much speculation about what alternative processes might be occurring [33]. Similarly, with accumulating evidence that children with ASD had normal HCs at birth that became enlarged between 3 and 4 years of age, it was hypothesized that accelerated brain growth was occurring [21]. Using T2 relaxation to assess the temporal progression of development in ASD children, we hypothesized that decreased water relaxation might support this theory. Instead, our observations of elevated water T2 relaxation measures were more consistent with normal or delayed brain growth at this developmental interval [82]. Combined, these MRS and quantitative water T2 relaxation findings in ASD at the 2- to 4-year-old age range suggest atypical cellular features that may be characterized by reduced synapse density [96, 97], column density/packing abnormalities [90], and/or active inflammatory processes such as gliosis and edema [91]. Multivariate assessment of several modalities (phenotypes, imaging parameters, immunology, genetics) and description of changes across the developmental course in future work will be critical to extend and refine these observations.

References

1. *Diagnostic and Statistical Manual of Mental Disorders*, 4th ed. Washington, DC: American Psychiatric Association; 1994.
2. Lord C, Risi S, DiLavore PS, Shulman C, Thurm A, Pickles A: Autism from 2 to 9 years of age. *Arch Gen Psychiatry* 2006;63:694–701.
3. Zwaigenbaum L, Bryson S, Rogers T, Roberts W, Brian J, Szatmari P: Behavioral manifestations of autism in the first year of life. *Int J Dev Neurosci* 2005;23:143–152.

4. Landa RJ, Holman KC, Elizabeth Garrett-Mayer E: Social and communication development in toddlers with early and later diagnosis of autism spectrum disorders. *Arch Gen Psychiatry* 2007;64:853–864.
5. Rapin I, Katzman R: Neurobiology of autism. *Ann Neurol* 1998;43:7–14.
6. DiCicco-Bloom E, Lord C, Zwaigenbaum L, Courchesne E, Dager SR, Schmitz C, Schultz RT, Crawley J, Young LJ: The developmental neurobiology of autism spectrum disorder. *J Neurosci* 2006;26:6897–6906.
7. Gillberg C, Coleman M: Autism and medical disorders: a review of the literature. *Dev Med Child Neurol* 1996;38:191–202.
8. Dawson G, Munson J, Estes A, Osterling J, McPartland J, Toth K, Carver L, Abbott R: Neurocognitive function and joint attention ability in young children with autism spectrum disorder versus developmental delay. *Child Dev* 2002; 73:345–358.
9. Bachevalier J: Medial temporal lobe structures and autism: a review of clinical and experimental findings. *Neuropsychologia* 1994;32:627–648.
10. Chakrabarti S, Fombonne E: Pervasive developmental disorders in preschool children: confirmation of high prevalence. *Am J Psychiatry* 2005 Jun;162(6):1133–1141.
11. Volkmar FR, Nelson DS: Seizure disorders in autism. *J Am Acad Child Adolesc Psychiatry* 1990;29:127–129.
12. Giovanardi R, Posar A, Parmeggiani A: Epilepsy in adolescents and young adults with autistic disorder. *Brain Dev* 2000;22:102–168.
13. Dawson G, Webb S, Schellenberg G, Dager SR, Friedman SD, Aylward E, Richards T: Defining the broader phenotype of autism: genetic, brain, and behavioral perspectives. *Dev Psychopathol* 2002;14:581–611.
14. Lyon GR, Rumsey JM: *Neuroimaging Studies of Autism: Neuroimaging: A Window to the Neurological Foundations of Learning and Behavior in Children*. Baltimore: Paul H. Brookes Publishing Co; 1996.
15. Courchesne E, Karns CM, Davis HR, Ziccardi R, Carper RA, Tigue ZD, Chisum HJ, Moses P, Pierce K, Lord C, Lincoln AJ, Pizzo S, Schreibman L, Haas RH, Akshoomoff NA, Courchesne RY: Unusual brain growth patterns in early life in patients with autistic disorder: an MRI study. *Neurology* 2001;57:245–254.
16. Sparks BF, Friedman SD, Shaw DW, Aylward EH, Echelard D, Artru AA, Maravilla KR, Giedd JN, Munson J, Dawson G, Dager SR: Brain structural abnormalities in young children with autism spectrum disorder. *Neurology* 2002;59:184–192.
17. Hazlett HC, Poe M, Gerig G, Smith RG, Provenzale J, Ross A, Gilmore J, Piven J: Magnetic resonance imaging and head circumference study of brain size in autism: birth through age 2 years. *Arch Gen Psychiatry* 2005;62:1366–1376.
18. Aylward EH, Minshew NJ, Field K, Sparks BF, Singh N: Effects of age on brain volume and head circumference in autism. *Neurology* 2002;59:175–183.
19. Lainhart JE, Piven J, Wzorek M, Landa R, Santangelo SL, Coon H, Folstein SE: Macrocephaly in children and adults with autism. *J Am Acad Child Adolesc Psychiatry* 1997;36:282–290.
20. Dementieva YA, Vance DD, Donnelly SL, Elston LA, Wolpert CM, Ravan SA, DeLong GR, Abramson RK, Wright HH, Cuccaro ML: Accelerated head growth in early development of individuals with autism. *Pediatr Neurol* 2005;32(2):102–108.
21. Courchesne E, Carper R, Akshoomoff N: Evidence of brain overgrowth in the first year of life in autism. *JAMA* 2003;290:337–344.
22. Dawson G, Munson J, Webb SJ, Nalty T, Abbott R, Toth K: Deceleration in rate of head growth and decline in skills in the second year of life in autism. *Biological Psychiatry* 2007;61(4):458–464.
23. Fukumoto A, Hashimoto T, Ito H, Nishimura M, Tsuda Y, Miyazaki M, Mori K, Arisawa K, Kagami SJ: Growth of head circumference in autistic infants during the first year of life. *J Autism Dev Disord* 2007 Jul 24; [Epub ahead of print].

24. Redcay E, Courchesne E: When is the brain enlarged in autism ? A meta-analysis of all brain size reports. *Biol Psychiatry* 2005 Jul 1;58(1):1–9.
25. Kanner L: Autistic disturbance of affective contact. *Nerv Child* 1943;2:217–250.
26. Bailey A, Luthert P, Dean A, Harding B, Janota I, Montgomery M, Rutter M, Lantos P: A clinicopathological study of autism. *Brain* 1998; 121(Pt 5):889–905.
27. Piven J, Arndt S, Bailey J, Havercamp S, Andreasen NC, Palmer P: An MRI study of brain size in autism. *Am J Psychiatry* 1995;152:1145–1149.
28. Aylward EH, Minshew NJ, Goldstein G, Honeycutt NA, Augustine AM, Yates KO, Barta PE, Pearlson GD: MRI volumes of amygdala and hippocampus in non-mentally retarded autistic adolescents and adults. *Neurology* 1999;53:2145–2150.
29. Hardan AY, Minshew NJ, Mallikarjuhn M, Keshavan MS: Brain volume in autism. *J Child Neurol* 2001;16:421–424.
30. Lotspeich LJ, Kwon H, Schumann CM, Fryer SL, Goodlin-Jones BL, Buonocore MH, Lammers CR, Amaral DG, Reiss AL: Investigation of neuroanatomical differences between autism and Asperger syndrome. *Arch Gen Psychiatry* 2004;61(3):291–298.
31. Herbert MR: Large brains in autism: the challenge of pervasive abnormality. *Neuroscientist* 2005;11:417–440.
32. Courchesne E, Redcay E, Kennedy DP: The autistic brain: birth through adulthood. *Curr Opin Neurobiol* 2004;17:489–496.
33. Friedman SD, Shaw DW, Artru AA, Richards TL, Gardner J, Dawson G, Posse S, Dager SR: Regional brain chemical alterations in young children with autism spectrum disorder. *Neurology* 2003;60:100–107.
34. Bauman M, Kemper TL: Histoanatomic observations of the brain in early infantile autism. *Neurology* 1985;35:866–874.
35. Gadian DF: *Nuclear Magnentic Resonance and its Applications to Living Systems*. New York: Oxford University Press, 1982.
36. Ross B, Michaelis T. Clinical applications of magnetic resonance spectroscopy. *Magn Reson Med* 1994;10:191–247.
37. Friedman SD, Brooks WM, Jung RE, Chiulli SJ, Sloan JH, Montoya BT, Hart BL, Yeo RA. Quantitative proton MRS predicts outcome after traumatic brain injury. *Neurology* 1999;52:1384–1391.
38. Friedman SD, Brooks WM, Stidley CA, Hart BL, Sibbitt WL Jr. Brain injury and neurometabolic abnormalities in systemic lupus erythematosus. *Radiology* 1998;209:79–84.
39. Dager SR, Friedman SD, Heide A, Layton ME, Richards T, Artru A, Strauss W, Hayes C, Posse S: Two-dimensional proton echo-planar spectroscopic imaging of brain metabolic changes during lactate-induced panic. *Arch Gen Psychiatry* 1999;56:70–77.
40. Dager SR, Friedman SD, Parow A, Demopulos C, Stoll AL, Lyoo IK, Dunner DL, Renshaw PF: Brain metabolic alterations in medication-free patients with bipolar disorder. *Arch Gen Psychiatry* 2004;61:450–458.
41. Kreis R, Ernst T, Ross BD: Development of the human brain: in vivo quantification of metabolite and water content with proton magnetic resonance spectroscopy. *Magn Reson Med* 1993;30:424–437.
42. Dager SR, Steen RG: Applications of magnetic resonance spectroscopy to the investigation of neuropsychiatric disorders. *Neuropsychopharmacology* 1992;6:249–266.
43. Lyoo IK, Renshaw PF. Magnetic resonance spectroscopy: current and future applications in psychiatric research. *Biol Psychiatry* 2002;51:195–207.
44. Tallan HH, Moore S, Stein WH: N-Acetyl-L-aspartic acid in brain. *J Biol Chem* 1956;219:257–264.
45. Birken DL, Oldendorf WH. N-acetyl-l-aspartic acid: a literature review of a compound prominent in 1H NMR spectroscopic studies of brain. *Neurosci Biobehav Rev* 1989;13:23–31.
46. Coyle JT, Schwarcz R: Mind glue: implications of glial cell biology for psychiatry. *Arch Gen Psychiatry* 2000;57:90–93.

47. Baslow MH: Functions of N-acetyl-L-aspartate and N-acetyl-L-aspartylglutamate in the vertebrate brain: role in glial cell-specific signaling. *J Neurochem* 2000;75:453–459.

48. Neale JH, Bzdega T, Wroblewska B: N-Acetylaspartylglutamate: the most abundant peptide neurotransmitter in the mammalian central nervous system. *J Neurochem* 2000;75:443–452.

49. van der Knaap MS, van der Grond J, van Rijen PC, Faber JA, Valk J, Willemse K: Age-dependent changes in localized proton and phosphorus MR spectroscopy of the brain. *Radiology* 1990;176:509–515.

50. Huppi PS, Posse S, Lazeyras F, Burri R, Bossi E, Herschkowitz N: Magnetic resonance in preterm and term newborns: 1H-spectroscopy in developing human brain. *Pediatr Res* 1991;30:574–578.

51. Kimura H, Fujii Y, Itoh S, et al. Metabolic alterations in the neonate and infant brain during development: evaluation with proton MR spec- troscopy. *Radiology* 1995;194:483–489.

52. Moore CM, Breeze JL, Kukes TJ, Rose SL, Dager SR, Cohen BM, Renshaw PF: Effects of myo-inositol ingestion on human brain myo-inositol levels: a proton magnetic reso-nance spectroscopic imaging study. *Biol Psychiatry* 1999;45:1197–1202.

53. Jung RE, Yeo RA, Love TM, Petropoulos H, Sibbitt WL, Jr., Brooks WM: Biochemical markers of mood: a proton magnetic resonance spectroscopy study of normal human brain. *Biol Psychiatry* 2002;51:224–229.

54. Friedman SD, Brooks WM, Jung RE, Hart BL, Yeo RA: Proton MR spectroscopic findings correspond to neuropsychological function in traumatic brain injury. *AJNR Am J Neuroradiol* 1998;19:1879–1885.

55. Behar KL, den Hollander JA, Stromski ME, Ogino T, Shulman RG, Petroff OA, Prichard JW: High-resolution 1H nuclear magnetic resonance study of cerebral hypoxia in vivo. *Proc Natl Acad Sci USA* 1983;80:4945–4948.

56. Dager SR, Strauss WL, Marro KI, Richards TL, Metzger GD, Artru AA: Proton magnetic resonance spectroscopy investigation of hyperventilation in subjects with panic disorder and comparison subjects. *Am J Psychiatry* 1995;152:666–672.

57. Dager SR, Layton ME, Strauss W, Richards TL, Heide A, Friedman SD, Artru AA, Hayes CE, Posse S: Human brain metabolic response to caffeine and the effects of tolerance. *Am J Psychiatry* 1999;156:229–237.

58. Chugani DC, Sundram BS, Behen M, Lee ML, Moore GJ. Evidence of altered energy metabolism in autistic children. *Prog Neuropsychopharmacol Biol Psychiatry* 1999;23:635–641.

59. Otsuka H, Harada M, Mori K, Hisaoka S, Nishitani H. Brain metabo- lites in the hippocampus-amygdala region and cerebellum in autism: an 1 H-MR spectroscopy study. *Pediatr Neuroradiol* 1999;41:517–519.

60. Hisaoka S, Harada M, Nishitani H, Mori K. Regional magnetic reso-nance spectroscopy of the brain in autistic individuals. *Neuroradiology* 2001;43:496–498.

61. Mori K, Hashimoto T, Harada M, Yoneda Y, Shimakawa S, Fujii E, Yamaue T, Miyazaki M, Saijo T, Kuroda Y. Proton magnetic resonance spectroscopy of the autistic brain [in Japanese]. *No To Hattatsu* 2001;33:329–335.

62. Murphy DG, Critchley HD, Schmitz N, McAlonan G, Van Amelsvoort T, Robert- son D, Daly E, Rowe A, Russell A, Simmons A, Murphy KC, Howlin P. Asperger syndrome: a proton magnetic resonance spectroscopy study of brain. *Arch Gen Psychiatry* 2002;59:885–891.

63. Levitt JG, O'Neill J, Blanton RE, Smalley S, Fadale D, McCracken JT, Guthrie D, Toga AW, Alger JR: Proton magnetic resonance spectroscopic imaging of the brain in child-hood autism. *Biol Psychiatry* 2003;54:1355–1366.

64. Fayed N, Modrego PJ. Comparative study of cerebral white matter in autism and attention- deficit / hyper activity disorder by means of magnetic resonance spectroscopy. *Acad Radiol* 2005;12:566–569.

65. Zeegers M, van der Grond J, van Daalen E, Buitelaar J, van Engeland H: Proton magnetic resonance spectroscopy in developmentally delayed young boys with or without autism. *J Neural Transm* 2007;114(2):289–295.

66. Friedman SD, Shaw DW, Artru AA, Dawson G, Petropoulos H, Dager SR: Gray and white matter brain chemistry in young children with autism. *Arch Gen Psychiatry* 2006;63:786–794.

67. Page LA, Daly E, Schmitz N, Simmons A, Toal F, Deeley Q, Ambery F, McAlonan GM, Murphy KC, Murphy DG: In vivo 1H-magnetic resonance spectroscopy study of amygdala-hippocampal and parietal regions in autism. *Am J Psychiatry* 2006;163(12):2189–2192.

68. DeVito TJ, Drost DJ, Neufeld RW, Rajakumar N, Pavlosky W, Williamson P, Nicolson R: Evidence for cortical dysfunction in autism: a proton magnetic resonance spectroscopic imaging study. *Biol Psychiatry* 2007;Feb 15;61(4):465–473.

69. Endo T, Shioiri T, Kitamura H, Kimura T, Endo S, Masuzawa N, Someya T: Altered chemical metabolites in the amygdala-hippocampus region contribute to autistic symptoms of autism spectrum disorders. *Biol Psychiatry* 2007;Jul 11; [Epub ahead of print].

70. Kleinhans NM, Schweinsburg BC, Cohen DN, Muller RA, Courchesne E. N-acetyl aspartate in autism spectrum disorders: Regional effects and relationship to fMRI activation. *Brain Res* 2007;1162:85–97.

71. Posse S, Schuknecht B, Smith ME, van Zijl PC, Herschkowitz N, Moonen CT: Short echo time proton MR spectroscopic imaging. *J Comput Assist Tomogr* 1993;17:1–14.

72. Posse S, Dager SR, Richards TL, Yuan C, Ogg R, Artru AA, Muller-Gartner HW, Hayes C: In vivo measurement of regional brain metabolic response to hyperventilation using magnetic resonance: proton echo planar spectroscopic imaging (PEPSI). *Magn Reson Med* 1997;37:858–865.

73. Hashimoto T, Tayama M, Miyazaki M, et al. Differences in brain metabolites between patients with autism and mental retardation as detected by in vivo localized proton magnetic resonance spectroscopy. *J Child Neurol* 1997;12:91–96.

74. Polleux F, Lauder JM: Toward a developmental neurobiology of autism. *Ment Retard Dev Disabil Res Rev* 2004;10;303–317.

75. Shinohe A, Hashimoto K, Nakamura K, Tsujii M, Iwata Y, Tsuchiya KJ, Sekine Y, Suda S, Suzuki K, Sugihara GI, Matsuzaki H, Minabe Y, Sugiyama T, Kawai M, Iyo M, Takei N, Mori N: Increased serum levels of glutamate in adult patients with autism. *Prog Neuropsychopharmacol Biol Psychiatry* 2006;30, 30(8):1472–1477.

76. Barkovich AJ: Magnetic resonance techniques in the assessment of myelin and myelination. *J Inherit Metab Dis* 2005;28:311–343.

77. Ding XQ, Kucinski T, Wittkugel O, Goebell E, Grzyska U, Görg M, Kohlschütter A, Zeumer H: Normal brain maturation characterized with age-related T2 relaxation times: an attempt to develop a quantitative imaging measure for clinical use. *Invest Radiol* 2004;39:740–746.

78. Inder T, Huppi, P. In vivo studies of brain development by magnetic resonance techniques. *Ment Retard Dev Disabil Res Rev* 2000;6:59–67.

79. Paus T, Collins DL, Evans AC, Leonard G, Pike B, Zijdenbos A: Maturation of white matter in the human brain: a review of magnetic resonance studies. *Brain Res Bull* 2001;54(3):255–266.

80. Dietrich RB, Bradley WG, Zaragoza EJ 4th, Otto RJ, Taira RK, Wilson GH, Kangarloo H: MR evalutaion of early myelination patterns in normal and developmentally delayed infants. *Am J Roentgenol* 1988; 150(4): 889–896.

81. Baratti C, Barnett AS, Pierpaoli C: Comparative MR imaging study of brain maturation in kittens with T1, T2, and the trace of the diffusion tensor. *Radiology* 1999; 210:133–142.

82. Petropoulos H, Friedman SD, Shaw D, Artru A, Dawson G, Dager SR: T2 relaxometry reveals evidence for gray matter abnormalities in Autism Spectrum Disorder. *Neurology* 2006;67:632–636.

83. Palmen SJ, Hulshoff Pol HE, Kemner C, Schnack HG, Durston S, Lahuis BE, Kahn RS, Van Engeland H. Increased gray-matter volume in medication-naive high-functioning children with autism spectrum disorder. *Psychol Med* 2005;35:561–570.

84. Santangelo S, Tsatsanis K: What is known about autism – genes, brain, and behavior. *Am J Pharmacogenomics* 2005;5:71–92.

85. Hendry J, DeVito T, Gelman N, Densmore M, Rajakumar N, Pavlosky W, Williamson R, Thompson P, Drost D, Nicolson R: White matter abnormalities in autism detected through transverse relaxation time imaging. *NeuroImage* 2006;29:1049–1057.

86. Araghi-Niknam M, Fatemi SH: Levels of Bcl-2 and P53 are altered in superior frontal and cerebellar cortices of autistic subjects. *Cell Mol Neurobiol* 2003;23(6):945–952.

87. Fatemi SH, Snow AV, Stary JM, Araghi-Niknam M, Reutiman TJ, Lee S, Brooks AI, Pearce DA. "Reelin signaling is imparied in autism." *Biol Psychiatry* 2005;57:777–787.

88. Whitaker-Azmitia, P: Behavioral and cellular consequences of increasing serotonergic activity during brain development: a role in autism ? *Int J Dev Neurosci* 2005;23:75–83.

89. Chugani, D: Serotonin in autism and pediatric epilepsies. *Ment. Retard. Dev Disabil Res Rev* 2004;10:112–116.

90. Casanova MF, Buxhoeveden DP, Switala AE, Roy E. Minicolumnar pathology in autism. *Neurology* 2002;58:428–432.

91. Vargas DL, Nascimbene C, Krishnan C, Zimmerman AW, Pardo CA: Neuroglial activation and neuroinflammation in the brain of patients with autism. *Ann Neurol* 2005;57:67–81.

92. Laurence, J. and S. Fatemi: Glial fibrillary acidic protein is elevated in superior frontal, parietal, and cerebellar cortices of autistic subjects. *Cerebellum* 2005;4:206–210.

93. Jyonouchi H, Sun S, Le Hoa. Proinflammatory and regulatory cytokine production associated innate and adaptive immune responses in children with autism spectrum disorders and developmental regression. *J Neuroimmunol* 2001;120:170–179.

94. Jyonouchi H, Sun S, Itokazu N. Innate immunity associated with inflammatory responses and cytokine production against common dietary proteins in patients with autism spectrum disorder. *Neuropsychobiology* 2002;46:76–84.

95. Jyonouchi H, Geng L, Ruby A, Zimmerman-Bier B. Dysregulated innate immune responses in young children with autism spectrum disorders: their relationship to gastrointestinal symptoms and dietary intervention. *Neuropsychobiology* 2005;51:77–85.

96. Fatemi SH, Halt AR: Altered levels of Bcl2 and p53 proteins in parietal cortex reflect deranged apoptotic regulation in autism. *Synapse* 2001;42:281–284.

97. Mukaetova-Ladinska EB, Arnold H, Jaros E, Perry R, Perry E: Depletion of MAP2 expression and laminar cytoarchitectonic changes in dorsolateral prefrontal cortex in adult autistic individuals. *Neuropathol Appl Neurobiol* 2004;30:615–623.

Chapter 18
Information Processing, Neural Connectivity, and Neuronal Organization

Nancy J. Minshew, Diane L. Williams, and Kathryn McFadden

Abstract Twenty years ago we proposed a neurobiological model of autism as a widespread disorder of association cortex and the development of connectivity of neocortical systems. We have subsequently provided substantial behavioral and neurofunctional evidence in support of this model. Neuropsychological studies with high functioning children and adults reveal the selective involvement of higher order cognitive processing across a broad range of domains and functions. Studies of face and object processing relate these processing deficits to the difficulty with automatic integration of multiple features, in contrast to the integrity of performance relying on individual features. The results of systematic functional magnetic resonance imaging studies of many social, cognitive, and language systems have added to our original model with evidence of functional underconnectivity of neural systems and disturbances in cortical specialization. Our model is also supported by the neuroimaging and neuropathology findings from other research groups that provide evidence of disturbances in cortico-cortical white matter (WM) connections but not interhemispheric or long tract WM connections with an increase in local connectivity that supports basic cognitive processes. Recent studies of neocortical growth and neuronal organization are consistent with our model and provide further details about potential mechanisms that complete the sequence from genes to behavior. Data continue to accumulate that support the basic tenets of this information processing, neural connectivity, and neuronal organization model and the specific neurodevelopmental mechanisms from which neurobiologically based, behaviorally defined disorder of autism arises.

Keywords Autism · brain development · cognition · functional connectivity · fMRI · information processing · neuroimaging

N.J. Minshew
Webster Hall, Suite 300, 3811 O'Hara Street, Pittsburgh, PA 15213, USA
e-mail: minshewnj@upmc.edu

A.W. Zimmerman (ed.), *Autism*, DOI: 10.1007/978-1-60327-489-0_18,
© Humana Press, Totowa, NJ 2008

Introduction

This chapter presents a model of autism that has been steadily built over the past 20 years within a neurological perspective. From the beginning, we have viewed autism first and foremost as a neurologic disorder originating in the brain and secondarily as a behavioral disorder [1]. This may appear to be a subtle distinction in orientation, but it has proven to be an important one, because it immediately resulted in different conceptualizations of the pathophysiology from that of other research groups. These groups have recently adopted major aspects of conceptualizations about autism that we have long proposed, such as the disturbances in cortical connectivity particularly involving frontal connectivity in high functioning individuals [2, 3, 4, 5, 6, 7, 8]; however, they continue to lack full appreciation of the pathophysiology that comes from an integrated neurologic perspective of the disorder. Secondly, we also recognized from the outset that idiopathic autism has its origins in developmental neurobiological mechanisms, specifically neuronal organizational events, whose processes would be expressed at the molecular level (superimposed on the background of individual familial inheritance) [1]. Additionally, because programs for brain development are under genetic control, we appreciated that further clarification of the syndrome would be found at the genetic level. Hence, our view has been that the ultimate explanatory principles for the syndrome of autism will be found at the genetic and molecular levels, not at the cognitive level. That is to say, we expect that descriptions of the disorder at levels above the molecular and genetic level will be inadequate or incomplete. However, we also appreciate that, as autism-associated genes and potential developmental neurobiologic mechanisms have been reported, identification of the genetic bases of autism alone will not result in the insights into the clinical syndrome that will achieve the goals of restoring brain integrity and predicting risk before conception. To accomplish these clinical goals, one must understand the expression of these genetic defects in brain function and structure and the connection to cognition and behavior. Recognizing these tenets, we have attempted throughout our research to identify overall constructs with validity at each level of the pathophysiologic sequence (Table 18.1) that would carry over and inform adjacent levels in the sequence.

Table 18.1 Pathophysiologic sequence of a neurodevelopmental disorder

Abnormalities in genetic code for brain development
↓
Abnormal mechanisms of brain development
↓
Structural and functional abnormalities of brain
↓
Cognitive and neurologic abnormalities
↓
Behavioral syndrome

It is important to bear in mind that models for autism can be developed for different purposes. The overall purpose of the model described in this chapter is to understand the pathophysiology of autism from gene to behavior and eventually the influence of etiologies on this sequence with the ultimate purpose of supporting the development of interventions at multiple levels of the pathophysiologic sequence. However, models that focus on isolated levels of the pathophysiologic sequence can also be developed, usually for the purpose of guiding an intervention. These clinical models often focus on behavior and its immediate or proximate "cause(s)." Such models have a different view or concept of "cause" than a model of the entire pathophysiologic sequence. When focusing on the translation of our research findings to behavioral interventions, we focus on social-emotional, communication, and reasoning impairments as other models do—not neurological or genetic impairments. In this context, the model is being related in terms that allow the translation to behavioral intervention. It is not meant to express the pathophysiology of autism from gene to brain to behavior.

Complex Information Processing-Disconnectivity-Neuronal Organization Model of Autism

We variously refer to our model of autism as a "complex information processing disorder" [9,10], a "connectivity disorder" [11], a "disconnectivity disorder" [12], a "disorder of cortical development" [13], an "underconnectivity disorder" [14], and a "neuronal organization disorder" [11, 15] depending on the context in which we use it and the evolution of our findings. Each term emphasizes a different level of the pathophysiology or aspect of the findings from which the model was derived, but all are part and parcel of the same model and pathophysiologic sequence.

Our model proposes that at the *cognitive-neurologic level*, autism is a disorder of information processing that disproportionately impacts complex or higher order processing with intact or enhanced simple information processing. The complex information processing characterization that we ascribe to the cognitive and neurologic deficits is not meant to denote a single primary deficit or "A" specific deficit but to identify a general characteristic of deficits that transcends domains, modalities, and methodologies. Fundamentally, impairments are exhibited in testable individuals with autism relative to age, IQ, and gender-matched controls when tasks require integration of multiple features rather than reliance on one or two individual features, speed of processing, processing of large amounts of information, or processing of novel material. All of these task demands tax information processing or integration capacity and reveal the limitation in processing in autism.

At the brain level, our model proposes that the observed information processing limitations in testable individuals with autism are the consequence of underdevelopment of neocortical systems that particularly affects connections of unimodal cortices with heteromodal cortex and therefore disproportionately impacts connectivity with frontal cortex. The model further proposes that autism involves alterations in local cortical connectivity. In individuals with high functioning autism (HFA), some alterations constitute an increase in local connectivity that supports basic cognitive processes, especially visual processes. We would further propose that these are reciprocally linked developmental disturbances of connectivity[1] in that the increased local circuitry is linked developmentally to the failure of neural systems connectivity to emerge.

We propose two clarifications to the issue of connectivity. First, functional underconnectivity of neural systems is not synonymous with the volume of white matter (WM) tracts ("structural over-connectivity"); functional studies are likely to be more reflective of the functional status of the dendritic tree and synapses, whereas WM volume reflects the long tracts. In fact, it could be that the over-development of WM tracts is a consequence of the functional underconnectivity. That is, because there is a disruption in effective cortical development, attempts at compensation result in the over-development of WM tracts. Second, in more severely affected individuals with autism, functional connectivity is proportionately more affected such that there is essentially no functional connectivity of neural systems, for example, no connectivity to frontal or parietal (heteromodal) cortices and little to no functional connectivity between primary sensori-motor and uni-modal association cortex. It would be difficult to get the profound mental retardation with disproportionate loss of social, communication, and reasoning abilities and unusual preservation of attention to detail in low functioning individuals without the pervasive impact on association cortex connectivity, for example, a frontal connectivity model is insufficient for the spectrum of autism. We propose that these connectivity disturbances are the consequence of abnormalities in selective aspects of the *developmental neurobiologic* processes referred to as *neuronal organization* and in turn abnormalities in *genes or epigenetic* processes that control these developmental neurobiologic mechanisms. These same gene disturbances are also responsible for the *multi-organ involvement* observed in autism (e.g., gastrointestinal and immune systems).

Principles of Neurological Disorders

One of the basic principles of neurological disorders that has guided our research from its inception is that brain dysfunction results in a *constellation* of signs and symptoms whose pattern reflects the impact of the underlying

---

[1] As used in this context, connectivity refers to the functional coordination or synchronization of cortical regions. It also embraces in principle structural disturbances in white matter and in dendritic and synaptic structure or function though white matter volume may be inversely or directly related to volume depending on the pathologic defect.

pathophysiologic process or mechanism on the brain. The pathophysiologic process or mechanism can be vascular, infectious, traumatic, neoplastic, toxic, degenerative, or developmental, to name the most common, or combinations of these when etiology is considered. None of these processes or mechanisms results in a single neurologic sign or symptom but rather a constellation of signs and symptoms and a characteristic mode of onset. Consequently, neurologists do not view disorders or diseases in terms of single primary deficits. Such a search is incompatible with how the brain is organized and impacted by disease processes. Hence, when investigating brain disorders, neurologists search for a constellation of signs and symptoms and then for a shared characteristic in this constellation that will identify the underlying pathophysiologic process. A related practice is to define intact abilities, because what is not impacted confirms conclusions about what is impacted (pertinent negatives), and what is not affected is equally important to understanding the effect of the disease process on the patient's capacity to function. The neurologic examination also provides specification regarding localization.

General Principles of Developmental Neurobiologic Disorders

When a neurologic disorder appears to be of developmental origin, the first consideration is to establish to the extent possible, the aspect of brain development that is involved. The tools for making such determinations from clinical evidence are limited, but clues are present that provide constraints and guidelines. The periods of brain development are divided grossly between the first two, during which the basic form of the nervous system is established and subsequent events, during which the major structures and cytoarchitechture of the nervous system are established. Table 18.2 provides a list of these events for cerebral structures. Some of the obvious clinical manifestations of disturbances in each of these events have been described and provide guidelines for clinical neurologic appraisals of developmental disorders [16]. The description of neuronal organization events as those producing the circuitry that supports the abilities that are most uniquely human makes this category of events highly relevant for autism.

The cellular and molecular events involved in neuronal organization and the genetics and epigenetic control over its unfolding during development are

Table 18.2 Developmental processes

Organogenesis
Neuronal proliferation
Glial proliferation, migration
Neuronal migration
Neuronal organization
Myelination

incredibly complex and incompletely understood but will eventually provide the greater level of specificity regarding the disturbances in brain development that underlie autism. (The most relevant progress in this area is discussed at the end of the chapter after presentation of the evidence supporting the earlier levels of our model.) Similarly, the gap in knowledge between autism candidate gene reports and knock-out animal models and the autism syndrome is wide. This leaves enormous amounts of research to be done—with minimal translational value—until the complex sequence of events establishing the connection with autism can be confirmed. The gap will be closed by research studies that will define how candidate genes influence behavior in response to environmental pressures or how genetic variability is related to anatomic variability and to variability in outcome, or how genetic variability is related to variability in medication response and to behavior and outcome. All of these efforts are in progress with the ultimate goal of an integrated understanding of the full sequence from gene (of which there may be many genetic and epigenetic variants) and developmental neurobiology (and its many variants on the same theme) in a common pathophysiologic brain to behavior sequence. If the brain-behavior sequence were not common in most ways across autism individuals, then there would not be a syndrome of autism. That does not mean there is not inter-individual variability. However, this variability is likely to reflect variability not only in the autism genetic abnormality but also the large familial variability that occurs in typically developing siblings born within a family.

Forming a Neurologically Based Model for Autism: Origins of Our Model

From a clinical perspective, idiopathic autism is characterized by deficits in complex behavior and higher order cognition, mental retardation in 60–70%, and seizures in 30%. In neurology, this constellation is classic evidence of dysfunction of cortical gray matter and specifically association cortex. The absence of blindness and deafness in idiopathic cases and the lack of functionally significant hypotonia, spasticity, cerebral palsy, and Babinski signs indicated sparing of primary sensory and motor cortices and of WM long tracts. Based on the preceding observations and neurological conventions, Minshew hypothesized in 1986 that autism is a disorder that broadly involves association cortex with sparing of sensorimotor cortex and WM long tracts [17]. She further hypothesized that in high functioning individuals, prefrontal cortex, and specifically dorsolateral prefrontal cortex, mediated a number of the cognitive deficits and behaviors associated with autism [1].

There were two notable observations that amplified this brain-behavior hypothesis. It became clear while studying preschoolers with HFA that they did not have the signs of focal cortical brain dysfunction common in children seen in child neurology clinics, for example, visuospatial deficits and difficulty acquiring words. In fact, the children with HFA are precocious in these areas. This led to the added specification that autism is a distributed neocortical systems disorder, for example, the abnormality is in the connections that constitute neural systems, not in focal dysfunction of association cortex [18]. A second insight was based on the early capacity of preschool children with autism to repeat words or phrases they heard without the ability to comprehend or spontaneously use those words in an original way (immediate echolalia or, in neurologic terms, mixed transcortical aphasia). The key question was what do comprehension and productive use of language have in common at a brain level that does not affect repetition of language? The answer appeared to be lack of connections to adjacent association cortex. Minshew then began to think of abnormalities in brain development and the behavioral manifestations of autism in terms of disturbances in the development of local connectivity between primary sensorimotor cortex and association cortex, followed by disturbances in the development of the long distance connections that constitute neural systems, and then the connections between neural systems. The most severe cases of autism were conceptualized as little to no development of functional connections between sensorimotor cortex and association cortex, and thus no meaning was being attached to information. This conceptualization also led to the hypothesis that connectivity with heteromodal cortex, particularly frontal cortex, would be more greatly impacted than connectivity with unimodal cortex in high functioning individuals [1].

This conceptualization of autism as a disorder of association cortex and its connectivity explains a number of other clinical observations. Reduced development of local connections explains why a child with autism could point to a picture of a shovel when asked using its name but not if his mother says "Point to the thing you dig with." He has the name but not the meaning, which requires greater connectivity. This also explains why nouns are acquired more readily than abstract words like verbs.

Other clinical observations suggest that there is a failure in the development of connections between neural systems. The failure of connections to develop between the language and social neural systems explains why some children with autism speak when no one is in the room to hear them, not recognizing that language requires a listener to accomplish communication. Reduced connections between the memory and executive systems explain the impairments documented in memory related to the failure to utilize strategies to support recall [19]. It also explains the presence of impairments in concept formation across multiple domains (memory, story recall, rule learning-concept formation) [9]. All of these impairments were ultimately demonstrated by fMRI studies to be the result of under-development of connections of relevant brain regions with frontal lobe. Thus, additional clinical modifiers to the original model involves the development of connections within and then between neural systems and the greater number of connections required for abstract (non-visualizable) constructs than for constructs that could be visualized (1:1 correspondence). A map of Brodmann's area that divides cortex into primary

Fig. 18.1 A map of Brodmann's areas that divides the cortex into primary sensory and motor cortices, and association cortex into unimodal and heteromodal cortices and assigns numbers to specialized regions. The figure displays both a lateral and a medial view. From Fundamentals of Human Neuropsychology, 5/e by Bryan Kolb and Ian Q. Whishaw © 2003 by Worth Publishers. Used with permission (*see* Color Plate 8)

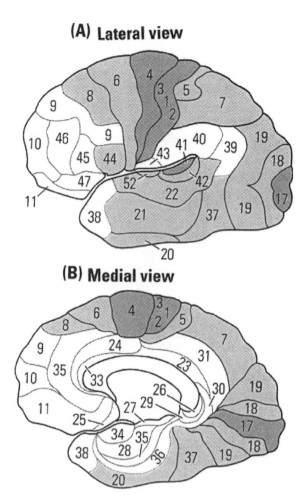

sensory and motor cortices and association cortex into unimodal and hetero-modal cortices and assigns numbers to specialized regions is a common and constant referent for our evolving model (Fig. 18.1 and see Color Plate 8, following p. XX).

Empiric Evidence for Our Model: Disordered Complex Information Processing

Following the principles of neurology, the purpose of our first research study was to define the deficits in the major domains of neuropsychologic functioning and the intact abilities. Because tests of social cognition for high functioning individuals had not yet been developed, the social domain could not be assessed other than with the Autism Diagnostic Observation Schedule [20].[2] We studied 33 individually matched pairs of high functioning adolescents and adults with and without autism [9], assessing a broad range of abilities within each of the domains from the most basic skills to higher order skills. This battery was designed not only to test our hypothesis but also to test competing theories of autism at that time. Our hypothesis was that autism broadly involved association cortex and therefore higher order abilities across domains, including higher cortical sensory and motor abilities. We also wanted to confirm that deficits in these higher order abilities were not secondary to deficits in basic abilities and that higher order abilities were selectively affected in individuals with HFA. Other contemporary theories proposed primary deficits in basic abilities like elementary sensory perception, various aspects of attention, and basic associative memory abilities.

On completion of the study, we found a subtle deficit in higher cortical sensory perception and significant deficits in skilled motor movements, memory abilities dependent on a cognitive strategy, and the concept formation aspects of abstraction [9]. Intact or enhanced abilities were found in the areas of attention, elementary sensory perception, elementary motor abilities, associative memory abilities, formal language, and the rule-learning and attribute identification aspects of abstraction. The deficits in higher cortical sensory perception, skilled motor movements, and cognitive memory exhibit the same simple-complex dissociation as found in other domains, providing evidence that they are caused by the same neurobiologic mechanism as the other deficits documented and thus are integral parts of autism. Additionally, the profile is characterized by deficits in higher order abilities AND by preserved or enhanced basic abilities. These features appear to be two related or linked parts of a developmental neurobiologic process, as clinical observations of patients over time reveals subtle losses in their enhanced basic abilities with subtle gains in higher order skills. Thus, it appears as though excess local connectivity diminishes in the process of gaining systems connectivity.

We subsequently completed a study in 56 children with autism and 56 controls aged 8–15 years [10] that replicated the overall pattern with some interesting differences related to development. The child study revealed deficits in skilled motor movements, higher cortical sensory perception, memory for material that requires an organizing strategy or detection of an organizing strategy, and higher order language meaning. Intact abilities were found in the domains of attention, elementary motor abilities, elementary sensory abilities, basic associative memory abilities, formal language skills (vocabulary, spelling, fluency, decoding), and abstraction. The differences from the adult

---

[2] Note: All subjects participating in our studies meet criteria for autism (not autism spectrum disorder) on all domains of the Autism Diagnostic Interview [21] including evidence of abnormal development before 3 years of age; meet autism cutoffs on the communication and the reciprocal social interaction domains and total scores of the ADOS; and, meet DSM IV criteria for autism based on expert clinical opinion.

study are more pronounced deficits in higher cortical sensory perception, the absence of the superior formal language abilities observed in the adults with autism, and the absence of a significant deficit in concept formation abilities in the children. In the case of abstract reasoning, the absence of significant deficits for the domain is related to the poor performance by both the autism and the control children. These are abilities that are related to the maturation of frontal circuitry in the second decade in typically developing individuals. Therefore, no significant deficits in performance are evident in the children, but significant deficits in performance are documented when studying adolescents and adults with autism [9].

The analysis of this constellation of findings with adults and children with autism led to several conclusions. First, the deficits are not in the acquisition aspects of information. These children, adolescents, and adult individuals with HFA learn word meanings, how to read, how to spell, and how to do arithmetic, sometimes at superior levels to matched peers. Interestingly, deficits are consistently present in the same domains as intact or enhanced skills, thus appearing to reflect a biologic boundary or process. The impaired abilities are those that place the highest demand on information processing within the affected domains, and the domains with the most prominent symptoms are those that have the highest information processing demands. Therefore, we propose that the common denominator of the impaired abilities is their demands on integrative or "complex" information processing. Neurologists associate "complex or higher order information processing" with cerebral cortical architecture, dendrites, synapses, and neural systems; this term is also used in neurophysiology and cognitive neuroscience. This shared characteristic, together with the developmental mode of onset and neurologic examination, suggests that autism is the result of a disturbance in the development of neocortical circuitry and involved neuronal organizational events.

The intact and enhanced abilities, as shown in the profile studies, are all basic or elementary abilities with low information processing demands that often have a visual component [10, 22]. One might, however, argue that spelling is a complex skill involving the holding of information on line (the auditory production of to-be-spelled word) while the speller retrieves previously stored information about the word and the relationship between phonemes and graphemes. However, the very nature of spelling may lend itself well to the way in which the brain with autism handles information. There is only one way to spell a word (not withstanding English versus American spellings). Therefore, decisions do not have to be made between multiple meanings nor does situational and cultural information have to be considered. The most complex part of spelling is the phoneme–grapheme relationship. In English, there are a number of graphemic representations for a particular phoneme (read, red, for instance). This might necessitate a level of analysis that would support an argument that spelling is "complex information." However, if the spelling is accomplished by retrieval of a visual representation of the word without consideration of the phoneme–grapheme representation, the task becomes much less complex. One word, spelled one way, has one visual representation. All of this processing can be accomplished within a fairly restricted area of the brain alleviating the need for the coordination of diverse brain regions to accomplish the task.

A related issue is why any language task is not more complex than any motor or sensory task. The response is that the nervous system is arranged by systems and modalities and thus tasks cannot be judged independent of systems. Complexity is judged within systems. As mentioned earlier, the domains most severely impacted in autism are those with the highest demands for integration of information.

The next major finding of these profile studies is that there are several affected domains not considered to be integral parts of the autism syndrome, specifically the sensori-perceptual, motor, and memory domains. Hence, the evidence suggests that autism involves the brain far more generally than the DSM-IV diagnostic triad indicated and that the common denominator is a dependence on the degree of integration at the information processing level. This suggests that there might be a cytoarchitectural feature common to all neural systems that is designed for higher level processing or integration demands regardless of domain. Subsequent fMRI studies confirmed this inference beginning with an fMRI study of spatial working memory (oculomotor delayed response task) that should have activated left dorsolateral prefrontal cortex but failed to do so in the non-retarded adults with autism [23]. Instead, the individuals with autism demonstrated greater activation bilaterally of occipital parietal regions consistent with the neuropsychological profile findings of reliance on basic cognitive abilities and impaired higher order abilities.

Concept Formation: A Common Theme and Predictor of Underconnectivity with Frontal Cortex

One additional observation of the profile of neuropsychological function in individuals with autism is that deficits in concept formation are present in multiple domains, for example, in the use of or detection of a cognitive theme to support memory, in the telling of a logical story or detecting the theme of a story, and in forming concepts. Subsequently, we have found deficits in learning motor concepts (unpublished data). These findings have not only led to the extension of our research into category formation as a possible common underlying mechanism but also the hypothesis that underdevelopment of cortical connections with frontal lobe is a common neurobiologic basis for these deficits. Subsequent fMRI studies (reviewed below) support this hypothesis; underconnectivity with frontal cortex is a common finding across studies regardless of the cognitive domain and brain region studied.

Visual Processing: Another Example of Generalized Integrative Processing Deficits

At the time of our original profile study, we observed deficits in memory for complex visual stimuli but not for copying or for simple visual stimuli (see results for Rey Osterreith Complex Visual Figure and stylus mazes of varying difficulty) [9]. However, we failed to include challenging tests in the visuospatial domain. The traditional neuropsychological test of complexity in the visuospatial domain is face recognition. Ultimately, we documented severe face recognition deficits (gender, identity, emotion) in about one-third of individuals with HFA and increased reaction times in the remaining two-thirds consistent with a 10–15% slowing of processing speed [24]. The latter slowing is a substantial impairment given the speed of on-line processing required for much of adaptive function. In addition, cross-sectional studies of face gender, emotion, and identity were completed for ages 5–7, 8–12, 13–17 years and adults. These studies furthermore used typical and atypical stimuli and included or excluded hair cues. Typical children were near adult levels of competency on all stimuli by 12 years of age; in other words, it took about 10 years or so for them to reach full competency. However, the individuals with HFA were slower to develop skills and topped out at about the level of 10 year olds and required 10–15% more time to make their assessments. These results are of major interest because optimal performance is based on the brain's automatic capacity for integrating multiple features and making comparisons to previously created prototypes. Thus, even the least impaired individuals with autism displayed limitations with integrative processing under these conditions. This difficulty with automatic integrative processing and category formation was also found to extend to objects [24, 25]. Hence, the deficits in face processing are not unique because they are social but are merely one class of complex stimuli that require expertise (automatic processing mediated by prototypes that are used to integrate multi-dimensional information). This limitation in integrating multi-dimensional information proves helpful in understanding the cognitive limitations in autism more broadly. This limitation in integrating multi-dimensional information is identical to that in the abstraction studies in which rule learning and attribute identification are intact but concept formation, the foundation for adaptive function, is impaired [26]. This would also appear to be the basis for the flexible application of information. That is, the full encoding of information enables it to be applied variably with changing circumstances; perhaps the underlying mechanism for flexibility in use of other concepts (rules) is a prototype, just as it is for faces or objects.

Face recognition deficits are widely documented in autism and are attributed by some to deficits in high spatial frequency detection and in turn to a selective deficit in the magnocellular visual system [27]. In a recent study from our group, Behrmann and colleagues [25] investigated spatial frequency thresholds and face recognition abilities and found that subjects with autism performed within the normal boundaries in detecting low- and high-frequency gratings despite impairments in face recognition abilities. Thus, observed deficits in face processing are not attributable to deficits in high spatial frequency detection but are more likely the result of deficits in higher order perceptual processes. Furthermore, the abnormalities in fusiform face area are not attributable to abnormalities in the magnocellular system as hypothesized. The disturbed cognitive mechanism in this context may be the deficiency in the automatic formation of prototypes that provide for integration of multiple features or elements, whereas the neural mechanism may be the specialization of the fusiform face area and its specialized circuitry [28].

A Dual-Task Experiment: Support for the Complex Information Processing Model

When deciding how complexity might be defined in our model, we considered not only number of elements and intrinsic complexity of stimulus material to be processed but also the *amount* of information to be processed in a unit of time. One study had already investigated this issue in autism using a dual task design [29]. In this study, average IQ individuals with and without autism were asked to do two simple tasks (digit span, motor tracking) alone and at the same time. When the autism group performed either task alone, their performance was identical to the typical group. When the typical group performed both tasks simultaneously, there was no decrement in their performance for either task. When the autism group performed both tasks together, their performance declined about 40% on both tasks. These tasks were simple and the individuals with autism were of average IQ. The findings of this study support our conclusions about the presence of substantial constraints on information processing capacity in autism. From a computational modeling perspective, Marcel Just proposes that processing constraints in autism are comparable to a limitation in bandwidth [30]. That is, the neural connections in autism are compromised such that information is transmitted at a slower rate or in a degraded form as a result of the inability to process all of the information. We have therefore clarified the cognitive aspect of our model to specify capacity limits on number of tasks and speed of performance as parameters in defining task complexity. The above dual task findings also explain the common clinical observations that individuals with autism have one speed for doing things (and react negatively when others try to rush or hurry them) and may only be able to attend to one modality of information at a time.

Postural Control: Evidence of Abnormal Connectivity and Broad Involvement of the Brain

Problems with integration of information are observed in autism in areas that are outside of the cognitive domains. In a study of 79 individuals with autism and 61 typical controls between 5 and 52 years, we reported abnormalities in postural control involving delayed maturation and failure to achieve adult levels of function [31]. Impairments are not related to the motor system or to the function of the cerebellum, rather they are related to inadequacy of multi-modal sensory (vestibular, visual, position senses) integration. This study provides evidence that the difficulty the brain of individuals with autism has with complex processing can be extended to completely unexpected (based on DSM-IV criteria for autistic disorders) areas of brain function. As has been seen for the other domains, these areas are ones that depend on the integration of large amounts of information.

These findings are noteworthy for their support of disturbances in neural connectivity. More importantly, they reiterate the more pervasive involvement of the brain in autism and the resulting need to re-conceptualize the disorder and the DSM-IV criteria for the autism spectrum disorders to accommodate such deficits. Additionally, the study also demonstrates that the brain continues to develop into adulthood following a trajectory that paralleled the normal curve. Such parallel but lower developmental curves suggest that the brain in autism is organized similarly to the normal brain, is plastic, and thus theoretically has the potential for rehabilitation, restoration, or rescue of circuitry through intervention.

The Neural Basis of Autism: The Origin of Functional Underconnectivity

The availability of modern functional neuroimaging makes it possible for us to observe the brain activity of individuals with autism as they perform cognitive tasks. These studies help to establish a neural basis for the disturbances in information processing documented by our cognitive and neurologic studies described above and to build a brain-behavior connection. Prior to the availability of this technology, we relied upon oculomotor studies to help establish localization [32]. These oculomotor studies are very helpful in determining that the so-called shifting attention deficit is an executive function deficit arising from dysfunction of dorsolateral prefrontal cortex. However, functional magnetic resonance imaging (fMRI) contributes greatly to understanding the brain and the mind of autism. Functional MRI makes it possible to gain insight into when and how individuals with autism have difficulty with processing information, specifically when dealing with increasing complexity. An initial fMRI study examined brain activation during the reading of active and passive sentences in older adolescents and adults with HFA [14]. The autism group had increased activation in Wernicke's area, relative to the age and IQ-matched controls (possibly increased local connections) that corresponded to more extensive processing of the meanings of the individual words. The autism group also had relatively decreased activation in Broca's area (possibly decreased long-distance connections) corresponding to reduced integration of individual words into conceptually and syntactically coherent sentences.

Furthermore, functional connectivity, or the degree of synchronization or correlation of the time series of the activation between the various participating cortical areas, was lower in the autism group between key areas of the cortical language network. These results suggest that individuals with autism have a lower degree of information integration and synchronization across the large-scale cortical network for language processing.

Reduced functional connectivity for the autism group has been found in a number of different neuroimaging studies. In an earlier PET study of adults with autism or Asperger syndrome performing a theory of mind task, Castelli and colleagues [33] reported reduced functional connectivity between the extrastriate cortex and the superior temporal sulcus at the temporo-parietal junction. In a study using sentences requiring either high or low imagery, reduced functional connectivity occurred for the HFA group between frontal and parietal areas, key regions for performing visuospatial imagery [34]. Reduced frontal-parietal connectivity was also reported during executive processing using the Tower of London task in adults with HFA [30]. During a response inhibition task, adults with HFA had lower levels of synchronization between the inhibition network (anterior cingulate gyrus, middle cingulate gyrus, and insula) and the right middle and inferior frontal and right inferior parietal regions [35]. In an *n*-back working memory task using photographic facial stimuli, the HFA group had activation in a slightly different area of the fusiform gyrus with networks that were smaller and less synchronized, with lower functional connectivity with frontal areas but no underconnectivity among posterior cortical regions [36].

These findings of reduced functional connectivity in autism during cognitive processing are the basis of the *underconnectivity theory* of autism [14]. According to this theory, autism is a cognitive and neurobiological disorder marked and caused by underfunctioning integrative circuitry that is needed for tasks that require coordination of processing at a high level, regardless of the task domain. The theory predicts that any facet of psychological or neurological function that is dependent on the coordination or integration of brain regions is susceptible to disruption, particularly when the computational demand of the coordination is large. The underconnectivity theory is a neurobiological correlate to our complex information processing model of autism.

Additional findings from fMRI studies are consistent with our complex information processing model of autism and give further insight into the difficulties individuals with autism have during cognitive processing. Two fMRI studies investigated the brain activation of individuals with HFA during a verbal working memory [37] and a spatial working memory [12] task. The verbal working memory task was a classic *n*-back letter task with three conditions: 0 back, 1 back, and 2 back.[3] Behaviorally, the autism group performed

---

[3] In a 0-back task, the participant indicates by pressing a button whenever they see a pre-designated letter appear on the screen. In a 1-back task, the button is pressed when two letters in a row are the same. In a 2-back task, the button is pressed when a letter is the same as the one that appeared two letters earlier, e.g. A-H-A.

similarly to the control group with error rates of 9.8 and 11.1%, respectively, in the 2-back condition. However, examination of the brain activation patterns revealed significant differences between the two groups. The control group had the expected left prefrontal and left parietal verbal working memory network. This network is thought to represent the conversion of the letters to verbal-phonological codes that reduce the working memory load. However, the autism group had a right frontal and right parietal working memory network. Thus, the autism group appeared to use non-verbal, visually oriented processing that resulted in the retention of the letters as visuo-graphical codes. The fMRI study of spatial working memory [12] used an oculomotor delayed response task. In the control group, there was substantial activation of left frontal cortex and left posterior visual and parietal cortices. However, in the autism group, there was almost no activation in left frontal cortex, but there was bilateral activation of posterior cortical regions indicating near total reliance on visuospatial areas for performance of this task. Thus, in both working memory tasks, there was a failure of the autism group to activate higher order circuitry underlying executive functions and a reliance on lower order circuitry of visual cortex for task performance. This occurred even when the tasks did not appear to be behaviorally demanding.

Functional MRI studies of higher order language tasks provide additional evidence that, in autism, the brain has relatively more difficulty than in individuals with typical development as the processing demands of a task increase. The brain activation of high functioning children and adolescents with autism or Asperger syndrome was examined in a study of the comprehension of irony [38]. Three conditions were used to examine the use of contextual knowledge and prosodic cues during the interpretation of ironic utterances. Behaviorally, the children and adolescents with ASD differed significantly in the two conditions that required the processing of contextual information and contextual information with prosodic cues, but not in the condition that required only the processing of the prosodic information. The challenge seen in the behavioral performance for the ASD group was evidenced by increased activation in right inferior frontal gyrus in the contextual information only condition and bilaterally in inferior frontal gyrus in the condition that required integration of contextual and prosodic information. Overall, the ASD group activated prefrontal and temporal regions more strongly than the typically developing children, a result that the authors interpret as indicative of the response of the brain to increased task difficulty.

An fMRI study of discourse examined processing during three different types of reading passages [39]. The passages required either inferring the physical causation, internal motivation of a protagonist, or emotional reaction of a protagonist. The results showed that the right hemisphere (RH) activation was substantially greater for all sentence types in the HFA group than in the control group, suggesting decreased left hemisphere (LH) capacity in autism resulting in a *spillover* of processing to RH homologs. This is a pattern that is consistent with greater processing effort by the autism group. In addition, the control group showed a differential response to the three different types of inferencing; they had the expected pattern of results with greater activation in medial frontal gyrus and bilateral temporo-parietal junction while processing intentional inferences than for physical or emotional inferences. However, the autism group activated the RH language and theory of mind network (which includes the right temporo-partietal junction) for all three types of passages. That is, during the reading of the physical causation (or least demanding) passages, the individuals with autism were already using cognitive resources that the control group would not use until they were processing the more demanding emotional inference passages. Furthermore, the autism group had lower functional connectivity within the theory of mind network and also between the theory of mind and LH language network. The results of this study gives some indication of why an increase in complexity results in a degradation in performance for individuals with autism sooner than would be predicted based on their overall level of function.

In summary, fMRI studies with individuals with autism indicate that the cortical networks underlying basic abilities are intact. When processing language, individuals with autism activate LH language areas; when processing visual information, they activate occipital areas. However, these networks are relied upon to perform tasks usually performed with more highly integrative networks. For individuals with autism, functional underconnectivity occurs during processing of complex information, and dynamic recruitment of neural resources as processing demands increase is compromised. Equivalent behavioral performance may occur by using a cognitive processing pattern that is different from that used by typically developing individuals. Even when individuals with autism successfully perform a task behaviorally, they may be using the upper limits of their information processing resources.

Disturbances in the Development of Neural Circuitry: A Second Disruption in Development

A study of volitional saccades in adolescents and adults with and without autism demonstrated deficits in frontal cortex but not cerebellum as the basis for deficits in "shifting attention" [32]. A subsequent study applying the same methods to children with and without autism revealed no deficits, possibly because neither group had developed the frontal abilities [40]. When a sample of sufficient size was assembled to look at age differences, it became apparent that deficits emerged in autism in the second decade because of the maturation of frontal lobe abilities in the typical population but not the autism group [41]. Another study of children through adults with and without autism using the Rey Osterreith Complex Figure (unpublished data) also revealed no deficits in the children because neither group had developed these higher order abilities; whereas, by the second decade the controls were using executive abilities and deficits emerged in the autism group who had failed to develop these skills. These two studies indicate that the second decade of life is an important period of brain maturation that impacts the expression of deficits in autism. The failure of these frontally dependent abilities to mature may explain why individuals with HFA have unexpectedly poor outcomes despite their high IQ scores. That is, they appear to have good cognitive potential, frequently excel in school even attaining college degrees, yet lack good function in life situations. However, they are unable to solve problems, create structure and order for themselves, or respond flexibly to life by considering the numerous variables that dictate variation in responses to situations, making independent functioning in adulthood difficult.

Social Impairments in Autism

Our model views social impairments in autism from the perspective of deficits in the underlying mechanisms involved in processing social information. As described earlier, this class of information is not viewed as special or distinct from other information by virtue of being social, only by virtue of its complexity. The discovery of "theory of mind" deficits was a significant advance in connecting social difficulties to underlying problems in cognition. However, the social cognitive deficits that exist in autism go far beyond theory of mind. Not well understood at a mechanistic level are a plethora of difficulties in the processing of social information, including problems in face perception and recognition, the perception and understanding of emotions, the understanding and interpretation of communicative efforts, and the comprehension of context. Analogous issues also exist with the processing of information from eye gaze and prosody. Progress is clearly being made in understanding the cognitive and neural mechanisms of these deficits, though once such mechanisms are understood at one level, underlying levels remain to be conquered until the molecular and genetic levels are reached. Thus, our model is concerned with understanding social impairments, but from the perspective of seeking to understand underlying information processing and neural mechanisms of social processing as opposed to just motivational or reward differences.

Even less well understood than the cognitive aspects of the social impairments are the deficits in the emotional aspects of social function in autism. It is clear that individuals with autism have disturbances in their connections with others and in empathy. Despite autism originally being termed a "disturbance of affective contact" [42], there is little actual understanding of the mechanisms underlying the affective or emotional impairments typical of autism. Some insight into these differences has been generated by studies of the amygdala [43] and of the mirror neuron system [44], but clearly there is much left to be discovered.

Autism as a Disorder of Cortical Development

As we discussed earlier, according to our model, to be fully understood, autism must be described at each pathophysiologic level. Behavioral differences and disturbances in cognitive and neurological function must ultimately be related to underlying genetic and developmental neurobiologic disturbances. The pathogenesis of a central nervous system (CNS) developmental disorder such as autism may be constructed as a cascade of events in which an insult, genetic or environmental, initiates a sequence of molecular, cellular, and architectural alterations that evolve over time. The final phenotype is a patterned constellation of morphologic and functional characteristics resulting from this abnormal trajectory. Traditionally, developmental CNS disorders are classified according to the major event most impacted, for example neurulation, forebrain induction, neuronal migration, and so on. Wherever the genetic etiology is adequately understood (e.g., some forms of lissencephaly), there is a shift to using genotype as the basis of classification [45]. Although complicated, this approach may have greater explanatory power when confronted with certain complex phenotypic patterns, as any single gene or genetic pathway may impact multiple developmental processes at different time points.

Recent morphometric and neuropathologic studies implicate perturbations in the processes of neocortical growth and organization in the development of an autistic phenotype. Studies examining head circumference (HC) and brain volume (by MRI) in infants and children with autism demonstrate abnormally accelerated neocortical growth in the first few years of life [46,47,48,49]. Mean HC and brain volume are not significantly different from normal at birth [50]. Soon after there occurs a period of abnormally accelerated brain growth in the first 2–4 years of life until brain volume is significantly larger than normal [46,47,48,49]. Following this period of rapid growth, the rate of growth declines significantly relative to normal controls so that brain volumes of individuals with autism in early adolescence are again no larger than average [49].

There is a marked rostral-caudal gradient in this altered trajectory so that the frontal cortical grey matter (GM) and WM show the most enlargement. The temporal GM and WM and the parietal GM exhibit milder enlargement, and the occipital GM and WM and parietal GM tend not to vary significantly from normal [51]. Within the frontal lobes, the areas most abnormally enlarged are the dorsolateral and mesial prefrontal cortex [52]. Similarly, the WM most involved appears to be the radiate compartment immediately underlying the prefrontal cortex [53], whereas the sagittal/bridging WM, particularly the corpus callosum and internal capsule, are not enlarged. This suggests overgrowth of later-myelinating, intrahemispheric (cortico-cortical) connections (originating from layer III neurons) with relative sparing of the interhemispheric (layer III) and cortical to subcortical (layers V and VI) connections [53].

Recent examinations of cortical microarchitecture have found subtle but pervasive abnormalities. Although most of these analyses are not rigorously stereologic, the impression is of increased numbers of narrow minicolumns with increased densities of neurons [54, 55, 56] (see Chapter 16 by Casanova). Here, minicolumns appear to correspond to ontogenetic columns of vertically oriented cords of cells resulting from the migration of neurons along radial glial fibers. Minicolumn packing is most pronounced in the dorsolateral prefrontal cortex. Buxhoeveden similarly found narrower minicolumns in subjects with autism predominately in the dorsal and orbital regions of the frontal cortex [57]. No difference in minicolumn width/density has been found between subjects with autism and controls in the visual cortex.

In summary, volumetric and histologic studies of subjects with autism show a substantially altered postnatal cortical growth trajectory associated with an abnormal underlying microarchitecture. Furthermore, there is agreement between these two sets of analysis concerning the anatomic distribution of change. The observed abnormalities occur in a rostrocaudal gradient so that later developing anterior structures associated with complex processing are disproportionately affected. The overgrowth is observed during the first 2 years of life—a period of active neuronal cell body growth, axonal/dendritic elaboration, and synaptogenesis. However, the constellation of increased cell density, packed radial units, and a rostrocaudal gradient indicates the initial processes affected are prenatal ones, namely cortical regionalization, neuronal precursor proliferation, and lamination. Interestingly, it has been increasingly recognized that master transcription factors, expressed in a rostral-caudal (or reverse) gradient, control most of these molecular cascades underlying these processes.

The embryonic telencephalic neuroepithelium is initially regionalized along a rostral-caudal axis and into initial dorsal and ventral domains by extracellular signaling molecules (e.g., Shh, BMPs, FGFs and Wnts) emitted from major organizing centers [58, 59, 60]. The dorsal progenitor domain or pallium eventually gives rise to the cerebral cortex and the ventral subpallium, consisting of the lateral and medial ganglionic eminences (LGE and MGE) and gives rise to the basal ganglia. This early dorsal-ventral patterning specifies, among other things, the genesis of neurons that express different classes of neurotransmitters. Cortical germinal zones produce glutamatergic projection neurons, the LGE produces GABAergic interneurons, and the MGE produces GABAergic and cholinergic neurons [61].

Within the dorsal pallium, these early extrinsic patterning signals establish regional graded expression domains of transcriptional factors. These regionally restricted genes serve to specify the identity of the cortical area in which they are expressed. In rodents and primates, the transcription factors Emx2 and Pax6 regulate much of the rostral-caudal arealization of the cortex. Pax6 specifies a rostral identity to cortical progenitor cells, whereas Emx2 confers caudal identity [62]. Pax6 additionally confers a dorsal identity in the developing cortex by mutually cross-repressive interactions with Gsh1/2 and by directly activating Neurogenin 2 (Ngn2), which in turn acts to repress ventral specification [63]. The orphan nuclear receptor tailless (Tlx), also expressed in a rostral-caudal gradient, interacts with Pax6 (and probably Ngn2) to help establish the pallio-subpallial boundary [64].

Initial cortical expansion occurs by symmetrical divisions of neuroepithelial cells in the germinal ventricular zone (VZ). As neurogenesis progresses, these transform into radial glial cells (RGCs) that give rise to glutamatergic projection neurons by undergoing asymmetric cell divisions. Daughter neurons migrate radially along RGC processes to form cortical lamina in an inside-out fashion so that earlier born neurons contribute to deeper cortical layers and vice versa. In later neurogenesis, a distinct population of basal or intermediate progenitors (derived from neuroepithelial and RGCs) form the subventricular zone (SVZ). Here they undergo symmetrical cell division to form two neurons. The SVZ is the source of cortical layers II–IV [65]. Concurrently, the majority of inhibitory GABAergic neurons are born in the ganglionic eminences and migrate to the neocortex where they integrate into the circuitry as interneurons [66]. Pax6, acting through Ngn1/2, specifies a glutamatergic phenotype [67] and is also required for the production and maintenance of proper numbers of neurons by preventing premature differentiation [68]. The synergistic interaction of Pax6 with Tlx is important for the specification of later born neurons, particularly those produced by symmetrical division in the SVZ [65, 67].

These regulatory molecules are excellent candidates for the root cause processes mediating the pathophysiology of autism. They are expressed in the right places and orchestrate the developmental events we are concerned with. It is doubtful, however, that an autistic phenotype results from a high impact mutation (e.g., knock-out). Rather, more subtle changes causing less drastic misregulation should be sought. Because of the degree of interdependence and cross-regulation among these molecules, it is likely a similar phenotype could be obtained by perturbing the function of more than one. Furthermore, individual genetic background alleles would be more likely to influence the impact of these perturbations.

Summary

Two decades ago we proposed that autism is a widespread disorder of association cortex and the development of its connectivity. In low-ability individuals with autism, we propose that the disturbance involves the failed development of functional connections between primary sensorimotor cortices and unimodal association cortices, which results in the inability to associate elementary meanings with information. In high-ability individuals, we propose that autism is the consequence of disturbances in the development of the connectivity of neocortical systems, particularly involving heteromodal cortex and, thus, especially connections with frontal cortex; the consequence is broad impairments in higher order cognitive and neurologic abilities. Our subsequent research systematically provides substantial evidence in support of these hypotheses including evidence of the selective involvement of higher order cognitive and neurologic abilities across a broad range of domains and functions; systematic fMRI evidence of functional underconnectivity of neural systems related to many social, cognitive, and language systems; and disturbances in cortical specialization. Studies of face and object processing impairments highlight the relationship of information processing deficits to the difficulty with automatic integration of multiple features, as opposed to the integrity of performance relying on individual features. fMRI and cognitive studies suggest that more basic mechanisms may involve the process of cortical specialization resulting from a disturbance in automatic processing dependent on prototype mechanisms and the supporting neural circuitry.

Our model is supported by other neuroimaging and neuropathology findings that provide evidence of broad, though not even, involvement of the circuitry of the cerebral hemispheres and of the cerebral cortex [15]. Of particular note is the evidence of disturbances in cortico-cortical WM connections but not interhemispheric or long tract WM connections [53] and of abnormalities in cortical connectivity [56, 69, 70].

The terms we chose for our model—autism as a disorder of complex information processing-connectivity-neuronal organization—are not intended as endpoints but as overarching concepts with the expectation that research would provide greater specification of mechanisms as time passed. The validity of our model is reinforced in that other research groups have recently assumed similar versions of what we originally proposed. More importantly, data have continued to accumulate that support the basic tenets of this model and elaborate specific neurodevelopmental mechanisms. Hopefully, future research efforts will rapidly provide the details of the mechanisms that will complete the sequence from gene to behavior that will enable the development of definitive neurobiologically based interventions as well as prediction of risk for inception and intervention from birth.

References

1. Minshew NJ, Payton JB. New perspectives in autism, Part II: The differential diagnosis and neurobiology of autism. *Curr Probl Pediatr* 1988; 18: 613–694.
2. Baron-Cohen S, Belmonte MK. Autism: A window onto the development of the social and the analytic brain. *Ann Rev Neurosci* 2005; 28: 109–126.
3. Belmonte MK, Allen G, Beckel-Mitchener A et al. Autism and abnormal development of brain connectivity. *J Neurosci* 2004; 24: 9228–9231.
4. Courchesne E, Pierce K. Why the frontal cortex in autism might be talking only to itself: Local over-connectivity but long-distance disconnection. *Curr Opin Neurobiol* 2005; 15: 225–230.
5. Geschwind DH, Levitt P. Autism spectrum disorders: Developmental disconnection syndromes. *Curr Opin Neurobiol* 2007; 17: 103–111.
6. Hughes JR. Autism: The first firm finding = underconnectivity? *Epilepsy Behav* 2007; 11: 20–24.

7. Mizuno A, Villalobos ME, Davies MM, Dahl BC, Müller RA. Partially enhanced thalamocortical functional connectivity in autism. *Brain Res* 2006; 1104: 160–174.

8. Villalobos ME, Mizuno A, Dahl BC, Kemmotsu N, Müller RA. Reduced functional connectivity between V1 and inferior frontal cortex associated with visuomotor performance in autism. *Neuroimage* 2004; 25: 916–925.

9. Minshew NJ, Goldstein G, Siegel DJ. Neuropsychologic functioning in autism: Profile of a complex information processing disorder. *J Int Neuropsychol Soc* 1997; 3: 303–316.

10. Williams DL, Goldstein G, Minshew NJ. Neuropsychologic functioning in children with autism: Further evidence for disordered complex information-processing. *Child Neuropsychol* 2006; 12: 279–298.

11. Minshew NJ. Brief report: Brain mechanisms in autism: Functional and structural abnormalities. *J Autism Dev Disord* 1996; 26: 205–209.

12. Luna B, Minshew NJ, Garver KE et al. Neocortical system abnormalities in autism: An fMRI study of spatial working memory. *Neurology* 2002; 59: 834–840.

13. Minshew NJ, Sweeney JA, Luna B. Autism as a selective disorder of complex information processing and underdevelopment of neocortical systems. *Mol Psychiatry* 2002; 7: S14–S15.

14. Just MA, Cherkassky VL, Keller TA, Minshew NJ. Cortical activation and synchronization during sentence comprehension in high-functioning autism: Evidence of underconnectivity. *Brain* 2004; 127: 1811–1821.

15. Minshew NJ, Williams DL. The new neurobiology of autism. *Arch Neurol* 2007; 64: 945–950.

16. Volpe JJ. *Neurology of the Newborn*, 2nd ed., Philadelphia: W.B. Saunders Company, 1987: 33–68.

17. Minshew NJ, Payton JB, Sclabassi RJ. Cortical neurophysiologic abnormalities in autism. *Neurology* 1986; 36(Suppl 1): 194.

18. Minshew NJ, Goldstein G, Maurer RG, Bauman ML, Goldman-Rakic PS. The neurobiology of autism: An integrated theory of the clinical and anatomic deficits. *J Clin Exp Neuropsychol* 1989; 11: 66–85.

19. Minshew, NJ, Goldstein, G. Is autism an amnesic disorder? Evidence from the California Verbal Learning Test. *Neuropsychology* 1993; 7: 1–8.

20. Lord C, Risi S, Lambrecht L et al. The autism diagnostic observations schedule-generic: A standard measure of social and communication deficits associated with autism spectrum disorder. *J Autism Dev Disord* 2000; 30: 205–223.

21. Lord C, Rutter M, Le Couteur A. Autism diagnostic interview—revised: a revised version of a diagnostic interview for caregivers of individuals with possible pervasive developmental disorders. *J Autism Dev Disord* 1994;24:659-685.

22. Minshew N, Webb SJ, Williams DL, Dawson G. Neuropsychology and neurophysiology of autism spectrum disorders. In Moldin SO, Rubenstein JLR, eds. *Understanding Autism*, Boca Raton, FL: Taylor & Francis, 2006: 379–415.

23. Luna B, Minshew NJ, Garver KE et al. Neocortical system abnormalities in autism. *Neurology* 2002; 59: 834–840.

24. Gastgeb HZ, Strauss MS, Minshew NJ. Do individuals with autism process categories differently? The effect of typicality and development. *Child Dev* 2006; 77: 1717–1729.

25. Behrmann M, Avidan G, Leonard GL et al. Configural processing in autism and its relationship to face processing. *Neuropsychologia* 2006; 44: 110–129.

26. Minshew NJ, Meyer J, Goldstein G. Abstract reasoning in autism: A dissociation between concept formation and concept identification. *Neuropsychology* 2002; 16: 327–334.

27. Schultz R. Developmental deficits in social perception in autism: The role of the amygdala and fusiform face area. *Int. J. Devl Neurosci* 2005; 23: 125–141.

28. Wilson TW, Rojas DC, Reite ML Teale PD, Rogers SJ. Children and adolescents with autism exhibit reduced MEG steady-state gamma responses. *Biol Psychiatry* 2007; 62: 192–197.

29. García-Viallamisar D, Della Sala S. Dual-task performance in adults with autism. *Cognit Neuropsychiatry* 2002; 7: 63–74.
30. Just MA, Cherkassky VL, Keller TA, Kana RK, Minshew NJ. Functional and anatomical cortical underconnectivity in autism: Evidence from an fMRI study of an executive function task and corpus callosum morphometry. *Cereb Cortex* 2007; 17: 951–961.
31. Minshew NJ, Sung K, Jones BL, Furman JM. Underdevelopment of the postural control system in autism. *Neurology* 2004; 63: 2056–2061.
32. Minshew NJ, Luna B, Sweeney JA. Oculomotor evidence for neocortical systems but not cerebellar dysfunction in autism. *Neurology* 1999; 52: 917–922.
33. Castelli F, Frith C, Happé F, Frith U. Autism, asperger syndrome and brain mechanisms for the attribution of mental states to animated shapes. *Brain* 2002; 125: 1839–1849.
34. Kana RK, Keller TA, Cherkassky VL, Minshew NJ, Just MA. Sentence comprehension in autism: Thinking in pictures with decreased functional connectivity. *Brain* 2006; 129: 2484–2493.
35. Kana RK, Keller TA, Minshew NJ, Just MA. Inhibitory control in high-functioning autism: Decreased activation and underconnectivity in inhibition networks. *Biol Psychiatry* 2007; 62: 198–206.
36. Koshino H, Kana RK, Keller TA et al. fMRI investigation of working memory for faces: Visual coding and underconnectivity with frontal areas. *Cereb Cortex* 2007; [E pub 5/20/2007].
37. Koshino H, Carpenter PA, Minshew NJ et al. Functional connectivity in an fMRI working memory task in high-functioning autism. *Neuroimage* 2005; 24: 810–821.
38. Wang AT, Lee SS, Sigman M, Dapretto M. Neural basis of irony comprehension in children with autism: The role of prosody and context. *Brain* 2006; 129: 932–943.
39. Mason RA, Williams DL, Kana RK, Minshew NJ, Just MA. Theory of mind disruption and recruitment of the right hemisphere during narrative comprehension in autism. *Neuropsychologia* 2008; 46: 269–280.
40. van der Geest JN, Kemner C, Camfferman G, Verbaten MN, van Engeland H. Eye movements, visual attention, and autism: A saccadic reaction time study using the gap and overlap paradigm. *Biol Psychiatry* 2001; 50: 614–619.
41. Luna B, Doll SK, Hegedus SJ, Minshew NJ, Sweeney JA. Maturation of executive function in autism. *Biol Psychiatry* 2006; 61: 474–481.
42. Kanner L. Autistic disturbances of affective contact. *Nerv Child* 1943; 2: 217–250.
43. Amaral DG, Corbett BA. The amygdala, autism and anxiety. *Novartis Found Symp* 2003; 251: 177–187.
44. Dapretto, M, Davies, MS, Pfeifer, JH et al. Understanding emotions in others: Mirror neuron dysfunction in children with autism spectrum disorders. *Nat Neurosci* 2005; 9: 28–30.
45. Barkovich AJ, Kuzniecky RI, Jackson GD et al. A developmental and genetic classification for malformations of cortical development. *Neurology* 2005; 65: 1873–1887.
46. Courchesne E, Karns CM, Davis HR et al. Unusual brain growth patterns in early life in patients with autistic disorder: An MRI study. *Neurology* 2001; 57: 245–254.
47. Sparks BF, Friedman SD, Shaw DW et al. Brain structural abnormalities in young children with autism spectrum disorder. *Neurology* 2002; 59: 184–192.
48. Aylward EH, Minshew NJ, Field K et al. Effects of age on brain volume and head circumference in autism. *Neurology* 2002; 59: 175–183.
49. Redcay E, Courchesne E. When is the brain enlarged in autism? A meta-analysis of all brain size reports. *Biol Psychiatry* 2005; 58: 1–9.
50. Courchesne E, Pierce K. Brain overgrowth in autism during a critical time in development: Implications for frontal pyramidal neuron and interneuron development and connectivity. *Int J Dev Neurosci* 2005; 23: 153–170.
51. Carper RA, Moses P, Tigue ZD, Courchesne E. Cerebral lobes in autism: Early hyperplasia and abnormal age effects. *Neuroimage* 2002; 16: 1038–1051.

52. Carper RA, Courchesne E. Localized enlargement of the frontal cortex in early autism. *Biol Psychiatry* 2005; 57: 126–133.
53. Herbert MR, Ziegler DA, Makris N et al. Localization of white matter volume increase in autism and developmental language disorder. *Ann Neurol* 2004; 55: 530–540.
54. Casanova MF, Buxhoeveden DP, Switala AE, Roy E. Minicolumnar pathology in autism. *Neurology* 2002; 58: 428–432.
55. Casanova MF, Buxhoeveden DP, Switala AE, Roy E. Neuronal density and architecture (gray level index) in the brains of autistic patients. *J Child Neurol* 2002; 17: 515–521.
56. Casanova MF, van Kooten IA, Switala AE et al. Minicolumnar abnormalities in autism. *Acta Neuropathol (Berl)* 2006; 112: 287–303.
57. Buxhoeveden DP, Semendeferi K, Buckwalter J et al. Reduced minicolumns in the frontal cortex of patients with autism. *Neuropathol Appl Neurobiol* 2006; 32: 483–491.
58. Crossley PH, Martinez S, Ohkubo Y, Rubenstein JL. Coordinate expression of Fgf8, Otx2, Bmp4, and Shh in the rostral prosencephalon during development of the telencephalic and optic vesicles. *Neuroscience* 2001; 108: 183–206.
59. Fukuchi-Shimogori T, Grove EA. Neocortex patterning by the secreted signaling molecule FGF8. *Science* 2001; 294: 1071–1074.
60. Hebert JM, Mishina Y, McConnell SK. BMP signaling is required locally to pattern the dorsal telencephalic midline. *Neuron* 2002; 35: 1029–1041.
61. Wilson SW, Rubenstein JL. Induction and dorsoventral patterning of the telencephalon. *Neuron* 2000; 28: 641–651.
62. Bishop KM, Rubenstein JL, O,Leary DD. Distinct actions of Emx1, Emx2, and Pax6 in regulating the specification of areas in the developing neocortex. *J Neurosci* 2002; 22: 7627–7638.
63. Schuurmans C, Guillemot F. Molecular mechanisms underlying cell fate specification in the developing telencephalon. *Curr Opin Neurobiol* 2002; 12: 26–34.
64. Stenman J, Yu RT, Evans RM, Campbell K. Tlx and Pax6 co-operate genetically to establish the pallio-subpallial boundary in the embryonic mouse telencephalon. *Development* 2003; 130: 1113–1122.
65. Tarabykin V, Stoykova A, Usman N, Gruss P. Cortical upper layer neurons derive from the subventricular zone as indicated by Svet1 gene expression. *Development* 2001; 128: 1983–1993.
66. Parnavelas JG, Anderson SA, Lavdas AA et al. The contribution of the ganglionic eminence to the neuronal cell types of the cerebral cortex. *Novartis Found Symp* 2000; 228: 129–139; discussion 139–147.
67. Schuurmans C, Armant O, Nieto M et al. Sequential phases of cortical specification involve neurogenin-dependent and -independent pathways. *EMBO J* 2004; 23: 2892–2902.
68. Quinn JC, Molinek M, Martynoga BS et al. Pax6 controls cerebral cortical cell number by regulating exit from the cell cycle and specifies cortical cell identity by a cell autonomous mechanism. *Dev Biol* 2006; 302: 50–65.
69. Friedman, SD, Shaw, DWW, Artru, AA et al. Gray and white matter brain chemistry in young children with autism. *Arch Gen Psychiatry* 2006; 63: 786–794.
70. Petropoulous, H, Friedman, SD, Shaw, DWW et al. Gray matter abnormalities in autism spectrum disorder revealed by T2 relaxation. *Neurology* 2006; 67: 632–636.

Part VI
Environmental Mechanisms and Models

Chapter 19
Evidence for Environmental Susceptibility in Autism

What We Need to Know About Gene × Environment Interactions

Isaac N. Pessah and Pamela J. Lein

Abstract Research into the pathophysiology and genetics of autism may inform the identification of environmental susceptibility factors that promote adverse outcomes in brain development. Conversely, understanding how low-level chemical exposure influences molecular, cellular, and behavioral outcomes relevant to the development of autism will enlighten geneticists, neuroscientists, and immunologists about autism's complex etiologies and possibly yield novel intervention strategies. The inherent imbalances in neuronal connectivity in children at risk for autism are likely to provide the biological substrate for enhanced susceptibility to environmental triggers that are known to target signaling systems that establish the basic patterns of connectivity, from early neuronal migration and axonal pathfinding to postnatal refining of neuronal connections. Three examples of gene × environment interactions that likely contribute to autism risk are illustrated: pesticides that interfere with (1) acetylcholine (ACh) and (2) γ-aminobutyric acid (GABA) neurotransmission; and (3) the persistent organic pollutants that directly alter Ca^{2+} signaling pathways and Ca^{2+}-dependent effectors. One fundamental way in which heritable genetic vulnerabilities can amplify the adverse effects triggered by environmental exposures is if both factors (genes and environment) converge to dysregulate the same neurotransmitter and/or signaling systems at critical times during development.

Keywords Autism · genes · environment · susceptibility · pollution

Introduction

Autism is a heterogeneous neurodevelopmental disorder defined by core deficits in social reciprocity, communication, and restrictive/repetitive patterns of

I.N. Pessah
Department of Molecular Bioscience, School of Veterinary Medicine and Center for Children's Environmental Health Sciences and Disease Prevention, University of California, Davis, CA 95616, USA
e-mail: inpessah@ucdavis.edu

A.W. Zimmerman (ed.), *Autism*, DOI: 10.1007/978-1-60327-489-0_19,
© Humana Press, Totowa, NJ 2008

interest and behavior [1]. Generally accepted estimates of prevalence range from 1:750 for the narrowest diagnostic criteria to 1:150 for autism spectrum disorder (ASD) [2, 3]. Although autism may be one of the most heritable complex disorders [4, 5], the genes linked to autism risk do not segregate in a simple Mendelian manner [6]. Results from over ten genome-wide autism screens indicate that potential susceptibility genes are spread across the entire genome. Estimates of the number of genes involved in autism range from as few as 3 to as many as 100 [6, 7, 8, 9]. However, no single locus alone appears to be sufficient to cause the full clinical phenotype [10, 11]. Evaluation of a broadly defined autistic phenotype that included communication and social disorders increased concordance from 60 to 92% in monozygotic twins and from 0 to 10% in dizygotic pairs [12]. These results indicate that interactions among multiple genes are likely to contribute to autism, and heritable epigenetic factors and/or non-heritable environmental exposures may significantly influence susceptibility and variable expression of autism and autism-related traits. A major challenge in the field is to identify environmental factors of relevance to autism. The relative importance of environmental factors in autism is suggested by recent studies linking viral infections and chronic inflammation early in development to ASD (reviewed in this book by Van de Water et al., Chapter 12; Pardo, Chapter 15) and evidence suggesting that sex hormones mediate the well-documented gender differences in autism susceptibility [13] via gene-specific epigenetic modifications of DNA and histones [14]. Reports that exposure to teratogens such as thalidomide, valproic acid, and misoprostol during critical periods of development can cause autism-related traits in children [15, 16] have raised interest regarding the role of chemical exposures in autism.

This chapter will focus on what we know and what we need to know about environmental chemicals and autism susceptibility. From a toxicologist's perspective, the identity of "defective" genes and signaling pathways linked to autism provides important clues about exposures to environmental chemicals that influence autism susceptibility, severity, and/or treatment outcomes. One fundamental way by which heritable genetic vulnerabilities can amplify the adverse effects triggered by environmental exposures is if both factors (genes and environment) converge to dysregulate the same neurotransmitter and/or signaling systems at critical times of development. A framework for understanding how such gene $\times$ environment interactions could influence adverse outcomes in the developing child is depicted by the hypothetical quantal dose–response relationship shown in Fig. 19.1. Any population will have a fraction of susceptible and resistant individuals that respond adversely below and above the exposure level at which the population mean (TD_{50}) exhibits a defined adverse effect, respectively (Fig. 1A). When applied to the entire autism spectrum in which the adverse effect is a specified autism-related phenotype, it may be difficult to measure a significant shift in the population mean relative to the general "typically developing" population because autism spectrum is likely to be a common set of behavioral deficits that can arise from several etiological processes. Thus, for ASD, the distribution about the quantal dose–response relationship for an adverse response is likely to

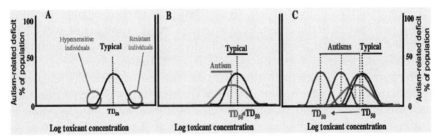

Fig. 19.1 (**A**) Depiction of how a typically developing population responds to increasing concentrations of a hypothetical environmental toxicant or combination of toxicants. In every population, there are hypersensitive individuals that, due to their genetic makeup, timing of exposure, and/or coincidental illnesses (*e.g.*, viral or bacterial infection), are particularly vulnerable to adverse effects associated with low-level exposures. The TD_{50} is the exposure dose of a chemical(s) that produces an autism-related adverse effect in 50% of the population. Resistant individuals are likely to possess more robust metabolic defenses and repair mechanisms. (**B**) An example of how autism spectrum, grouped as a single population, could influence the mean sensitivity to an environmental exposure. Because of autism's heterogeneity, the distribution about the TD_{50} is likely to be broader, thereby adversely affecting a larger fraction of the autism population. (**C**) Stratification of autism into definable endophenotypes possessing distinct but overlapping sets of susceptibility genes could result in definable subpopulations having different sensitivity to environmental modifiers of autism-related deficits, severity, and/or treatment outcomes

be broader. However, a broader variance about the mean predicts that a larger fraction of the autism population may respond adversely to exposures below the *no observable effect level* (NOEL) for the general population (Fig. 1B). Current research is being directed at defining distinct neurological endophenotypes in autism that may possess overlapping sets of susceptibility genes [17]. It is not, therefore, unreasonable to expect that as subgroups of autistic children are better defined based on common biological abnormalities, distinct sensitivity to environmental modifiers of autism risk, severity and/or treatment outcomes will be identified (Fig. 1C). Each distribution shown in Fig. 1C is meant to portray a distinct subgroup within the autism spectrum and their respective sensitivity to the adverse effect of a chemical (or chemical mixture). This hypothetical example is intended to underscore the concept that combinations of heritable factors contributing to autism risk can influence the sensitivity of the developing brain to the adverse effects of specific types of environmental exposure(s). The degree of environmental susceptibility can differ among autism's endophenotypes. In Fig. 1C, each of these subgroups is depicted as having a measurable difference in the mean and/or variance about the mean for an adverse outcome resulting from exposure to xenobiotic chemical(s), and these can be thought of as distinct "xenotypes." Some of these subgroups may in fact represent a very small fraction of the total autistic population, but in fact are the xenotypes that may most benefit from interventions that limit exposure.

Current efforts to identify clinical endophenotypes within the autism spectrum are therefore likely to help our understanding of the constellations of genes that confer differential sensitivity to distinct environmental exposures during gestational and neonatal development. Such approaches will likely prove useful in defining subgroups of children that differ in susceptibility to environmental exposures that promote autism risk and severity.

The purpose of this chapter is to illustrate how research into the pathophysiology and genetics of autism may inform the identification of environmental susceptibility factors that promote adverse outcomes in brain development. Conversely, understanding how low-level chemical exposure influences molecular, cellular, and behavioral outcomes relevant to autism will better inform geneticists, neuroscientists, and immunologists about autism's complex etiologies. Recent genetic findings coupled with emerging histological, electrophysiological, and functional imaging studies suggest that autism results from an imbalance in the ratio of excitatory and inhibitory neurons within the developing brain [18] and the neural networks they form [19]. Such imbalances are likely to result in the failure of connections to form normally between brain regions involved in higher order associations [20]. The inherent imbalances in neuronal connectivity present in children at risk for autism are likely to provide the biological substrate for enhanced susceptibility to environmental triggers that are known to target signaling systems that establish the basic patterns of connectivity from early neuronal migration and axonal pathfinding to postnatal events that function to refine neuronal connections, such as dendritic growth, synapse formation, and pruning [19, 20, 21]. The toxicological literature includes ample evidence that many of the environmental chemicals of concern to human health either directly or indirectly affect the very same signaling systems that are impaired in autism. For example, several pesticides of historical and current importance are in fact known to interfere with acetylcholine (ACh) and γ-aminobutyric acid (GABA) neurotransmission, which are neurotransmitter systems demonstrated to be altered in autism, and these will serve as examples. A third example proposed in this chapter is the possible role of persistent organic pollutants that directly alter Ca^{2+} signaling pathways and Ca^{2+}-dependent effectors and are thus likely to impact a broad range of developmental processes, especially in individuals susceptible to autism.

Developmental Neurotoxicants and Autism

Cholinergic Agents

In the United States alone, more than a billion pounds of organophosphate (OP) insecticide is used for agricultural and domestic purposes annually [22]. OPs are potent inhibitors of the enzyme acetylcholinesterase (AChE) that is responsible for rapid hydrolysis of ACh at nicotinic and muscarinic synapses. Inhibition of AChE by OPs has been amply documented to cause a plethora of toxicological effects that result from initially overstimulating, and consequently desensitizing, cholinergic transmission. More recent research suggests that some OPs cause developmental neurotoxicity by mechanisms independent of AChE inhibition [23, 24, 25, 26, 27]. For example, the OP chlorpyrifos has been shown to enhance the phosphorylation of the Ca^{2+}/cAMP response element binding protein CREB in neuronal cultures at concentrations well below those that inhibit AChE [24]. Tight regulation of CREB activity by Ca^{2+}-dependent phosphorylation is critical for normal neural progenitor proliferation and differentiation, dendritic development, and cognitive function [28, 29]. CREB activity in turn regulates the expression of many nerve-specific genes including NF-l, TH, chromogranin, pEJ, and a number of immediate early genes that encode transcription factors (e.g., c-Jun). Slotkin and co-workers recently showed that neonatal exposure of rats to two widely used OPs, chlorpyrifos and diazinon, produced marked changes in the pattern of expression of specific isoforms of fibroblast growth factor genes (*fgf*) and their receptor (*fgfr1*) at doses below those that cause systemic toxicity or growth impairment resulting from AChE inhibition [27]. Although the molecular mechanisms mediating these developmental effects are unknown, they may have a common origin as the nuclear accumulation of FGFR1 protein is activated by changes in cell contacts and by stimulation of cells with growth factors, neurotransmitters, and hormones. Stachowiak and co-workers proposed that FGFR1 represents a key convergence point for integrating signaling cascades initiated by specific membrane receptors that transmit signals to sequence-specific transcription factors through co-activation of CREB [30]. Investigation into the adverse toxicological consequences of OPs via AChE-dependent and AChE-independent mechanisms and how they can be synergized by inborn errors in Ca^{2+} signaling such as Timothy syndrome (see below), which has a 60% autism rate, are needed to permit more meaningful risk assessments of susceptible populations.

In addition to OPs, a newer class of insecticides modeled after the chemical structure of nicotine mimics the effects of nicotine at nicotinic cholinergic receptors (nAChRs). Neonicotinoids such as imidacloprid are commonly used in domestic applications. Because of their purported selectivity toward nAChR isoforms found in target insects, the use of neonicotinoids has been generally considered to have a wide margin of safety [31]. Nevertheless, neonicotinoids have not been directly tested for their potency toward all the nAChR subtypes expressed in mammalian brain that are composed of pentameric combinations of nine alpha subunits ($\alpha2$–$\alpha10$) and three beta subunits ($\beta1$–$\beta3$) [32].

Relevance to Autism

Neurotransmission at cholinergic synapses is essential for functional aspects of cognition, reward, motor activity, and analgesia. In addition, dysregulation of cholinergic neurotransmission has been associated with several pathological conditions including autism. To date, no differences in the expression of brain muscarinic cholinergic receptors or choline acetyltransferase (ChAT) have been associated with autism. By contrast, regional differences in the expression of specific nAChR subunits have been documented in postmortem adult brains obtained from autistic and non-autistic individuals. A decrease in the density of [^{3}H]epibatidine-binding sites within the parietal cortex, cerebellum, and thalamus of autistic brains was correlated to decreases in the levels of α3, α4, and β2 subunits by western blotting [33, 34, 35]. In the cerebellum, the [^{3}H]epibatidine-binding density to α3 and α4 subunits was reduced by 40–50% in the granule cell, Purkinje, and molecular layers in the autistic group compared with the non-autistic group. Cerebella from autistic patients possessed approximately threefold higher binding of [^{125}I]α-bungarotoxin to the α7 subunit in the granule cell layer compared to the general population comparison group, and this increase was not observed to the same extent in the other mental retardation groups examined in the study [35]. How these differences in nAChR expression arise is not known. However, these findings raise important questions as to whether deficits in nAChR neurotransmission identified in at least a subset of autistic children might confer a particularly high sensitivity to the adverse effects of OP and/or neonicotinoid exposure. Allelic variants of the gene that encodes for paraoxonase (PON), a metabolic enzyme that is critical to the detoxification of OPs, have been associated with autism in North America, and these variants appear to have lower metabolic activity in vitro [36]. Lower PON activity was also recently associated with autism in a small Romanian study [37]. Thus, reduced metabolic capacity to detoxify and excrete OPs, in conjunction with deficits in nAChR neurotransmission could provide a framework for future studies of gene × environment interactions relevant to autism risk. Environmental epidemiological studies of possible associations between pesticide exposures and the major triad of deficits seen in autism (social reciprocity, language, and stereotyped behavior) have only begun to receive attention [38]. Eskanazi and co-workers recently reported evidence of a positive association between levels of dialkylphosphate metabolites (biomarkers of OP exposure) in maternal and child urine with risk of pervasive developmental disorder at the age of 24 months in a longitudinal birth cohort of primarily Latino farm worker families in California [39]. Although these new associations between OP exposure and PDD should not be interpreted as evidence of causation, clearly more research is needed to understand how long-term low-level exposure to OPs, neonicotinoids, and other chemicals that interfere with cholinergic signaling influence adverse developmental outcomes, especially in children in whom these signaling systems are genetically impaired.

Chemicals that Interfere with GABA Neurotransmission

Several classes of pesticides are known to interfere with GABA-mediated neurotransmission because they bind to the type A family of GABA receptors (GABR) and block their ability to mediate chloride fluxes. Along with nAChRs discussed above, GABR channels are members of the ligand-gated ion channel superfamily of genes that code for proteins having high sequence homology but divergent spatiotemporal expression patterns and functions. In immature neurons, GABR primarily mediates excitatory responses, but during maturation and in mature neurons they mediate rapid inhibition of transmission [40]. This "switch" in function is the result of a reversal in direction of the chloride gradient across the neuronal membrane with the maturation-dependent expression of the potassium-chloride co-transporter KCC2 that lowers the internal Cl^- concentration below that found in the extracellular fluid [41]. The physiological relevance of these different functional GABR responses during neuronal maturation remains unclear. What is clear is that altered expression and function of these chloride channels have been implicated in autism and associated co-morbidities including sleep disturbances, anxiety, and epilepsy.

Organochlorine (OC) insecticides were widely used for both large-scale agriculture and domestic pest control beginning in the early 1940s. The vast majority of OC insecticides were banned from use in the United States beginning in the late 1970s due to their persistence in the environment, adverse effects on avian reproduction, and their ability to promote hepatic tumors in rodent studies. OC insecticides that possess polychloroalkane structures are known to bind to GABR in the mammalian brain and potently block their ability to conduct Cl^- with many having nanomolar affinity for their receptor-binding site [42, 43]. In the United States, toxaphene represents one of the most heavily used pesticides and was banned in 1990, whereas heptachlor and dieldrin were banned in the late 1980s. Examples of polychloroalkane insecticides that are currently being used in the United States include endosulfan and lindane. Dicofol, an OC structure similar to DDT, is also currently registered for use on agricultural crops. Because of their chemical stability, global distribution from countries that continue to use these compounds, and their propensity to bioaccumulate, exposures to OC insecticides continue to be a significant concern to human health worldwide. Yet, relatively little is known about their developmental neurotoxicity, and the long-term consequences of low-dose exposures have not been adequately evaluated [44].

Another class of insecticide that interferes with GABA neurotransmission and has attained broad domestic and commercial use is the 4-alkyl-1-phenylpyrazoles. Although these compounds do not persist in the environment to the extent observed with OCs, they are heavily used within the home and for commercial pest control. For example, fipronil is formulated as a topical for control of fleas and ticks on pets. Fipronil's insecticidal activity is mediated primarily through its actions as a non-competitive antagonist of GABR. In this regard, fipronil was initially developed as an insecticide because of its higher selectivity for insect forms of GABR relative to mammalian GABR [45]. However, results from recent studies indicate that fipronil, like endosulfan and lindane, is indeed a high-affinity noncompetitive antagonist for the mammalian β_3-homopentameric GABR (GABRβ_3) [46, 47]. In fact, the GABRβ_3 recognizes fipronil with nanomolar affinity in a manner indistinguishable from its interaction with insect GABA receptors. These more recent findings raise questions about the high degree of selectivity once attributed to 4-alkyl-1-phenylpyrazoles. More importantly, the finding that a diverse group of widely used insecticides converge on a common molecular target within mammalian GABRβ3 has important implications for human risk, especially in those individuals with heritable impairments in $GABA_A R$ signaling pathways.

Relevance to Autism

The hypothesis of an imbalance between excitation and inhibition within developing CNS circuits as a common etiological factor contributing to autism susceptibility is supported by several reports of GABR abnormalities in autism. Of particular relevance to the environmental exposures to OC insecticides and 4-alkyl-1-phenylpyrazoles discussed above is the significantly lower level of $GABR\beta_3$ expression measured in postmortem brain samples obtained from children diagnosed with autism, Rett syndrome, and Angelman syndrome [48]. The gene encoding for $GABR\beta_3$ is located in a non-imprinting region of chromosome 15q11–13 suggesting an overlapping pathway of gene dysregulation common to all three disorders that may stem from abnormal regulation of DNA methylation by the methyl CpG-binding protein MeCP2 [49]. Quantitative immunoblot analyses of the brains of Mecp2-deficient mice, a model of human Rett syndrome, are deficient in the $GABR\beta_3$ subunit [48], further supporting the working hypothesis that epigenetic mechanisms conferring gene dysregulation within 15q11–q13 in Angelman syndrome and autism may significantly contribute to heightened susceptibility to compounds that potently block the chloride channel of $GABR\beta_3$. Autism has been linked to polymorphisms within additional genes that encode GABR, including $GABR\gamma_1$ located on 15q11–13 [50, 51]. Complex epistatic interactions between genes that encode $GABR\alpha_4$ and $GABR\beta_1$ within 4q12 have also been reported in autistic probands [52]. Collectively, these data implicate GABR dysregulation as a major contributor to imbalances between excitation and inhibition in the autistic brain and may provide the best lead for studying gene × environment interactions that enhance autism susceptibility.

Is there any epidemiological evidence linking exposure to chemicals that interfere with GABR neurotransmission and autism risk? Roberts and coworkers recently published the results of a study that examined possible association between maternal residence near agricultural pesticide applications during key periods of gestation and development of autism [53]. Of nearly 250 hypotheses tested, children of mothers living within 500 m of field sites with the highest non-zero quartile of OC (primarily endosulfan and dicofol) poundage had a risk factor for autism that was 6.1 times higher than that of mothers not living near agricultural fields. Autism risk increased with poundage of OC applied and decreased with distance from field sites [53]. Although small in sample size, these new findings underscore the critical need to understand how heritable imbalances in GABR signaling and their associated pathways influence susceptibility to chemicals that are known to directly target GABA receptors, especially GABRβ_3, during gestation and postnatal development. However, predicting the functional consequences of exposure to these chemicals is difficult because we do not have sufficient scientific data regarding the specificity and functional outcome of interactions between toxicants that target GABR and the 16 known GABR subunits that oligomerize as heteropentamers or homotetramers to give rise to the diverse functional roles of these receptors in regulating CNS excitability. Development of relevant models in mice will be essential to understanding the mechanisms by which complex gene × environment interaction at the level of GABR signaling influence autism risk and to better predict susceptible endophenotypes.

Chemicals that Interfere with Calcium Signalling

Foreseeing the extraordinary importance of the calcium ion (Ca^{2+}) in biology and medicine, the Nobel laureate Otto Loewi stated "Ja, Kalzium das ist alles!" (yes, calcium is everything!). Changes in localized and global intracellular Ca^{2+} concentration represent one of the most common ways in which cells regulate cell cycle, terminal differentiation, migration, and death. Moreover, intracellular Ca^{2+} orchestrates hundreds, if not thousands, of biochemical processes essential for metabolism, transport, secretion, and regulation of gene transcription and translation in most mammalian cell types. Thus, Ca^{2+} is a fundamental regulator of most biological processes. Although Ca^{2+}-regulated processes vary according to cell type and the chemical and physical environment in which the cell finds itself, significant progress has been made toward understanding the mechanisms by which cells generate spatially and temporally segregated Ca^{2+} signals that are context-specific [54]. Both the nAChR and GABR that are implicated in autism contribute directly or indirectly to the regulation of Ca^{2+}-dependent pathways and are themselves regulated by Ca^{2+} dependent processes. For example, GABR activation in neurons initiates at least two distinct signal transduction pathways, one in which the Cl^- current is activated and another which involves the elevation of intracellular Ca^{2+} through functional coupling with voltage activated Ca^{2+} channels [55].

Moreover, several of the candidate genes for autism encode proteins whose primary role is to generate intracellular Ca^{2+} signals or are themselves tightly regulated by local fluctuations in Ca^{2+} concentrations (Table 19.1).

A large number of priority chemicals of concern to human environmental health have been shown to affect the integrity of cellular Ca^{2+} signals. Prominent examples are the polyaromatic hydrocarbons that mediate their toxicity through the arylhydrocarbon receptor (AhR), such as dioxin, or through selective interactions with ryanodine receptors (RyR) in the brain, such as

Table 19.1 Examples of Ca^{2+} regulating and Ca^{2+} regulated genes linked to autism

Gene (map)	Function	Mutation (Dysfunction)
CACNA1C (12p13.3)	L-type voltage-dependent Ca^{2+} channel (CaV1.2)	G406R-delayed inactivation (Timothy Syndrome)
CACNA1H (16p13.3)	T-type voltage-dependent Ca^{2+} channel (CaV3.2)	R212C; R902W,W962C, A1874V- altered activation (autism)
SLC25A12 (2q24)	Ca^{2+}-dependent mitochondrial aspartate/glutamate carrier	SNPs (autism)
KCNMA1 (10q22.3)	Ca^{2+}-activated K^+ channel (BK_{Ca}^{2+})	Balanced 9q23/10q22 translocation
PTEN (10q23.3)	Ca^{2+}-regulated PI-3-phosphatase; regulates CaV1.2	H93R, D252G, F241S- (macrocephaly; autism)
MECP2 (Xq28)	DNA methylation (Ca^{2+}-dependent phosphorylation)	Down regulated/mutations- altered DNA methylation (autism, RETT syndrome)
MET (7q21.1)	Tyrosine receptor kinase for hepatocyte growth factor coupled to IP3 production	Polymorphism-down regulation (autism)
CADPS2 (7q31–q32)	Ca^{2+}-dependent activator protein for secretion	Aberrant alternative splicing lacks exon 3 (autism)
NL-1; NL-3; 3q26.31; Xq13.1	Neuroligin-synapse formation/ function EF-hands	NL-1 R476C (autism) NL-3 R471C (autism)

non-coplanar polychlorinated biphenyls (PCBs). With respect to the former, Dale and Eltom [56] recently showed a major role for Ca^{2+}/calpain in AhR transformation, transactivation, and subsequent down-regulation. Interactions of non-coplanar PCBs with RyR greatly sensitize the release of Ca^{2+} from microsomal intracellular stores [57, 58]. Sensitization of RyRs by PCBs has been shown to be responsible for a host of toxicological outcomes in vitro including amplification of NMDA-mediated excitation and Ca^{2+}/caspase-mediated apoptosis [59, 60, 61]. RyRs are physically and functionally linked to voltage-gated Ca^{2+} channels at the surface of the neuron where they form Ca^{2+} release units responsible for generating microdomains of signaling (Ca^{2+} microdomains) [54]. Of relevance to autism susceptibility, a gain-of-function missense mutation in the L-type Ca^{2+} channel CaV1.2 causes Timothy syndrome which has a 60% autism rate [62]. Could inherent abnormalities in Ca^{2+} signaling pathways represent a major point of convergence in the pathobiology of autism? Could environmental factors that dysregulate the spatial and temporal fidelity of Ca^{2+} microdomains amplify inborn weaknesses in generating and decoding Ca^{2+}-dependent processes leading to heightened susceptibilities to autism risk and severity and its associated co-morbidities? Considering that defective neuronal connectivity is now being considered a common etiological feature in most forms of autism and its phenotypic manifestation is closely linked to imbalances in brain circuitry, the hypothesis of gene × environment as it relates to Ca^{2+} signals requires more attention.

Ca^{2+} and Neuronal Connectivity: How Chemicals can Make a Good Plan go Bad

A principal determinant of neuronal connectivity is activity, which increases intracellular Ca^{2+} levels and activates downstream Ca^{2+}-dependent signaling pathways. Spontaneous brain activity influences early neurodevelopmental events that determine the initial architecture of sensory and motor systems, and patterned electrophysiological activity triggered by sensory experience is critical for refining neural circuits to form functionally integrated sensory and motor networks [63–67]. Emerging genetic evidence has linked defects in activity-dependent signaling with autism [68] and recent reports indicate that environmental factors that alter Ca^{2+} signaling also disrupt neuronal connectivity [69, 70]. These observations suggest that genetic and environmental factors that modulate Ca^{2+}-dependent signaling pathways may interact either in parallel or in series to determine susceptibility to autism. Here, we illustrate this hypothesis in the context of dendritic morphogenesis.

Dendritic morphology is a critical determinant of neuronal connectivity. The size of the dendritic arbor and the density of dendritic spines determine the total synaptic input a neuron can receive [71, 72] and influence the types and distribution of these inputs [73, 74, 75]. Dendritic morphogenesis can be separated into two, often concurrent, phases: (1) overall growth typified by the addition and elongation of dendritic processes and (2) refinement or patterning of the dendritic arbor through localized outgrowth and branching, spine formation, and selective maintenance and retraction of terminal dendritic branches and spines [76]. Both phases of dendritic growth are strongly influenced by neuronal activity as evidenced by the remarkable effect of experience on the development and refinement of synaptic connections, which not only patterns neural circuitry during development (reviewed by [77, 78, 79]) but also underlies associative learning [80, 81, 82, 83].

The effects of neuronal activity on dendritic growth are mediated primarily, if not exclusively, by changes in intracellular Ca^{2+} (reviewed in [84, 85]). Ca^{2+}-dependent signaling affects dendritic development globally by controlling the size of the arbor and density of spines and locally by shaping branching patterns and regulating fine structural changes in spines [86]. A generalization emerging from studies of activity-dependent dendritic growth is that Ca^{2+} exerts bimodal effects on dendritic structure. Thus, increased intracellular Ca^{2+} has been linked to both dendritic growth and to dendritic retraction (reviewed in [84, 85, 87]). Two possibilities have been proposed to explain why Ca^{2+} signaling stimulates dendritic growth in some cases while inhibiting dendritic plasticity in others. First, Segal et al. [84] suggest that moderate and/or transient increases in intracellular Ca^{2+} cause growth of dendritic branches and spines, whereas large Ca^{2+} increases cause destabilization and retraction of these dendritic structures. Consistent with this possibility, relatively sustained increases in intracellular Ca^{2+} versus transient changes in Ca^{2+} influx activate different Ca^{2+}-dependent signaling pathways with distinct effects on dendrites [87, 88]. The second possibility is that dendritic responses to Ca^{2+} differ with neuronal maturation, such that early in development, increased Ca^{2+} promotes dendritic growth, while later in development, increased Ca^{2+} functions to stabilize dendritic structure [85]. Thus, factors that alter Ca^{2+} signaling could give rise to diverse and heterogeneous patterns of altered dendritic arborization depending on the timing, the direction of change, and the level of endogenous activity.

Activity increases intracellular Ca^{2+} by activating neurotransmitter receptors resulting in $Ca2^+$ influx through Ca^{2+}-permeant ionotropic receptors, Ca^{2+} entry via voltage-gated Ca^{2+} channels, or release of Ca^{2+} from intracellular stores (e.g., Ca^{2+}-induced Ca^{2+} release or CICR). Somatic or global Ca^{2+} changes influence general dendritic growth via modulation of gene transcription, whereas local changes in Ca^{2+} within dendrites or spines alter dendritic patterning via translation of mRNA localized within postsynaptic densities or non-genomic regulation of the actin cytoskeleton [84, 85]. Two signaling pathways linking neuronal activity to transcription of gene products that regulate dendritic growth have recently been described. One pathway involves the nuclear translocation of the C-terminal fragment of $Ca_v1.2$, an L-type voltage-gated calcium channel (LTC) to regulate transcription [89]. Nuclear calcium channel associated transcription regulator (CCAT) expression is developmentally regulated in the brain, with peak levels coinciding temporally with the period of most rapid dendritic growth and synaptogenesis. Overexpression of CCAT increases dendritic arborization in cultured cortical neurons, and this effect requires an intact transcriptional activation domain. However, the concentration of nuclear CCAT in cortical neurons is decreased by experimental manipulations that increase intracellular Ca^{2+}. While this latter observation raises questions about the functional relationship between activity-dependent and CCAT-dependent dendritic growth, it is of interest in light of recent studies showing that a gain-in-function mutation in the gene encoding $Ca_v1.2$ is linked to autism [62]. This mutation increases intracellular Ca^{2+} levels, which would be expected to decrease nuclear levels of CCAT resulting in attenuated dendritic growth. Reduced dendritic arborization has been observed in some cases of autism [21, 90].

The second pathway linking activity to dendritic growth is more clearly delineated: NMDA receptor activation triggers dendritic growth and spine formation in cultured hippocampal neurons via the sequential activation of CaM-dependent kinase kinase (CamKK), CamKI, and MEK/ERK to enhance CREB-mediated transcription of Wnt-2 [91]. In other studies, activity-dependent activation of CREB-mediated transcription has been shown to upregulate expression of BDNF [92] as has activity-dependent phosphorylation of MECP2 [93, 94]. BDNF increases dendritic growth in part by activating the phosphoinosidtide-3-kinase (PI3K) and protein kinase B (Akt) signaling pathway, which modulates local protein synthesis via the serine-threonine protein kinase mammalian target of rapamycin (mTOR) [95, 96, 97, 98]. The PI3K-AKT-mTOR signaling pathway is negatively regulated by the phosphatase and tensin homolog on chromosome ten (PTEN). Defects in these signaling pathways have been linked to altered dendritic arborization in animal models and to autism. Mice deficient in *Mecp2* display delayed neuronal maturation in the cerebral cortex with reduced dendrite growth and dendritic spine density [99–100], whereas conditional mouse knockouts of *Pten* exhibit hypertrophic and ectopic dendritic growth and increased spine density coincident with abnormal social interactions and exaggerated responses to sensory stimuli [101]. Mutations in *Wnt-2* [102], *Mecp2* [49, 103], and *Pten* [104, 105] have all been linked to autism.

While environmental factors that specifically target $Ca_v1.2$, Wnt-2, MECP2, and PTEN have not been described, environmental factors that alter the same Ca^{2+}-dependent pathways that regulate dendritic growth and plasticity have been identified. PCBs [106, 107, 108] and redox damage [109] have been shown to modulate the expression and function of RyR-dependent calcium release units. This is significant because RyRs are expressed in both pre- and postneuronal synapses in virtually all brain areas [54] and contribute to fundamentally important aspects of neuronal excitability and synaptic plasticity [86, 110, 111, 112]. Repetitive or prolonged depolarization of hippocampal neurons, such as occurs during learning, activates signaling microdomains between L-type Ca^{2+} channels and RyR2, causing sequential activation of CaM kinases, CREB, and transcription of genes encoding Ca^{2+}-regulated proteins [113]. Ligands that directly modulate RyR, such as ryanodine, alter several functional aspects of neuroplasticity in the hippocampus, including long-term potentiation (LTP) [114] and long-term depression (LTD) [115]. Rapamycin and FK506, which deregulate RyR1 and RyR2 by dissociating the RyR2/FKBP12/12.3 complex also inhibit LTD [116]. In this regard, the activity of non-coplanar PCBs requires an intact RyR/FKPB12 complex [117]. Thus, RyRs appear to play a critical role in use-dependent plasticity that underlies the early stages of associative memory. RyR2 through its interaction with calexcitin may also alter Ca^{2+} signaling over a longer time frame, implying a critical role for RyRs in the consolidation phase of associative memory [118]. Most intriguing is work showing a tight correlation between acquisition of spatial learning and selective up-regulation of RyR2 in the dentate gyrus and CA3 [119], implicating RyR2 in storage phases of associative memory.

These observations raise the question of whether developmental PCB exposures known to increase intracellular Ca^{2+} levels via effects on RyR-dependent mechanisms also disrupt neuronal connectivity and synaptic plasticity. Recent evidence suggests this is the case. Studies in animal models demonstrate that developmental exposures to PCBs cause an imbalance between excitation and inhibition in the auditory cortex [69], a pronounced age-related increase in dendritic growth in pyramidal neurons of the hippocampus and cortex and in cerebellar Purkinje cells, and a decrease in experience-dependent dendritic plasticity [70, 120, 121]. Preliminary observations in cultured neurons suggest that PCB effects on dendritic growth are mediated by RyR-dependent Ca^{2+} signals that activate the CamK-MEK-Wnt signaling pathway (Lein, Pessah and

Wayman, unpublished observations). These findings suggest the possibility that exposure to environmental factors that influence Ca^{2+} signaling pathways important in neurodevelopment increase susceptibility to autism in populations with heritable defects in mutations in Ca^{2+} signaling.

While this discussion has focused on activity-dependent dendritic growth, it is important to note that dendritic growth and plasticity are also strongly influenced by activity-independent mechanisms, including hormonal status [122, 123] and interactions with glia [124, 125, 126, 127] or target tissues [72, 128, 129, 130, 131]. The molecular mechanisms by which these cues regulate dendritic growth are not well characterized, but specific molecules have been implicated, including neurotrophins, semaphorins, Notch, bone morphogenetic proteins and Reelin [132, 133, 134].

Thus, the possibility exists that environmental and genetic factors acting via activity-dependent and activity-independent pathways interact via convergence on the same neurodevelopmental endpoint to influence susceptibility to autism. Clearly, work is urgently needed to better predict which combination of defective genes and environmental exposures pose the greatest autism risk.

Acknowledgments Supported by NIH grants 1PO1 ES11269, 1 R01 NS046649, and 1 R56 ES014901-01, the US Environmental Protection Agency through the Science to Achieve Results (STAR) program (grant R829388), Cure Autism Now, and the UC Davis M.I.N.D. Institute. Additional support came from Superfund Basic Research Program (P42 ES04699).

References

1. American Psychiatric Association. Diagnostic and Statistical Manual of Mental Disorders DMS-IV-TR (Text Revision) Washington D.C., American Psychiatric Association, 2000.
2. Fombonne E. The prevalence of autism. *J Am Med Assoc* 2003;289:87–89.
3. Rice C. Prevalence of Autism Spectrum Disorders. *MMWR, Centers for Disease Control* 2000;56(SS01):1–11.
4. Wassink TH, Piven J. The molecular genetics of autism. *Curr Psychiatry Rep* 2000;2:170–175.
5. Veenstra-Vanderweele J, Christian SL, Cook EH Jr. Autism as a paradigmatic complex genetic disorder. *Annu Rev Genomics Hum Genet* 2004;5:379–405.
6. Pickles A, Bolton P, Macdonald H, Bailey A, Le Couteur A, Sim CH, Rutter, M. Latent-class analysis of recurrence risk for complex phenotypes with selection and measurement error: a twin and family history study of autism. *Am J Hum Genet* 1995;57:717–726.
7. Folstein SE, Rosen-Sheidley B. Genetics of autism: complex aetiology for a heterogeneous disorder. *Nat Rev Genet* 2001;2:943–955.
8. Risch N, Spiker D, Lotspeich L, Nouri N, Hinds D, Hallmayer J, Kalaydjieva L, McCague P, Dimiceli S, Pitts T, Nguyen L, Yang J, Harper C, Thorpe D, Vermeer S, Young H, Hebert J, Lin A, Ferguson J, Chiotti C, Wiese-Slater S, Rogers T, Salmon B, Nicholas P, Petersen PB, Pingree C, McMahon W, Wong DL, Cavalli-Sforza LL, Kraemer HC, Myers RM. A genomic screen of autism: evidence for a multilocus etiology. *Am J Hum Genet* 1999;65:493–507.
9. Pritchard JK. Are rare variants responsible for susceptibility to complex diseases? *Am J Hum Genet* 2001;69:124–137.

10. Trikalinos TA, Karvouni A, Zintzaras E, Ylisaukko-Oja T, Peltonen L, Jarvela I, Ioannidis JP. A heterogeneity-based genome search meta-analysis for autism-spectrum disorders. *Mol Psychiatry* 2005;11:29–36.

11. Maestrini E, Paul A, Monaco AP, Bailey A. Identifying autism susceptibility genes. *Neuron* 2000;28:19–24.

12. Muhle R, Trentacoste SV, Rapin I. The genetics of autism. *Pediatrics* 2004;113:472–486.

13. Mills JL, Hediger ML, Molloy CA, Chrousos GP, Manning-Courtney P, Yu KF, Brasington M, England LJ. Elevated levels of growth-related hormones in autism and autism spectrum disorder. *Clin Endocrinol (Oxf)* 2007;67:230–237.

14. Kaminsky Z, Wang SC, Petronis A. Complex disease, gender and epigenetics. *Ann Med* 2006;38:530–44.

15. Arndt TL, Stodgell CJ, Rodier PM. The teratology of autism. *Int J Dev Neurosci* 2005;23:189–199.

16. Miller MT, Stromland K, Ventura L, Johansson M, Bandim JM, Gillberg C. Autism associated with conditions characterized by developmental errors in early embryogenesis: a mini review. *Int J Dev Neurosci* 2005;23:201–219.

17. Muller RA. The study of autism as a distributed disorder. *Ment Retard Dev Disabil Res Rev* 2007;13:85–95.

18. Rubenstein JL, Merzenich MM. Model of autism: increased ratio of excitation/inhibition in key neural systems. *Genes Brain Behav* 2003;2:255–267.

19. Belmonte MK, Bourgeron T. Fragile X syndrome and autism at the intersection of genetic and neural networks. *Nat Neurosci* 2006;9:1221–1225.

20. Geschwind DH, Levitt P. Autism spectrum disorders: developmental disconnection syndromes. *Curr Opin Neurobiol* 2007;17:103–111.

21. DiCicco-Bloom E, Lord C, Zwaigenbaum L, Courchesne E, Dager SR, Schmitz C, Schultz RT, Crawley J, Young LJ. The developmental neurobiology of autism spectrum disorder. *J Neurosci* 2006; 26:6897–6906.

22. Donaldson D, Kiely T, Grube A. Pesticides Industry Sales and Usage 1998 and 1999 Market Estimates. Washington, DC:U.S. Environmental Protection Agency, Office of Prevention, Pesticides, and Toxic Substances, Office of Pesticide Programs, 2002.

23. Pope CN. Organophosphorus pesticides: do they all have the same mechanism of toxicity? *J Toxicol Environ Health B Crit Rev* 1999;2:161–181.

24. Schuh RA, Lein PJ, Beckles RA, Jett DA. Noncholinesterase mechanisms of chlorpyrifos neurotoxicity: altered phosphorylation of Ca^{2+}/cAMP response element binding protein in cultured neurons. *Toxicol Appl Pharmacol* 2002;182:176–185.

25. Jameson RR, Seidler FJ, Slotkin TA. Nonenzymatic functions of acetylcholinesterase splice variants in the developmental neurotoxicity of organophosphates: chlorpyrifos, chlorpyrifos oxon, and diazinon. *Environ Health Perspect* 2007;115:65–70.

26. Ricceri L, Venerosi A, Capone F, Cometa MF, Lorenzini P, Fortuna S, Calamandrei G. Developmental neurotoxicity of organophosphorous pesticides: fetal and neonatal exposure to chlorpyrifos alters sex-specific behaviors at adulthood in mice. *Toxicol Sci* 2006;93:105–113.

27. Slotkin TA, Seidler FJ, Fumagalli F. Exposure to organophosphates reduces the expression of neurotrophic factors in neonatal rat brain regions: similarities and differences in the effects of chlorpyrifos and diazinon on the fibroblast growth factor superfamily. *Environ Health Perspect* 2007;115:909–916.

28. Peltier J, O'neill A, Schaffer DV. PI3K/Akt and CREB regulate adult neural hippocampal progenitor proliferation and differentiation. *Dev Neurobiol* 2007;67:1348–1361.

29. Redmond L, Ghosh A. Regulation of dendritic development by calcium signaling. *Cell Calcium* 2005;37:411–416.

30. Stachowiak MK, Fang X, Myers JM, Dunham SM, Berezney R, Maher PA, Stachowiak EK. Integrative nuclear FGFR1 signaling (INFS) as a part of a universal "feed-forward-and-gate" signaling module that controls cell growth and differentiation. *J Cell Biochem* 2003;90:662–691.

31. Tomizawa M, Casida JE. Neonicotinoid insecticide toxicology: mechanisms of selective action. *Annu Rev Pharmacol Toxicol* 2005;45:247–268.

32. Gotti C, Zoli M, Clementi F. Brain nicotinic acetylcholine receptors: native subtypes and their relevance. *Trends Pharmacol Sci* 2006;27:482–491.

33. Martin-Ruiz CM, Lee M, Perry RH, Baumann M, Court JA, Perry EK. Molecular analysis of nicotinic receptor expression in autism. *Brain Res Mol Brain Res* 2004;123:81–90.

34. Lee M, Martin-Ruiz C, Graham A, Court J, Jaros E, Perry R, Iversen P, Bauman M, Perry E. Nicotinic receptor abnormalities in the cerebellar cortex in autism. *Brain* 2002;125:1483–1495.

35. Perry EK, Lee ML, Martin-Ruiz CM, et al Cholinergic activity in autism: abnormalities in the cerebral cortex and basal forebrain. *Am J Psychiatry* 2001;158:1058–1066.

36. D'Amelio M, Ricci I, Sacco R, Liu X, D'Agruma L, Muscarella LA, Guarnieri V, Militerni R, Bravaccio C, Elia M, Schneider C, Melmed R, Trillo S, Pascucci T, Puglisi-Allegra S, Reichelt KL, Macciardi F, Holden JJ, Persico AM. Paraoxonase gene variants are associated with autism in North America, but not in Italy: possible regional specificity in gene-environment interactions. *Mol Psychiatry* 2005;10:1006–1016.

37. Pasca SP, Nemes B, Vlase L, Gagyi CE, Dronca E, Miu AC, Dronca M. High levels of homocysteine and low serum paraoxonase 1 arylesterase activity in children with autism. *Life Sci* 2006;78:2244–2248.

38. Hertz-Picciotto I, Croen LA, Hansen R, Jones CR, van de Water J, Pessah IN. The CHARGE study: an epidemiologic investigation of genetic and environmental factors contributing to autism. *Environ Health Perspect* 2006;114:1119–1125.

39. Eskenazi B, Marks AR, Bradman A, Harley K, Barr DB, Johnson C, Morga N, Jewell NP. Organophosphate pesticide exposure and neurodevelopment in young Mexican-American children. *Environ Health Perspect* 2007;115:792–798.

40. Michels G, Moss SJ. GABAA receptors: properties and trafficking. *Crit Rev Biochem Mol Biol* 2007;42:3–14.

41. Lu J, Karadsheh M, Delpire E. Developmental regulation of the neuronal-specific isoform of K-Cl cotransporter KCC2 in postnatal rat brains. *J Neurobiol* 1999;39:558–568.

42. Cole LM, Casida JE. Polychlorocycloalkane insecticide-induced convulsions in mice in relation to disruption of the GABA-regulated chloride ionophore. *Life Sci* 1986;39:1855–1862.

43. Lawrence LJ, Casida JE. Interactions of lindane, toxaphene and cyclodienes with brain-specific t-butylbicyclophosphorothionate receptor. *Life Sci* 1984;35:171–178.

44. Slotkin TA, MacKillop EA, Ryde IT, Tate CA, Seidler FJ. Screening for developmental neurotoxicity using PC12 cells: comparisons of organophosphates with a carbamate, an organochlorine, and divalent nickel. *Environ Health Perspect* 2007;115:93–101.

45. Bloomquist JR. Chloride channels as tools for developing selective insecticides. *Arch Insect Biochem Physiol* 2003;54:145–156.

46. Chen L, Durkin KA, Casida JE. Structural model for gamma-aminobutyric acid receptor noncompetitive antagonist binding: widely diverse structures fit the same site. *Proc Natl Acad Sci U S A* 2006;103:5185–5190.

47. Sammelson RE, Caboni P, Durkin KA, Casida JE. GABA receptor antagonists and insecticides: common structural features of 4-alkyl-1-phenylpyrazoles and 4-alkyl-1-phenyltrioxabicyclooctanes. *Bioorg Med Chem* 2004;12:3345–3355.

48. Samaco RC, Hogart A, LaSalle JM. Epigenetic overlap in autism-spectrum neurodevelopmental disorders: MECP2 deficiency causes reduced expression of UBE3A and GABRB3. *Hum Mol Genet* 2005;14:483–492.

49. LaSalle JM, Hogart A, Thatcher KN. Rett syndrome: a Rosetta stone for understanding the molecular pathogenesis of autism. *Int Rev Neurobiol* 2005;71:131–165.

50. Vincent JB, Horike SI, Choufani S, Paterson AD, Roberts W, Szatmari P, Weksberg R, Fernandez B, Scherer SW. An inversion inv(4)(p12–p15.3) in autistic siblings implicates the 4p GABA receptor gene cluster. *J Med Genet* 2006;43:429–434.

51. Ashley-Koch AE, Mei H, Jaworski J, Ma DQ, Ritchie MD, Menold MM, Delong GR, Abramson RK, Wright HH, Hussman JP, Cuccaro ML, Gilbert JR, Martin ER,

Pericak-Vance MA. An analysis paradigm for investigating multi-locus effects in complex disease: examination of three GABA receptor subunit genes on 15q11–q13 as risk factors for autistic disorder. *Ann Hum Genet* 2006;70:281–292.

52. Ma DQ, Whitehead PL, Menold MM, Martin ER, Ashley-Koch AE, Mei H, Ritchie MD, Delong GR, Abramson RK, Wright HH, Cuccaro ML, Hussman JP, Gilbert JR, Pericak-Vance MA. Identification of significant association and gene-gene interaction of GABA receptor subunit genes in autism. *Am J Hum Genet* 2005;77:377–388.

53. Roberts EM, English PB, Grether JK, Windham GC, Somberg L, Wolff C. Maternal residence near agricultural pesticide applications and autism spectrum disorders among children in the California Central Valley. *Env Health Perspect* 2007;115:1482–1489.

54. Berridge MJ. Calcium microdomains: organization and function. *Cell Calcium* 2006;40:405–412.

55. Lyons HR, Land MB, Gibbs TT, Farb DH. Distinct signal transduction pathways for GABA-induced GABA(A) receptor down-regulation and uncoupling in neuronal culture: a role for voltage-gated calcium channels. *J Neurochem* 2001;78:1114–1126.

56. Dale YR, Eltom SE. Calpain mediates the dioxin-induced activation and down-regulation of the aryl hydrocarbon receptor. *Mol Pharmacol* 2006;70:1481–1487.

57. Pessah IN, Hansen LG, Albertson TE, Garner CE, Ta TA, Do Z, Kim KH, Wong PW. Structure-activity relationship for noncoplanar polychlorinated biphenyl congeners toward the ryanodine receptor-Ca^{2+} channel complex type 1 (RyR1). *Chem Res Toxicol* 2006;19:92–101.

58. Wong PW, Brackney WR, Pessah IN. Ortho-substituted polychlorinated biphenyls alter microsomal calcium transport by direct interaction with ryanodine receptors of mammalian brain. *J Biol Chem* 1997;272:15145–1553.

59. Gafni J, Wong PW, Pessah IN. Non-coplanar 2,2′,3,5′,6-pentachlorobiphenyl (PCB 95) amplifies ionotropic glutamate receptor signaling in embryonic cerebellar granule neurons by a mechanism involving ryanodine receptors. *Toxicol Sci* 2004;77:72–82.

60. Wong PW, Garcia EF, Pessah IN. Ortho-substituted PCB95 alters intracellular calcium signaling and causes cellular acidification in PC12 cells by an immunophilin-dependent mechanism. *J Neurochem* 2001;76:450–463.

61. Howard AS, Fitzpatrick R, Pessah I, Kostyniak P, Lein PJ. Polychlorinated biphenyls induce caspase-dependent cell death in cultured embryonic rat hippocampal but not cortical neurons via activation of the ryanodine receptor. *Toxicol Appl Pharmacol* 2003;190:72–86.

62. Splawski I, Timothy KW, Sharpe LM, Decher N, Kumar P, Bloise R, Napolitano C, Schwartz PJ, Joseph RM, Condouris K, Tager-Flusberg H, Priori SG, Sanguinetti MC, Keating MT. Ca(V)1.2 calcium channel dysfunction causes a multisystem disorder including arrhythmia and autism. *Cell* 2004;119:19–31.

63. Komuro H, Rakic P. Orchestration of neuronal migration by activity of ion channels, neurotransmitter receptors, and intracellular Ca^{2+} fluctuations. *J Neurobiol* 1998;37:110–130.

64. Aamodt SM, Constantine-Paton M. The role of neural activity in synaptic development and its implications for adult brain function. *Adv Neurol* 1999;79:133–144.

65. Cline HT. Dendritic arbor development and synaptogenesis. *Curr Opin Neurobiol* 2001;11:118–126.

66. Levitt P. Structural and functional maturation of the developing primate brain. *J Pediatr* 2003;143:S35–45.

67. Moody WJ, Bosma MM. Ion channel development, spontaneous activity, and activity-dependent development in nerve and muscle cells. *Physiol Rev* 2005;85:883–941.

68. Krey JF, Dolmetsch RE. Molecular mechanisms of autism: a possible role for Ca^{2+} signaling. *Curr Opin Neurobiol* 2007;17:112–119.

69. Kenet T, Froemke RC, Schreiner CE, Pessah IN, Merzenich MM. Perinatal exposure to a noncoplanar polychlorinated biphenyl alters tonotopy, receptive fields, and plasticity in rat primary auditory cortex. *Proc Natl Acad Sci U S A* 2007;104:7646–7651.

70. Lein PJ, Yang D, Bachstetter AD, Tilson HA, Harry GJ, Mervis RF, Kodavanti PR. Ontogenetic alterations in molecular and structural correlates of dendritic growth after developmental exposure to polychlorinated biphenyls. *Environ Health Perspect* 2007;115:556–563.

71. Engert F, Bonhoeffer T. Dendritic spine changes associated with hippocampal long-term synaptic plasticity. *Nature* 1999;399:66–70.

72. Purves D. *Body and Brain: A Trophic Theory of Neural Connections.* Cambridge, MA: Harvard University Press; 1988.

73. Miller JP, Jacobs GA. Relationships between neuronal structure and function. *J Exp Biol* 1984;112:129–145.

74. Schuman EM. Synapse specificity and long-term information storage. *Neuron* 1997;18:339–342.

75. Sejnowski TJ. The year of the dendrite. *Science* 1997;275:178–179.

76. Scott EK, Luo L. How do dendrites take their shape? *Nat Neurosci* 2001;4:359–365.

77. Grutzendler J, Gan WB. Two-photon imaging of synaptic plasticity and pathology in the living mouse brain. *NeuroRx* 2006;3:489–496.

78. Harms KJ, Dunaevsky A. Dendritic spine plasticity: looking beyond development. *Brain Res* 2007;1184:65–71 [Epub 2006 Apr 5]

79. Le Be JV, Markram H. Spontaneous and evoked synaptic rewiring in the neonatal neocortex. *Proc Natl Acad Sci U S A* 2006;103:13214–13219.

80. Sorra KE, Harris KM. Overview on the structure, composition, function, development, and plasticity of hippocampal dendritic spines. *Hippocampus* 2000;10:501–511.

81. Hering H, Sheng M. Dendritic spines: structure, dynamics and regulation. *Nat Rev Neurosci* 2001;2:880–888.

82. Pittenger C, Kandel ER. In search of general mechanisms for long-lasting plasticity: Aplysia and the hippocampus. *Philos Trans R Soc Lond B Biol Sci* 2003;358:757–763.

83. Leuner B, Shors TJ. New spines, new memories. *Mol Neurobiol* 2004;29:117–130.

84. Segal M, Korkotian E, Murphy DD. Dendritic spine formation and pruning: common cellular mechanisms? *Trends Neurosci* 2000;23:53–57.

85. Lohmann C, Wong RO. Regulation of dendritic growth and plasticity by local and global calcium dynamics. *Cell Calcium* 2005;37:403–409.

86. Korkotian E, Segal M. Release of calcium from stores alters the morphology of dendritic spines in cultured hippocampal neurons. *Proc Natl Acad Sci USA* 1999;96:12068–12072.

87. Redmond L, Kashani AH, Ghosh A. Calcium regulation of dendritic growth via CaM kinase IV and CREB-mediated transcription. *Neuron* 2002;34:999–1010.

88. Wilson MT, Kisaalita WS, Keith CH. Glutamate-induced changes in the pattern of hippocampal dendrite outgrowth: a role for calcium-dependent pathways and the microtubule cytoskeleton. *J Neurobiol* 2000;43:159–172.

89. Gomez-Ospina N, Tsuruta F, Barreto-Chang O, Hu L, Dolmetsch R. The C terminus of the L-type voltage-gated calcium channel Ca(V)1.2 encodes a transcription factor. *Cell* 2006;127:591–606.

90. Courchesne E, Redcay E, Morgan JT, Kennedy DP. Autism at the beginning: microstructural and growth abnormalities underlying the cognitive and behavioral phenotype of autism. *Dev Psychopathol* 2005;17:577–597.

91. Wayman GA, Impey S, Marks D, Saneyoshi T, Grant WF, Derkach V, Soderling TR. Activity-dependent dendritic arborization mediated by CaM-kinase I activation and enhanced CREB-dependent transcription of Wnt-2. *Neuron* 2006;50:897–909.

92. Ou LC, Gean PW. Transcriptional regulation of brain-derived neurotrophic factor in the amygdala during consolidation of fear memory. *Mol Pharmacol* 2007;72:350–358.

93. Chen WG, Chang Q, Lin Y, Meissner A, West AE, Griffith EC, Jaenisch R, Greenberg ME. Derepression of BDNF transcription involves calcium-dependent phosphorylation of MeCP2. *Science* 2003;302:885–889.

94. Zhou Z, Hong EJ, Cohen S, Zhao WN, Ho HY, Schmidt L, Chen WG, Lin Y, Savner E, Griffith EC, Hu L, Steen JA, Weitz CJ, Greenberg ME. Brain-specific phosphorylation of MeCP2 regulates activity-dependent Bdnf transcription, dendritic growth, and spine maturation. *Neuron* 2006;52:255–269.

95. Takei N, Inamura N, Kawamura M, Namba H, Hara K, Yonezawa K, Nawa H. Brain-derived neurotrophic factor induces mammalian target of rapamycin-dependent local activation of translation machinery and protein synthesis in neuronal dendrites. *J Neurosci* 2004;24:9760–9769.

96. Jaworski J, Spangler S, Seeburg DP, Hoogenraad CC, Sheng M. Control of dendritic arborization by the phosphoinositide-3'-kinase-Akt-mammalian target of rapamycin pathway. *J Neurosci* 2005;25:11300–11312.

97. Kumar V, Zhang MX, Swank MW, Kunz J, Wu GY. Regulation of dendritic morphogenesis by Ras-PI3K-Akt-mTOR and Ras-MAPK signaling pathways. *J Neurosci* 2005;25:11288–11299.

98. Gong R, Park CS, Abbassi NR, Tang SJ. Roles of glutamate receptors and the mammalian target of rapamycin (mTOR) signaling pathway in activity-dependent dendritic protein synthesis in hippocampal neurons. *J Biol Chem* 2006;281:18802–18815.

99. Kishi N, Macklis JD. MECP2 is progressively expressed in post-migratory neurons and is involved in neuronal maturation rather than cell fate decisions. *Mol Cell Neurosci* 2004;27:306–321.

100. Fukuda T, Itoh M, Ichikawa T, Washiyama K, Goto Y. Delayed maturation of neuronal architecture and synaptogenesis in cerebral cortex of Mecp2-deficient mice. *J Neuropathol Exp Neurol* 2005;64:537–544.

101. Kwon CH, Luikart BW, Powell CM, Zhou J, Matheny SA, Zhang W, Li Y, Baker SJ, Parada LF. Pten regulates neuronal arborization and social interaction in mice. *Neuron* 2006;50:377–388.

102. Wassink TH, Piven J, Vieland VJ, Huang J, Swiderski RE, Pietila J, Braun T, Beck G, Folstein SE, Haines JL, Sheffield VC. Evidence supporting WNT2 as an autism susceptibility gene. *Am J Med Genet* 2001;105:406–413.

103. Moretti P, Zoghbi HY. MeCP2 dysfunction in Rett syndrome and related disorders. *Curr Opin Genet Dev* 2006;16:276–281.

104. Goffin A, Hoefsloot LH, Bosgoed E, Swillen A, Fryns JP. PTEN mutation in a family with Cowden syndrome and autism. *Am J Med Genet* 2001;105:521–524.

105. Butler MG, Dasouki MJ, Zhou XP, Talebizadeh Z, Brown M, Takahashi TN, Miles JH, Wang CH, Stratton R, Pilarski R, Eng C. Subset of individuals with autism spectrum disorders and extreme macrocephaly associated with germline PTEN tumour suppressor gene mutations. *J Med Genet* 2005;42:318–321.

106. Schantz SL, Seo BW, Wong PW, Pessah IN. Long-term effects of developmental exposure to 2,2',3,5',6-pentachlorobiphenyl (PCB 95) on locomotor activity, spatial learning and memory and brain ryanodine binding. *Neurotoxicology* 1997;18:457–467.

107. Wong PW, Joy RM, Albertson TE, Schantz SL, Pessah IN. Ortho-substituted 2,2',3,5',6-pentachlorobiphenyl (PCB 95) alters rat hippocampal ryanodine receptors and neuroplasticity in vitro: evidence for altered hippocampal function. *Neurotoxicology* 1997;18:443–456.

108. Wong PW, Pessah IN. Ortho-substituted polychlorinated biphenyls alter calcium regulation by a ryanodine receptor-mediated mechanism: structural specificity toward skeletal – and cardiac-type microsomal calcium release channels. *Mol Pharmacol* 1996;49:740–751.

109. Pessah IN, Kim KH, Feng W. Redox sensing properties of the ryanodine receptor complex. *Front Biosci* 2002;7:a72–a79.

110. Kennedy MB. Signal-processing machines at the postsynaptic density. *Science* 2000;290:750–754.

111. Matus A. Actin-based plasticity in dendritic spines. *Science* 2000;290:754–758.

112. Segal M. New building blocks for the dendritic spine. *Neuron* 2001;31:169–171.

113. Deisseroth K, Heist EK, Tsien RW. Translocation of calmodulin to the nucleus supports CREB phosphorylation in hippocampal neurons. *Nature* 1998;392:198–202.

114. Wang Y, Wu J, Rowan MJ, Anwyl R. Ryanodine produces a low frequency stimulation-induced NMDA receptor-independent long-term potentiation in the rat dentate gyrus in vitro. *J Physiol* 1996;495:755–767.

115. Wang Y, Rowan MJ, Anwyl R. Induction of LTD in the dentate gyrus in vitro is NMDA receptor independent, but dependent on Ca2+ influx via low-voltage-activated Ca2+ channels and release of Ca2+ from intracellular stores. *J Neurophysiol* 1997;77:812–825.

116. Li ST, Kato K, Mikoshiba K. Effect of calcineurin inhibitors on long-term depression in CA1 rat hippocampal neurons. *28th Annu Meet Soc Neurosci Abs* 1998;24:1815.

117. Wong PW, Pessah IN. Noncoplanar PCB 95 alters microsomal calcium transport by an immunophilin FKBP12-dependent mechanism. *Mol Pharmacol* 1997;51(5):693–702.

118. Alkon DL, Nelson TJ, Zhao W, Cavallaro S. Time domains of neuronal Ca^{2+} signaling and associative memory: steps through a calexcitin, ryanodine receptor, K+ channel cascade. *Trends Neurosci* 1998;21:529–537.

119. Cavallaro S, Meiri N, Yi CL, Musco S, Ma W, Goldberg J, Alkon DL. Late memory-related genes in the hippocampus revealed by RNA fingerprinting. *Proc Natl Acad Sci USA* 1997;94:9669–9673.

120. Yang D, Kim KH, Phimister A, Girouard J, Ward T, Bachstetter A, Anderson KA, Kodavanti PRS, Stackman RW, Wisniewski AB, Klein S, Mervis R, Pessah IN, Lein PJ. PCBs alter dendritic plasticity coincident with disruptions of spatial learning in weanling rats. *Env Health Perspect* (submitted).

121. Kim KH, Inan SY, Berman RF, Pessah IN. Inhibitory deficits synergize hippocampus excitotoxicity of *Ortho*-substituted polychlorinated biphenyls and enhances seizure susceptibility. *Toxicol Appl Pharmacol* (submitted).

122. Kapfhammer JP. Cellular and molecular control of dendritic growth and development of cerebellar Purkinje cells. *Prog Histochem Cytochem* 2004;39:131–182.

123. Cooke BM, Woolley CS. Gonadal hormone modulation of dendrites in the mammalian CNS. *J Neurobiol* 2005;64:34–46.

124. Le Roux PD, Reh TA. Regional differences in glial-derived factors that promote dendritic outgrowth from mouse cortical neurons in vitro. *J Neurosci* 1994;14:4639–4655.

125. Prochiantz A. Neuronal polarity: giving neurons heads and tails. *Neuron* 1995;15:743–746.

126. Guo X, Metzler-Northrup J, Lein P, Rueger D, Higgins D. Leukemia inhibitory factor and ciliary neurotrophic factor regulate dendritic growth in cultures of rat sympathetic neurons. *Brain Res Dev Brain Res* 1997;104:101–110.

127. Bauch H, Stier H, Schlosshauer B. Axonal versus dendritic outgrowth is differentially affected by radial glia in discrete layers of the retina. *J Neurosci* 1998;18:1774–1785.

128. Blaser PF, Catsicas S, Clarke PG. Retrograde modulation of dendritic geometry in the vertebrate brain during development. *Brain Res Dev Brain Res* 1990;57:139–142.

129. Andrews TJ. Autonomic nervous system as a model of neuronal aging: the role of target tissues and neurotrophic factors. *Microsc Res Tech* 1996;35:2–19.

130. Brehmer A, Beleites B. Myenteric neurons with different projections have different dendritic tree patterns: a morphometric study in the pig ileum. *J Auton Nerv Syst* 1996;61:43–50.

131. Cowen T, Gavazzi I. Plasticity in adult and ageing sympathetic neurons. *Prog Neurobiol* 1998;54:249–288.

132. Higgins D, Burack M, Lein P, Banker G. Mechanisms of neuronal polarity. *Curr Opin Neurobiol* 1997;7:599–604.

133. McAllister AK. Cellular and molecular mechanisms of dendrite growth. *Cereb Cortex* 2000;10:963–973.

134. Herz J, Chen Y. Reelin, lipoprotein receptors and synaptic plasticity. *Nat Rev Neurosci* 2006;7:850–859.

Chapter 20
An Expanding Spectrum of Autism Models

From Fixed Developmental Defects to Reversible Functional Impairments

Martha R. Herbert and Matthew P. Anderson

Abstract In this review, we contrast previous models of autism pathogenesis with newer models inspired by some recently appreciated and previously minimally considered pathological and clinical features of the disease. Autism has conventionally been viewed as an incurable behavioral disorder resulting solely from genetic defects impacting brain development. However, emerging evidence suggests that autism affects many organ systems beyond the brain and that some neuropathological and somatic pathophysiological processes are active even into adulthood. We incorporate these newer observations into a model of how this systemic and persistent disease process might impact brain *function* and ultimately impair behavior through potentially reversible mechanisms. In particular, observations of substantial transient and sometimes enduring increases in function and even losses of the diagnosis challenge researchers to identify pathophysiological mechanisms consistent with this dynamic course and potential plasticity. This broadened appreciation of the disease phenomenology and prognosis of autism calls for mechanistic models that encompass the full range of its features. To this end, our review contrasts several models of autism pathophysiology and lays out their differing underlying assumptions regarding mechanisms. First, we compare models of autism that are based on different underlying biological and experimental perspectives to addressing the question of autism pathogenesis. We contrast a "bottom-up, modular, genes–brain–behavior" model with a more inclusive "middle-out, multi-system biology" model. We then contrast different models that consider autism's development over time. Beginning with a purely genetic prenatal brain development model, we expand the framework to include early environmental influences, epigenetics, later and ongoing environmental influences, and features consistent with chronic encephalopathy. The implications of these models, particularly the last, are spelled out through a discussion of the functional

M.R. Herbert
Assistant Professor of Neurology, Harvard Medical School, Massachusetts General Hospital, Pediatric Neurology, Center for Morphometric Analysis, Martinos Center for Biomedical Imaging, 149, 13th Street, Room 10018, Charlestown, MA 02129, USA
e-mail: mherbert1@partners.org

A.W. Zimmerman (ed.), *Autism:*, DOI: 10.1007/978-1-60327-489-0_20,
© Humana Press, Totowa, NJ 2008

consequences of one prominent chronic feature, persistent immune activation, and its impact on neural–glial interactions. The implications of these newer models on potential treatments are also discussed.

Keywords Autism · models · pathogenesis · function · brain · plasticity · epigenetics · environment

Introduction

For many years, autism was considered to be an incurable behavioral syndrome resulting from genetically determined in utero alterations in brain development. There is now a growing shift to an appreciation of autism as a heterogeneous whole-body, multi-system set of conditions that may begin during development, but whose pathophysiologic features may remain active and impact brain function well into adulthood [1]. These interrelated dimensions of reconceptualization are necessary responses to emerging evidence in clinical, pathophysiological, and epidemiological fields that challenge older models. Clinically, the high prevalence of gastrointestinal and immune symptomatology challenges the idea that autism, or what many are coming to call "*autisms*," is purely "brain and behavior" disorders [1, 2, 3, 4, 5, 6, 7, 8, 9, 10]. The description of pathophysiological features such as persistent immune responses, oxidative stress, and mitochondrial dysfunction in multiple tissues suggests systemic rather than brain-specific perturbation [11, 12, 13, 14]. A further clinical dimension is the emerging appreciation that these persistent components of autisms, when present, may be medically treatable. Etiologically, there is growing evidence that both genetics and environmental factors contribute [15]. Epidemiologically, the increased incidence of autisms may be due in part to increased awareness or broadening of diagnostic criteria; but those do not exclude the possible impact of environmental factors such as the growing numbers of and complex interactions among new-to-nature chemical agents present in the environment [16]. Finally, prognostically, substantial improvements and even loss of rigorously ascertained autism diagnoses are being reported in some autistic individuals [17, 18]; this observation is becoming an object of study, because validation of this phenomenon would challenge both the presumption of incurability and the neurobiological models based on that presumption.

Reconceptualizing autisms has provoked a re-review of the research findings and clinical phenomenology with fresh eyes. It is leading to a shift in the perceptual frames within which evidence is considered and according to which figure and ground are distinguished from each other. It is also reorganizing the models and narratives that organize research programs and clinical approaches. In this article, we will review some of the conceptual reorganization that is occurring. We will discuss schematized versions of how these models are being reconstructed, and will show how many emerging findings, when viewed from this new framework, point toward novel but plausible pathophysiologic mechanisms that may have major clinical and treatment implications. We will especially focus on immune activation as a theme with which to illustrate the points of our argument; this choice is based on our belief that the identification of immune, autoimmune, and inflammatory processes in autism are of fundamental importance to modeling the disease process. A similar discussion could be constructed around the metabolic or energetic disturbances that are also prevalent in the disorder, which may in many cases be at least equally important. We must remember that autism is defined by a specific set of behavioral abnormalities in the DSM manual, but just like motor impairments, abdominal pain, and even social phobias, autism is not a single disease, but instead a condition caused by multiple etiologies. Consequently, although we lay out a set of general arguments that we hope help frame how we think about the complexity of autisms, the examples we choose in our discussions below might only be relevant to a subset of all possible autism etiologies and pathophysiologies.

This conceptual reorganization is being fleshed out both up and down the biological hierarchy and across the lifespan along the temporal axis. Identification of immune activation and other emerging pathophysiological findings inspires us to consider new levels of the biological hierarchy in addition to genes, brain, and behavior. The persistent postnatal and chronic changes and processes suggest the disease process is dynamic over time. The increasing number of affected individuals and substantial intra-individual variability in behavioral symptoms in at least some people with autism suggest a neuromodulatory control of the phenotype. The reports of improvement with fever and cases of remission suggest the disease might at least in part be contingent and potentially partly or even completely reversible if the pathophysiologic processes are inhibited or overcome. Depending on the mechanisms that come to more comprehensively explain the various autisms, the significance of these mechanisms may relate to other "neurodevelopmental disorders" beyond autism (e.g., schizophrenia).

Viewing the behavioral impairments that define autism as reflecting an *ongoing effect* of biological dysregulations rather than a fixed neurodevelopmental defect carries the implication that autism includes more than the traditional "triad" of deficits that define it in the psychiatric DSM-IV manual [19]. From the broader whole-body vantage point, approaches that define autism in purely behavioral terms can be seen as limiting treatment targets, and is vulnerable to the criticism of some members of the autism community who see treatment as focusing upon those symptoms that are troubling to caregivers (e.g., behaviors seen as "inappropriate" by neurotypical individuals), whereas a whole-body approach can address a further range of symptoms (as well as their underlying mechanisms) that are uncomfortable, painful, or troublesome to those with the diagnosis.

Reflection on Autism Models

In what follows, we will present and contrast two pairs of models of autism:

1. In the first pair of models (Fig. 20.1), we focus on contrasting formulations of the biological hierarchy. This involves comparing an older genetic reductionist model in which the biological hierarchy is populated by a "bottom-up" sequence leading from causal genes through brain to behavior on the one hand with a newer and more inclusive "middle-out" model that shifts the focus toward dysregulated immunologic and biochemical processes and their impairment of neural circuit function, and grounds the investigations of upstream gene/environment/epigenetic contributors and downstream behaviors in an underlying dysregulated cellular biologic process, which is a "middle level" from which causes and consequences emerge.

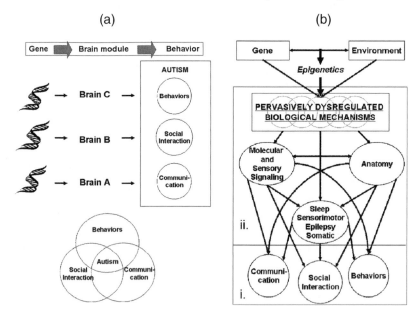

Fig. 20.1 Modular and Systems Approaches to Autism. (**a**) *Modular Approach to Autism.* This figure illustrates the "modular, gene–brain–behavior" model of autism, which is a "bottom-up" approach in which genes change the brain, and these brain alterations yield behavioral deficits. In this version of the model, each behavioral deficit is the result of its own genetically driven brain abnormality. The combination of three "gene–brain–behavior" modules yields the behavioral syndrome we call autism. Individuals with one or two of these modules would share features with autism but would not meet full criteria for autism disorder. (**b**) *Systems Biology Approach to Autism.* This figure depicts a "middle-out" approach with a set of interacting pervasively dysregulated biological mechanisms at the core, driven from upstream by a variety of gene–environment–epigenetic interactions, and cascading downstream into altered signaling, anatomic, neural and somatic functions, and behaviors. Figure 20.1b-i indicates the conventional behavioral characterization, whereas Fig. 20.1b-ii indicates the underlying biological abnormalities that may in the future provide the basis for biology- and signaling-based nosologies that use pathophysiological mechanisms to parse autism's heterogeneity

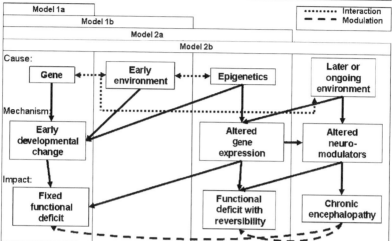

Fig. 20.2 Expanding the temporal axis of autism. This figure depicts a series of temporal models of autism, each of which adds to the one before at one or more of the cause, mechanism, and impact levels sketched in this schematic. Model 1a: *Static encephalopathy, genetically caused:* In this model, autism is a genetically determined disorder of early brain development with fixed functional deficits. Model 1b: *Static encephalopathy, caused by gene–environment interactions:* In this model, autism results from gene–environment interactions, with early environmental exposures contributing to developmental injury or alteration that leads to fixed functional deficits. Model 2a: *Potentially reversible epigenetically altered gene expression:* In this model, autism results from gene–environment–epigenetic–timing interactions, and involves a variable combination of early developmental injury or alteration with persistent but conceivably reversible alterations of gene expression. Model 2b: *Chronic encephalopathy sustained by later or ongoing environmental factors:* In this model, autism results from gene–environment–epigenetic–timing interactions, but environmental influences now include later and ongoing as well as early influences. These may affect not only gene expression but may also chronically dysregulate signaling and chemistry factors that may serve as neuromodulators, and may in this fashion impact function to create a modest or substantial component of a chronic encephalopathy. This raises the possibility that alterations of cellular structure, brain volume, and brain tissue parameters may be downstream consequences of persistent, chronic pathophysiologic abnormalities

2. In the second set of models (Fig. 20.2), we focus on the dimension of time in autism, contrasting an exclusively prenatal model with a model that adds and emphasizes the potentially fragile postnatal period when major synaptic remodeling occurs and an additional model in which ongoing chronic and dynamic physiological changes of brain circuit function are triggered by persistent disturbances in immunity and metabolism.

Although the multiple biological levels of integration are separable from the axis of developmental time at a conceptual level, these two axes are intertwined in the autism phenomenology that is driving this reconceptualization. In particular, several interrelated features that stretch the temporal axis well past the prenatal period raise compelling questions that need to be addressed in a systems biological fashion that integrates across levels of biological hierarchy. That is, we are seeing physical and functional changes over the life course in at least some of the autisms and not just prenatally, including postnatal rapid brain enlargement [20, 21], ongoing neuroinflammation and systemic immune dysregulation [22, 23], persistent systemic redox chemistry abnormalities [11, 12, 24, 25, 26, 27, 28], and fluctuating changes in brain functional connectivity [29]. These changes all have features that cannot be explained within a gene–brain–behavior model but instead demand a more comprehensive research agenda that features metabolic and glial–neuronal interactions in the disease pathophysiology [30].

As an example, one of the most consistent anatomical changes in autism is the larger than average brain size in young autistic individuals (with 20% over the 97th percentile, and most above average—although some quite small, supporting heterogeneity) that occurs during early postnatal periods—maximally during the first 2 years of life, a period of maximum axosynaptic remodeling [20, 21]. This observation raises the possibility that brain circuitry may be altered during postnatal rather than exclusively during the prenatal periods as previously assumed. More recent work has identified the presence of immune activation and signs of oxidative stress in the brain tissue of autistic individuals [14, 31, 32], and in particular involving glial alterations in a distribution that appears to follow the distribution of white matter enlargement, with recent evidence suggesting that astroglial hypertrophy may preferentially occur in the later-myelinating subcortical (radiate) white matter where macroscopic volumetric enlargement has been documented but not in the earlier-myelinating deep or bridging white matter where this macroscopic white matter enlargement does not occur (Pardo C, 2006, personal communication) [33]. Glial reactions were however also observed in gray matter such as cerebellum and cingulate cortex. Immune activation was also observed to be substantially more pronounced in brain tissue from a 3-year old than in brains from autistic individual ages 5–44 years (Pardo C, 2006, personal communication) . These observations suggest the brain and white matter volume increases may spatially and temporally co-localize with the immune activation.

Do these tissue changes have functional consequences pertinent to the behavioral syndrome we call autism? Altered interregional coordination or functional connectivity has been measured in a widespread distribution [29, 30, 34, 35]. Placing the tissue, volume and connectivity findings side by side, the question arises whether the connectivity abnormalities are a manifestation at the functional level of the same underlying pathophysiological mechanisms as the physical changes. If so, it becomes central to address at the cellular and molecular level the mechanisms by which these tissue reactions might impact functional connectivity. The confluence of brain developmental trajectories, brain tissue changes, functional decrements, and symptom evolution timetables points toward an integrated model of autism pathophysiologies crossing multiple levels of biological hierarchy and tracking far along the temporal axis of brain maturation. The implications of this confluence are the motivation for this discourse; the underlying details will be described in the following analysis of the evolution of concepts in the field.

From Behavioral to Multi-Leveled Nosology

A Modular "Bottom-Up" Approach: Genes–Brain–Behavior

Figure 20.1 addresses the question of biological hierarchy, illustrating the differences in structure between the commonly held "modular, gene–brain–behavior" models and the emerging more inclusive "systems biology model" of autism. In the conventional "modular" approach (Fig. 20.1a), the mechanisms by which genes affect the brain are often not of central interest—instead this model, which focuses on gene–behavior correlations, moves the underlying biology into the background. In one variant of this model, the three impaired behavioral domains (communication, social interaction, restrictive-repetitive behavior) are assumed to be distinct and to relate to disturbances in distinctive neural systems that are each controlled by distinct genes. This leads to models in which the disturbances in the three behavioral domains are considered to be independent but co-occurring; autism is thus thought to result from multiple separate gene defects coexisting in the same individual. Disorders that share a subset of autism's behavioral features (e.g., repetitive or obsessive behaviors) are thought to share a subset of autism's gene–brain–behavior alterations. This version of the modular model presumes a congruence or continuity—i.e., a tight coupling—across the gene–brain–behavior levels within each module. Findings that behavioral features of autism can travel separately [36] are cited to support this line of reasoning. Critics would, however, argue that behavioral fractionation does not prove a unique gene–behavior link, because not only genes but also non-genetic features (such as in utero exposure to maternal autoantibodies, infectious agents, or environmental toxins affecting gene products at both pre- and post-transcription levels) may affect specific behaviorally pertinent brain regions and cells during vulnerable time periods.

A less modular variant of this gene–brain–behavior view is that genetic mechanisms are valuable to identify not because they are applicable across the entire autistic population, but instead because even rare genetic variants can point toward a pathway that contributes even in differently inherited or non-inherited forms of the condition. Identifying these biologic pathways may then permit rational treatment design. This version of a gene–brain–behavior model begins to consider neurobiological mechanisms, but is still often heavily weighted toward genetics and would typically put biological factors outside the nervous system into background.

The Critical Nature of Models and Diversity in Scientific Investigation

This conventional modular model biases research programs toward presuming that the brain and behavior phenotype data will provide sufficient correlates for researchers to identify the genes that are currently presumed to cause autism. The agenda it sets for researchers is to improve the characterization of behaviors and to obtain ever larger samples of subjects characterized across the levels of genetics, brain, and behavior so that sample size can be sufficiently large to permit correlates across the three levels. However, if this model of autism pathophysiology is incorrect or incomplete, it sets an agenda that is bound to fail. We cannot stress enough how critical it is for experimentalists to work under accurate mechanistic models. One author's experiences in cystic fibrosis research are instructive. When the mutated gene in this disease was first identified, a viewpoint announced in *Nature* argued that the cystic fibrosis gene cannot encode a chloride channel, but instead must encode a transporter of some metabolite that secondarily regulates channels to impair apical membrane chloride permeability and lung fluid transport [37]. Although this public statement may have at first seemed innocuous, surprisingly, most laboratories pursuing cystic fibrosis research abandoned their ongoing research programs that had evolved through careful evaluation of the literature in the field and their own focused thoughts on potential mechanisms to pursue this "novel" hypothesis. Unfortunately, this shift led to unproductive research, failed research training experiences, and financial losses. The hypothesis announced in *Nature* was later shown to be incorrect [38], but not until after many laboratories had abandoned their prior agendas and fully invested into the new idea [39]. Three major lessons in research emerge from these experiences: (1) carefully examine and re-examine all evidence in the field before holding fast to any single model; (2) as an individual, be cautious in blindly following "novel" hypotheses made by others in the field that ignore existing information; and (3) as a funding agency, maintain a diverse portfolio in a research field; do not let individuals or small groups dominate the research agenda for the whole field. These rules may be difficult to follow during a period of heavy financial constraints and intense competition for research funding, but in times of scarce resources it is all the more important to invest wisely.

Limits of a "Gene–Brain–Behavior" Model

The gene–brain–behavior model setting the current autism research agenda is incomplete in that it does not encompass the other known system-wide endophenotypes or the potential additional utility of metabolic endophenotyping methods to characterize its subjects. The hope of this traditional model is that this gene–brain–behavior correlation process will identify causative or risk promoting genes, and then once such genes are identified, treatments can be developed. The shadow of this hope is the idea that without such gene-based mechanisms rational treatments of autism will not be possible; in the eyes of some proponents of this model, treatment should therefore be deferred until these genes and their associated mechanisms are identified.

Non-Neurological Features of Autism

One problem with this conventional autism model is that many prominent and common signs, symptoms, and pathophysiologies occur outside the confines of the central nervous system in probably the majority of individuals with autism. Interestingly, these somatic and systemic symptoms of autism are seen by some as recent and perhaps inappropriate interlopers in the discussion of this "neuropsychiatric" disorder. But we argue that characterizing autism as a purely behavioral disorder was probably overly restrictive from the start. The initial article by Leo Kanner that first identified the phenomenon of autism characterized it as a behavioral syndrome, and years of subsequent effort have continued to focus on refining its behavioral criteria. Yet, even in this very first case series, one finds that almost every individual manifested some symptoms of abnormal feeding, vomiting, constipation, diarrhea or nutritional malabsorption, and/or immune dysregulation. Many cases also included recurrent infections [40, 41]. Belief in the idea that autism is a pure brain disorder and neglect of these systemic findings may have persisted in part because of the general misperception that the blood-brain barrier is impermeable, in part due to the lack of broad knowledge of the now accumulating evidence for brain-immune-gastrointestinal cross-communication, and in part from the long-standing placement of somatic findings into background rather than foreground by most key opinion leaders in the field of autism pathogenesis and in psychiatry more broadly.

Epidemiology and Non-Genetic Contributors

A second problem with the "modular, gene–brain–behavior" formulation is that epidemiological evidence points away from autism being a purely genetic disorder. Twin studies in autism are intrinsically confounded, because twins share not only genetic material but also an intrauterine as well as family environment. One could even argue that the unusually high concordance rate of autism reported in identical twins results from a double contingency of a genetic susceptibility concurrent with an intrauterine environmental insult. In fact, there is no intrinsic reason why genetic and environmental influences need to be considered mutually exclusive. A monozygotic concordance rate of 60% can coexist with a 60% or even a 100% role for environmental contributors—these numbers do not have to add up to 100% (Hertz-Picciotto I, 2007, personal communication). In other words, a particular genotype or set of genotypes can be a precondition for environmental susceptibility, but without that environmental influence the disease phenotype may not manifest. If the environmental trigger is common, there could be substantial overlap of genetic and environmental factors across a population. Thus, high recurrence rates even within individual families could result from autoimmune mechanisms or environmental factors (chemical, infectious or other) rather than pure genetics. However, in our formulation of the pathogenesis of at least some autisms, genetic alleles that influence immunologic, metabolic, or even growth responses are likely to strongly influence the expression and severity of an environmentally induced autism disorder. Increasing autism rates may drive increased awareness as well as be driven by them, and there is no rigorous proof that the tenfold increase in autism rate is purely artifactual or solely because of broadened diagnostic criteria or diagnostic substitution [42]. On this account, the autism research and treatment agendas need to face the possibility that new environmental stressors are contributing, potentially strongly so, to the increasing number of autism cases.

Non-Genetic Brain Pathophysiology

A third problem with a modular gene–brain–behavior approach is the short-comings of the idea that one need only consider genes in accounting for neurobiological abnormalities. This is not proving to be the most useful way to approach autism. First, the overwhelming majority of brains of autistic individuals that have been examined *by* neuroimaging or neuropathology do not show signs of migrational or other classically described developmental abnormalities, nor do they except for a small minority (e.g., tuberous sclerosis) show signs of previously described neurogenetic disorders. When examined by magnetic resonance imaging (MRI), most brains of autistic individuals look clinically normal. Quantitative volumetric methodologies have been required to identify more subtle abnormalities that, aside from frequently being inconsistent, cannot be assumed to derive from genetic causes [43]. However, when specific techniques are used, a different and very prominent neuropathological process becomes apparent: marked immune activation (see chapter 15 by C. Pardo). Brain samples from a series of autistic individuals showed activation of microglia and astroglia along with increased levels of numerous chemokines and cytokines in the brain tissues and cerebrospinal fluid [14]. Surprisingly, these changes were found even in the brains of older individuals with autism diagnoses. Chronic inflammatory reactions are also found in the gastrointestinal system of at least some individuals with autism, along with a variety of immune system abnormalities, which may reflect a system-wide inflammatory response and immune dysregulatory process [22, 23, 44].

Immune activation is found in the central nervous system (CNS) even in strongly genetic neurodegenerative disorders such as Huntington's disease in the focal regions of neuronal cell death and in Alzheimer's disease where neuritic plaques have formed. Neuroinflammatory activation of astroglia and microglia is also well known to occur as a consequence of assorted environmental (infectious and toxic) exposures, autoimmune conditions (e.g., Rasmussen's encephalitis), and in response to traumatic and ischemic/hypoxic injuries. Consequently, this finding does not point to a specific etiologic agent responsible for autism, but it does undermine the assumption that autism pathogenesis must involve a static defect of brain development, because these disturbances of glial–neuronal interactions are ongoing and could underlie disturbances in brain function that could contribute to the autism behavioral syndrome. Not all of these brain conditions are known to be associated with the autism-like behavioral condition, but a few pro-inflammatory factors (e.g., maternal anti-brain antibodies, pre- or perinatal Cytomegalovirus (CMV) infection) have been well documented in cases of infantile autism. These observations suggest that either differences in the *distribution* (cell types or brain regions affected) or the *timing* (in utero, postnatal developmental periods, or adulthood) of the pro-inflammatory challenge might account for some of the heterogeneity of clinical presentation seen amongst individuals in the autism spectrum as well as amongst other conditions involving immune activation.

Brain-Body Cross-Talk

Until recently the brain was considered immune-privileged, so that inflammatory events occurring within the body would be considered irrelevant to local environment and ongoing functions of the brain. Recent studies, however, have identified ongoing cross-talk between the body and the brain. To illustrate with just a few examples, adipocytes (fat cells) synthesize the peptide hormone leptin, which travels through the blood stream into the brain to regulate neurons within the forebrain, hypothalamus, and brainstem that control food-seeking behavior, cellular metabolism, and autonomic nervous system activity [45]. The regulation of activities within these brain regions provides feedback signaling to peripheral organs to control blood flow and energy consumption with the goal of achieving energy homeostasis. Similarly, inflammation in the body signals the brain to promote not only the fever response, but also increased cardiac output and an immune response, and suppressed social interaction [46]. Thus, we see that there can be mutual influence between peripheral and brain immune responses.

Furthermore, the rationale for dismissing peripheral influences on the brain because of the belief that the brain is protected from the events that impinge on the peripheral organs is being progressively undermined. We now know that "barrier" is an over-statement regarding the blood–brain barrier, because it does not mature until substantially after birth, and its permeability is significantly modulated even in adulthood by factors such as fever and circulating cytokines.

This brings the question of the neurobehavioral pertinence of somatic infections and inflammation into the foreground. Placed alongside the documentation of immune activation in postmortem autistic brain tissue and in CSF, the relationship between peripheral and central immune dysregulation becomes an important area of investigation. Thus, the accumulating literature on the substantial prevalence of gastrointestinal and immune symptoms and disease processes in autism is moving from a parallel and subordinate track—a study of coincidental "comorbidities"—to an integrated study of the condition that takes into account each of these diverse manifestations in order to provide a common or at least interacting set of underlying pathophysiological mechanisms, as depicted in Fig. 20.1b.

A Systems Pathophysiology "Middle-Out" Approach: Dysregulated Biology

In light of the systemic pathophysiology that is being repeatedly documented in brain and peripheral tissues, it is now imperative that any model of autism pathogenesis incorporate mechanisms at additional physiological and signaling levels to explain the many findings that do not fit into the simpler "genes–brain–behavior" paradigm. Fig. 20.1b illustrates a "systems biology" or "systems pathophysiology" approach, which expands the dimensions under consideration beyond those that are highlighted in the "modular" approach. It also shifts the focus. The core of this model is a widely distributed dysregulated cell biology that goes beyond just the neuron. Beginning with the level of dysregulated biology represents a "*middle-out*" approach, where one works upstream to pathogenesis and downstream to a cascade of phenotypic manifestations that includes behavior but also much more.

- Upstream pathogenesis includes not only genes but also environmental factors and epigenetics. Even within genetics, this model calls for moving from the current narrow focus only on genes that specifically regulate neural development to a broader approach that also includes genes that regulate the metabolic and immune responses to environmental (infectious and xenobiotic) insults and the autoimmune response. These immunologic and metabolic perturbations would be upstream of impaired neural circuit and neural system functions in this model [15]. Additionally many genetic mutations may well target a substantially smaller set of cell, molecular, and biochemical mechanisms.
- Downstream of the core dysregulated biology are the components of the various autism phenotypes at the levels of biology, information processing, somatic symptoms, and behavior. Rather than assuming that each distinct behavioral domain has its own genetic basis, this model allows for the more probable interactions among systems and especially of emergent systems properties [47, 48], with multiple disease and behavioral manifestations

that *appear* distinct, but in fact, arise from common underlying pathophysiological mechanisms. Autism's substantial heterogeneity would result at least in part from differences in the severity, distribution, and timing of these systemic events, potentially relating to specific physical properties of the insult and the distinct biological responses to these insults.

Because phenotypic studies so far have been largely behaviorally oriented and correlated not to "mechanism," but rather to epidemiological data, and a limited number of "biomarkers," a fresh research agenda is needed to pursue the implications of this model.

Tight coupling between specific metabolic endophenotypes and specific behaviors might exist. This is an area badly in need of further study, and at this point one can only guess which potential mechanisms connect the pervasive metabolic, biochemical, and signaling dysregulation to the changes in function of specific target organs, brain regions, neuromodulatory systems, and behaviors, given that some kind of preferential targeting would need to be involved for locally enhanced effects to occur. However, we should note that the need to study underlying pathophysiology at the level of metabolism, biochemistry, immunology, and neural circuit levels is strongly defensible independent of the precise impact of these perturbations on the specific components of the autism behavioral syndrome.

It is in relation to how the levels in Fig. 20.1b relate to each other that the "systems model" also raises the question of the nature of autism. Autism has been defined behaviorally, i.e., in terms of the communication, social interaction, and behavioral domains included in box "i." But if autism rests on underlying biological factors, i.e., the components in box "ii", one aim of research might be to parse out subgroups within the larger population of individuals (i) who meet behavioral criteria and (ii) who are further characterized by having distinctive biological features. Such subgroups might be better defined, understood, and treated based on their specific underlying biological dysregulation rather than based on their behavioral deficits.

The "systems biology" model more sharply brings to the forefront critical questions of heterogeneity and of "final common pathways" regarding multiple mechanisms that may eventuate in common outcomes [49]. It also raises the question of where the bottleneck may be—i.e., at what level there may be the least heterogeneity. That is, are there few or many things *sufficient* to cause the autistic syndrome? And are there any things at all that are always *necessary*? Given the likely roles of epigenetics and environment, we can no longer assume that genes are the sole causal agents. Consequently, the question of the levels at which heterogeneity is both greatest and most restricted becomes pertinent, and poses a challenge to the research community in developing coordinated research strategies explicitly aimed at answering this class of questions. For example, how do we collaborate to determine whether and how multiple genes, environmental, and epigenetic factors may contribute to only one or a few types of dysregulated biology? Do all autistic individuals have immune activation? And if some do not, what does their biology look like and by what mechanisms do these noninflammatory forms lead to autistic behaviors? Is there convergence between the genetic targets in the rare familial cases and the inflammatory targets that may exist in the more prevalent forms of autism? Do all autistic individuals have reduced or otherwise altered functional connectivity? And if some do not, how can their processing patterns be characterized? Overall, what is it that different mechanisms leading to autistic behaviors have in common?

Analogies with multiple sclerosis

Concerning the issue of heterogeneity of etiologies and mechanisms in autism, another inflammatory neurologic disease with clear heterogeneity is multiple sclerosis. In this disease, demyelinating plaques affect different regions of the white matter in each individual. Although autism differs from multiple sclerosis in the type of immune reaction—innate vs. adaptive, respectively—it may still be similar in showing variable distribution and intensity from case to case and over the time course of the disease. Importantly, in MS, despite many studies, no correlation has been found between CNS [cerebrospinal fluid (CSF)] and peripheral immune markers, suggesting that it may be equally difficult to make such correlations between behavioral deficits and immune responses in autism. The heterogeneity in autisms, as in MS, may not result from the level of CSF (or systemic) immune markers, but instead from the brain regions targeted by the immune cells. Addressing this idea will require methods of labeling immune cells with radioisotopes combined with high-resolution imaging of these markers across different gray matter structures within the brain. The development of behavioral assays with quantifiable end points in humans will also be important in the analysis of these relationships. Ultimately, such methods will be necessary to monitor therapeutic responses to drugs. Embracing a systems model of autism with variable etiologies and disease penetrance will improve experimental design aimed at understanding the disease in these fashions.

Links Between Expanded Biological Hierarchy and Expanded Temporal Axis

The addition of environmental and epigenetic factors to the level of pathogenesis brings to the foreground the question of temporality. Although temporality in genetics may be implicit in that many genetically based conditions may not become symptomatic until substantially after birth, the time dimension is much more explicit regarding epigenetic mechanisms, which are by definition not inborn in the same way as are genes. Moreover, environmental exposures, whether chemical, infectious, or other, may influence critical periods of brain and somatic development *by* epigenetic or pretranscriptional mechanisms (e.g., may act as teratogens), but they may also accumulate or occur at subsequent points and yet still have an impact on brain *function*. This means that rather than taking for granted an exclusively early genetic determination of the autism brain wiring diagram, it is now becoming necessary to additionally consider contributors that may enter the system later and/or influence it in an ongoing fashion.

Presumptions about exclusively prenatal determination have been challenged by the phenomenon of what is called "autistic regression"—i.e., loss of acquired milestones, by the presence of chronic pathophysiology as discussed, and more recently also by the phenomenon of improvement, loss of diagnosis, and even reports of full recovery from autism. These challenges support revising the temporal narrative about autism.

Delayed Onset of Autism

Not so long ago, parental reports of later onset autism (beyond 2 years of age) were considered to be based on the inability of these parents to recognize early signs of autism. However, in recent years, the phenomenon of later onset autism or "regressive autism" has been validated, particularly by retrospective analysis of videotapes of children for signs of autism before the clear emergence of autism syndrome features [50]. Currently, it is estimated that upwards of 25% of autistic children develop the condition during the second year of postnatal life, with subtle or even absent signs of prior abnormality.

Neurodegeneration

The question of a neurodegenerative component in autism is being discussed, based on the identification of pathophysiological features such as ongoing systemic and CNS redox abnormalities and inflammation, and the evidence of brain volume [51] and neuron [52] loss, albeit modest, that are features of neurodegenerative conditions. Although there are no clear clinical correlates of continued decline in function as might be expected in a progressive neurodegenerative process (with the caveat that adult autism is poorly studied), it is commonly thought that early intervention is more effective than intervention even in mid to late childhood, suggesting a reduction of potential plasticity over time. Whether this preferential early responsiveness to therapy reflects the time course of the natural critical periods of human neurologic development or whether it identifies early postnatal development as a uniquely fragile period remains an open question for future research.

Improvement and Recovery

Finally, if current reports on the internet, in film, and in popular books and magazines chronicling loss of diagnosis and recovery [53] can be rigorously validated, the question arises whether and how at least some underlying patho-physiological mechanisms may at least partly be reversible. If even some cases of substantial recovery are validated, this is another important direction for future research. Moreover, reports of improvement and recovery have involved a variety of interventions ranging from standard behavioral therapies to treat-ment of somatic symptoms. This suggests either (a) a heterogeneity in potential treatment targets, (b) the possibility that the pathophysiology is a systems problem held in place by multiple simultaneous reinforcers, so that a potential may exist for self-correction even when only a subset of stressors are reduced, or (c) both.

From a Static Defect in Early Circuit Development to a Chronic Ongoing Disturbance of Circuit Remodeling and Function

A Model of Static Encephalopathy (Fig. 20.2)

The sequence in Fig. 20.2 addresses the contrast between models framing autism as a fixed, static encephalopathy with those including chronic and potentially reversible features. We elaborate the expansion of the temporal dimension in autism to include other key etiological levels and timetables of pathophysiology beyond a fixed deficit genetically caused entirely in utero.

Genetic Static Encephalopathy (Fig. 20.2, Model 1a)

In Fig. 20.2, Model 1a schematizes the standard narrative about autism cause and mechanisms. In this model, genes cause alterations in early brain development that permanently alter brain function. This model temporally instantiates the gene–brain–behavior version of the biological hierarchy. Along the temporal axis, autism in this model results from a genetically deter-mined disturbance of early brain development, and autism is a static encephalo-pathy, i.e., a fixed trait.

Static Encephalopathy Caused by Gene-Environment Interactions (Fig. 20.2, Model 1b)

In Model 1b, environmental and in utero immunologic factors enter the picture, and these, along with genes interacting with them, lead to a disturbance of early brain development that causes autism. But although the causal matrix is more complex, autism is still a static encephalopathy, a fixed trait. This model has been dominant in developmental neurotoxicology. Various chemicals, cytokines, and other substances have been shown to target morphogenetic signals, transduction pathways, and developmental events critical for multiple brain development processes [54]. The impact of insults at this level is arguably indelible and permanent. Immune activation and neuroinflammation could have such indelible effects, as reflected in a growing body of literature demonstrating immune influences on cellular proliferation and vulnerability as well as other features of brain development [55].

A Model of Autism as a Chronic Condition (Fig. 20.2, Model 2)

Epigenetic Contributors and Potential Plasticity (Fig. 20.2, Model 2a)

In Model 2a, one considers that not only might there be changes in the structural architecture of the brain's neural circuitry and glial cell distribution and numbers, but there may also be changes in nuclear histone-DNA architectures that dictate the levels and patterns of gene expression that must also be carefully adjusted to create a biologic system where all cells provide the appropriate quantities of molecular machineries (synaptic signaling, synapse-action potential coupling, oligodendroglial myelination, and glutamate transport) that cooperate to achieve a finely tuned neural circuit that transmits signals with high fidelity and precision. As an example, the use of valproate to treat epilepsy during pregnancy is associated with an increased risk of an autism spectrum disorder in the offspring [56]. This compound is a potent histone deacetylase inhibitor [57] and could therefore produce a lasting disturbance to the nuclear architecture that is critical to the tuning of neural circuit function [58]. Deficiencies of the methyl–CpG binding protein, MECP2, in Rett syndrome leads to a disturbance in the expression of a variety of proteins (e.g., decreased brain-derived neurotrophic factor (BDNF) [59] and increased Dlx5 [60]), suppressed layer V pyramidal neuron activity because of an increased inhibitory to excitatory synaptic input ratio [61], and behavioral deficits resembling autism. These observations suggest that disturbing the nuclear DNA architecture during development might be sufficient to produce autistic behavioral deficits. Importantly, some of the behavioral defects of the mouse model of Rett syndrome are reversible after development through conditional genetic rescue experiments [62]. Recent work also implicates alterations in chromatin remodeling in memory formation, recovery following neuronal damage, and a variety of inherited neurodevelopmental disorders (e.g., Rubinstein–Taybi syndrome, Rett syndrome) [63, 64, 65, 66]. We speculate that inflammation in utero or during early postnatal development, whether because of autoimmune mechanisms, xenobiotics, or infection, may also disturb nuclear DNA architecture to produce long-term effects on gene expression, some of which may be reversible but not easily so.

Ongoing Environmental Contributors to Chronic Encephalopathy (Fig. 20.2, Model 2b)

Model 2b formulates a hypothetical model in which the possible ongoing impact of accumulating environmental factors extends beyond the period of early development. Toxicants, autoantibodies, and infectious agents may not go away after the exposures during critical periods of development, and furthermore may continue to accumulate, or even arrive in the early postnatal period for the first time. For example, heavy metals such as lead, cadmium, mercury, or other neurotoxicants (such as polychlorinated biphenyls, pesticides, and air pollutants) can penetrate the nervous system. Once there, it is possible for them, by various mechanisms (e.g., organic mercurials that are de-ethylated or demethylated in microglia in the brain, yielding inorganic mercury) [67, 68, 69, 70, 71, 72, 73], to promote an oxidative response in these cells to stimulate a cytokine/chemokine inflammatory response within the brain. This effect may be long lasting if the toxin also impairs the function of cells (macrophages or microglial) that would normally try to clear the toxin. Various metals and persistent organic pollutants also accumulate in a number of body compartments, such as fatty tissue and liver, where they can have chronic metabolic impact such as inhibition of mitochondrial or hepatic enzymes. Viral infections might also contribute to chronic metabolic alterations [74]. The later or ongoing presence of such environmental influences can lead to a chronically dysregulated neuroglial environment, even if brain development is already substantially complete in all but the most subtle respects, and such dysregulation may not always require a prenatal initiating process.

Autism as Trait vs. Autism as Trait Modulated by State vs. Autism as State

The spectrum of mechanisms we are describing here are not mutually exclusive, as is illustrated in Fig. 20.2 in which each successive model includes additional features beyond the prior one but still includes the simpler model. It is likely that the heterogeneity of autism is substantially related to variability in several or all of the parameters involved in the gene–environment–epigenetics–timing interactions [75]. Although autism has been considered a "trait"—i.e., a fixed condition, it may also have elements of a "state"—i.e., contingent condition.

Autism as Trait Modulated by State

Autism in some cases may derive from a disturbance of early brain development whose impact is worsened by chronic or intermittent environmental influences (e.g., the child who gets worse with exposure to certain allergens or dietary peptides). In this model, autism is a static encephalopathy aggravated chronically or intermittently by disturbed intermediary metabolism and a dimension of metabolic encephalopathy—so it is still a trait, but has some malleability at the level of state changes that may lead to variability of severity within an individual. Support for this model of exacerbations of the neuroinflammatory state with systemic inflammatory challenge is supported by recent work showing that there is a magnified central pro-inflammatory cytokine response to a systemic inflammation trigger (lipopolysaccharide) when there is a long-standing weak stimulus of the brain's innate immune system [76]. Even a brief systemic proinflammatory insult to a 2-month-old rodent was shown to cause a persistent central elevation of the pro-inflammatory agent TNF-α 10 months later [77]. We propose that these observations suggest that an inflammatory insult in utero (e.g., maternal infection) might be a set-up for a marked central inflammatory response to a peripheral pro-inflammatory stimulant during early childhood (e.g., immunizations or viral illness) [78, 79, 80, 81]—that is, a low level of early onset immune activation may be the primary developmental condition that sets up a child for further injury later on (See also Patterson et al., Chapter 13).

Autism as Predominantly State Rather than Trait

For other autistic individuals, the interaction of gene–environment–epigenetics–timing may be not so much to *cause* autism as to increase vulnerability to subsequent contingent environmental stressors whose impacts are predominantly functional. In these individuals the tipping point into chronic immune activation, redox or energetic abnormalities, for example, may be reached more easily than in other individuals, leading to a chronically and multiply reinforced disturbance of intermediary metabolism, a state more than a trait—which can on occasion be reversed. Examples of such reversal include improvement on allergy medication; the striking amelioration of autistic features that has been anecdotally observed by various clinicians and families in some children who are on clear fluids or total parenteral nutrition, with a reversion to autism when enteral feeding is resumed; the transient but sometimes marked amelioration of autistic features in association with fever [82]; and the intermittent character of autistic features in some children diagnosed with mitochondrial disorders, whose autistic behaviors appear during fatigued but are absent when energetic (Korson M, 2007, personal communication). Further examples include instances of loss of diagnosis mentioned previously [17], which are poorly documented academically because of, among other things, the lack of inclusion until recently of useful indicators of improvement and recovery in outcomes research, presumably because of the assumption of autism as "trait," i.e., of incurability [83].

Mistaking State for Trait

It is important to note that chronicity may masquerade as incurability because the pathophysiology of encephalopathy secondary to disturbed intermediary metabolism in many cases may be quite complex. Inflammation may interact with redox abnormalities, recurrent and/or chronic infection, allergies, malabsorption and/or self-restricted diet leading to nutrient insufficiencies, disordered sleep and heightened stress responses to create a highly mutually reinforcing set of pathological feedback loops. The resultant physiological "gridlock," observed by some metabolically oriented clinicians (and much in need of more systematic study) may surface as a range of somatic and systemic symptoms that are just now moving from background to foreground in autism research and clinical practice. Because clinical phenotyping, which to date has skipped over these levels of autism phenomenology, is finally starting to move beyond behavior to include whole-body features (Fein D, 2008, personal communication) [1], the nature and prevalence of these features as well as the model that diverse somatic features may be linked by a shared set of cellular, signaling, and metabolic features can finally be explored and tested. If these chronic somatic and systemic features are core components of autism pathophysiology rather than secondary accompaniments, then their treatment and reversal may lead to reversal at the level of brain and behavior as well. Although it may require great ingenuity to unlock such chronicity, and although without such ingenuity and persistence this chronicity may be for all intents and purposes "incurable," the recalcitrance of such a state is not sufficient proof that the condition arises exclusively from early developmental mechanisms (although those may also be present at the same time as ongoing chronic effects in some or many individuals). With newer perspectives on previous findings in autism research [84] and the newer findings of a chronic ongoing pathophysiologic process characterized by inflammation and disturbed metabolism, autism should now be viewed as possibly resulting from a dynamic metabolic and physiologic disturbance rather than simply an incurable developmental injury.

Chronic Mechanisms in Functional Impairment in Autism

Considering the astroglial and microglial immune response now shown to be present in brains of autistic children [14], in this section we discuss potential mechanisms whereby activated glia could drive brain malfunction in autism. We must emphasize that this model is erected solely to guide experimental design, and that much of the evidence must still be acquired. The models will surely evolve as data are accumulated. We focus on mechanisms related to chronicity, i.e., to Model 2b in Fig. 20.2 as described just above. We will confine ourselves to this model not because it is the only possibility but rather because it has received the least attention to date, even though it includes plausible mechanisms that can be tested experimentally and that would have broad implications if they were to be identified as operating in autism.

Disturbances of Astroglial Function

Environmental xenobiotics may alter glial–neuronal interactions through effects on astrocyte electrolyte homeostasis. One effect of neurotoxins is astroglial swelling, which may be caused by inhibiting mitochondrial respiration, by altering passive chloride or cation conductances, or by impairing active transport of ions. Astrocyte swelling leads to amino acid efflux. This solute efflux in part serves to recover normal astrocyte volume, but more importantly, can produce major effects on neuronal synaptic transmission and intrinsic excitability as described below. Beyond these functional effects, glutamate and glycine release can promote excitotoxic electrolyte disturbances (swelling and calcium overload) in the neurons, which may ultimately lead to loss of axonal or dendritic processes or cause outright cell death. Some xenobiotics may block anion channels, which could reduce excitotoxicity at the expense of prolonging the astroglial swelling. Although not a major focus of current research programs, toxicity to astroglia themselves (process collapse or cell death) may also be a critical component of the neural circuit dysfunction because these cells play critical roles in promoting the formation and function of synapses [85, 86]. Another poorly appreciated effect of astroglial toxicity is the glial–vascular signaling that enables changes in neuron metabolism to couple to changes in cerebral vascular blood flow [87, 88]. From a purely structural perspective, astroglial swelling could also cause a substantial decrease in average capillary lumen volume, impeding blood flow that could cause ischemia in neurons possibly analogous to the cognitive defects that can arise in hypertensive or severe diabetic cerebrovascular disease. This may be a mechanism pertinent to multiple published reports of reduced cerebral perfusion in autism [89, 90, 91, 92]. If so, ameliorating some of the physiological reinforcers of this reduced perfusion could plausibly improve the level of functioning in autism. Additionally, astroglia provide the major energy source to neurons in the form of lactate, and therefore, impaired astroglial cell metabolism can have major secondary effects on neuron energy metabolism [93]. Astroglial foot processes also create the functional blood–brain barrier, and defective astroglia could impair this barrier, leading to increases of extracellular potassium or other metabolites that could impair neuron functions [94]. A potential link between defects in the blood–brain barrier and autism is inspired by the observations that neuroligin 3 is mutated in a family with X-linked autism [95] and that the Drosophila homolog of neuroligin 3, gliotactin, plays a critical role in formation of the blood–brain barrier [96]. Interestingly, this same protein may also contribute to formation of epithelial barriers based on its effects in Drosophila suggesting a possible link to the gastrointestinal problems and inflammation found in autism [2, 3, 4, 5, 6, 7, 8, 9, 97]. Finally, impaired diffusion of substrates and waste products between the intravascular compartment and the neuron could have major consequences for neuronal and glial metabolic efficiency [98].

Neuronal and Neural Circuit Changes from Astroglial Metabolic and Structural Changes Deriving from Chronic Inflammation

Model 2b in Fig. 20.2 regarding ongoing contributors to chronic encephalopathy, described above, may significantly describe the impact of immune activation and inflammation on autism. There are a variety of potential mechanisms. Components of the inflammatory response may normally reduce neuronal metabolic activity to protect neurons during self-limited disease (e.g., viral infection) and to promote neural circuit growth and repair after disease-induced damage. In autism, however, the condition is no longer self-limited; neural circuit silencing appears to be too widespread and too persistent, perhaps related to the chronicity of the immune activation. Although resting astroglia promote synaptic signaling, activated glia may impact synaptic signaling and neuronal functioning differently. Little is currently known about the effects of activated glia on neuronal circuit function during a chronic innate immune response as observed in autism. However, a sequence of changes can be proposed for investigation. It appears likely that chronic immune activation leads to metabolic changes in astroglia, which in turn might inhibit neuronal and neural circuit functioning in a series of ways. Primary impairments in glial metabolism could readily impair neural circuitry through their failure to provide metabolic support to neurons (i.e., lactate) and to clear glutamate from the synaptic cleft. Glia could *release* inhibitory neuromodulators, *retract* glial processes supporting synaptic or axonal transmission, or *dismantle* synapses, and even *collapse* dendrites.

Release of Inhibitory and Excitatory Neuromodulators

The release of inhibitory neuromodulators such as ATP, lipid metabolites, chemokines, cytokines, and growth factors by glia could lead to acute changes in neuronal function. Astroglia also release glutamate, GABA, taurine, and D-serine [99, 100, 101]. These latter agents might produce acute excitation or inhibition through group III metabotropic glutamate receptors, tonic GABA A receptors, glycinergic receptors, or NMDA receptors (on GABA interneurons for example), respectively.

A further potential mechanism of inhibition is the release of lipids such as *arachidonate metabolites* by astroglia [102, 103, 104, 105, 106, 107, 108]. Lipoxygenase metabolites inhibit synaptic transmission [106, 107, 108]. In the lateral amygdala, they also inhibit action potential firing by activating α-dendrotoxin-sensitive, low-threshold K^+ channels [108].

Inhibition at this level could impact neural systems functioning. Purinergic agonist (ATPγS) stimulation of glia in the retina causes them to release *ATP that is metabolized into adenosine* to inhibit neurons *by* adenosine R1 receptors [109]. We speculate that similar mechanisms might occur in the brains of individuals with autism because of innate immune activation. Chronic immune activation in autism might cause glia to release ATP to inhibit neural activity. In the retina, acute stimulation of glia causes them to release ATP that is sequentially metabolized into adenosine by ecto-ATPase and ectonucleotidases. Adenosine binds to adenosine A1 receptors to activate G-protein-coupled inwardly rectifying K^+ channels (GIRK) and inhibit action potential firing of retinal ganglion cells [109]. The thalamus contains a very high density of adenosine A1 receptor A_1R [110]. Adenosine activates adenosine A1 receptors to hyperpolarize thalamic relay neurons of lateral geniculate nucleus, converting the action potential firing mode from tonic to burst [111]. Burst firing by thalamic neurons is suspected to help achieve a stable sleep state [112]; consequently, inflammation-induced activation of burst firing in the thalamus of an individual who is awake might put affected regions of the thalamus into their sleep mode of firing and prevent normal sensory transmission, which is thought to utilize the tonic firing mode. Hyperpolarization and inhibition of tonic firing is explained by the GIRK K^+ currents activated by adenosine A1 receptors [111, 113].

Haznedar et al. [114] reported severely reduced [18F]-fluorodeoxyglucose uptake in the thalamus, striatum, and frontal cortex in autism. We suspect that effects of the innate immune activation on neural circuit function may be responsible. These effects could be due to depressed glutamatergic synaptic transmission, enhanced GABAergic synaptic transmission, or impaired neuronal excitability, all of which could be induced by neuromodulators released during chronic innate immune activation. The effects could occur within the local circuits or they could result from defects in neuromodulatory systems projecting to these circuits (e.g., the basal forebrain cholinergic system or the hypothalamic preoptic area that inhibits monoaminergic systems to initiate the sleep state). Alternatively, the suppressed activity measured by Haznedar's group might result from impaired synchronization of firing within populations of neurons that project to these regions. For example, BDNF is released by activated microglia. BDNF decreases the KCC2 K^+-Cl-cotransporter, causing the high intracellular chloride gradient to collapse. This could result in defective GABAergic synaptic transmission. The basal forebrain GABAergic system generates a synchronous GABAergic synaptic input at theta frequencies, which temporally synchronizes firing across broad regions of the cerebral cortex and hippocampus. Loss of inhibitory post-synaptic potentials in cortical pyramidal neurons because of BDNF down-regulation of KCC2 could impair pyramidal neuron synchronization and consequently lead to a failure to synchronously excite down-stream target sites. Synchrony is achieved in part through the broadly projecting GABAergic systems of the basal forebrain that temporally couple the firing of large populations of neurons within the theta and gamma frequency bands. These mechanisms of circuit suppression could also interact with defects in synaptic connections because of developmentally altered axonal targeting, growth, or pruning. An outcome of these mechanisms could be the underconnectivity measured through fMRI or an alteration of coherence as measured by EEG in autism [30, 34, 35].

Retraction of Glial Processes

Persistent innate immune activation could progress from a neuromodulatory and neurophysiological to a neurostructural level of impact depending on the duration or intensity of the inflammation and its temporal relationship to postnatal developmental time windows. *Retraction of glial processes* supporting synaptic or axonal signal transmission in neurons has been implicated in the supraoptic nucleus of the hypothalamus during lactation [99]. This has functional consequences: as the glial covering of a synapse is lost, glutamate released at the synapse that is usually taken up by the glial transporters now escapes into the extrasynaptic space where it can bind to presynaptic metabotropic glutamate receptors that will inhibit synaptic transmission. Similar mechanisms may be at work during autism-associated inflammation in the thalamus and elsewhere.

Dismantling and Collapse of Synapses and Dendrites

Dismantling and collapse of synapses and dendrites could silence neuron activity during chronic inflammation. Lehnardt et al. in 2006 [115] showed that bacterial meningitis causes a neurodegenerative process through toll-like receptor 2 (TLR2) possibly mediated by microglia. Although collapse would seem to be a pathologic process, it might also be adaptive because it preserves the neuron soma and might permit reconnection of the circuit later once the inflammatory condition has been cleared. This possibility of reversible impairments even in structural connectivity is highly pertinent to interpreting and pursuing reports of improvement and recovery phenomena in autism, which, to be adequately studied, require longitudinal repeated measures and a model of autism that is dynamic rather than static. Endotoxin, which activates TLR4, exacerbates neurodegeneration [76]. It may be pertinent to the trajectory of mild volume loss that appears to occur in adolescence and adulthood, after the volume increase seen in early childhood [20, 51].

Glial Calcium Oscillations

Inflammatory glial activation might have a more direct neurophysiological impact. One possible consequence is a strengthening of glial–neuronal coupling. Parri and Crunelli [116, 117] found that a minor subset of resting astroglia (4%) in the thalamus undergo spontaneous *calcium oscillations* (approximately 1/60 s). Although it has not yet been investigated, we hypothesize that immune activation might increase the number of astrocytes undergoing calcium oscillations. Increases of intracellular calcium during each cycle of the oscillation could promote pulsatile release of cytokines and chemokines.

We further speculate that intracellular calcium oscillations and waves could be promoted by the combination of gap junction coupling between astrocytes and calcium-induced calcium release mechanisms through ryanodine receptors. Under baseline conditions retinal *glial–neuronal inhibition* is relatively weak [109]: amongst retinal ganglion cells, 35.5% showed little or no adenosine-mediated slow hyperpolarization events (<0.2 mV), 52.2% showed moderate hyperpolarizations (0.2–5 mV), and only 12.3% showed large hyperpolarizations (>5 mV). We speculate that immune activation might increase the number or incidence of large hyperpolarizing events because of either more frequent large calcium oscillations or altered ATP release mechanisms, or altered adenosine responses of the neuron. These are all important issues we will investigate to gain a better understanding of the potential mechanisms of immune activation-induced alteration of neuronal activity in the brain of autism patients.

Sustaining Inflammation

What sustains the immune activation? The presence of *neuroglial synapses* has only recently been recognized [118], and their function remains a complete mystery. We speculate that neuroglial synapses might couple neural activity to glial release of inhibitory neuromodulators under conditions of immune activation. This type of mechanism could target the glial release of chemokines and cytokines to neuronal activity, providing feedback inhibition only when and where neuronal activity is occurring. Such a mechanism could conserve metabolic energy consumed by the immune-activated glia. We further speculate that this mechanism might promote greater inflammatory changes in resting-state or default-mode regions of the brain in autism (cingulate cortex, anterior thalamus) because these are the regions with the highest activity during the resting state [119].

Summary of Glial Impacts: The Foregoing as Examples of New Classes of Mechanisms to Pursue

The above speculations arise from the need for models to guide experiments that address the ongoing effects of immune activation on neuroglial interactions and on circuit dysfunction. Mechanisms such as we sketch in discussing Model 2b of Fig. 20.2 bear consideration as possible explanations for the reports of improvement and loss of diagnosis that have emerged in recent years and are recently coming to be studied. These mechanisms are not meant to be an exhaustive survey of possibilities, but rather to illustrate some levels at which investigation could profitably be pursued. Although it is also important to investigate how immune activation might additionally disrupt the normal pattern of neural circuit remodeling that occurs during early postnatal development (as in the Models 1a and 1b in Fig. 20.2), our focus has been on the later and chronic functional impacts of chronic innate immune activation, because these mechanisms of disruptioning circuit function have received almost no attention in autism research and might represent reversible defects that could be targeted by therapies.

Implications for Heterogeneity, Reversibility, Autism Symptoms, and Development–Chronicity Interactions

Heterogeneity of autisms may map to differences in the relative weightings of links in this set of causal chains, as mentioned above. There may also be heterogeneity in reversibility because among the various ongoing mechanisms, some can apparently reverse quickly (e.g., improvement during fever or abstinence from solid foods as mentioned earlier) although others, such as the dismantling and collapse of synapses and dendrites mentioned above, may take longer to reverse. Identifying such potentially reversible mechanisms needs to become an important part of the autism research agenda. Reversibility, particularly short-term marked improvement, raises the further question of whether recovery involves gains of skills or loss of inhibition (or some of both), because even transient markedly improved capabilities suggests more underlying soundness of circuitry than previously suspected.

It is also worth noting that anecdotal reports from individuals with autism suggest that many "autistic behaviors" result from sensory dysregulation and represent ways of coping with problems such as sensory overload. Such behaviors include self-injurious and so-called self-stimulatory behaviors and compulsions in addition to sometimes crippling anxiety [120]. Insofar as such sensory dysregulation is a consequence of functional impairment of neuroglial functioning through mechanisms such as those described above, such sensory overload could be lowered by reducing the chronic reinforcers of this impaired neuroglial functioning, thereby reversing behaviors [121, 122].

The alteration of neuronal functioning in Model 2 of Fig. 20.2 cannot be described as "developmental" in the classic sense, because it may occur at any point in the life course, even remote from the period of exuberant brain development in early life. Constructing the autism state may depend on an interaction of these chronic mechanisms with specific windows of developmental vulnerability. This may account for the apparent contradiction between the substantial overlap in dysregulated biology between autism and a variety of other multisystem conditions (such as autoimmune diseases, metabolic syndromes, and other neurodegenerative disorders) on the one hand and the uniquely autistic neurobehavioral profile on the other. Thus, the timing parameter in genes–environment–timing interactions may be exquisitely important.

The Challenge of Generic Mechanisms in a Specifically Described Disorder

Nevertheless, although understanding autism's uniqueness may be of great interest, it should be remembered that substantial advances in the well-being of individuals with autism may be achieved by addressing more mundane and even generic features of the condition's dysregulated biology, such as immune dysregulation, inflammation, and oxidative stress. What makes autism unique may be upstream or, perhaps more likely, downstream of pathophysiological mechanisms that may be more generic but that nevertheless may provide more leveraged treatment targets. Moreover, the specific features of the disease at the behavioral level may simply be a function of neural systems that are impacted by these mechanisms. The instigating mechanisms could be quite varied and yet lead to a common set of neural system defects related to common features of the pathological response (e.g., immune cell activation, cytokine release).

Summary and Conclusion

The foregoing discussion has sketched the structure of a reconceptualization of autism from an incurable genetically determined brain disorder into a heterogeneous whole-body, gene-environment, complex, and dynamic multisystem condition with multiple potential treatment possibilities. Our exposition of a range of autism models illustrating various forms and stages of this reconceptualization is meant to bring into full view the underlying structures of argument that motivate hypothesis formation and research agendas in autism. Such assumptions often go unstated. To articulate unstated assumptions is to allow them to be subjected to systematic and deliberate reflection. This allows us to move beyond the confusion generated when unstated assumptions are left merely implicit.

We have argued that we are moving to a systems approach that incorporates the complexities of many levels of the biological hierarchy in addition to genes, brain, and behavior, particularly including dysregulated biology and signaling. And we have argued that we are moving to an expanded temporal model of the autism syndrome as a dynamic condition with ongoing as well as developmental environmental modulation. Within this framework, it is possible that by correcting dysregulated biology and signaling we can ameliorate the severity of the condition, sometimes substantially. These systems and expanded temporal models of autism have several critical implications for the autism research agenda.

1. The logical outcome of the systems approach is that research in autism at any level needs to aim for some linkage to the underlying dysregulated biology and cellular signaling pathways.
2. The logical outcome of the expanded temporal approach is that the autism phenotype may be more fruitfully addressed as a modulated condition rather than an inborn defect with genetic and environmental determinants. A concept is proposed that the disease state can represent an ongoing mutually reinforcing physiological and signaling gridlock that may look incurable but that actually may be approached as a puzzle (as opposed to a mystery) to be untangled and addressed stepwise and systematically.
3. The logical outcome of integrating the systems and the expanded temporal models is that change is possible in autism, and that we will be most successful at optimizing opportunities and outcomes if we move beyond exclusively considering just genes, brain and behavior to inclusively considering all levels and interactions in the biological hierarchy. We posit that autism research and treatment need to address genes, brain, and behavior in relation to dysregulated biology, somatic symptoms, and clinical pathophysiology. The "black-box" approach to empirical therapy may be helpful, but it is reasonable to suggest that a more thorough and whole-body systems understanding of the disease and therapeutic responses will lead in the future to greater treatment efficacy.
4. The nature of systems dynamics and the idea that behavioral deficits may be emergent properties of perturbed systems function suggest that reducing dysfunction or stress at any level may have effects that cascade through multiple levels of the system. Thus, an analytic approach to identifying treatment targets, however mundane or seemingly removed from the "autism" as behaviorally defined, may improve quality of life and reduce the aspect of autism that involves suffering. Moreover, an expanded, more inclusive modeling of autism will help us to articulate precisely what aspects of autism involve suffering for the affected individual, which will enable practitioners to more effectively address the concerns of those members of the autism community who feel that their autism gives them many strengths and do not

wish to be subjected to treatments they feel are aimed at making them neurotypically "normal."

The models discussed herein have substantial implications for the research agenda in autism. The field needs to substantially upgrade attention and resources allocated to measuring the dysregulated intermediary biology, which to date has been sorely understudied in comparison with the efforts invested in genetics. It needs to consider autism dynamically, which means focusing more centrally upon change and in particular on multidisciplinary physiological measures in the analysis of treatment response and on what such data can teach us about subgroups and mechanisms. It means developing methodologies for analyzing complex datasets derived from subjects who are in many respects heterogeneous. It means looking for different underlying biological mechanisms in responders versus non-responders in treatment trials rather than averaging away the significance of treatments that may be effective for biologically distinct subgroups. And it means testing models as well as hypotheses, and welcoming a test of fresh models proactively. These implications, taken as goals, all follow from pursuing treatment, improvement, and recovery with thoroughness and expectation of success.

Acknowledgments Cure Autism Now Foundation, Nancy Lurie Marks Family Foundation, Bernard Fund for Autism Research, National Alliance for Autism Research, Autism Speaks, the National Institute of Mental Health, the National Institute of Neurologic Disease and Stroke, the Burroughs Wellcome Fund.

References

1. Herbert MR. Autism: A Brain disorder or a disorder that affects the brain? *Clin Neuropsychiatry* 2005; 2:354–79.
2. Horvath K, Perman JA. Autistic disorder and gastrointestinal disease. *Curr Opin Pediatr* 2002; 14:583–7.
3. Jyonouchi H, Geng L, Ruby A, Reddy C, Zimmerman-Bier B. Evaluation of an association between gastrointestinal symptoms and cytokine production against common dietary proteins in children with autism spectrum disorders. *J Pediatr* 2005; 146:605–10.
4. Jyonouchi H, Sun S, Itokazu N. Innate immunity associated with inflammatory responses and cytokine production against common dietary proteins in patients with autism spectrum disorder. *Neuropsychobiology* 2002; 46:76–84.
5. Lucarelli S, Frediani T, Zingoni AM, Ferruzzi F, Giardini O, Quintieri F, et al. Food allergy and infantile autism. *Panminerva Med* 1995; 37:137–41.
6. Valicenti-McDermott M, McVicar K, Rapin I, Wershil BK, Cohen H, Shinnar S. Frequency of gastrointestinal symptoms in children with autistic spectrum disorders and association with family history of autoimmune disease. *J Dev Behav Pediatr* 2006; 27: S128–36.
7. Jass JR. The intestinal lesion of autistic spectrum disorder. *Eur J Gastroenterol Hepatol* 2005; 17:821–2.
8. Afzal N, Murch S, Thirrupathy K, Berger L, Fagbemi A, Heuschkel R. Constipation with acquired megarectum in children with autism. *Pediatrics* 2003; 112:939–42.

9. Torrente F, Ashwood P, Day R, Machado N, Furlano RI, Anthony A, et al. Small intestinal enteropathy with epithelial IgG and complement deposition in children with regressive autism. *Mol Psychiatry* 2002; 7:375–82, 334.

10. Hornig M, Mervis R, Hoffman K, Lipkin WI. Infectious and immune factors in neurodevelopmental damage. *Mol Psychiatry* 2002; 7: S34–5.

11. James SJ, Melnyk S, Jernigan S, Cleves MA, Halsted CH, Wong DH, et al. Metabolic endophenotype and related genotypes are associated with oxidative stress in children with autism. *Am J Med Genet B Neuropsychiatr Genet* 2006; 141:947–56.

12. Chauhan A, Chauhan V, Brown WT, Cohen I. Oxidative stress in autism: increased lipid peroxidation and reduced serum levels of ceruloplasmin and transferrin – the antioxidant proteins. *Life Sci* 2004; 75:2539–49.

13. Ashwood P, Wills S, Van de Water J. The immune response in autism: a new frontier for autism research. *J Leukoc Biol* 2006; 80:1–15.

14. Vargas DL, Nascimbene C, Krishnan C, Zimmerman AW, Pardo CA. Neuroglial activation and neuroinflammation in the brain of patients with autism. *Ann Neurol* 2005; 57:67–81.

15. Herbert MR, Russo JP, Yang S, Roohi J, Blaxill M, Kahler SG, et al. Autism and environmental genomics. *Neurotoxicology* 2006; 27:671–84.

16. Newschaffer CJ, Falb MD, Gurney JG. National autism prevalence trends from United States special education data. *Pediatrics* 2005; 115:e277–82.

17. Kelley E, Paul JJ, Fein D, Naigles LR. Residual language deficits in optimal outcome children with a history of autism. *J Autism Dev Disord* 2006; 36:807–28.

18. Fein D, Dixon P, Paul J, Levin H. Pervasive developmental disorder can evolve into ADHD: case illustrations. *J Autism Dev Disord* 2005; 35:525–34.

19. American Psychiatric Association. *Diagnostic and Statistical Manual of Mental Disorders.* 4th edn. (DSM IV). Washington, DC: APA, 1994.

20. Redcay E, Courchesne E. When is the brain enlarged in autism? A meta-analysis of all brain size reports. *Biol Psychiatry* 2005; 58:1–9.

21. Herbert MR. Large brains in autism: the challenge of pervasive abnormality. *Neuroscientist* 2005; 11:417–40.

22. Ashwood P, Van de Water J. Is autism an autoimmune disease? *Autoimmun Rev* 2004; 3:557–62.

23. Ashwood P, Van de Water J. A review of autism and the immune response. *Clin Dev Immunol* 2004; 11:165–74.

24. McGinnis WR. Could oxidative stress from psychosocial stress affect neurodevelopment in autism? *J Autism Dev Disord* 2007; 37:993–4.

25. MacFabe DF, Cain DP, Rodriguez-Capote K, Franklin AE, Hoffman JE, Boon F, et al. Neurobiological effects of intraventricular propionic acid in rats: possible role of short chain fatty acids on the pathogenesis and characteristics of autism spectrum disorders. *Behav Brain Res* 2007; 176:149–69.

26. Kern JK, Jones AM. Evidence of toxicity, oxidative stress, and neuronal insult in autism. *J Toxicol Environ Health B Crit Rev* 2006; 9:485–99.

27. Yao Y, Walsh WJ, McGinnis WR, Pratico D. Altered vascular phenotype in autism: correlation with oxidative stress. *Arch Neurol* 2006; 63:1161–4.

28. Ming X, Stein TP, Brimacombe M, Johnson WG, Lambert GH, Wagner GC. Increased excretion of a lipid peroxidation biomarker in autism. *Prostaglandins Leukot Essent Fatty Acids* 2005; 73:379–384.

29. Just MA, Cherkassky VL, Keller TA, Kana RK, Minshew NJ. Functional and anatomical cortical underconnectivity in autism: evidence from an fMRI study of an executive function task and corpus callosum morphometry. *Cereb Cortex* 2006; 17:951–61.

30. Murias M, Webb SJ, Greenson J, Dawson G. Resting state cortical connectivity reflected in EEG coherence in individuals with autism. *Biol Psychiatry* 2007; 62:270–3.

31. Vargas DL, Bandaru V, Zerrate MC, Zimmerman AW, Haughey N, Pardo CA. Oxidative stress in brain tissues from autistic patients: increased concentration of isoprostanes. IMFAR 2006, 2006; Poster PS2.6.

32. Perry G, Nunomura A, Harris P, siedlak S, Smith M, Salomon R. Is autism a disease of oxidative stress? Oxidative Stress in Autism Symposium, New York State Institute for Basic Research in Developmental Disabilities, Staten Island, NY, 2005; p. 15.

33. Herbert MR, Ziegler DA, Makris N, Filipek PA, Kemper TL, Normandin JJ, et al. Localization of white matter volume increase in autism and developmental language disorder. *Ann Neurol* 2004; 55:530–40.

34. Just MA, Cherkassky VL, Keller TA, Minshew NJ. Cortical activation and synchronization during sentence comprehension in high-functioning autism: evidence of underconnectivity. *Brain* 2004; 127:1811–21.

35. Rippon G, Brock J, Brown C, Boucher J. Disordered connectivity in the autistic brain: challenges for the 'new psychophysiology'. *Int J Psychophysiol* 2007; 63:164–72.

36. Happe F, Ronald A, Plomin R. Time to give up on a single explanation for autism. *Nat Neurosci* 2006; 9:1218–20.

37. Hyde SC, Emsley P, Hartshorn MJ, Mimmack MM, Gileadi U, Pearce SR, et al. Structural model of ATP-binding proteins associated with cystic fibrosis, multidrug resistance and bacterial transport. *Nature* 1990; 346:362–5.

38. Anderson MP, Gregory RJ, Thompson S, Souza DW, Paul S, Mulligan RC, et al. Demonstration that CFTR is a chloride channel by alteration of its anion selectivity. *Science* 1991; 253:202–5.

39. Anderson MP, Rich DP, Gregory RJ, Smith AE, Welsh MJ. Generation of cAMP-activated chloride currents by expression of CFTR. *Science* 1991; 251:679–82.

40. Kanner L. Autistic disturbances of affective contact. *Nerv Child* 1943; 10:217–50.

41. Jepson B, Johnson J, Wright K. *Changing the Course of Autism.* Boulder, co, 2007.

42. Rutter M. Incidence of autism spectrum disorders: changes over time and their meaning. *Acta Paediatr* 2005; 94:2–15.

43. Herbert MR, Ziegler DA. Volumetric Neuroimaging and Low-Dose Early-Life exposures: loose coupling of pathogenesis-brain-behavior links. *Neurotoxicology* 2005; 26:565–72.

44. Ashwood P, Anthony A, Pellicer AA, Torrente F, Walker-Smith JA, Wakefield AJ. Intestinal lymphocyte populations in children with regressive autism: evidence for extensive mucosal immunopathology. *J Clin Immunol* 2003; 23:504–17.

45. Badman MK, Flier JS. The gut and energy balance: visceral allies in the obesity wars. *Science* 2005; 307:1909–14.

46. Elmquist JK, Scammell TE, Saper CB. Mechanisms of CNS response to systemic immune challenge: the febrile response. *Trends Neurosci* 1997; 20:565–70.

47. Morton J, Frith U. Causal modelling: a structural approach to developmental psychopathology. In: Cicchetti D, Cohen DJ, editors. *Manual of Developmental Psychopathology.* New York: John Wiley, 1995:357–390.

48. Karmiloff-Smith A. The tortuous route from genes to behavior: a neuroconstructivist approach. *Cogn Affect Behav Neurosci* 2006; 6:9–17.

49. Li Z, Dong T, Proschel C, Noble M. Chemically diverse toxicants converge on Fyn and c-Cbl to disrupt precursor cell function. *PLoS Biol* 2007; 5:e35.

50. Werner E, Dawson G. Validation of the phenomenon of autistic regression using home videotapes. *Arch Gen Psychiatry* 2005; 62:889–95.

51. Aylward EH, Minshew NJ, Field K, Sparks BF, Singh N. Effects of age on brain volume and head circumference in autism. *Neurology* 2002; 59:175–83.

52. Bauman ML, Kemper TL. Neuroanatomic observations of the brain in autism: a review and future directions. *Int J Dev Neurosci* 2005; 23:183–7.

53. Edelson SM, Rimland B. *Recovering Autistic Children.* San Diego: Autism Research Institute, 2006.

54. Jensen KF, Catalano SM. Brain morphogenesis and developmental neurotoxicology. In: Slikker W, Chang LW, editors. *Developmental Neurotoxicology*. San Diego, CA: Academic Press, 1998: 3–41.
55. Hagberg H, Mallard C. Effect of inflammation on central nervous system development and vulnerability. *Curr Opin Neurol* 2005; 18:117–23.
56. Moore SJ, Turnpenny P, Quinn A, Glover S, Lloyd DJ, Montgomery T, et al. A clinical study of 57 children with fetal anticonvulsant syndromes. *J Med Genet* 2000; 37:489–97.
57. Phiel CJ, Zhang F, Huang EY, Guenther MG, Lazar MA, Klein PS. Histone deacetylase is a direct target of valproic acid, a potent anticonvulsant, mood stabilizer, and teratogen. *J Biol Chem* 2001; 276:36734–41.
58. Zhang MM, Yu K, Xiao C, Ruan DY. The influence of developmental periods of sodium valproate exposure on synaptic plasticity in the CA1 region of rat hippocampus. *Neurosci Lett* 2003; 351:165–8.
59. Chang Q, Khare G, Dani V, Nelson S, Jaenisch R. The disease progression of Mecp2 mutant mice is affected by the level of BDNF expression. *Neuron* 2006; 49:341–8.
60. Horike S, Cai S, Miyano M, Cheng JF, Kohwi-Shigematsu T. Loss of silent-chromatin looping and impaired imprinting of DLX5 in Rett syndrome. *Nat Genet* 2005; 37:31–40.
61. Dani VS, Chang Q, Maffei A, Turrigiano GG, Jaenisch R, Nelson SB. Reduced cortical activity due to a shift in the balance between excitation and inhibition in a mouse model of Rett syndrome. *Proc Natl Acad Sci USA* 2005; 102:12560–5.
62. Guy J, Gan J, Selfridge J, Cobb S, Bird A. Reversal of neurological defects in a mouse model of Rett syndrome. *Science* 2007; 315:1143–7.
63. Guan Z, Giustetto M, Lomvardas S, Kim JH, Miniaci MC, Schwartz JH, et al. Integration of long-term-memory-related synaptic plasticity involves bidirectional regulation of gene expression and chromatin structure. *Cell* 2002; 111:483–93.
64. Alarcon JM, Malleret G, Touzani K, Vronskaya S, Ishii S, Kandel ER, et al. Chromatin acetylation, memory, and LTP are impaired in CBP +/− mice: a model for the cognitive deficit in Rubinstein-Taybi syndrome and its amelioration. *Neuron* 2004; 42:947–59.
65. Korzus E, Rosenfeld MG, Mayford M. CBP histone acetyltransferase activity is a critical component of memory consolidation. *Neuron* 2004; 42:961–72.
66. Fischer A, Sananbenesi F, Wang X, Dobbin M, Tsai LH. Recovery of learning and memory is associated with chromatin remodelling. *Nature* 2007; 447:178–82.
67. Charleston JS, Body RL, Bolender RP, Mottet NK, Vahter ME, Burbacher TM. Changes in the number of astrocytes and microglia in the thalamus of the monkey Macaca fascicularis following long-term subclinical methylmercury exposure. *Neurotoxicology* 1996; 17:127–38.
68. Garg TK, Chang JY. Methylmercury causes oxidative stress and cytotoxicity in microglia: attenuation by 15-deoxy-delta 12, 14-prostaglandin J2. *J Neuroimmunol* 2006; 171:17–28.
69. Kim SH, Johnson VJ, Sharma RP. Mercury inhibits nitric oxide production but activates proinflammatory cytokine expression in murine macrophage: differential modulation of NF-kappaB and p38 MAPK signaling pathways. *Nitric Oxide* 2002; 7:67–74.
70. Zurich MG, Eskes C, Honegger P, Berode M, Monnet-Tschudi F. Maturation-dependent neurotoxicity of lead acetate in vitro: implication of glial reactions. *J Neurosci Res* 2002; 70:108–16.
71. Campbell A. Inflammation, neurodegenerative diseases, and environmental exposures. *Ann N Y Acad Sci* 2004; 1035:117–32.
72. Shanker G, Aschner JL, Syversen T, Aschner M. Free radical formation in cerebral cortical astrocytes in culture induced by methylmercury. *Brain Res Mol Brain Res* 2004; 128:48–57.
73. Filipov NM, Seegal RF, Lawrence DA. Manganese potentiates in vitro production of proinflammatory cytokines and nitric oxide by microglia through a nuclear factor kappa B-dependent mechanism. *Toxicol Sci* 2005; 84:139–48.

74. Munger J, Bajad SU, Coller HA, Shenk T, Rabinowitz JD. Dynamics of the cellular metabolome during human cytomegalovirus infection. *PLoS Pathog* 2006; 2:e132.

75. Lipkin I, Hornig M, Gorman J. The 'Three Strikes' concept of autism. *Autism Advocate* 2006; 45:45.

76. Cunningham C, Wilcockson DC, Campion S, Lunnon K, Perry VH. Central and systemic endotoxin challenges exacerbate the local inflammatory response and increase neuronal death during chronic neurodegeneration. *J Neurosci* 2005; 25:9275–84.

77. Qin L, Wu X, Block ML, Liu Y, Breese GR, Hong JS, et al. Systemic LPS causes chronic neuroinflammation and progressive neurodegeneration. *Glia* 2007; 55:453–62.

78. Mallard C, Hagberg H. Inflammation-induced preconditioning in the immature brain. *Semin Fetal Neonatal Med* 2007;12(4):280–6.

79. Shi L, Fatemi SH, Sidwell RW, Patterson PH. Maternal influenza infection causes marked behavioral and pharmacological changes in the offspring. *J Neurosci* 2003; 23:297–302.

80. Patterson PH. Maternal infection: window on neuroimmune interactions in fetal brain development and mental illness. *Curr Opin Neurobiol* 2002; 12:115–8.

81. Fatemi SH, Earle J, Kanodia R, Kist D, Emamian ES, Patterson PH, et al. Prenatal viral infection leads to pyramidal cell atrophy and macrocephaly in adulthood: implications for genesis of autism and schizophrenia. *Cell Mol Neurobiol* 2002; 22:25–33.

82. Curran L, Newschaffer C, Lee L, Crawford S, Johnston M, Zimmerman A. Behaviors associated with fever in children with autism spectrum disorders. *Pediatrics* 2007; 120(6): e1386–92.

83. Kelley E, Paul JJ, Fein D, Naigles LR. Residual language deficits in optimal outcome children with a history of autism. *J Autism Dev Disord* 2006; 36(6):807–28.

84. Bauman M. Beyond behavior – Biomedical diagnoses in autism spectrum disorders. *Autism Advocate* 2006; 45:27–29.

85. Anderson M, Hooker B, Herbert M. Bridging from cells to cognition in autism pathophysiology: biological pathways to defective brain function and plasticity. *Am J Biochem Biotechnol* 2008; 4(2):167–76.

86. Pfrieger FW, Barres BA. Synaptic efficacy enhanced by glial cells in vitro. *Science* 1997; 277:1684–7.

87. Ullian EM, Sapperstein SK, Christopherson KS, Barres BA. Control of synapse number by glia. *Science* 2001; 291:657–61.

88. Takano T, Tian GF, Peng W, Lou N, Libionka W, Han X, et al. Astrocyte-mediated control of cerebral blood flow. *Nat Neurosci* 2006; 9:260–7.

89. Zonta M, Angulo MC, Gobbo S, Rosengarten B, Hossmann KA, Pozzan T, et al. Neuron-to-astrocyte signaling is central to the dynamic control of brain microcirculation. *Nat Neurosci* 2003; 6:43–50.

90. Mountz JM, Tolbert LC, Lill DW, Katholi CR, Liu HG. Functional deficits in autistic disorder: characterization by technetium-99m-HMPAO and SPECT. *J Nucl Med* 1995; 36:1156–62.

91. Zilbovicius M, Boddaert N, Belin P, Poline JB, Remy P, Mangin JF, et al. Temporal lobe dysfunction in childhood autism: a PET study. Positron emission tomography. *Am J Psychiatry* 2000; 157:1988–93.

92. Chiron C, Leboyer M, Leon F, Jambaque I, Nuttin C, Syrota A. SPECT of the brain in childhood autism: evidence for a lack of normal hemispheric asymmetry. *Dev Med Child Neurol* 1995; 37:849–60.

93. Pellerin L. How astrocytes feed hungry neurons. *Mol Neurobiol* 2005; 32:59–72.

94. Simard M, Nedergaard M. The neurobiology of glia in the context of water and ion homeostasis. *Neuroscience* 2004; 129:877–96.

95. Jamain S, Quach H, Betancur C, Rastam M, Colineaux C, Gillberg IC, et al. Mutations of the X-linked genes encoding neuroligins NLGN3 and NLGN4 are associated with autism. *Nat Genet* 2003; 34:27–9.

96. Auld VJ, Fetter RD, Broadie K, Goodman CS. Gliotactin, a novel transmembrane protein on peripheral glia, is required to form the blood-nerve barrier in Drosophila. *Cell* 1995; 81:757–67.

97. Schulte J, Charish K, Que J, Ravn S, MacKinnon C, Auld VJ. Gliotactin and Discs large form a protein complex at the tricellular junction of polarized epithelial cells in Drosophila. *J Cell Sci* 2006; 119:4391–401.

98. Aschner M, Allen JW, Kimelberg HK, LoPachin RM, Streit WJ. Glial cells in neuro-toxicity development. *Annu Rev Pharmacol Toxicol* 1999; 39:151–73.

99. Oliet SH, Piet R, Poulain DA. Control of glutamate clearance and synaptic efficacy by glial coverage of neurons. *Science* 2001; 292:923–6.

100. Wang CM, Chang YY, Kuo JS, Sun SH. Activation of P2X(7) receptors induced GABA release from the RBA-2 type-2 astrocyte cell line through a Cl(-)/HCO(3)(-)-dependent mechanism. *Glia* 2002; 37:8–18.

101. Fields RD, Burnstock G. Purinergic signalling in neuron-glia interactions. *Nat Rev Neurosci* 2006; 7:423–36.

102. Petroni A, Blasevich M, Visioli F, Zancocchia B, Caruso D, Galli C. Arachidonic acid cycloxygenase and lipoxygenase pathways are differently activated by platelet activating factor and the calcium-ionophore A23187 in a primary culture of astroglial cells. *Brain Res Dev Brain Res* 1991; 63:221–7.

103. Vahter ME, Mottet NK, Friberg LT, Lind SB, Charleston JS, Burbacher TM. Demethylation of methyl mercury in different brain sites of Macaca fascicularis monkeys during long-term subclinical methyl mercury exposure. *Toxicol Appl Pharmacol* 1995; 134:273–84.

104. Ji KA, Yang MS, Jou I, Shong MH, Joe EH. Thrombin induces expression of cytokine-induced SH2 protein (CIS) in rat brain astrocytes: involvement of phospholipase A2, cyclooxygenase, and lipoxygenase. *Glia* 2004; 48:102–11.

105. Won JS, Im YB, Khan M, Singh AK, Singh I. Involvement of phospholipase A2 and lipoxygenase in lipopolysaccharide-induced inducible nitric oxide synthase expression in glial cells. *Glia* 2005; 51:13–21.

106. Vaughan CW, Ingram SL, Connor MA, Christie MJ. How opioids inhibit GABA-mediated neurotransmission. *Nature* 1997; 390:611–4.

107. Feinmark SJ, Begum R, Tsvetkov E, Goussakov I, Funk CD, Siegelbaum SA, et al. 12-lipoxygenase metabolites of arachidonic acid mediate metabotropic glutamate receptor-dependent long-term depression at hippocampal CA3-CA1 synapses. *J Neurosci* 2003; 23:11427–35.

108. Faber ES, Sah P. Opioids inhibit lateral amygdala pyramidal neurons by enhancing a dendritic potassium current. *J Neurosci* 2004; 24:3031–9.

109. Newman EA. Glial cell inhibition of neurons by release of ATP. *J Neurosci* 2003; 23:1659–66.

110. Ochiishi T, Chen L, Yukawa A, Saitoh Y, Sekino Y, Arai T, et al. Cellular localization of adenosine A1 receptors in rat forebrain: immunohistochemical analysis using adenosine A1 receptor-specific monoclonal antibody. *J Comp Neurol* 1999; 411:301–16.

111. Pape HC. Adenosine promotes burst activity in guinea-pig geniculocortical neurones through two different ionic mechanisms. *J Physiol* 1992; 447:729–53.

112. Anderson MP, Mochizuki T, Xie J, Fischler W, Manger JP, Talley EM, et al. Thalamic Cav3.1 T-type Ca2+ channel plays a crucial role in stabilizing sleep. *Proc Natl Acad Sci USA* 2005; 102:1743–8.

113. Wetherington JP, Lambert NA. Differential desensitization of responses mediated by presynaptic and postsynaptic A1 adenosine receptors. *J Neurosci* 2002; 22:1248–55.

114. Haznedar MM, Buchsbaum MS, Hazlett EA, LiCalzi EM, Cartwright C, Hollander E. Volumetric analysis and three-dimensional glucose metabolic mapping of the striatum and thalamus in patients with autism spectrum disorders. *Am J Psychiatry* 2006; 163:1252–63.

115. Lehnardt S, Henneke P, Lien E, Kasper DL, Volpe JJ, Bechmann I, et al. A mechanism for neurodegeneration induced by group B streptococci through activation of the TLR2/MyD88 pathway in microglia. *J Immunol* 2006; 177:583–92.

116. Parri HR, Crunelli V. The role of Ca2+ in the generation of spontaneous astrocytic Ca2+ oscillations. *Neuroscience* 2003; 120:979–92.

117. Parri HR, Crunelli V. Pacemaker calcium oscillations in thalamic astrocytes in situ. *Neuroreport* 2001; 12:3897–900.

118. Lin SC, Bergles DE. Synaptic signaling between neurons and glia. *Glia* 2004; 47:290–8.

119. Gusnard DA, Raichle ME, Raichle ME. Searching for a baseline: functional imaging and the resting human brain. *Nat Rev Neurosci* 2001; 2:685–94.

120. Grandin T. *Thinking in Pictures*. NY: Vintage, 1996.

121. Iversen P. *Strange Son*. Riverhead trade, NY: Penguin, 2007.

122. Mottron L, Mineau S, Martel G, Bernier CS, Berthiaume C, Dawson M, et al. Lateral glances toward moving stimuli among young children with autism: Early regulation of locally oriented perception? *Dev Psychopathol* 2007; 19:23–36.

Index

Printed in the United States of America